G. Kessler

PROLIFERATION-PROOF URANIUM/PLUTONIUM AND THORIUM/URANIUM FUEL CYCLES

Safeguards and Non-Proliferation

2nd revised edition

Proliferation-Proof Uranium/Plutonium and Thorium/Uranium Fuel Cycles

Safeguards and Non-Proliferation

by
G. Kessler

2nd revised edition

KIT Scientific Publishing

This book was written without any support by private institutions or government organizations.

Impressum

KIT Scientific Publishing

Karlsruher Institut für Technologie (KIT)
KIT Scientific Publishing
Straße am Forum 2
D-76131 Karlsruhe

www.ksp.kit.edu

Print on Demand 2017 – Gedruckt auf FSC-zertifiziertem Papier

ISBN 978-3-7315-0516-7
DOI 10.5445/KSP/1000054131

To Lotte,
Birgit, Anne and Jürgen
for their encouragement and patience.

Preface (second edition)

Chapters 9-14 of the first edition dealt with plutonium from the commercial nuclear fuel cycle as a proliferation problem. As a result of scientific analyses these chapters provided ranges for the isotopic concentration of Pu-238 above which reactor grade plutonium is **proliferation-proof**.

This second edition deals, in addition, with the possibility of re-enrichment of <20% U-235 enriched uranium of research and test reactors to weapons type high enriched uranium in relatively small cascades of gas ultra-centrifuges. Kryuchkov et al. of Moscow Engineering Physics Institute (MEPHI) in a publication in Nuclear Science and Engineering Vol. **162**, p.208-213 (2009) suggested to dope the <20% U-235 enriched fuel with 0.1% of the uranium isotope U-232. This uranium isotope U-232 has a very high alpha-particle activity and, therefore, a higher internal heat source than Pu-238. It also has a relatively high spontaneous neutron activity. In addition, the U-232 isotope is by three atomic mass units lighter than the isotope U-235. During re-enrichment in gas ultra-centrifuges it, therefore, would be enriched from 0.1% U-235 enrichment to 11%, whereas U-235 is enriched from <20% U-235 enrichment to about 70%. As will be shown in chapters 15 and 16 this will make <20% U-235 **proliferation-proof** in several ways.

U-233 originating from the conversion/breeding process in the thorium/U-233 fuel cycle is always contaminated by the uranium isotope U-232. Enrichment of <12% U-233 uranium in gas ultra-centrifuges to weapons type high enriched U-233 (proliferation problem) would similarly lead to higher enrichment of the contaminant U-232. However, U-232 is only one atomic mass unit lighter than U-233. Therefore, this factor of U-232 enrichment differs not as strongly as in the above case of U-235.

For Light Water Reactors operating in the thorium/U-233 fuel cycle the contamination of their fuel by U-232 is sufficiently high after a fuel burnup of 20,000-30,000 MWd/t to make such fuel **proliferation-proof** against enrichment in gas ultra-centrifuges. This is also valid for the fuel of Fast Breeders having a burnup > 100,000 MWd/t. This is shown in chapters 17 and 18.

INTERNET links in chapters 15 through 18 refer to the publishing year 2016

The author would like to thank:

- Dr. E. Kiefhaber for critical comments and proof reading; in particular he contributed in prevealing the mistake in Table 3.5.
- Dr. A. Rineiski for advice on critical masses
- Dipl.-Ing. A. Veser for correcting and arranging the manuscript of this second edition
- Mrs. M. Wettstein for assistance in literature search

Again the author hopes that this book will make a helpful contribution to the advancement of the difficult future scientific and political discussion of the nuclear proliferation problem.

G. Kessler

Karlsruhe, 2016

Preface (first edition)

This book was written in an effort to present a comprehensive overview of the findings and proposed solutions elaborated after 2004 by a Karlsruhe group of retired scientists of the former Institute for Neutron Physics and Reactor Technology of the former Karlsruhe Nuclear Research Center about the problem of "Plutonium Proliferation of Nuclear Power". These findings were published in two scientific journals between 2007 and 2010 as solutions to subproblems of plutonium proliferation.

From the beginnings of the civil use of nuclear power there have been fears that such use could produce fissile nuclear material (highly enriched uranium with up to 93% U-235 or more than 12% U-233, or weapon-grade plutonium or neptunium), which could be passed on or used directly to build nuclear weapons.

All official nuclear weapon states (NWSs) so far (USA, Russia, UK, France, China, India, Pakistan, North Korea), however, conducted the development and construction of their nuclear weapons in special military programs and started the civil use of nuclear power at a later date.

The International Atomic Energy Agency (IAEA) was founded in Vienna in 1956, and the Non-proliferation Treaty (NPT) was presented for signature in 1968, to solve the proliferation problem. By 2003, the NPT had been signed and ratified by 186 nations of the world.

From around 1970 onward, the IAEA published a series of statements in INFCIRC reports and scientific and technical publications. According to these rules, IAEA inspectors may monitor the inventories of fissile materials in the facilities of the nuclear fuel cycle (enrichment plants, fuel fabrication plants, nuclear power plants, reprocessing plants, waste disposal) of the non-nuclear weapon states (NNWSs) running civil nuclear power programs. Scientific equipment and analytical methods for measuring fissile material inventories are now available.

However, three events in 1975-1980 caused further international discussions and restrictions, especially in the United States of America:

- The Indian initiation of a nuclear explosion in 1974.
- The finding that the spent fuel elements of civil nuclear reactors had accumulated more plutonium than had existed in the nuclear weapon arsenals of the nuclear weapon states.

- The realization in the FORD-MITRE study (1977), that reactor-grade plutonium was good for nuclear weapons which, though unreliable and only able to generate relatively low explosion energies, could still be dangerous nuclear explosives.

The United States then gave up reprocessing of spent nuclear fuel elements and the technology of recycling plutonium in nuclear reactors approximately in 1978-80. Also the development of breeder reactors based on the uranium/plutonium fuel cycle was abandoned. “Direct disposal” of spent fuel elements was proposed instead. Only a few states with programs to utilize nuclear power followed this proposal after some delay.

Roughly around 1995, the USA and Russia, as part of their obligations under the NPT, decided to transfer to the civil nuclear fuel cycle and use a total of approx. 50 t of their weapon-grade plutonium and several 100 tons of their highly U-235-enriched weapon-grade uranium. The UK is the only other nuclear weapon state to follow that example. In addition, the nuclear weapon arsenals of the USA and Russia were reduced by other disarmament agreements.

At the same time scientific organizations in the USA (US-Department of Energy, American Physical Society, and American Nuclear Society) reiterated the assertion that it was possible to use reactor-grade plutonium for nuclear weapons. Since 1972 the IAEA provisions continue to be upheld (INFCIRC/173) that all plutonium of the civil nuclear fuel cycle was to be treated like weapon-grade plutonium (with the exception of plutonium containing more than 80% of the Pu-238 isotope).

As a consequence of this situation, large reprocessing plants for spent nuclear fuel have been built and commissioned so far only in NWS (France, UK, Russia). Japan is an exception with its 800 t/a plant of Rokkasho-mura. The reasons are the very restrictive conditions imposed by the IAEA and the limited accuracy in measuring the plutonium inventories of large reprocessing plants.

However, the civil use of nuclear power has progressed further since 1990. Around 2010, approx. 430 GW(e) of civil nuclear power plants were operated in the world, and another 35-40 GW(e) were being planned or under construction. The quantity of plutonium, neptunium, and americium in spent fuel elements has accumulated to approx. 2300 tonnes of reactor-grade plutonium, some 90 tonnes of neptunium, and 150 tonnes of americium. Only some 30% of these fuel elements were reprocessed in NWSs and recycled as uranium/plutonium mixed oxide fuel especially in Europe and Japan.

It has become clear in the meantime that all plutonium and neptunium (except for small residues of approx. 0.1% in the chemical steps of reprocessing plants and in refabrication) can be destroyed by nuclear fission. Direct disposal as planned in the United States (Yucca Mountain repository) has suffered a setback. The licensing applications filed for that repository were withdrawn by the U.S. Department of Energy.

In this general situation, the "Karlsruhe Group" queried the statement, in scientific and technical terms, that reactor-grade plutonium of any composition could be used to make nuclear explosives. Limits were worked out above which the share of Pu-238 isotopes in plutonium renders the use in nuclear explosives technically impossible (proliferation-proof).

Moreover, options are indicated for the uranium-plutonium fuel cycle which allow plutonium with a sufficiently high content of Pu-238 to be produced. This Pu-238-isotope content can be maintained even after repeated recycling. Nuclear weapon-grade neptunium can be avoided in these nuclear fuel cycle options. The use of americium allows also the plutonium generated in breeder blankets to be kept always above the proposed limit of the Pu-238 isotope content. In this way, also the construction of nuclear explosives with blanket plutonium of fast breeders becomes impossible.

In this scenario, future breeder technology also would permit complete use of U-238 for nuclear fission and employ only so-called proliferation-proof plutonium. In this way, the exploitation of the uranium resource could be increased by a factor of 100.

The present status of the civil use of nuclear power with the U/Pu nuclear fuel cycle and the associated IAEA safeguards is described briefly in chapters 1-8. This was deemed to be necessary to explain the background to the previous debates about safeguards and the proliferation problem.

The analysis of assertions that reactor-grade plutonium could be used for nuclear explosives, and the very restrictive regulations by the IAEA, are described in sections 9-11. New scientific solutions with denatured or proliferation-proof reactor-grade plutonium to run future Light Water Reactors and Fast BreederReactors are covered in sections 12-14. These new technical and scientific approaches at the same time allow plutonium, neptunium, and americium to be incinerated through nuclear fission.

The author had the good fortune to work on the solution of these problems with outstanding members, almost all now retired, of the former Institute of Neutron Physics and Reactor Technology of the former Karlsruhe Research Center. These are

Prof. Dr. W. Seifritz (retired)	Dr. B. Goel (retired)
Dipl.Math. W. Höbel (retired)	Dr. C.H.M. Broeders (retired)
Dr. D. Wilhelm (retired)	Dr. A. Rineiski

He wishes to dedicate chapters 9-14 to these excellent scientists.

The scientific findings described in sections 9-14 were discussed at international workshops and covered in seven publications in "Nuclear Science and Engineering" and "Nuclear Engineering and Design."

Prof. Dr. Y. Fuji-ie (Emeritus Commissioner of the Japan Atomic Energy Commission) first suggested that, roughly about the year 2005, both actinide incineration and the solution to the nuclear proliferation problem might be linked. The international workshops were organized by

Prof. Dr. Saito, Tokyo Institute of Technology, Japan,

Dr. Ch. Ganguly, International Atomic Energy Agency, Austria:

- International Seminar on "Advanced Nuclear Energy System Toward Zero Release of Radioactive Wastes," Susono, Japan, November 6-9, 2000,
- IAEA Consultancy Meeting on "Protected Plutonium Production," IAEA, Vienna, Austria, June 19-20, 2003,
- International Science and Technology Forum on "Protected Plutonium Utilization for Peace and Sustainable Prosperity," Tokyo Institute of Technology, Tokyo, Japan, March 1-3, 2004,
- IAEA Consultancy Meeting on "Protected Plutonium Production," IAEA, Vienna, Austria, June 15-16, 2006,
- COE Satellite Technical Meeting on Non-proliferation and Protected Plutonium Production, Tokyo Institute of Technology, Tokyo, Japan, December 1, 2006,
- International Science and Technology Forum on "Protected Plutonium Utilization for Peace and Sustainable Prosperity," Tokyo Institute of Technology, Tokyo, Japan, September 16-19, 2008.

Prof. Dr. V. Artisyuk, Obninsk State University and SCICET (Rosatom) in Obninsk, Russia, directed these workshops:

- Special Session on "Nonproliferation of Nuclear Materials" at the 10th International Conference on Nuclear Power Safety and Nuclear Education, October 1-7, 2007, Obninsk, Russia,
- International Workshop on Non-Proliferation of Nuclear Materials, September 29 to October 3, 2008, Obninsk, Russia,
- International Workshop on Non-Proliferation of Nuclear Materials, September 29 to October 3, 2009, Obninsk, Russia.

The author would like to thank the following scientists for assisting him with critical discussions and suggestions in the analysis of the proliferation problem:

- Dr. E. Kiefhaber (retired scientist of the same Institute as the author). His help and his critical comments were of inestimable value in writing and publishing this book.
- Prof. W. Häfele, former Director of the Research Centers of Jülich and Dresden, Germany, for his interest and critical suggestions.

- Prof. H.H. Hennies, one of the former Directors of the Karlsruhe Research Center, for his support and advice.
- Prof. Dr. jur. Burckhardt Jähnke, Vice President of the Federal Supreme Court, Karlsruhe, Germany, for his advice on German publication law.
- Dr. G. Schumacher (retired scientist of the same Institute as the author) for his continued interest.
- A number of former staff members for their suggestions in numerical mathematics and information technology.

The author hopes that this book will make a helpful contribution to the advancement of the difficult future scientific and political discussion of the nuclear proliferation problem.

G. Kessler

Karlsruhe, December 15, 2010

Summary

A brief outline of the historical development of the proliferation problem is followed by a description of the uranium-plutonium nuclear fuel cycle with uranium enrichment, fuel fabrication, the light-water reactors mainly in operation, and the breeder reactors still under development. The next item discussed is reprocessing of spent fuel with plutonium recycling and the future possibility to incinerate plutonium and the minor actinides: neptunium, americium, and curium. Much attention is devoted to the technical and scientific treatment of the IAEA surveillance concept of the uranium-plutonium fuel cycle. In this context, especially the physically possible accuracy of measuring U/Pu flow in the fuel cycle, and the criticism expressed of the accuracy in measuring the plutonium balance in large reprocessing plants of non-nuclear weapon states are analyzed.

The second part of the book initially examines the assertion that reactor-grade plutonium could be used to build nuclear weapons whose explosive yield cannot be predicted accurately, but whose minimum explosive yield is still far above that of chemical explosive charges. Methods employed in reactor physics are used to show that such hypothetical nuclear explosive devices (HNEDs) would attain too high temperatures in the required implosion lenses as a result of the heat generated by the Pu-238 isotope always present in reactor plutonium of current light-water reactors. These lenses would either melt or tend to undergo chemical auto-explosion. Limits to the content of the Pu-238 isotope are determined above which such hypothetical nuclear weapons are not feasible on technical grounds. This situation is analyzed for various possibilities of the technical state of the art of making implosion lenses and various ways of cooling up to the use of liquid helium. The outcome is that, depending on the existing state of the art, reactor-grade plutonium from spent fuel elements of light-water reactors with a burnup of 35 to 58 GWd/t cannot be used for making nuclear weapons. This statement does not apply to reactor-grade plutonium from fuel elements of lower burnup of less than 30 GWd/t (heavy-water reactors, older gas-graphite reactors or researach reactors), as their plutonium contains too little of the Pu-238 isotope. Today's light-water reactors, however, attain fuel burnups in excess of 50 GWd/t. In the future, fuel burnups of more than 60 GWd/t are aimed at.

In the next part of the book, nuclear fuel cycle options are examined which allow larger shares of the Pu-238 plutonium isotope (up to more than 10%) in reactor-grade plutonium to be achieved. This is easily possible by using re-enriched reprocessed uranium (RRU) arising in reprocessing spent fuel, whose low contents of U-235 and U-236 can be enriched. Moreover, the minor actinides, neptunium and americium, can be added to the fresh fuel.

It is shown that reactor-grade americium produced in spent fuel cannot be used to build nuclear weapons for similar reasons as reactor-grade plutonium. The Am-241 isotope always present in reactor-grade americium generates so much heat as a result of alpha decay that any use in making hypothetical nuclear weapons becomes technically unfeasible. The nuclear physics properties of the neptunium minor actinide, however, are such that it can be used directly as a metal to build nuclear weapons. This leaves the only possibility to prevent neptunium in future nuclear fuel cycles. Such fuel cycle options are analyzed in the last but one chapter of the book.

I a next chapter of the book a proposal is made of a transition phase leading to a future proliferation-proof civil use of nuclear power. This employs the IAEA proposal henceforth to use multilateral fuel cycle centers which are multinational. As today's large enrichment plants and reprocessing facilities are operated almost exclusively in nuclear weapon states, and as plutonium recycling is most advanced there as well, these are also the places where existing light-water reactors could produce plutonium with a higher Pu-238 content using neptunium and re-enriched reprocessed uranium. This is done by chemical co-separation of plutonium and neptunium in reprocessing. This plutonium, which has a higher content of Pu-238, is proliferation-proof, i.e. cannot be abused to make nuclear weapons. It can be used and burnt in non-nuclear weapon states under surveillance by the IAEA.

To hold this plutonium always at the required (proliferation-proof) content of Pu-238, several percent of (proliferation-proof) americium must be added to the fresh fuel. In this way, it is possible in a future proliferation-proof uranium-plutonium fuel cycle incorporating light-water reactors and breeders with a fast neutron spectrum to use U-238 and so produce energy over very long periods of time (thousands of years) and incinerate all the existing plutonium and minor actinides. This can be achieved by a sophisticated change in the uranium-plutonium fuel cycle and by the production of reactor-grade proliferation-proof plutonium with higher contents of Pu-238.

For the future use of thorium in a proliferation-proof U-233/thorium fuel cycle similarly the isotope U-232 can be applied .

In the next part of the book it is shown that a small cascade of gas-ultra-centrifuges can be used to re-enrich the fresh fuel with <20% U-235 enriched uranium of present research and test reactors or future <20% U-235 enriched uranium fuel of light-water-reactors operating in the U/Th fuel cycle to high enriched uranium (HEU) which could be misused to design nuclear weapons.

A technical solution for this proliferation danger of <20% U-235 enriched uranium is the admixture of about 0.1% of the isotope U-232. It is shown that during re-enrichment in gas ultra-centrifuges this U-232 admixture is enriched to 11% while the U-235 is enriched to 72%.Tis leads to a number of technical consequences which make such high enriched uranium **proliferation-proof**.

Similarly the U-233/U-235 fuel of future light-water-reactors operating in the U/Th fuel cycle can become **proliferation-proof** against re-enrichment in a cascade of gas ultra-centrifuges as this fuel produces a certain contamination by U-232 from nuclear reactions of neutrons with Th-232 during its burnup in the reactor core.

Table of contents

1 Nuclear Proliferation and IAEA-Safeguards

1.1 Historical Development

Unfortunately, the first application of nuclear energy occurred for military use in 1945. The plutonium for this military application had been produced in special graphite moderated gas cooled reactors. Some years later pressurized water reactors were first used for nuclear submarine propulsion. Civil application of nuclear energy with electricity generating nuclear power reactors did not start until 1955-1958.

Therefore, nuclear technology is considered to be a dual use technology which allows both peaceful and military applications. From the peaceful use of nuclear energy technologies, nuclear materials and nuclear facilities have been disseminated all over the world.

Nuclear weapons were developed and manufactured before and independently of the peaceful exploitation of nuclear energy. This has been borne out by historical developments so far in nuclear weapon countries, e.g. the USA (1945), USSR (1949), UK (1953), France (1960), China (1964) as well as in the de facto nuclear weapon states: India (1974), Pakistan (1998), Israel and North Korea (2008). All these Nuclear Weapons States (NWSs) produced their fissile nuclear materials: highly enriched uranium (≥93% U-235) or weapon-grade plutonium by military programs and not through the peaceful use of nuclear energy. Accordingly, the proliferation of nuclear weapons cannot simply be prevented by restrictions on the peaceful uses of nuclear energy.

The inherent proliferation risk in the use of nuclear energy was recognized at the very beginning of the development of nuclear power, and a number of proposals have been made and measures taken in the course of time to prevent proliferation. The period up until 1953 can be regarded as a phase of complete classification of any kind of utilization of nuclear power. As early as 1945/46, the idea originated in the USA to make the peaceful utilization of nuclear power accessible to other states while, at the same time, preventing the proliferation of nuclear weapons. The proposal contained in the so-called Acheson-Lilienthal report provided for the establishment of an "international atomic development authority," which was to manage or possess all nuclear activities, i.e., an international body to monopolize the field of nuclear power utilization. In 1946, the USA submitted a proposal to the Atomic Energy Commission of the United Nations. That proposal, which became known as the "Baruch Plan," failed because it called for a far-reaching surrender of national sovereignty; and therefore classification was maintained [1].

The worldwide utilization of nuclear power began with the "Atoms for Peace" program announced by US President Eisenhower before the General Assembly of the United Nations in December 1953, under which a promotion of nuclear power utilization was planned in conjunction with control measures. This initiative also lead to the 1954 Geneva United Nations Conference on the peaceful uses of nuclear energy. One major constituent of the plan was the establishment of an "International Atomic Energy Agency (IAEA)," which was to promote and, at the same time, monitor all international cooperation in the field of nuclear technology. After a series of negotiations, the IAEA Statute was submitted for signature in October 1956. Article II of that Statute reads, inter alia: "The Agency ensures, so far as it is able, that assistance provided by it or at its request or under its supervision or control is not used in such a way as to further any military purpose" (IAEA Statute) [2].

Controls were defined in agreements between the IAEA and the countries it supports. The guideline used was the INFCIRC/66 document, "The Agency's Safeguards System" [3]. It provided for controls of plants and materials. A system of records, reports and inspections was created. The IAEA's only verifying compliance were agreements by the signatory states, i.e., to not abuse nuclear power for military purposes.

In 1957, the European Community established in the frame of the Euratom Treaty a nuclear material control system. Euratom safeguards were designed to ensure that nuclear materials were not diverted from their intended use and to guarantee that the Community complies with its international obligations concerning the supply and use of nuclear materials. Supply agreements with Euratom employed Euratom safeguards in recognition of the multinational character of its safeguards system. After the full development of IAEA safeguards, special arrangements and cooperative mechanisms between Euratom and IAEA inspections were worked out and continue to evolve.

The safeguards system at that time had been designed for the surveillance of small reactor plants below 100 MW(e) power output. It was soon found to be inadequate for a quickly expanding commercial nuclear power reactor technology. In 1963 the first civil nuclear power reactors were ordered in the USA, Russia, Canada, the UK, France and other European countries. All countries built their own commercial nuclear power reactors with several 100 MW(e) output.

It remained at the discretion of these countries to build and operate nuclear facilities without IAEA controls. As a consequence, the Treaty of Tlatelolco (UN Treaty Series No. 9068) [4] was concluded in 1967 for the Latin American countries, and the Non-Proliferation Treaty (NPT) was negotiated and signed in 1968. It entered into force in May 1970 (INFCIRC/140) [5]. This is a summary of its contents:

Each non-nuclear weapon state (NNWS) that becomes party to the NPT binds itself not to acquire nuclear weapons or other nuclear explosives (Article II). It also binds itself to conclude an agreement with IAEA for the application of safeguards to all its peaceful nuclear activities with a view to verifying the fulfilment of its obligations under the treaty (Article III).
In return, the treaty recognizes the right of all parties to participate in the fullest possible exchange of equipment, materials, and scientific and technological information for the peaceful uses of nuclear energy; in other words, all parties are guaranteed full access to peaceful nuclear technology (Article IV).
The parties also undertake to pursue negotiations in good faith towards nuclear disarmament (Article VI) and reaffirm their determination to achieve the discontinuance of all tests of nuclear weapons (Preamble); these latter commitments apply principally to the nuclear weapon states (NWSs) themselves.
The safeguards required under the NPT shall be applied to all sources and special fissionable material in all peaceful nuclear activities within the territory of such state, under its jurisdiction, or carried out under its control anywhere.

The structure and the contents of a verification agreement are contained in a recommendation by IAEA, which was to constitute the basis of negotiations, but in fact represents the contents of all agreements. This was documented in INFCIRC/153 Corrected (1972) [6]. This document became the basis for all Comprehensive Safeguards Agreements between NPT member states and the IAEA. These agreements have a number of important features. One is the requirement to place under safeguards all nuclear materials in peaceful uses in the state, which would later prove to have significance in determining the Agency's authority to search for undeclared nuclear materials and activities. A second feature is the requirement for states to establish so-called State's System of Accounting and Control (SSACs) to track domestic inventories of nuclear materials and provide reports to the IAEA. In many countries, these SSACs are also the national authorities regulating nuclear activities including domestic safeguards and security. A third feature is that the agreement obligates the IAEA to apply safeguards with all states that have such agreements. Part II of INFCIRC/153 Corrected outlines detailed procedures for the application of IAEA safeguards under the agreement.

India's nuclear test explosion in 1974 shocked the nuclear non-proliferation community. It initiated greater interest in controlling the nuclear trade (nuclear fuel and technology) and lead to the association of states exporting nuclear technology called Nuclear Suppliers Group. This group agreed to enforced rules requiring special commitments to non-proliferation criteria from recipient states.

In 1975 M. Willrich and Th. Taylor [7] warned of possible theft of nuclear materials and stated that reactor-grade plutonium could be misused for crude and inefficient nuclear explosive devices. The steadily increasing amounts of spent fuel from the growing nuclear industry and the plans for reprocessing and plutonium recycling in reactors lead to even more serious concerns. As a consequence the results of the FORD/MITRE [8] report published in 1977 became the basis for the declaration by US-president J. Carter that the USA would refrain from civil reprocessing of spent fuel, plutonium recycling and breeder technology. The US Nuclear Non-proliferation Act of 1978 (NAPA) strengthened international control and security measures to avoid further proliferation of nuclear materials and knowledge.

The burnup of spent fuel in gas cooled reactors and heavy water reactors – using natural uranium as fresh fuel – was about 7 GWd/t at that time and for light water reactors the burnup of the spent fuel was increased up to about 30 GWd/t. Table 1.1 shows the plutonium isotopic compositions of such reactor-grade plutonium and compares them with weapon-grade plutonium (the sum for isotopic compositions of weapon-grade plutonium does not add up fully to unity).

Table 1.1: Isotopic composition (weight fraction) of plutonium separated from gas cooled reactors (MAGNOX), Heavy Water Reactors (CANDUs) and Pressurized Water Reactors (PWRs) around 1975 in comparison with weapon-grade plutonium as defined by US-DOE and US-NRC [14].

Reactor type	Burnup GWd/t	Plutonium isotopic composition				
		Pu-238	Pu-239	Pu-240	Pu-241	Pu-242
MAGNOX	5	~0	0.685	0.25	0.053	0.012
CANDU	7.5	~0	0.668	0.265	0.055	0.012
PWR	30	0.016	0.565	0.238	0.128	0.053
Weapon-grade plutonium (US-DOE, US-NRC)	very low	0.00012	0.938	0.058	0.0035	0.00022

Table 1.1 explains that there were already considerable differences for reactor-grade plutonium and weapon-grade plutonium around 1975. These differences increased considerably until 2010. The burnup of LWR spent fuel of LWRs was increased for economical reasons up to 55 GWd/t in 2010 (Section 9).

The following international debate on civil reprocessing and reactor-grade plutonium recycling lead to a two years International Fuel Cycle Evaluation (INFCE) [9] under the guidance of the IAEA in Vienna. The result of these studies was that under the considered burnup conditions of the spent fuel there are no fuel cycle options which could guarantee absolute proliferation resistance. Therefore, it was recommended that safeguard concepts should be further developed and more institutional concepts, such as collocation of fuel reprocessing and re-fabrication plants or international spent fuel storage facilities and other technical measures, e.g. co-processing, should be developed.

Denaturing, i.e., dilution of fissile isotopes by non-fissile isotopes to such an extent that the fissile material can not directly be used for nuclear weapons, was proposed as a technical measure.

For uranium fuel it was proposed keeping U-235/U-238 mixtures below 20% of U-235 and U-233/U-238 mixtures below 12% of U-233 (Section 8.1.1).

Denaturing of reactor-grade plutonium by the isotope Pu-238 to contents higher than about 5% in plutonium in order to increase the proliferation-resistance of reactor-grade plutonium, was proposed in the scientific literature by Campbell and Gift (1978) [10], Heising-Goodman (1980) [11], as well as Massey and Schneider (1982) [12]. Unfortunately these scientific proposals were not pursued further.

Although US-president Reagan lifted the ban on commercial reprocessing in the USA in 1981. US-president Clinton discouraged again the nuclear industry from reprocessing spent nuclear fuel in a policy statement in 1993. In addition, the US National Academy of Sciences declared nuclear fuel reprocessing as impractical and too costly in1996 and the US National Research Council found no economic justification in developing nuclear fuel reprocessing facilities in 2007.

The US administration decided the Nuclear Waste Policy Act in 1982 which defines a nuclear waste policy allowing only direct spent fuel disposal in deep geological repositories. France, the UK, Russia and Japan did not follow this once-through fuel cycle strategy. They built civil reprocessing facilities (LaHague, in France, Windscale known also as Sellafield in the UK and Rokkasho-mura in Japan) to chemically reprocess their own as well as foreign spent fuel. Sweden and Finland decided to follow the once-through fuel cycle concept that provides permanent spent fuel storage in deep geological repositories. Germany and Switzerland allowed both lines (reprocessing with plutonium recycling or direct spent fuel disposal). But Germany refrained from the reprocessing strategy in 2005.

The US spent nuclear fuel disposal site Yucca Mountain was closed in 2009 after the US-Department of Energy had proposed to eliminate all funding awaiting to hear a final decision of the US Nuclear Regulatoty Commission.

The international safeguards system was extended and improved considerably during the time period 1980 - 2006. Among the research and development efforts were the destructive and nondestructive assay methods used for independent measurements by IAEA inspectors. The concepts of material balance areas (MBAs) with key measurement points

(KMPs), as well as near real time accountancy (NRTA) and the containment and surveillance (C/S) concepts were developed and demonstrated for nuclear reactors, reprocessing and fuel refabrication plants. Continuous monitoring of nuclear facilities (unattended monitoring systems) allows more cost effective safeguards surveillance for the future (Section 8.1.7 and 8.1.8). The concept of physical protection for nuclear materials (INFCIRC/225 Rev. 4) [13] was revised in 1999. The so-called Additional Protocol to the NPT (INFCIRC/540) [15] was introduced in 1997. This allows IAEA inspectors access to information and locations in a state (not only those with declared nuclear materials) to follow up on evidence of safeguard violations.

1.2 Safeguards Implementation

By the end of 2003, 189 states including five nuclear weapon states (USA, UK, USSR, France and China) had signed the NPT. Four de facto nuclear weapon states are not parties to the NPT: India, Israel, Pakistan and North Korea. Out of these Israel is widely believed to possess nuclear weapons, but did not openly declare it. India (1974), Pakistan (1998) and North Korea (2006/2009) have first openly tested and then declared that they possess nuclear weapons. North Korea had acceded to the NPT, violated it, and withdrew from it in 2003.

By 2009, there were safeguard agreements in force in more than 145 countries. Only Iran was found in noncompliance with its IAEA safeguards agreement in 2005.

Almost the entire known nuclear industry outside the NWSs is thus under the safeguards control of IAEA. By late 2009, some 229 power reactors, 153 research reactors and critical assemblies, 18 uranium conversion plants, 46 fuel fabrication plants, 13 reprocessing plants, 17 enrichment plants, 118 separate storage facilities, and 76 others (mostly research and development facilities) were under safeguards [26].

1.3 Arms reduction initiatives

As a result of arms reduction initiatives, the USA and Russia agreed in 1991 and 2010 to reduce their nuclear armaments. A considerable amount of highly enriched uranium (HEU) was provided for its use in civilian power reactors. HEU can be downblended with natural uranium to form low enriched uranium (LEU) fuel for nuclear power reactors.

In 1993, Russia agreed to downblend 500 t of HEU into low enriched uranium and sell part of it to the USA for commercial use in nuclear reactors. In 1994, the USA declared, 174 t of its HEU stocks to be excess of military purposes and designated 85% of it to be downblended and converted into low enriched fuel for commercial nuclear reactors. In

2005, the USA announced it would remove another 200 t of HEU from its weapons stockpile [1].

In 1995, both the USA and Russia (former USSR) declared 50 t of their weapon-grade plutonium as surplus to their national security needs. Both the USA and Russia agreed to dispose 34 t of weapon-grade plutonium in 2000 [27]. The UK declared 3 t of its weapon-grade plutonium as surplus to its security needs. Studies by the National Academy of Sciences and other organisations in the USA led to the decision to transform this metallic weapon-grade plutonium into mixed oxide uranium plutonium (MOX) fuel for irradiation or burnup in Light Water Reactors (LWRs). After a burnup period of about 5 years the weapon-grade plutonium will become reactor-grade plutonium. After removal from LWR cores, this spent MOX fuel can be assigned the lowest level "E" for attractiveness in weapons use (US-DOE Safeguards Categories). It will not be useable for weapons purposes (NAS, 1995) [14].

Table 1.2: Estimates of stocks of HEU and weapon grade plutonium [16].

Nuclear Weapon States (NWS)	t of HEU*	t of weapon-grade Pu**
USA***	654	92
Russia***	985±300	145±25
UK***	23.4	7.6
France	36.5	5
China	20	4
De facto Nuclear Weapon States		
India	0.2	0.52
Israel	not known	0.45
Pakistan	1.3	0.064
North Korea	not known	0.035
Non Nuclear Weapon States (NNWSs)	10	-----

* HEU (≥93% enriched in U-235)

** Weapon-grade plutonium ≥94% Pu-239

*** This takes into account the already implemented reductions of HEU by downblending into low enriched uranium for civil nuclear reactors until 2007.

The remaining amounts of HEU and weapon-grade plutonium in the NWSs and in NNWSs are shown in Table 1.2. The respective data collected by the Institute for Science and International Security (ISIS) [25] differ only slightly from those of Table 1.2.

1.4 Amounts of reactor-grade plutonium in the world

The amount of reactor-grade plutonium in spent fuel elements of civil nuclear reactors in the world will be about 2300 t by the year 2010 [18]. Fig. 1.1 shows the data and projection until 2030 for reactor-grade plutonium in the world. Almost one third of this spent fuel was reprocessed and used for the refabrication of MOX fuel which is recycled in MOX fueled LWRs.

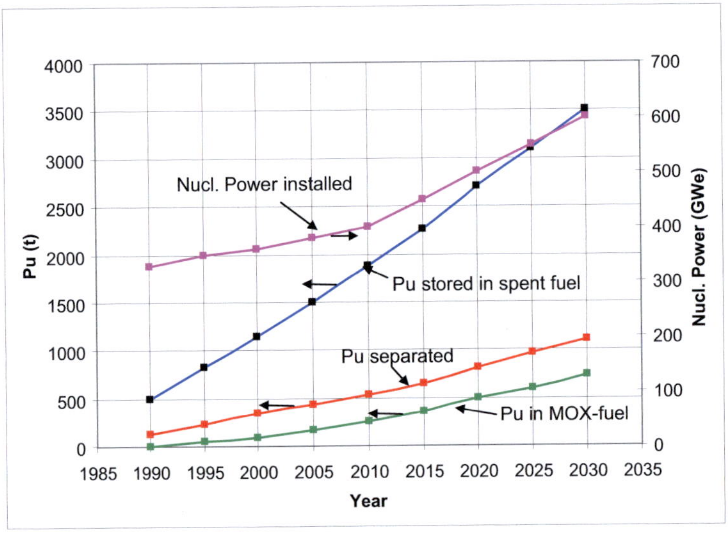

Figure 1.1: Nuclear power installed as well as plutonium stored in spent fuel elements, plutonium separated and plutonium in MOX fuel until 2030 [18].

1.5 Amounts of reactor-grade americium and neptunium in the world

There is growing interest in countries with large nuclear energy programs to separate and incinerate neptunium and americium in order to minimize the radioactive inventories of these long-lived nuclides in deep geological waste repositories. However, separated neptunium and americium have long been of concern in proliferation discussions. The IAEA has, therefore, began to consider a program to monitor also neptunium and americium.

1.5.1 Neptunium and americium

Neptunium is considered useable in nuclear explosive devices. It has a bare critical mass of 57±4 kg [21]. A reflector, e.g., beryllium can reduce the critical mass to approximately 45 kg. Neptunium produces no alpha heat and has a low spontaneous fission neutron rate of 0.11 n/kg·s, which is lower than that of U-235 (0.29 n/kg·s) [22].

Americium generated in nuclear reactors is a mixture of Am-241, Am-242m, and Am-243. Am-241 without admixture of other americium isotopes can originate from the decay of Pu-241. The critical mass of Am-241 was calculated to be approximately 34-45 kg, that of Am-243 between 111 and 193 kg, where both calculations used steel as reflector [23].

The critical mass of Am-242m is as low as 3.7-5.6 kg when reflected by steel [23]. However, Am-242m amounts to less than 1% in the americium of spent LWR fuel, and approximately 4% in americium of spent fuel of fast reactors (FRs). Spontaneous fission neutron emission of Am-241, Am-242m and Am-243 is relatively high. Am-241 has a high alpha particle heat output of 110 W/kg.

The amount of neptunium and reactor-grade americium in spent fuel elements will be, respectively, about 90 t and about 150 t by 2010. Figs. 1.2 and 1.3 display the data and projection until 2030 for reactor neptunium and americium [18].

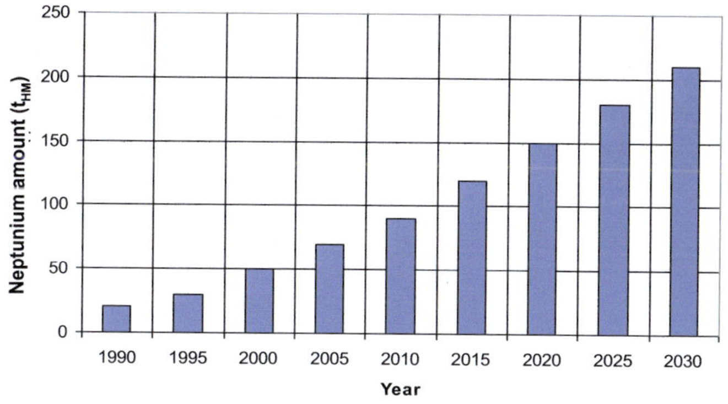

Figure 1.2: World wide neptunium stored in spent fuel elements or high active waste until 2030 [18].

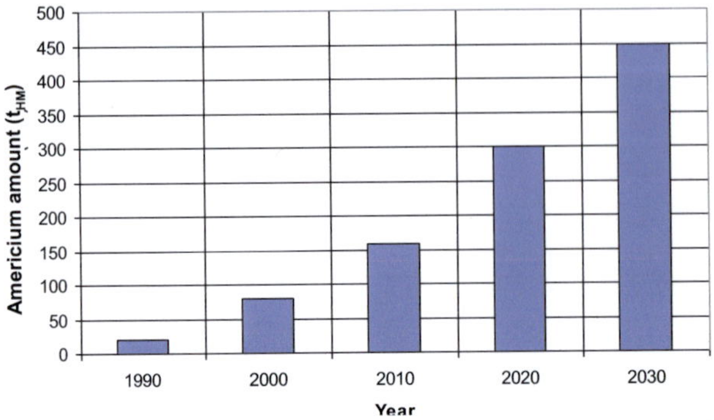

Figure 1.3: World wide americium stored in spent fuel elements or in high active waste until 2030 [18].

1.6 Nuclear fuel cycle concepts

The two fuel cycle concepts that are being pursued are as follows:

- The once-through fuel cycle followed by direct spent fuel disposal in deep repositories (USA, Sweden, Finland), and
- the closed fuel cycle with reprocessing of the spent fuel followed by recycling and incineration of the plutonium (France, Japan etc.).

Technical difficulties were discussed between 2008 and 2010 regarding the US national repository Yucca Mountain after certain temperature limits were set by the US regulatory agencies for the local areas surrounding the waste packages.

In 2010 the US-DOE withdraw the license application for the Yucca Mountain high level waste repository [24].

A new international initiative on the proliferation resistance of future generation nuclear reactors and fuel cycles was initiated around 2005 (INFCIRC/640) [20].

In February 2006 the US government announced the Global Nuclear Energy Partnership (GNEP), which envisions a close coupling of nonproliferation measures with new developments in nuclear energy technology. Among the nonproliferation measures are the possibilities of a small number of states which possess fuel cycle facilities employing advanced technologies. These states could ensure fuel cycle services, e.g. fresh fuel supply and spent fuel back services to other states which have forgone sensitive fuel cycle technologies, but still operate nuclear reactors.

In June 2007 Russia announced an initiative which offers enrichment services to other states and proposes to create a Global Nuclear Power Infrastructure [17].

1.7 New scientific results and further developments

New results from scientific analysis within the international community became apparent after about 2005:

- The burnup of spent fuel in LWRs was about 55 GWd/t around 2010 and will increase to about 70 GWd/t in the future;
- The reactor-grade plutonium in spent fuel of civil reactors after reprocessing and re-fabrication as MOX fuel can be multi-recycled and be incinerated in LWRs, fast neutron reactors (FRs), or accelerator driven systems (ADSs);
- The minor actinides (MAs) can be chemically partitioned by different aqueous chemical processes. They can be used in the future for refabrication of fuel elements and be incinerated by multi-recycling in LWRs, FRs, and ADSs. Another approach is applying pyroprocessing keeping the actinides together and multi-recycle them in FRs;
- Incineration of reactor-grade plutonium and the main important minor actinides (neptunium, americium) will drastically lower the long term radiotoxicity caused by the waste disposal in deep geological repositories;
- The isotopic content of the isotope Pu-238 in the reactor-grade plutonium can be increased to above 5% by different fuel cycle options (re-enriched reprocessed uranium (RRU) or MOX fuel with small addition of minor actinides). This denatured reactor-grade plutonium can be considered proliferation-proof as it will make so-called Hypothetical Nuclear Explosive Devices (HNEDs) technically unfeasible;
- The denatured and proliferation-resistant reactor-grade plutonium can be fully incinerated by multi-recycling in LWRs, FRs or ADSs;
- During multi-recycling steps, denatured plutonium can remain denatured by small additions of reactor-grade americium (not useable for nuclear weapons). At the same time the production of neptunium (useable for nuclear weapons) can be avoided.

After earlier suggestions [19], in 2005 the IAEA proposed the so called multilateral fuel cycle centers, which should be built and operated by multinational industrial companies (INFCIRC/640) [20]. This opens the innovative possibility of combining the above scientific findings with the new IAEA proposal of 2005:

During a transition period of several decades, the fuel reprocessing and refabrication for the production of denatured proliferation-resistant reactor grade plutonium could be done in existing reprocessing and refabrication centers in the NWSs (LaHague in France and Sellafield in the UK). They would provide denatured, proliferation-proof reactor-grade MOX fuel for LWRs and FRs in the NNWSs (Sections 12 - 14).

In addition, two NWSs (USA and Russia) refabricate their weapon-grade plutonium for irradiation in MOX-LWRs (USA) or MOX-FRs (Russia).

In a second phase, the NNWSs could irradiate the denatured proliferation-proof plutonium in own LWRs or FRs, and reprocess, refabricate and recycle the proliferation-proof plutonium in Multilateral Fuel Cycle Centers (with IAEA safeguards) of NNWSs. Neptunium would virtually be avoided. In this second phase only denatured proliferation-proof reactor-grade plutonium would exist and be maintained in NNWSs. Neptunium would be avoided. All existing and future originating proliferation-proof reactor-grade plutonium would be incinerated in the future (Sections 13 and 14).

- The addition of small amounts, e.g. 0.1% U-232 to the <20% U-235 enriched uranium fuel of present research and test reactors or to the <20% U-235 enriched U/Th fuel of future Light Water Reactors operating in the U/Th fuel cycle will make such fuel **proliferation-proof** against re-enrichment in a small cascade of ultra-centrifuges.
- Similarly the contaminant U-232 originating from nuclear reactions of neutrons with Th-232 in the U/Th fuel can make this fuel **proliferation-proof** against re-enrichment in small cascades of gas ultra-centrifuges.

The present safeguards and surveillance concept of the IAEA would still be needed in its present form, e.g. to prevent concealed misuse of nuclear facilities. However, many presently discussed problems of IAEA safeguards survey would no longer exist. This will be discussed in chapters 8 to 14.

References Section 1

[1] Tape, J., and Pilat, J., Nuclear safeguards and the security of nuclear materials. In J.E. Doyle: Nuclear Safeguards, Security and Nonproliferation. Butterworth and Heinemann, Elsevier (2008).

[2] The Agency's Statute (as amended up to 1 June 1973). Vienna: International Atomic Energy Agency (1980).

[3] The Agency's Safeguards System (1965, as provisionally extended in 1966 and 1968). Vienna: International Atomic Energy Agency. INFCIRC/66/Rev. 2 (1968).

[4] UN Treaty Series No. 9068: Treaty for the Prohibition of Nuclear Weapons in Latin America, 14 February 1967. New York: United Nations (1967).

[5] The Treaty on the Non-Proliferation of Nuclear Weapons; London, Moscow, Washington, 1 July 1968. Vienna: International Atomic Energy Agency, INFCIRC/140 (1970).

[6] The Structure and Content of Agreements between the Agency and States Required in Connection with the Treaty on the Non-Proliferation of Nuclear Weapons. Vienna: International Atomic Energy Agency, INFCIRC/153 Corrected (1972).

[7] Taylor, Th., Nuclear Safeguards. In: Annual Review of Nuclear Science, 25, pp. 406-21. Palo Alto, Cal.: Annual Reviews Inc. (1975).

[8] Nuclear power, issues and choices. Report of the Nuclear Energy Policy Study Group sponsored by the Ford Foundation. Administered by MITRE cooperation. Ballinger, Cambridge, Mass. (1977).

[9] INFCE (International Nuclear Fuel Cycle Examination). Report of the First Plenary Conference of the International Nuclear Fuel Cycle Examination (INFCE). International Atomic Energy Agency, Vienna (1980).

[10] D.O. Campbell, E.M. Gift, Proliferation Resistant Nuclear Fuel Cycles, ORNL/TM-6392, Oak Ridge National Laboratory (1978).

[11] C.D. Heising-Goodman, An Evaluation of the Plutonium Denaturing Concept in an Effective Safeguards Method, Nucl. Technol., 50, 242 (1980).

[12] J.V. Massey, A. Schneider, The Role of Plutonium-238 in Nuclear Fuel Cycles, Nucl. Technol., 56, 55 (1982).

[13] IAEA (International Atomic Energy Agency). The Physical Protection of Nuclear Materials and Nuclear Facilities. IAEA-INFCIRC/225/Rev. 4 (Corrected), IAEA, Vienna (1999).

[14] NAS (National Academy of Sciences), Committee on International Security and Arms Control, Panel on Reactor-Related Options. Management and Disposition of Excess Plutonium: Reactor-Related Options. National Academy Press, Washington, D.C. (1995).

[15] Model protocol additional to the agreements between states and the IAEA for the application of safeguards. INFCIRC/540 Corrected (1997).

[16] Sokova, E. et al., Elimination of excess fissile material. In J.E. Doyle: Nuclear Safeguards, Security and Nonproliferation. Butterworth and Heinemann, Elsevier (2008).

[17] Russian initiative for creation of global nuclear power infrastructure. INFCIRC/708 (2007).

[18] Fukuda, K. et al., Prospects of Inventories of Uranium, Plutonium and Minor Actinides and Mass Balance, presented at 2nd Consultancy Meeting Protected Plutonium Production Project, Vienna, June 15-16, 2006, International Atomic Energy Agency.

[19] Haefele, W., Energy in a finite world, Harper and Row (1982).

[20] IAEA, Multilateral approaches to the nuclear fuel cycle: Expert group report submitted to the director general of the IAEA, INFCIRC/640 (2005).

[21] Loiza, D. et al., Results and analysis of spherical neptunium-237 critical experiments surrounded by highly enriched uranium hemispherical shells. Nucl. Science and Engineering, 152, 65-75 (2006).

[22] Holden, N. et al., Spontaneous fission half lives for ground state nuclei. Pure Appl. Chem., 72, 8, 1525-1562 (2000).

[23] Diaz, M. et al., Critical mass calculations for Am-241, Am-242m and Am-243, Int. Conf. Nucl. Criticality Safety (ICNC 2003), Tokai-mura, Japan.

[24] US-DOE withdraws repository license application, Nuclear News, 53, 4, p. 63 (2010).

[25] Institute for Science and International Security (ISIS), Global Plutonium and Highly Enriched Uranium Stocks (2004), www.isis-online.org

[26] http://www.iaea.org/publications/Reports/Anrap2009/table_a5.pdf

[27] Agreement between the government of the United States of America and the government of the Russian Federation concerning the management and disposition of plutonium designated as no longer required for defense purposes and related cooperation U.S./Russia (2000).

2 Technical applications of nuclear power reactors

The majority of nuclear power reactors built and operated today is used for electricity generation. Such nuclear power reactors are built in unit sizes up to 1300 and 1600 MW(e) and operated in the so called base load regime. Other potential applications are process heat generation for the substitution of oil and natural gas as primary sources of energy.

Smaller size nuclear power reactors are used for ship propulsion, mainly submarines, air craft carries etc. Early attempts for application of nuclear power for commercial ship propulsion, e.g. the ships Savannah (USA), Otto Hahn (Germany), Mutsu (Japan), were given up for economical reasons and because of difficulties to obtain permits to stay in international ports. Only Russian icebreakers, e.g. N.S. Lenin, are still operating. They serve to keep the Russian Arctic route open for ship traffic.

More than 80% of the electricity producing nuclear power reactors are light water moderated and cooled reactors (LWRs). Their heat produced by the fission process in the uranium fuel elements is used to achieve steam conditions of about 290 °C and 70 to 78 bars to drive the steam turbine and generator. This leads to a thermal efficiency of about 33 to 36% for electricity generation.

High temperature graphite moderated and helium gas cooled nuclear reactors or advanced gas cooled reactors (AGRs or HTRs) attain gas temperatures of 700-900 °C. In that range of temperature a number of technical processes requiring process heat are possible. If the gas temperatures are used for steam generation and electricity production a thermal efficiency of 42% can be attained.

A different class of nuclear power reactors are Liquid Metal cooled Fast Breeder Reactors (LMFBRs) with a fast neutron spectrum. Contrary to light water in LWRs or graphite in AGRs and HTRs the liquid metal does not thermalize the fission neutrons. As a consequence these nuclear power reactors can use the abundant non fissile U-238 by converting it to the fissile plutonium (breeding process) which is then utilized as fissile material. Also thorium can be converted to the fissile U-233 by this breeding process.

Other reactor types, e.g. heavy water moderated reactors (HWRs) will be described in Section 5.

2.1 Nuclear reactors operating in the world in 2008

In 2008, there were 439 nuclear reactors with a total electric capacity of 372 GW(e) operating in the world. They supplied about 16% of the world's electricity. The countries providing the largest nuclear reactor electricity generating capacities are presented in Table 2.1 [1,2]. Smaller countries operating only several GW(e) of nuclear powere are listed in [7]. Nuclear power is generated (Tab. 2.2) chiefly in

- light water reactors (LWRs) of which there are two types: the Pressurized Water Reactors (PWRs) and the Boiling Water Reactors (BWRs)
- the Heavy Water Reactors (HWRs or CANDUs)
- the Advanced Gas Cooled Reactors (AGRs) and
- the Light Water Cooled Graphite Moderated Reactors (LWGRs)
- the Liquid Metal Cooled Fast Breeder Reactors (LMFBRs).

Table 2.1: Worldwide nuclear generating capacity in June 2008 [1,2].

Country	Number of operating reactors	Generating capacity (GW(e)net)
United States	104	100,582
France	59	63,260
Japan	55	47,587
Russian Federation	31	21,743
Germany	17	20,470
Korea	20	17,451
Ukraine	15	13,107
Canada	18	12,589
United Kingdom	19	10,222
Sweden	10	9,014
China	11	8,572
Spain	8	7,450
Rest of the world	72	40,155
Total	439	372,202

Table 2.2: Fractions of different types of nuclear reactors [1,2].

Type	number	percentage %
PWR	265	60.36
BWR	94	21.41
CANDU (HWR)	44	10.02
AGR	18	4.10
LWGR	16	3.64
LMFBR	2	0.45

In 2016 the number of nuclear reactors in operation has changed slightly to 447. In addition, 60 nuclear reactors were under construction. Fig. 2.1 shows the projections for nuclear power capacity of IAEA 2006 and of OECD/NEA worldwide until 2040.

More than 40 additional nuclear power reactors were under construction by 2009. About 80% of these were again LWRs. In addition many countries including China, India, Japan, Korea and Russia announced ambitious plans to expand their nuclear power capacities.

The future market share of nuclear power in the world electricity market will depend on its competitiveness against other established sources of primary energy, e.g. hard coal, lignite, oil and natural gas or hydropower and renewable energies (wind, photovoltaics etc.). In addition other factors like public acceptance and the solution of the proliferation problem will play a dominant role.

Nuclear power reactors are operating with yearly energy availability factors of 85 to 90%. Their operating life time is increasing from 35 years at present to 60 years in the future.

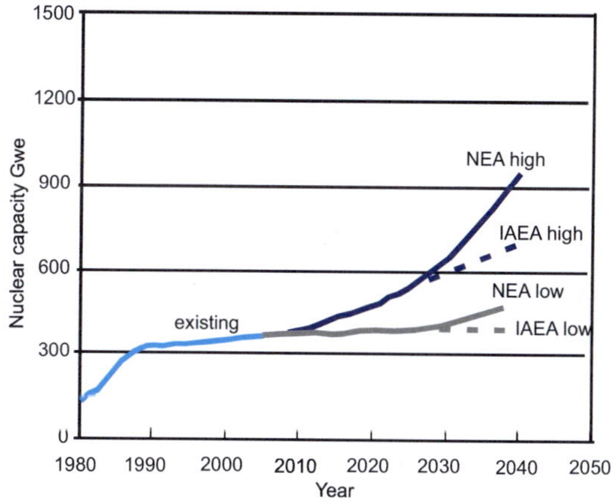

Figure 2.1: Existing and projected nuclear power reactor capacity [1,2].

2.2 The nuclear fuel cycle

Uranium ores contain 0.72% U-235 and 99.28% U-238 (disregarding the tiny amount of 0.0054% U-234). Only the isotope U-235 can be fissioned by thermalized neutrons in nuclear power reactors. After mining of uranium ores these are milled and chemically processed for conversion into the uranium oxides UO_2, UO_3 and U_3O_8 (Fig. 2.2) [3,4,5]. These are then chemically treated and converted into the only gaseous uranium compound

UF_6. In enrichment plants which presently apply either the gas diffusion or gas centrifuge technique the natural isotopic mixture of the UF_6 gas is enriched to about 5% U-235 and 95% U-238. This low enriched UF_6 is chemically re-converted to low enriched UO_2 fuel powder. The UO_2 fuel powder is pressed into cylindrical pellets which are sintered at more than 1600 °C. The pellets are filled into Zircaloy tubes which are assembled to fuel elements also called fuel assemblies

Fuel elements with metallic fuel are also used for LMFBRs (Section 6).

The fuel elements of LWRs produce about 50 to 60 GWd/t over an operation time of about 5 years. Partial loading, reloading and shuffling schemes after time periods of about 12 to 24 months guarantee an optimal economic operation of the nuclear power reactor. The unloaded spent fuel is stored for several years in on site spent fuel storage water pools or dry containers cooled by air. Afterwards the spent fuel is either sent to spent fuel conditioning for later direct fuel disposal in deep geological repositories **(open fuel cycle,** Fig. 2.2a) or sent to a spent fuel reprocessing plant for chemical separation of the uranium, of the generated plutonium and of the minor actinides. The separated fission products become high active waste. Separated plutonium and separated minor actinides can be further recycled as fissile fuel. The fission products are vitrified and – after further cooling over about 40-50 years – deposited in deep geological repositories **(closed fuel cycle)** (Fig. 2.2b).

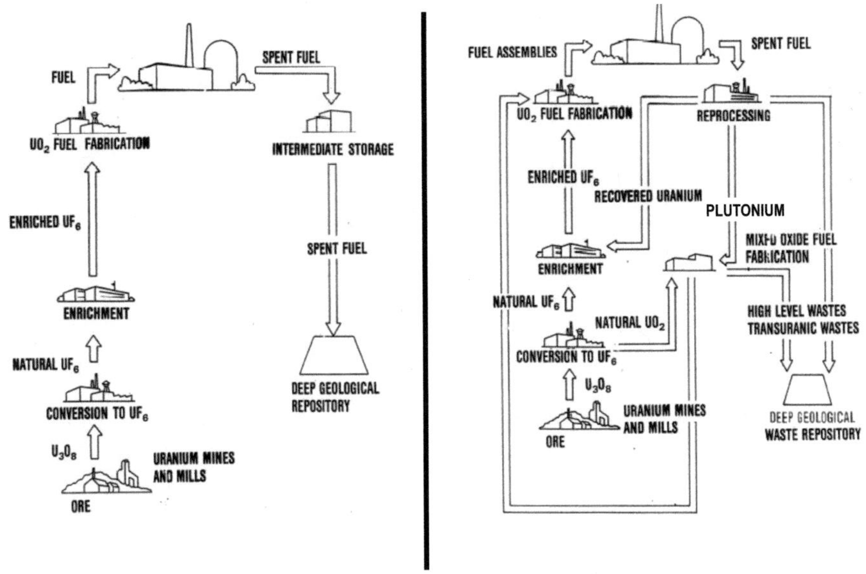

a) Once-through fuel cycle b) Closed fuel cycle

Figure 2.2: The nuclear fuel cycle (once through and closed fuel cycle) (NUREG).

The infrastructure needed for nuclear power reactors are uranium mining and milling, uranium conversion, uranium enrichment, uranium fuel fabrication and spent fuel reprocessing facilities.

Table 2.3 presents the number of commercial operating facilities of the uranium fuel cycle in the world.

Table 2.3: Number of fuel cycle facilities operating in 2008 [1,2,3].

Process	Number of facilities in commercial operation
Uranium mining and milling	37
Conversion	22
Enrichment	13
Uranium fuel fabrication	40
Spent fuel reprocessing	5

The largest uranium conversion plants are located in Canada, France, Russia and the USA. The largest uranium enrichment plants operate in the USA, Russia, France, the UK, Germany and the Netherlands. Only France, the UK, Russia and Japan operated large scale commercial reprocessing plants in 2016.

2.3 Natural uranium ores

Natural uranium is found in uranium ores in concentrations ranging from around fractions of a percent to several percent. It contains the fissile isotope U-235 found in nature in an isotopic concentration of 0.7204%. The remaining isotopes are 99.2742% of U-238 and 0.0054% U-234. Uranium ores are obtained from open pit or underground mining. In addition it can be produced by in situ leaching or it is found as a co-product or by-product during mining of other materials, e.g. phosphate rock, mineral sands etc.

2.3.1 Uranium resources

Uranium resources are classified on the basis of geological certainty and foreseeable costs of mining. Reasonably assured resources (RAR) are based on high confidence estimates which are compatible with decision-making standards for mining. Inferred resources are defined on a similar basis, but additional measurements are required before making decisions for mining.

Undiscovered resources (prognosticated and speculative) are expected to exist, based on geology or previously discovered resources. Both prognosticated and speculative resources still require significant efforts for exploration. The OECD nuclear energy agency

(NEA) and the IAEA collect the resource data of the different countries in the world on a yearly basis. These data are categorized on a **fictive** US dollar recovery cost base.

Table 2.4 shows the world uranium resources (reasonable assured and inferred) for three cost categories as published by OECD/NEA and IAEA in 2007 (other data given in the literature may vary slightly from these IAEA and OCD/NEA data).

Table 2.4: Reasonably assured and inferred natural uranium resources (tonnes) in the world as reported by 2007 [1,2].

	Cost ranges		
	<US dollar 40/kg U	<US dollar 80/kg U	<US dollar 130/kg U
Reasonably assured resources (RAR)	1,766,400	2,598,000	3,338,300
Inferred resources	1,203,600	1,858,400	2,130,600

Table 2.5 shows the distributions of the shares for natural uranium resources and natural uranium production in 2007 in the main uranium producing countries [2]. The total annual world production of uranium was about 39,000 t.

Table 2.5: Shares of uranium resources and production in some countries in 2007.

Country	% of resources	% of production
Australia	23.0	21
Canada	7.7	23
United States	6.2	4
Namibia	5.0	7
Niger	5.0	8
South Africa	8.0	1
Kazakhstan	14.9	16
Russian Federation	10.0	8
Uzbekistan	2.0	6
Ukraine	3.6	2

Undiscovered (prognosticated and speculative) resources were estimated to another 7,771,100 tonnes worldwide [1,2]. In addition, so called unconventional natural uranium resources in black shales, e.g. at Chattanooga (USA), with a total of 4.2 million tonnes of uranium resources as well as additional uranium resources associated with monazite, coastal sands and phosphorite deposits are mentioned by OECD/NEA and IAEA. The total global amounts are estimated to 22 million tonnes of natural uranium resources.

Seawater may also be regarded as a possible resource with very low concentration of uranium and high extraction costs.

2.3.2 Thorium resources

Thorium can also be used in combination with enriched uranium as a fertile nuclear fuel. It is found mainly as Th-232 and is considered to be three times more abundant in the earth's crust than uranium [8].

It is mainly recovered from monazite sand as a by-product and from deposits of titanium- zirconium- or tin-bearing minerals. Worldwide reasonably assured and inferred thorium resources were estimated in 2007 to a total of 6.08 million tonnes of thorium [2].

2.4 Concentration of uranium

Uranium ores must be separated from byproducts. They are concentrated by physical concentration methods, e.g. crushing, and gravity, magnetic or flotation types of separation. Roasting is applied to improve the solubility of the uranium. Leaching is performed by different agents depending on the type of uranium-bearing minerals. Agents can be either sulfuric, nitric and hydrochloric acids or sodium hydroxide and alkaline carbonates. Solvent extraction methods are preferred, if acid solutions are employed. In the solvent extraction process the active agent is an inorganic amine salt diluted in kerosene that can selectively extract the uranium ions. The uranium is finally precipitated as uranium diuranate and dried afterwards. The result is called yellow cake [3, 4, 5].

2.5 Purification of uranium

For purification the uranium concentrates are dissolved in nitric acid. The resulting uranyl-nitrate is then extracted by tributyl phosphate in kerosene. By this purification step some elements like boron, beryllium, cadmium, rare earths and other elements are removed down to a concentration of a few ppm. The end product after purification is one of the uranium oxides UO_2, UO_3 or U_3O_8.

2.6 Uranium conversion

The uranium oxides are chemically converted into uranium hexafluoride, UF_6. This is the only volatile or gaseous compound of uranium. It has a sublimation point of 56.5 °C at 1 bar. This gaseous compound UF_6 is necessary for the enrichment process in gaseous enrichment plants.

2.7 Natural uranium consumption and needs by the nuclear power industry

2.7.1 Natural uranium consumption by different reactor types

As already explained in Section 2.2 the natural uranium must be enriched, e.g to. 3.5 to 5% of U-235 for the low enriched fuel (LEU) of LWRs. For HWRs or CANDUs it can be either the natural enrichment or LEU with enrichment of about 1.5% U-235. For AGRs or HTRs the enrichment of the LEU fuel must be about 8% U-235 due to the higher burnup foreseen for their fuel.

The yearly need of natural uranium for the different reactor types operating about 85% of the year at full power is given for several examples in Table 2.6. Only for the LWR-LEU a second value for a load factor of 0.93 is given. This load factor is reached by many LWRs now.

Due to their better neutron economy (heavy water and graphite absorb less neutrons than light water) HWRs and AGRs consume less natural uranium than LWRs. Plutonium recycling can reduce the natural uranium consumption by about 35%, but it requires reprocessing of the spent fuel and MOX fuel refabrication (closed fuel cycle). FRs with plutonium fuel can make use of the 99.27% U-238 of the natural uranium. They are continuously breeding new plutonium resulting from neutron capture in U-238. Consequently their natural uranium or depleted uranium or depleted uranium consumption is very low with about 14 te/GWe·a. FRs can also operate on thorium as a fertile fuel.

Table 2.6: Annual natural uranium consumption [tonne] for different reactor types per GW(e) and year operating 85% of the year at full power.

Reactor type	Initial fuel enrichment [%]	Natural uranium consumption [t/GW(e)·y]
LWR-LEU	3.3	171*
LWR-LEU	3.3	156
HWR-U_{nat}	0.7	150
HWR-LEU	1.5	107
HTR-LEU	8	125
LWR U/Pu recycle	4%	110
FBR U/Pu breeder	20%	1.4**

* load factor 0.93

** depleted uranium or natural uranium

2.7.2 Future need for natural uranium

The future need for natural uranium can be illustrated by the following simple example: Assuming a constant installed nuclear reactor power capacity of 400 GWe over the next 60 to 80 years (which corresponds roughly to the projection of IAEA low in Fig. 2.1) then the natural uranium consumption in future decades is shown by Fig. 2.3.

Fig. 2.3 shows that 400 GW(e) LWRs-LEU with a load factor of 0.93 operating during a lifetime of 60 years would consume about 4.1 million tonnes of natural uranium. HWRs-LEU or HTRs with a power capacity of 400 GW(e) would consume 2,57 and 3,0 million tonnes of natural uranium over 60 years, respectively. Uranium/plutonium recycling in LWRs would lead to similar savings of natural uranium. Comparing these numbers with the world uranium resources of Section 3 one may conclude that LWRs would consume the uranium resources in the cost range of <130 \$/kg U (reasonably assured and inferred, Table 2.4) either within 80 (LWR-LEU) or up to about 140 (LWR U/Pu recycle) years. For other reactor types (HWRs, AGRs or HTRs) this time period would be similar. If prognosticated and speculative or even unconventional uranium resources would be considered, these time periods could be extended. Future increases in the installed nuclear power capacity above the simple example of constant 400 GW(e) (see, e.g. INFCE high, Fig. 2.1) would shorten these time periods accordingly.

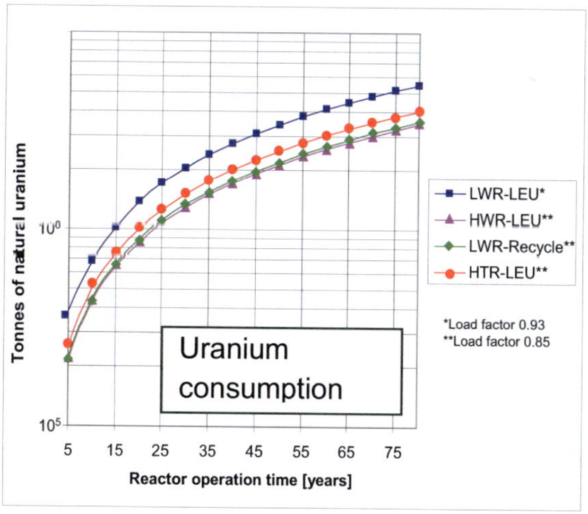

Figure 2.3: Natural uranium consumption (tonnes of natural uranium) for different reactor systems as a function of time.

Reprocessing and recycling of uranium and plutonium or better neutron economy of the different reactors can help to curb the natural uranium consumption. A major cutback of natural uranium consumption, however, can only be achieved by the introduction of breeder reactors with a fast neutron spectrum. Fig. 2.4 shows the natural uranium consumption for 3 cases in general form

- LWR-LEU in once through fuel cycle (OT) (load factor 0.93)
- HWR-U_{nat} in once trough fuel cycle (OT)
- LWR-U/Pu recycle in the closed fuel cycle
- LWR + LMFBR in the closed fuel cycle

Fig. 2.4 is based on scenarios for nuclear power installations which where assumed during INFCE. Although these scenarios might not be fully consistent with future developments, the form of the curves for the consumption of natural uranium of the different reactor scenarios is generally valid.

Fig. 2.4 shows that only the large scale introduction of fast neutron spectrum liquid metal cooled breeder reactors (LMFBRs with oxide, carbide, nitride or metallic Pu/U fuel) are able to curb the natural uranium consumption curves. A large scale introduction of LMFBRs around, e.g. 2050 could limit the total natural uranium consumption to about 6-7 mill. tonnes [6]. If the LMFBRs are introduced later into the energy market, then the total natural uranium consumption would be higher correspondingly or the low cost rescources would be exhausted at an earlier time.

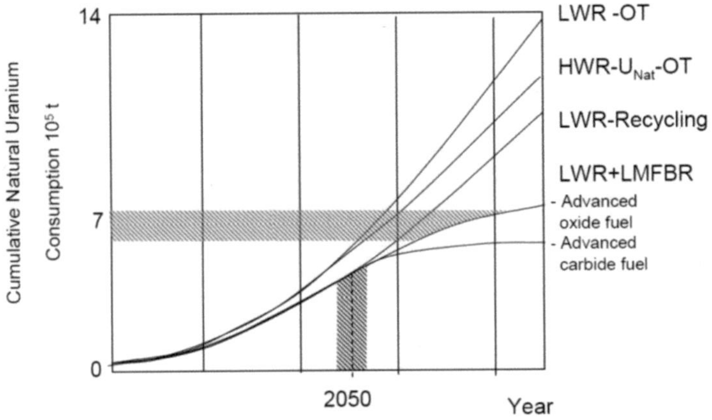

Figure 2.4: General projection of natural uranium consumption for different reactor types including fast neutron spectrum breeder reactors.

The technical feasibility of LMFBRs was already proven between the years 1970-2010 (Section 6). Prototype LMFBRs with electrical power of 250 to 800 MWe were operated safely over more than 30 years. The time and rate of introduction of LMFBRs depends mainly on the availability of reprocessing and refabrication plant capacities as well as political constraints (proliferation policy).

References Section 2

[1] Nuclear energy outlook, NEA No. 6348, OECD (2008).

[2] Uranium 2007: Resources, production and demand, A joint report by OECD/NEA and IAEA, NEA No. 6345, OECD (2008).

[3] Hardy, C.J., The Chemistry of Uranium Milling. Radiochimica Acta, 25, 121-134 (1978).

[4] Seidel, D.C., Extracting Uranium from its Ores. International Atomic Energy Agency Bulletin, 23 (2), 24-28 (1981).

[5] Benedict, M. et al., Nuclear Chemical Engineering. New York: McGraw-Hill (1981).

[6] Carré, F., et al., Overview of French nuclear fuel cycle strategy and transition scenarios, Proceedings of Global 2009, paper 9439, Paris (2009).

[7] http://web.archive.org/web/20080303234143/http://www.uic.com.an/reactors.htm

[8] Thorium based fuel options for the generation of electricity: Developments in the 1990s, IAEA-TECDOC-1155 (2000).

3 Uranium enrichment

3.1 Introduction

Naturally occuring uranium consists of three isotopes: U-234, U-235 and U-238 [1]. The amounts in weight percent of each isotope are shown in Table 1.

Table 3.1: Isotope concentration of natural uranium in weight percent [1].

U-234	0.0054%
U-235	0.7204%
U-238	99.2742%

Only the naturally occurring isotope U-235 is fissionable by thermal neutrons. Fission neutrons from spontaneous or neutron induced fission have an average kinetic energy of about 2 MeV. The probability of a neutron to fission U-235 is much higher for slow neutrons (thermal energies) compared to that for high energy fission neutrons. Therefore, the fission neutrons must be moderated by scattering processes from an average kinetic energy of about 2 MeV down to an average thermal energy of 0.025 eV. This degradation of neutron energy is called slowing down process.

Good moderators are light and heavy water as well as beryllium and graphite. However, only heavy water or high purity graphite together with carbon dioxide or helium gas and low neutron absorbing structural materials can be used in nuclear reactor cores together with the naturally occurring uranium in appropriate material arrangements in the reactor core. Such nuclear reactors are the heavy water moderated reactors (HWRs or CANDUs) and the gas cooled graphite moderated reactors. Their fuel is exchanged continuously at full power for attaining good neutron economy. However, only a low burnup can be attained with natural uranium fuel.

Light water as moderator and coolant requires enrichment of the natural uranium to about 3-5% U-235. Even higher enrichment of about 8% or more U-235 is needed for high temperature gas cooled reactors (HTGRs) which can attain higher burnup of the fuel (Section 5).

3.2 Enrichment technologies

There are a number of enrichment technologies which were tested at begin of the nuclear energy development. However, for large scale commerical enrichment only two methods: gaseous diffusion (USA , France) and gas centrifugation (Russia, Europe) became predominant with more than 90% of installed enrichment capacity in the world. In addition one enrichment plant using LASER enrichment – the so called SILEX process – is being built in the USA from 2010 on [2, 3, 4].

Other enrichment methods like aerodynamic methods (separation nozzle (Germany) or advanced vortex tube (Helikon in South Africa) were only tested and operated on pilot plant scale. LASER enrichment methods based on the atomic vapor laser excitation were given up around 1994 in the USA. A variation of the molecular laser isotope separation (MLIS) technology developed in Australia, is being applied on a large commercial scale as the SILEX (separation of isotopes by laser excitation) process for the first time in the USA [5,6,7].

Table 3.2 shows the world enrichment capacities by the year 2008 [4]. The enrichment capacity is given in kg or ton separative work units (SWU). The separative work unit is a measure of the amount of work necessary in the enrichment plant to produce a certain amount of enriched uranium. It has the dimension of mass in kg or ton SWU (Section 3.6). As an example, the annual reload of a 1 GW(e) PWR of 25 tons of 4.4% U-235 enriched UO_2 fuel requires about 175 t SWU.

About 53 million kg (SWU) are installed in operating enrichment plants, mainly in Russia, the USA and Europe. The large scale gaseous diffusion plants in the USA and France still represent about 42% of the world enrichment capacity. Gaseous diffusion enrichment will slowly be phased out in the future and be replaced by the more economic gas ultra-centrifuge technology. About 23 million kg SWU in gas ultra-centrifuge technology were under construction or in the planning phase in 2008. Again most of this gas ultra-centrifuge capacity is located in Russia, Europe and in the USA. China, South Korea and Japan – having large nuclear power reactor programs – will certainly make up with own enrichment capacities. In addition, the laser isotope separation technology may become an additional economically promising enrichment technology.

Table 3.2: Enrichment capacities installed in gaseous diffusion, gas centrifuge and LASER enrichment plants in the world [4].

Enrichment method	Enrichment capacity in million kg SWU/a		
	operating	under construction	pre-licensing planned
Gaseous diffusion			
USA	11.3		
France	10.8		
China	0.2		
Gas centrifuge			
Russia	20		
UK	3.7		
Netherlands	3.5		1.0
Germany	1.8		2.7
USA		3.0	6.8
France			7.5
Japan	0.3		1.2
China	1.0		0.5
Brazil		0.13	
Iran			0.25
India	0.01		
Pakistan	0.02		
Laser (SILEX)			
USA			3.5-6.0
TOTAL	52.63	3.13	23.45 – 26.45

Multinational enrichment companies like URENCO (UK, Netherlands, Germany, France) are favored by IAEA. Russia has offered his large industrial enrichment capacity for future multinational participation [4].

3.3 Enrichment and cascade theory

The smallest element of an enrichment plant is a separating unit (Fig. 3.1.). The feed, e.g. natural uranium (0.72% U-235, 99.27% U-238) with mass flow rate F and fissile atom fraction x_F in % is separated into a heads fraction (enriched in the desired isotope U-235) with mass flow rate P and atom fraction x_P in % and a tails fraction (depleted in the desired isotope U-235) with a mass flow rate W and atomic fraction x_W in %. The abundance ratio, ξ, often used instead of the atom fraction, x in %, is defined as $\xi = \frac{x}{1-x}$ [2,8].

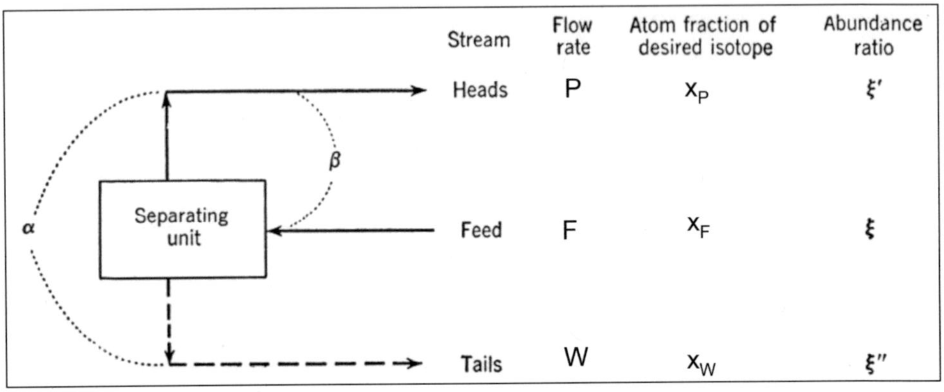

Figure 3.1: Separating unit and definitions.

The material balance on both isotopes is

$$F = P + W \tag{3.1}$$

$$F \cdot x_F = P \cdot x_P + W \cdot x_W \tag{3.2}$$

The ratio of head to feed is called the cut, θ

$$\theta = \frac{P}{F} \tag{3.3}$$

The stage separation factor, α, is defined as the abundance ratio of heads to tails

$$\alpha = \frac{x_P}{x_W}\frac{(1-x_W)}{(1-x_P)} \tag{3.4}$$

and the heads separation factor, ß, is defined as the abundance ratio of heads to feed

$$ß = \frac{x_P}{x_F}\frac{1-x_F}{1-x_P} \tag{3.5}$$

As the required enrichment cannot be achieved with one single separating unit, several or many separating units working in parallel are connected to a cascade. In a simple cascade the heads from the first row of parallel working separating units – called a stage – become the feed of the next stage of separating units (Fig. 3.2).

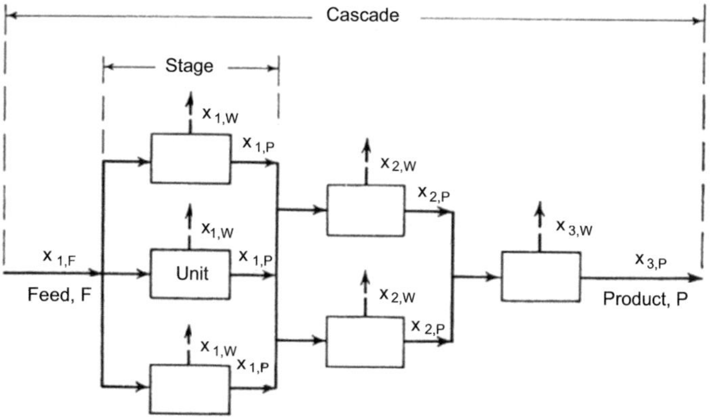

Figure 3.2: Separating unit, stage and cascade [2].

A countercurrent recycle cascade is used, if the partially depleted tails of a stage have sufficient value to warrant reprocessing. This countercurrent cascade scheme is applied for uranium enrichment. Fig.3.3 shows the scheme of a countercurrent recycle cascade.

The feed for each stage consists of the heads from the next lower stage and tails from the next higher stage.

The number of stages between the feed point (natural uranium feed, $x_F = 0.72\%$) and the product (4% enriched uranium, $x_P = 4\%$) is called the enriching section. The portion between the feed point (natural uranium) and the waste (tails) (0.2% depleted uranium, $x_W = 0.2\%$) is called the stripping section.

The aim of the enriching section is to produce enriched uranium. The purpose of the stripping section is to increase the recovery of the desired isotope U 235 from the feed and to reduce the amount of feed required. The stages in Fig. 3.3 are numbered from stage 1 at the waste end to stage n at the product end. The highest stage of the stripping section is numbered n_w.

As the flow in the enriching section becomes smaller the higher the product (enriched uranium) is enriched also the number of separating units as well as the mass flow per stages is lowered. This leads to so called tapered cascades. Tapered cascades can be optimized by selecting different cascade design parameters.

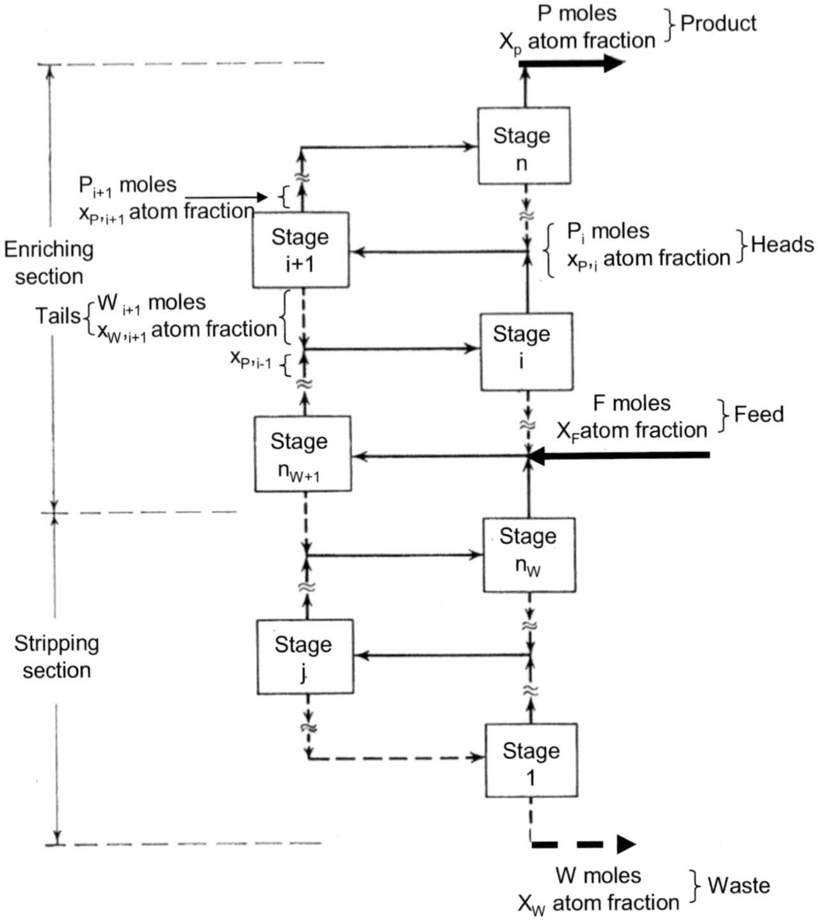

Figure 3.3: Countercurrent recycle cascade [8].

3.4 Ideal symmetric cascade

The theory of ideal symmetric cascades was originally developed by Dirac and Peierls in the UK and Cohen and Kaplan in the USA [2, 8, 9]. It leads to minimum total interstage mass flow and is approximated by all enrichement plants.

Table 3.3 shows the characteristic shape of an ideal cascade for a gas centrifuge plant (Section 3.8). Gas centrifuges have a relatively high stage separation factor α per separation unit and therefore a relatively low number of stages. Table 3.3 show the number of stages and the number of centrifuges (machines) per stage for enrichment of natural uranium to 3.02% U-235 and a tails assay of 0.29% U-235 in uranium [10].

Gaseous diffusion plants have only a separation factor of 1.0043 per separation unit. The number of stages is therefore about 1400 to attain a U-235 enrichment of about 4% [2]. For an ideal symmetric cascade the flow rates are adjusted such that head and tail streams feeding to a stage have the same compositions (Fig.3.3) [8]. In ideal cascade Fig.3.4 can be approximated by squares with equal stages. This leeds to a higher number of equal machines and lower capital costs [2,7,8]. An ideal cascade and its approximation by square is shown in Fig. 3.4.

Table 3.3: Number of centrifuges required in an approximated ideal cascade, assuming that each gas centrifuge has a separation factor 1.2 [10].

Number of machines in stage	Concentration % of feed to stage
10	3·02 product
23	2·53
38	2·12
55	1·77
76	1·48
192	1·24
132	1·03
168	0·86
146	0·72 feed
118	0·60
86	0·50
47	0·42
	0·35
	0·29 tails

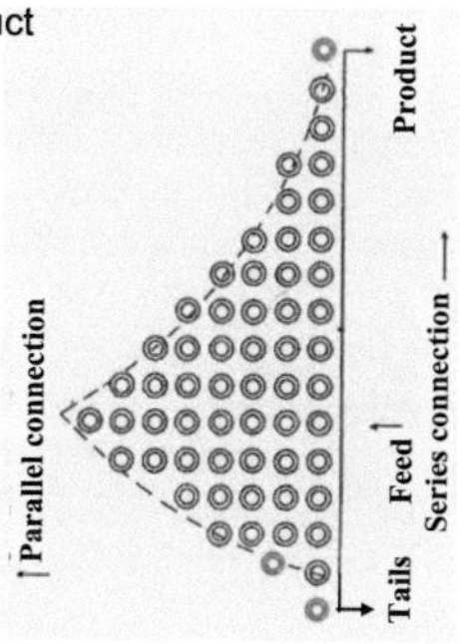

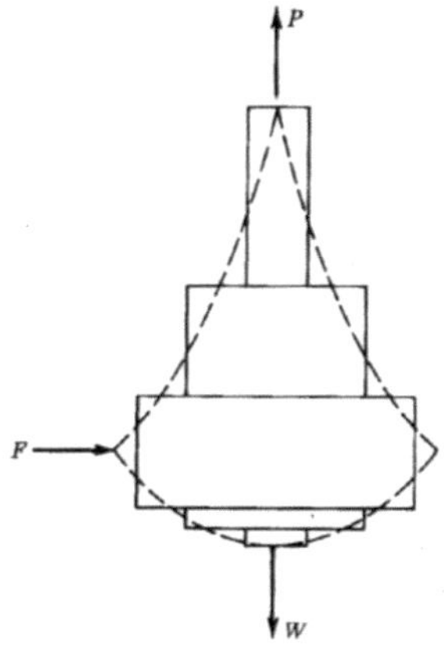

Figure 3.4: Ideal cascade approximation using squere sections.

3.4.1 Number of stages for an ideal symmetrie cascade [8]

The total number n of stages for an ideal cascade is given by

$$n = 2 \frac{\ln \dfrac{x_P \, (1-x_W)}{(1-x_P) \cdot x_W}}{\ln \alpha} - 1 \tag{3.6}$$

with

x_P atom fraction of final product
x_W atom fraction of tails depleted uranium
α stage separation factor

The number of stages in the stripping section is

$$n_W = \frac{\ln \dfrac{x_F \, (1-x_W)}{(1-x_F) \cdot x_W}}{\ln ß} - 1 \tag{3.7}$$

x_F atom fraction of feed
ß heads separation factor

For the ideal cascade the head separation factor is: $\beta = \sqrt{\alpha}$.

The number of stages in the enriching section

$$n - n_W = \frac{\ln \dfrac{x_P (1-x_F)}{(1-x_P) \cdot x_F}}{\ln ß} \tag{3.8}$$

Table 3.4 shows – as an example – the total number of stages required and the number of stages in the enrichment and stripping zone of an ideal cascade for an enrichment of natural uranium to $x_P = 4\%$ LEU and a tails assay of $x_W = 0.2\%$

- a gaseous diffusion enrichment plant with $\alpha = 1.0043$
- a gas centrifuge enrichment plant with $\alpha = 1.30$.

Table 3.4: Number of stages of an ideal cascade for gas diffusion and gas centrifuge enrichment.

	Gaseous diffusion $\alpha = 1.0043$	Gas centrifuge $\alpha = 1.30$
enrichment section	815	13
stripping section	598	9
total number of stages	1413	22

3.5 Inputs and Outputs of the enrichment process

The mass input to the enrichment process is the uranium feed F of natural uranium (atom fraction $x_F = 0.72\%$) or reprocessed uranium (0.8% U-235 and 0.6% U-236 as well as 98.6% U-238) in the form of uranium hexafluoride. The mass output in the enriched uranium product P with atom fraction x_P in % and the waste stream or tails W (atom fraction x_W of, e.g. 0.2%).

The ratio of the feed mass and the product mass is

$$\frac{F}{P} = \frac{x_P - x_W}{x_F - x_W} \tag{3.9}$$

and the ratio of waste mass to product mass is

$$\frac{W}{P} = \frac{x_P - x_F}{x_F - x_W} \tag{3.10}$$

3.6 Separative work of the enrichment process

The separative work is a measure of the amount of work necessary in the enrichment plant to produce a certain amount of enriched uranium. It has the dimension of mass and is indicated in kg SWU or tonne SWU.

The separative work S in SWU/kg can be expressed in terms of the different mass streams for feed, product and waste [8]. Applying the relations between these mass streams and their atom fractions, the relation for the separative work unit per product mass is given by

$$\frac{S}{P} = (2x_P-1) \ln \frac{x_P}{1-x_P} + \frac{x_P-x_F}{x_F-x_W} (2x_W-1) \ln \frac{x_W}{1-x_W} - \frac{x_P-x_W}{x_F-x_W} (2x_F-1) \ln \frac{x_F}{1-x_F}$$

product term waste term feed term

(3.11)

This relation is evaluated for P = 1 kg of product mass in Table 3.5 for different enrichment levels [7,8]. In the left column the different enrichment levels for low enriched uranium (LEU) fuel are shown. The second column indicates how many kg of natural uranium feed as UF_6 are needed to produce 1 kg of low enriched uranium

example: for P = 1 kg of 4% U-235 enriched uranium, a feed of F = 7.436 kg natural uranium as UF_6 must be provided.

The third column gives the quantity of U_3O_8 in pounds including 0.5% conversion losses between U_3O_8 and UF_6.

The fourth column gives the separative work needed to enrich the natural uranium as feed mass to a certain enrichment

example: for 1 kg of 4% U-235 enriched uranium 6.544 kg separative work are needed starting from natural uranium.

A 1.3 GW(e) PWR requires about 30 t of UF_6 per year with 4% U-235 enrichment. The separative work required would be roughly 30 x 6.544 = 196.3 t SWU (Table 3.5) for a tails assay of 0.2% wt% U-235. A 1 million kg SWU or 1000 t SWU enrichment plant would be able to enrich the fuel for about 5 PWRs of 1.3 GW(e).

For enrichment levels higher than 10% U-235 in uranium or other enrichments than given in Table 3.5 the above equation (3.11) must be applied.

The uranium industry in the USA uses different units, e.g. uranium concentrates are generally measured in pounds of U_3O_8. However, enrichment services are measured in kg U as UF_6. Only 714/842 (ratio of molecular weights) or 0.848 kg U are contained in U_3O_8. Taking into account 0.5% loss for the conversion of from U_3O_8 to UF_6 this ratio decreases to 0.8438. This leads to [12]:

1 lb U_3O_8 is equivalent to 0.383 kg U as UF_6
1 kg U as UF_6 is equivalent to 2.613 lb of U_3O_8

Table 3.5: Characteristic data: kg of natural uranium feed as UF6 or lb U_3O_8 required for 1 kg of enriched uranium product and corresponding kg SWU needed [7].

Enrichment,	kg of Natural U Feed Material to Enrichment Plant	lb of U_3O_8 to be Purchased[a]	kg Separative Work[b]
wt% ^{235}U	Per Kilogram of Enriched Uranium Product		
Nat. 0.72	1.000	2.613	0.000
0.8	1.174	3.068	0.104
0.9	1.370	3.580	0.236
1.0	1.566	4.092	0.380
1.2	1.957	5.114	0.698
1.4	2.348	6.136	1.045
1.6	2.740	7.160	1.413
1.8	3.131	8.182	1.797
2.0	3.523	9.206	2.194
2.1	3.718	9.716	2.397
2.2	3.914	10.228	2.602
2.3	4.110	10.740	2.809
2.4	4.305	11.250	3.018
2.5	4.501	11.762	3.229
2.6	4.697	12.274	3.441
2.7	4.892	12.784	3.656
2.8	5.088	13.296	3.871
2.9	5.284	13.808	4.088
3.0	5.479	14.318	4.306
3.1	5.675	14.830	4.526
3.2	5.871	15.342	4.746
3.3	6.067	15.854	4.968
3.4	6.262	16.364	5.191
3.5	6.458	16.876	5.414
3.6	6.654	17.398	5.638
3.7	6.849	17.898	5.864
3.8	7.045	18.410	6.090
3.9	7.241	18.922	6.316
4.0	7.436	19.432	6.544
5.0	9.393	24.544	8.851
10.0	19.178	50.112	20.863

[a] 0.5% U_3O_8 to UF_6 conversion losses included.

[b] Tails assay at 0.2 wt. % U-235

Footnote: As column 3 of Table 3.5 did not show units in /7/, kg had been erroneously assumed as units by the author in the first edition.

3.7 Gaseous Diffusion Technology

Gaseous diffusion is based on molecular diffusion through micropores in a membrane. In thermal equilibrium U-235- and U-238-molecules have the same average kinetic energy. Hence the lighter molecules of the isotope U-235 have a faster velocity and impinge on the wall of the membrane more often. If the wall is porous the lighter molecules will pass through the membrane at a slightly higher rate than the heavier molecules of U-238. More 235-UF_6 molecules will pass through the membrane than 238-UF_6 molecules. The theoretical separation factor per stage is equal to 1.0043.

The diameter of the pores in the membrane must be very small. Membrane barriers with pores of 10^{-6} cm have been reported [7]. Sintering of fine alumina or nickel powder was applied. The barriers are about 50 μm thick. They can be arranged in form of sintered nickel tubes assembled to tube bundles, housed in the diffuser or converter. The diffusion cell (diffuser) is divided into two compartments by the porous membrane. A compressor maintains the pressure at the inlet: A heat exchanger must remove the heat of compression from the UF_6 gas. Because of the corrosiveness of the UF_6 gas, the structural materials in the process vessels (converters) are nickel coated. Fig. 3.5 shows the converter units of a US gaseous diffusion plant with their motors for the compressors and connecting pipings and valves.

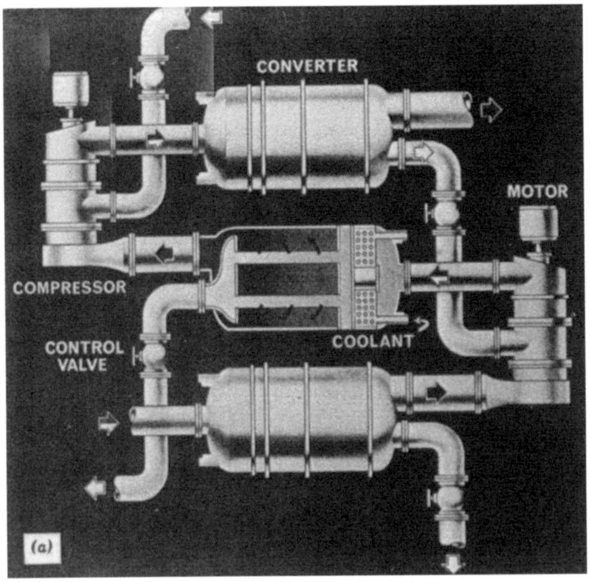

Figure 3.5: Basic stage equipment in a gaseous diffusion cascade [7].

Due to the low enrichment factor, gaseous diffusion plants typically need 1200-1400 stages to produce low enriched uranium for LWR fuel of 3 to 4% U-235 enrichment. Most gaseous diffusion plants are designed to operate between 65 and 110 °C at low pressures of about 0.35 bars. After enrichment of the 235-UF_6 product the gas is liquefied and put into storage cyclinders. The tails are also liquefied and put into containers. They are stored for further use e.g. in fast reactors (FRs). Large scale gaseous diffusion enrichment plants are operating at Paducah, USA with an enrichment capacity of 11.3 million kg SWU/year and at Tricastin, France with an enrichment capacity of 10.8 million kg SWU/kg (Table 3.2).

Gaseous diffusion enrichment plants have a high specific power consumption on the order of 2400 kWh/kg SWU. Therefore the power requirements of a large gaseous diffusion enrichment plant like at Tricastin is about 3000 MWe.

3.8 Gas Ultra-centrifuge

The gas ultra-centrifuge was developed in Europe and Russia by Zippe, Kistemaker, Whitley, Kamenev and others. The gas ultra-centrifuge process is based on the separation effect of UF_6 gas in a strong centrifugal field of a fast spinning rotating cyclinder. When the cylinder rotates at high speed, radial pressure differences cause the separation of the U-235 and U-238 molecules.

Table 3.6 gives the enrichment factors as a function of different peripheral speeds [2,10].

Table 3.6: Peripheral speeds and enrichment factors of a gas centrifuge [2].

peripheral speed m/s	300	400	500	600	700	800
enrichment factor	1.056	1.10	1.16	1.24	1.34	1.46

For rotor speeds of about 400 m/s at the outer radius of the centrifuge the local separation factor would be about 1.10. By creating a weak countercurrent flow inside the centrifuge the separative factor can be increased. The countercurrent flow can be achieved by traffles, molecular pumps or an axial temperature gradient. The result is that the U-238 diffuses to the bottom and the U-235 to the top of the centrifuge, where they are collected (see Fig. 3.6).

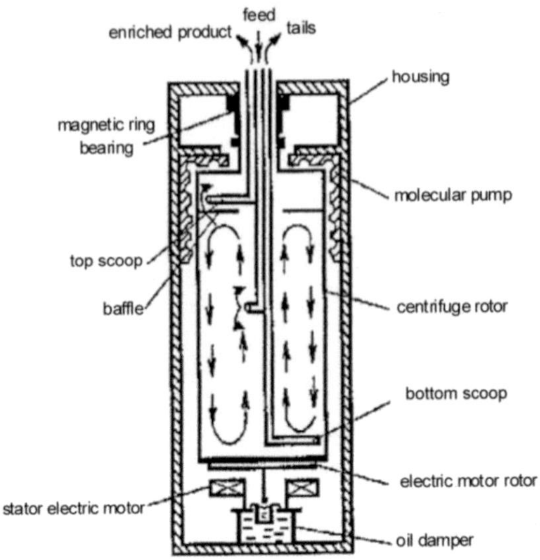

Figure 3.6: Russian gas centrifuge design according to Kamenev [11].

3.9 Gas Ultra-centrifuge technology

New materials have to be used to increase the peripheral speed of the centrifuges. Table 3.7 shows the tensile strength, the density and the approximate maximum peripheral speed these materials can attain and endure in a gas ultra-centrifuge.

Table 3.7: Maximum peripheral speeds for thin-walled cylinders [10].

Material	Tensile trength σ (kg/cm2)	Density ρ (g/cm3)	σ/ρ (cm)	approx. max. peripheral speed (m/s)
Al-alloy	5200	2.8	1900	425
Titanium	9200	4.6	2000	440
Maraging steel	22500	8.0	2800	525
Glass fibre	7000	1.9	3700	600
Carbon fibre	8500	1.7	5000	700

Gas ultra-centrifuges based on aluminum or steel can have a peripheral rotor speed in the range of 400 m/s. Composite materials allow peripheral velocities in the range of 600-700 m/s. For enrichment of natural uranium to 4% low enriched uranium for LWR fuel a total of 22 stages of gas centrifuges must operate in cascade if an enrichment factor of $\alpha = 1.3$ and a tails assay of 0.2% U-235 are assumed (Section 3.4).

3.10 Russian Ultra-centrifuge design

The highest installed capacity in gas ultra-centrifuges is located in Russia (Section 3.2). A detailed review of Russian centrifuge technology is given by Bukharin [11]. Some characteristic design data of Russian centrifuges designed around 2000-2010 are given in Table 3.8.

Table 3.8: Russian centrifuge characteristic data [11].

Generation	Separative capacity [SWU/a]	Peripheral speed
6	2.5	580 m/s
7	3.2	630 m/s
8	4.2	690 m/s

Modern designs of the centrifuge rotor consist of an inner metal sheet around which aramid and graphite fibers are wrapped to achieve the required high strength.

3.11 Rotor dynamics

No matter how carefully rotors are balanced, this balancing can never be perfect. This leads to vibrations at certain speed which become violent at a certain critical resonance speed. These critical (resonance) speeds are determined by the rotor design (length/diameter ratio or L/D) and characteristics of its materials.

Table 3.9: Critical peripheral speeds as a function of length/diameter L/D for aluminum rotors [10].

L/D	Critical speeds [m/s]			
	1st	2nd	3rd	4th
7	400			
11.6	145	400		
16.3	74	204	400	
21	45	123	242	400
25.5	30	83	162	269

A subcritical rotor with L/D = 7 would have no critical (resonance) speed as long as it runs at a peripheral speed of 350 m/s. A rotor with L/D = 11.6 would have to negotiate one critical speed at 145 m/s before it would experience another one at 400 m/s [Tab. 3.9].

A centrifuge operating below its first resonance is termed subcritical (short centrifuge). A centrifuge operating above the rotors vibrational resonances is termed supercritical (long rotor with damping measures). Modern design centrifuges with high stage separation factors and high separation capacity (Section 3.10) operate in the supercritical range. They have an energy consumption of about 100 kWh/SWU.

3.12 Laser enrichment [4,7]

Laser isotope enrichment is based on photo-excitation. The slight difference in electron energy levels of U-235 and U-238 is used for this enrichment technology. Laser light produced at the same wave length as these electron energy levels leads to resonances, enhanced ionization or molecular dissociations. Two main laser enrichment methods have developed:

Atomic vapor laser isotope separation (AVLIS)
Molecular laser isotope separation (MLIS)

3.12.1 The AVLIS enrichment technology

Two different ALVIS techniques were developed and tested at Livermore Laboratories in the USA.

- An oven at 2600 K produces uranium vapor. A xenon laser operating in the ultraviolet range excites only the U-235 atoms. These atoms are then ionized by a krypton laser. The ionized U-235 atoms are collected on a plate in an electric field.
- A sequence of three tunable dye lasers driven by a high-repetition copper laser excite the U-235 atoms. These laser photons produce auto-ionizing of the uranium atoms. The uranium ions can be collected from the vapor as described above.

Both AVLIS technologies were given up in 1994 and did not enter the commercialization phase.

3.12.2 Molecular Laser Isotope Separation (MLIS)

The MLIS process developed at Los Alamos, USA is based on the vibrational modes (vibrational and rotational frequencies) of a UF_6 molecule. The UF_6 gas together with hydrogen is expanded through a hypersonic nozzle to cool it down to about 30 K. This brings the molecules into the lowest vibrational ground state. In a next step an infrared laser operating at 16 μm wavelength selectively excites the 235-UF_6 molecules. The laser radiation in the ultraviolet part of the spectrum at 0.308 μm is used to dissociate the 235-UF_6 molecules into U-235-UF_5 and F_2. The U-235-UF_5 precipitates as a fine powder.

A variant of the MLIS process called SILEX originally developed in Australia is the basis of the first laser enrichment plant in USA (Section 3.2).

3.12.3 Conversion of UF_6 into UO_2 powder

The UF_6 coming from the enrichment plant must be converted into UO_2 powder in the UO_2 pellet fabrication plant. One of the chemical processes applied is the so-called AUC-process. The UF_6 is vaporized and mixed with water to form uranyl fluoride UO_2F_2. This is mixed with ammonia NH_3 and carbon dioxide CO_2 which react to ammonium uranyl-carbonate $(NH_4)_4\ UO_2\ (CO_3)_3$ which precipitates from the suspension. Further heating leads to thermal decomposition and uranium trioxide UO_3. This can be reduced by means of hydrogen to uranium dioxide UO_2.

After homogenization of the UO_2 powder, binders and lubricants, are added. This mixture is pressed to cylindrical pellets of 10 mm diameter and height. In a furnace the pellets are sintered at temperatures of 1600 to 1750 °C in hydrogen. Densities of the UO_2 pellets of about 10.4 g/cm^3 are achieved. Afterwards the pellets are ground to the required dimensions.

Finally the pellets are filled into the zircaloy tubes which are welded tight. These fuel rods are assembled to fuel assemblies (Section 5).

References Section 3

[1] Wood, M., Effects of separation processes on minor uranium isotopes in enrichment cascades, Science and Global Security, 16, 26-36 (2008).

[2] Uranium Enrichment (Villani, S., ed.). Topics in Applied Physics, Vol. 35, Berlin-Heidelberg-New York : Springer (1979).

[3] International Nuclear Fuel Cycle Evaluation, Enrichment Availability. Report of INFCE Working Group 2. Vienna: International Atomic Energy Agency (1980).

[4] Laughter, M., Profile of world uranium enrichment programs – 2007, ORNL/TM-2007/193 (2007).

[5] Becker, E.W., et al., Uranium Enrichment by the Separation Nozzle Method within the Framework of German/Brazilian cooperation. Nucl. Tech., 52, 105-114 (1981).

[6] Ehrfeld, W., Ehrfeld, U., Anreicherung von U-235. In: Gmelin Handbuch der Anorganischen Chemie, Uran, Ergänzungsband 2A, Isotope. Berlin-Heidelberg-New York: Springer (1980).

[7] Rahn, F., et al., A guide to nuclear power technology, John Wiley & Sons, New York (1984).

[8] Benedikt, M., et al., Nuclear Chemical Engineering. New York: McGraw-Hill (1981).

[9] Cohen, K., The theory of isotope separation as applied to the large-scale production of U-235. New York: McGraw-Hill (1951).

[10] Withley, S., The uranium ultracentrifuge, Phys. Technology, 10, 8-26 (1979).

[11] Bukharin, O., Russia's gaseous centrifuge technology and enrichment complex. Science and Global Security Program, Princeton, University (2004).

[12] Secretarial Determination for the Sale or Transfer of Uranium. Analysis of Potential Impacts of Uranium Transfers on the Domestic Uranium Mining, Conversion, and Enrichment Industries, US. Department of Energy, May 2015.

4 Neutron and reactor physics

Only some basic nuclear processes and data are presented in this Section. This shall help to increase the understanding of the subsequent Sections. For a deeper understanding of nuclear reactor physics the literature given in the reference list is recommended [1,2,3,4,5,6,7,8,9].

4.1 Fission process

If a neutron of a certain velocity (kinetic energy) is absorbed by a fissile heavy nucleus, e.g. U-233, U-235 or Pu-239, the resulting compound nucleus can become unstable and split (fission) into two or even three fragments. The fission fragments are created essentially according to a double humped yield distribution function with mass numbers between about 70 and 165. The mass yield distribution functions are similar for heavy nuclei fissioned by neutrons with kinetic energies up to about 2 MeV. They depend slightly upon the kinetic energy of the incident neutrons causing fission and on the type of heavy nuclei (U-233, U-235, Pu-239).

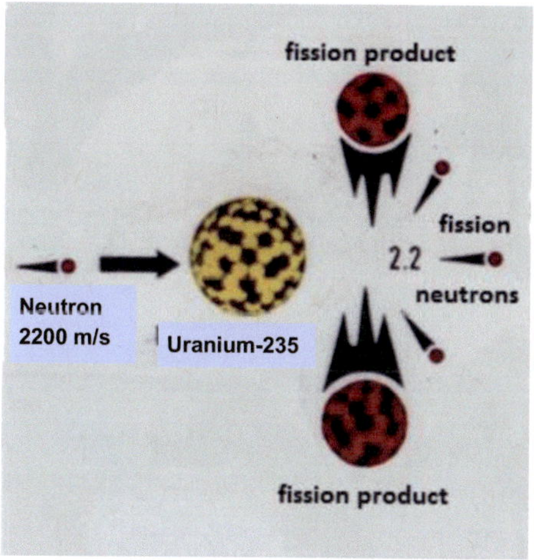

Figure 4.1: Fission of U-235 nucleus by a neutron.

In addition to the fission products (fragments), 2-3 prompt neutrons are emitted during the fission process. These prompt fission neutrons appear within some 10^{-14} s. They are created with different kinetic energies following a certain distribution curve. In some heavy nuclei, most times with even mass numbers, nuclear fission can only be initiated by incident neutrons with a certain, relatively high, threshold kinetic energy (Table 4.1).

Table 4.1: Threshold kinetic energy for incident neutrons causing substantial fission in different heavy nuclei [6].

Heavy Nucleus	Th-232	U-233	U-234	U-235	U-238	Pu-239
Incident neutron kinetic energy [MeV]	>1.3	0	>0.4	0	>1.1	0

The fission products can be solid, volatile or gaseous. Some of the fission products decay further emitting so-called delayed neutrons, β-particles, γ-rays and anti-neutrinos. The delayed neutrons resulting from the decay of particular fission products – called precursors – represent less than 1% of all neutrons. The fraction of delayed neutrons originating from fissioning by thermal neutrons of U-235 is 0.67%, and 0.22% from fissioning of Pu-239. They appear with decay constants of 0.01 to 3 s^{-1} for U-235 and 0.01 to 2.6 s^{-1} for Pu-239. These delayed neutrons are of absolute necessity for the safe control of nuclear fission reactors [11,12,13].

The energy release per fission Q_{tot} appears as kinetic energy of the fission products, E_f, of the prompt fission neutrons, E_n, as β-radiation, E_β, as γ-radiation, E_γ, or as neutrino radiation, E_v, (Table 4.2). The neutrino radiation does not produce heat in the reactor core. However, antineutrinos are relevant for safeguards and surveillance measures (Section 8.1.7). Table 4.2 also shows the total energy, Q_{tot}, and the thermal energy, Q_{th}, released during fission of a nucleus. Some of β-radiation and γ-radiation of the fission products is not released instantaneously but delayed according to the decay of the different fission products.

Table 4.2: Different components of energy release per fission of some heavy nuclei in MeV by incident neutrons of different kinetic energy (in eV or MeV) [10,11].

Heavy Nucleus	Incident neutron energy	E_f	E_n	E_β	E_γ	E_v	Q_{tot}	Q_{th}
U-235	0.025 eV	169.75	4.79	6.41	13.19	8.62	202.76	194.14
	0.5 MeV	169.85	4.80	6.38	13.17	8.58	202.28	193.7
U-238	3.10 MeV	170.29	5.51	8.21	14.29	11.04	206.24	195.82
Pu-239	0.025 eV	176.07	5.9	5.27	12.91	7.09	207.24	200.15
	0.5 MeV	176.09	5.9	5.24	12.88	7.05	206.66	199.61
Pu-240	2.39 MeV	175.98	6.18	5.74	12.09	7.72	206.68	198.94

On the average, about 194 MeV or 3.11×10^{-11} J are released per fission of one U-235 atom. Most of the fission energy is released instantaneously. However, a small fraction appears with some delay, the associated decay constants ranging between about 10^{-10} to 1 per s.

Since 1 g of U-235 metal contains 2.56×10^{21} atoms, the complete fission of 1 g of U-235 results in:

$$7.96 \times 10^{10} \text{ J or } 2.21 \times 10^{4} \text{ kWh} \quad \text{or} \quad 0.92 \text{ MWd}_{(th)} \text{ thermal energy}$$

For other fissile materials like U-233 or Pu-239 the energy release per fission is similar. Fission by neutrons with thermal energies (0.025 eV) or by energies of 0.5 MeV leads to almost equal energy releases.

Usually the thermal energy produced in the fuel of a fission reactor core is given in MWd(th) per tonne or GWd(th) per tonne of fuel which also corresponds roughly to the number of grams or kg, respectively, of U-235 fissioned in one tonne of fuel [2].

4.2 Neutron reactions

Neutrons produced in nuclear fission have a certain speed or kinetic energy and direction of flight. In a fission reactor core they may be scattered elastically or inelastically or absorbed by different atomic nuclei. In some cases the absorption of neutrons may induce nuclear fissions in heavy nuclei (U-235 etc.) so that successive generations of fission neutrons are produced and a fission chain reaction is established.

In addition to fission the scattering and absorption reactions also contribute to the energy release in a nuclear reactor.

4.2.1 Reaction rates

If $n(\vec{r}, v, \Omega)$ is the number of neutrons at point $\vec{r}$, with speed v and the direction of flight Ω, then these neutrons can react within a volume element dV with $N \cdot dV$ atomic nuclei (N being the number of atomic nuclei per cm^3 of reactor volume). The number of reactions per second, e.g. scattering or absorption, is then proportional to

$$v \cdot n(\vec{r}, v, \Omega) \quad \text{and to} \quad N \cdot dV$$

The proportionality factor $\sigma(v)$ is a measure for the probability of the nuclear reactions and is called microscopic cross section of the nucleus for a specific type of reaction. The microscopic cross section $\sigma(v)$ is measured in 10^{-24} $cm^2 \triangleq 1$ barn. It is a function of the speed or kinetic energy of the neutron and of the type of reaction and differs for every type of atomic nucleus. As for an absorption reaction the neutron can either remain captured or lead to fission of a heavy nucleus the relation (neglecting for simplicity of presentation less important reactions)

$$\sigma_a(v) = \sigma_c(v) + \sigma_f(v)$$

is valid with

$\sigma_a(v)$ microscopic absorption cross section
$\sigma_c(v)$ microscopic capture cross section
$\sigma_f(v)$ microscopic fission cross section

The reaction rate can be written

$$R(\vec{r},v) = \sigma(v)\cdot N(\vec{r})\cdot v\cdot n(\vec{r},v) = \Sigma(\vec{r},v)\cdot\phi(\vec{r},v).$$

The quantity $\Sigma(\vec{r},v) = N(\vec{r})\cdot\sigma(v)$ is called macroscopic cross section.

The quantity:

$\phi(\vec{r},v) = v\cdot n(\vec{r},v)$ is called the neutron flux.

Fig. 4.2 shows the microscopic fission cross section as a function of the neutron kinetic energy for the heavy nuclei U-235, U-238 and Pu-239. The fission cross sections for U-235 and Pu-239 increase with lower kinetic energies. In the energy region of about 0.1 eV to 10^3 eV this behavior is superposed by resonance cross sections.

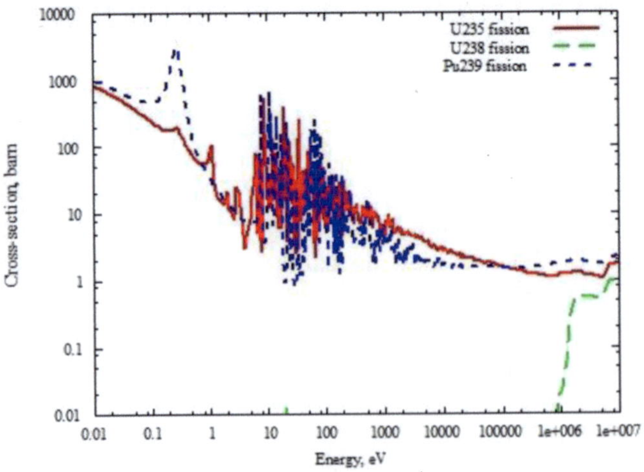

Figure 4.2: Microscopic fission cross section for U-235, U-238 and Pu-239.

The capture cross section for U-238 are shown in Fig. 4.3. Such microscopic cross sections are steadily compiled, supplemented and revised in nuclear data libraries, e.g. JEFF [19] etc.

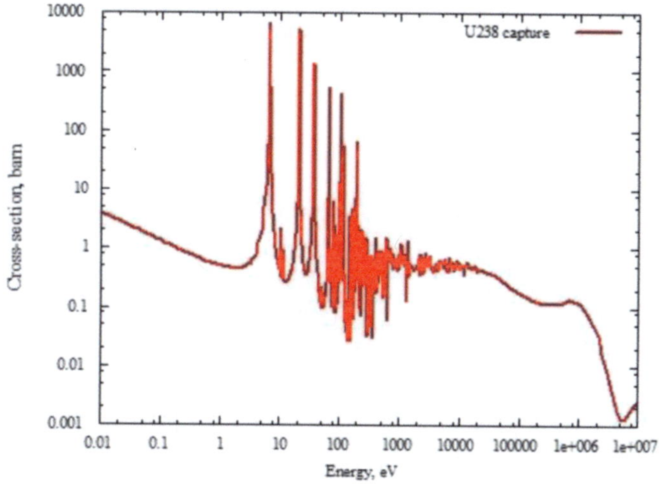

Figure 4.3: Microscopic capture cross section of U-238.

4.3 Spatial distribution of the neutron flux in the reactor core

The spatial distribution of the neutrons with a certain speed or kinetic energy and flight direction can be described by the Boltzmann neutron transport equation or by Monte Carlo methods [2,14]. For both cases, numerical methods in one-, two- or three-dimensional geometry were developed. Computer program packages (deterministic codes for the solution of the Boltzmann transport equation and Monte Carlo codes) are available for applications [14,15,16,17,18].

For many practical applications it is sufficient to solve the neutron diffusion equation which is an approximation to the Boltzmann neutron transport equation. The continuous energy range can be approximated by a subdivision into a number of energy groups with specifically defined microscopic group cross sections, e.g. JEFF [19], ENDF/B [20], or JENDL [21]. Group cross section sets for a given number of energy groups are, e.g. WIMS [22] and the ABBN set [23,25].

Fig. 4.4 shows the neutron energy spectra of a pressurized water reactor (PWR), of a sodium cooled fast reactor (SFR) with metal, oxide and nitride fuel as well as of a very high temperature reactor (VHTR) as neutron flux $\phi(u) = E \cdot \phi(E)$ per unit lethargy u. The lethargy u is defined as $\ln(E_o/E)$, where E_o is the upper limit of the energy scale. This logarithmic energy is suggested by the fact that the average logarithmic energy loss per elastic collision of a neutron with a nucleus is an energy independent constant.

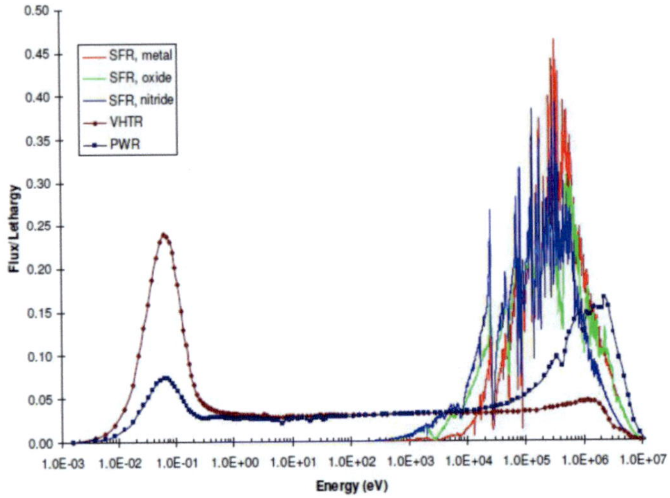

Figure 4.4: Neutron energy distribution (spectrum) in a PWR and VHTR and FRs [36].

Fig. 4.5 displays the spatial distribution of the neutron flux in the range of thermal energies for a PWR [24]. The absorber or control rods are partially inserted in axial direction in the PWR core. The control rods absorb neutrons and are responsible for the spatial distortions of the thermal neutron flux. They influence the criticality level and the spatial power distribution.

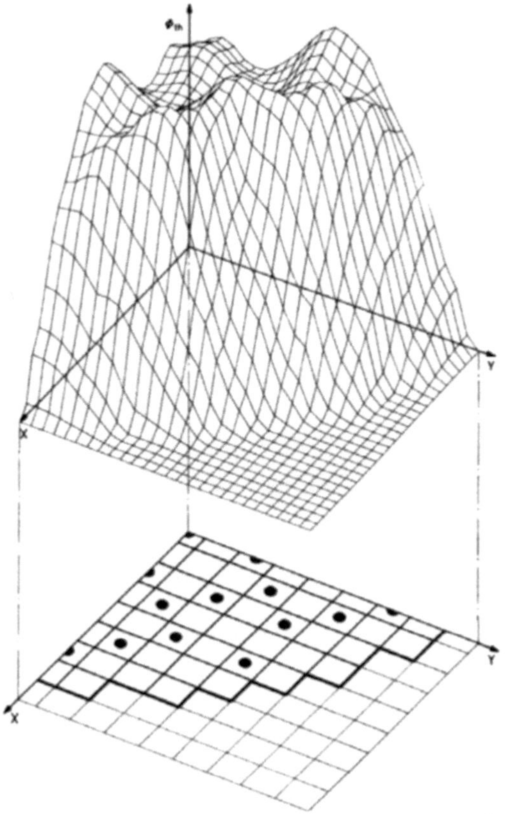

Figure 4.5: Spatial distribution of the thermal neutron flux, $\phi_{th}(\vec{r})$, in a PWR core with partially inserted control rods [24].

4.4 Criticality factor k_{eff} [1,2,3,4]

The ratio between the number of newly generated neutrons by fission and the number of neutrons absorbed in the reactor core or escaping from the reactor is called the criticality factor or effective multiplication factor, k_{eff}.

For a $k_{eff} = 1$ the reactor core is critical and can be operated in steady state. At $k_{eff} < 1$ the reactor core is subcritical, e.g. with the control or absorber rods fully inserted in the core. Boron, cadmium or gadolinium etc. can be used as absorber materials, either as metallic alloys in control rods or as burnable poisons in ceramic form in fuel rods and special rods or as a fluid, e.g. boric acid in the coolant of a PWR.

For $k_{eff} > 1$ the reactor is supercritical. More neutrons are produced than are absorbed in the reactor core or do escape from the core. The neutron chain reaction is ascending (reaction rates and the number of neutrons increase as a function of time).

The neutron flux in the reactor core can be controlled by moving or adding, e.g. absorber materials. This is done in a k_{eff}-range, where the delayed neutrons are dominating the transient behavior of the neutron flux. The delayed neutrons come into being in a time range of seconds. Therefore, the neutron flux in reactor cores can also be controlled safely by moving absorber materials in the time range of seconds.

4.5 Design of a reactor core

The choice of the type of fuel, e.g. natural or enriched uranium as well as plutonium or U-233 and the choice of the moderator, coolant as well as the structural materials determine the type of nuclear reactor. In addition the type of fuel can be used as metallic, oxide, carbide or nitride fuel. Coated particles used, e.g. in VHTRs and pebble bed HTR module reactors, or molten salts (fluorides or chlorides) have also been employed in the past or are still investigated. However, these reactor types are outside the scope of this Section. Table 4.3 shows the combination of fuel, moderator, coolant and structural materials for the most common types of reactors: high temperature reactors (HTR), heavy water reactors (HWR), light water reactors (LWR) and fast neutron reactors (FR) (Sections 5 and 6).

Table 4.3: Combinations of fuel, moderator, coolant and structural material in different reactor types.

Reactor type	Fuel	Moderator	Coolant	Structural material
HTR	Natural or slightly enriched uranium	Graphite	Gas: CO_2, He	Graphite, Mg-alloys
HWR	Natural or slightly enriched uranium	Heavy water	Light or heavy water	Zircaloy
LWR	Enriched uranium 4-5% U-235 + U-238	Light water	Light water	Zircaloy
Fast neutrons Reactor (FR)	20-30% Plutonium + U-238 15-25% U-233 + Thorium	No moderator	He-Gas, sodium, Lead or Lead-Bismuth	Stainless steel

The characteristic thermal properties of the coolant/moderator and of the structural material determine the power density in kW/l of the reactor core, e.g. 100 kW/l in a PWR and 300 kW/l in an FR. Assuming a certain power output of the reactor core leads to the required volume of the reactor core (radius and height of the reactor core). The thermo-hydraulic and thermal properties of the coolant together with assumed coolant inlet and outlet temperatures result in the required coolant mass flow for cooling the reactor core. With these rough design parameters, the necessary enrichment in fissile isotopes in the fuel can be determined by solving, e.g. the Boltzmann neutron transport equation or the multigroup diffusion equation. This is done for the given geometry of the reactor core by solving the equations for k_{eff}.

The fissile fuel enrichment is determined such that k_{eff} is slightly above 1 to allow (insertion of control material) for control of the reactor core over a full operation cycle i.e. account for the effects of decreasing concentrations of fissionable material, e.g. U-235, and of the accumulation of fission products or temperature changes.

For nuclear reactor cores with a thermal neutron spectrum the ratio of fuel volume to moderator volume is determined by the number of scattering collision required to slow down the fission neutrons to thermal energies. This leads to moderator/fuel volume ratios for the different reactor types as given in Table 4.4.

The spatial distribution of the neutron flux in fuel rods and surrounding moderator is determined by subdividing the reactor into cells [2,3,4].

Table 4.4: Moderator/Fuel volume ratio for different reactor types.

Reactor type (thermal)	Moderator/Fuel volume ratio
LWR	2-3
HWR	20
HTR	54

Fig. 4.6 shows such a cell arrangement for a square lattice as used for most Western LWRs. Triangular cells and hexagonal subassemblies are used, e.g. for Russian PWRs. These cells have a micro distribution of the neutron flux. This micro-distribution of the neutron flux is obtained by solving, e.g. the Boltzmann neutron transport equations in multigroup form for the geometry shown in Fig. 4.6 (fuel rod with cladding, surrounding moderator and the respective boundary conditions).

Fig. 4.6 shows a two group representation of the neutron flux within a lattice cell for, e.g. LWRs. The fast neutrons are generated by fission in the fuel rod. They move relatively fast out of the fuel to the moderator region, where they are slowed down to thermal energies. This avoids to a large extent undesired captures in the resonance energy region (see Figs. 4.2 and 4.3).

The thermal neutrons in the moderator region diffuse back into the fuel rod where they are absorbed by the fuel nuclei and cause fissions. The fast neutron flux is highest in the fuel rod and lowest in the moderator region, whereas the thermal neutron flux is highest in the moderator region. Most of the thermal neutrons are already absorbed in the outer regions of the fuel rod (spatial self-shielding effect). Therefore the radial fission power density generated is also somewhat higher in the outer region than in the center of the fuel rod.

The radial fission power density distribution is given by

$$P(\vec{r}) = 3.11 \cdot 10^{-11} \cdot \underset{g=1}{\overset{G}{\mathrm{Sum}}} \cdot \Sigma_{f,g}(\vec{r}) \cdot \phi_g(\vec{r}) \left[\mathrm{W/cm^3}\right] \quad (4.1)$$

with

$\Sigma_{f,g}(\vec{r})$ macroscopic fission cross section of energy group g

$\phi_{f,g}(\vec{r})$ neutron flux in energy group g

In the above example the number of energy groups is G = 2.

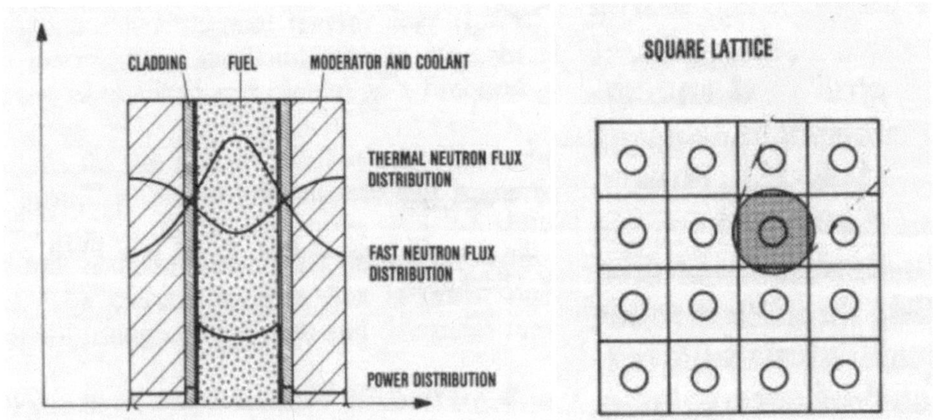

Figure 4.6: Square lattice cell with fuel rod, cladding, and moderator of a LWR-fuel element.

With these multigroup neutron flux distributions in the cell spatially averaged multigroup cross sections can be determined. With these spatially averaged cross sections the Boltzmann multigroup neutron transport equations can be solved numerically for the whole reactor core with boundary conditions. The real neutron flux then results in a superposition of the micro-distribution of the cell averaged over a subassembly, e.g. fuel rods with different fuel enrichment and burnable poison or water channels, and the macro-distribution over the entire core.

4.6 Fuel burnup and transmutation during reactor operation

During reactor operation over months and years the initially loaded U-235 in the low enriched uranium fuel will be consumed due to fission and neutron capture processes. As a consequence also the initial criticality factor k_{eff} decreases. Neutron capture in fertile isotopes, e.g. U-238 or Th-232 leads to buildup of new fissile isotopes, e.g. Pu-239 or U-233. This increases somewhat the criticality factor k_{eff}. Fission products originating from the fission of fissile isotopes decrease the criticality factor k_{eff} due to their absorption cross section. The combination of these three effects results in a time dependent change – usually a decrease – of the criticality factor k_{eff} during reactor operation.

This burnup effect on k_{eff} is accounted for by special design of the reactor core. The enrichment of the initially loaded fuel is increased such that k_{eff} becomes slightly >1. As the k_{eff} shall be equal 1 during the whole operation cycle, this is balanced by absorber materials in the core (movable absorber rods or special rods with burnable absorber material or burnable absorber materials dissolved in the coolant or mixed with the fuel). The accumulating fission products and the decreasing k_{eff} are counteracted e.g. by moving absorber rods slowly out of the core during reactor operation. At the end of the operation cycle the absorber rods are almost withdrawn out of the core and spent fuel must be unloaded and replaced by new fuel elements.

The calculation of the change in concentration of all fuel isotopes, actinides and fission products requires the knowledge of the microscopic cross section and decay constants of all isotopes as well as the yields of fission products during the fission process. Fig. 4.7 shows the uranium-plutonium reaction chain with α- and β^--decays and their decay half-lives. Only the most important nuclear reactions and isotopes are shown. More details, e.g. about the Am-isotopes will be discussed in Section 11.

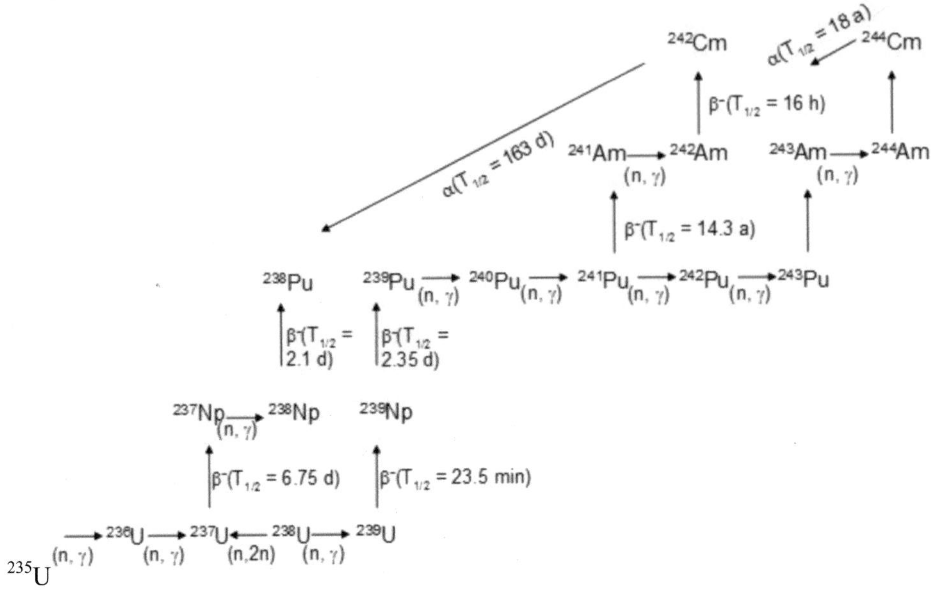

Figure 4.7: Uranium-plutonium reaction chain with buildup of neptunium, americium and curium.

The concentrations of the various isotopes in the reactor core can be described by a coupled set of ordinary differential equations which balance the production and destruction of the different isotopes [25,26]. This is explained in the following for only some fuel isotopes using the following indices (left side) and neglecting minor important contributions like the α-decay of the fairly stable uranium isotopes.

Index	Isotope	
24	U-234	$\frac{d}{dt}N_{24} = -\sigma_a^{24} \cdot \phi \cdot N_{24}$
25	U-235	$\frac{d}{dt}N_{25} = \sigma_c^{24} \cdot \phi \cdot N_{24} - \sigma_a^{25} \cdot \phi \cdot N_{25}$
26	U-236	$\frac{d}{dt}N_{26} = \sigma_c^{25} \cdot \phi \cdot N_{25} - \sigma_a^{26} \cdot \phi \cdot N_{26}$
27	U-237	$\frac{d}{dt}N_{27} = \sigma_c^{26} \cdot \phi \cdot N_{26} - \sigma_{n,2n}^{28} \cdot \phi \cdot N_{28} - \lambda^{27} \cdot N_{27}$

This set of coupled differential equations must be carried on for all fuel isotopes and fission products, structural material, absorber and coolant

- N represents the number of isotopes per cm^3,
- ϕ the neutron flux $n/(cm^2 \cdot s)$,
- σ the one group microscopic cross section in cm^2, averaged over the neutron energy spectrum for capture (index c), absorption (index a), or n,2n reactions (index n,2n)
- λ the decay constant.

The above coupled set of differential equations is solved numerically by computer programs for each point $\vec{r}$ in the reactor core assuming the initial concentrations of the isotope as initial conditions and a specified level of neutron flux or power density. Often this burnup chain has to be truncated appropriately accounting only for the most important isotopes [27,28,29]

One of the computer codes used internationally to predict the evolution of the fuel and fission product isotopes is ORIGEN-S developed at ORNL. Its modification, KORIGEN, [27] can treat 129 actinides, more than 1100 individual fission products and about 700 isotopes associated with coolant, absorber and structural materials.

The evolution chain in concentrations of all isotopes must be determined for each point of the reactor core with different spatial neutron flux. Therefore, the code ORIGEN is coupled with a code which solves the Boltzmann neutron transport equation. Such a code package is, e.g. SCALE developed by ORNL [30]. Another code package based on a MONTE CARLO code is MONTEBURNS [33].

Figs. 4.8a and 4.8b show the decrease of fissile U-235 nuclei (burnup), the buildup of U-236 and of the different plutonium isotopes as well as the buildup of Np-237, Am-243, Sr-90 and Cs-137 in one tonne of LWR fuel over a period of about five years of operation. It is seen that the isotopic distribution of plutonium (Pu-239, Pu-240, Pu-241, Pu-242) also changes as a function of burnup or time. While the enrichment of U-235 has dropped from 5% to roughly 0.77% or 7.7 kg in 1 tonne of fuel, i.e. similar but slightly higher than the U-235 concentration in natural uranium by the end of the period of operation (burnup of 60.000 MWd/t), the concentration of plutonium (sum of all plutonium isotopes) has risen to almost 0.9% over the same period. The plutonium buildup is caused by neutron capture processes. In this way the fissile isotopes Pu-239 and Pu-241, produced during operation make contributions to the heat production. From Fig. 4.8a it is understood that the sum of the concentrations of the two fissile plutonium isotopes exceed at a burnup of 60 GWd/t the concentrations of U-235.

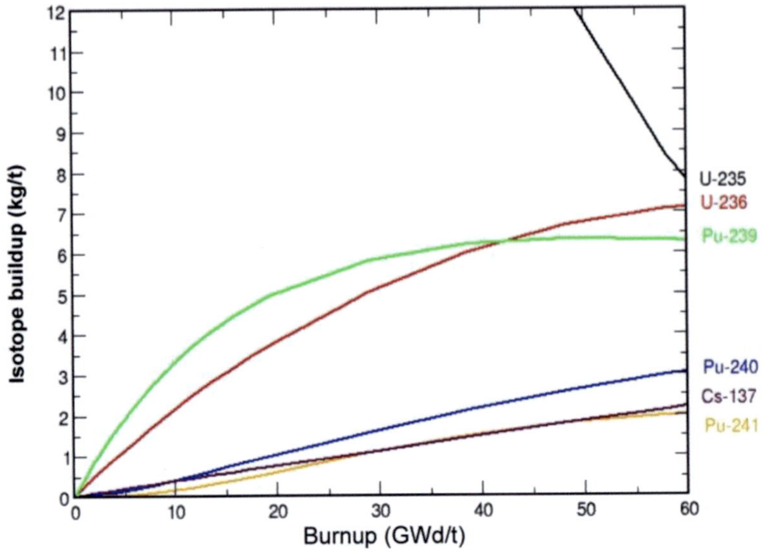

Figure 4.8a: Isotope concentrations for U-235, U-236, Pu-239, Pu-240, Pu-241 and Cs-137 as a function of burnup [GWd/to] of LWR fuel. Initial fuel enrichment: 5% U-235 [32].

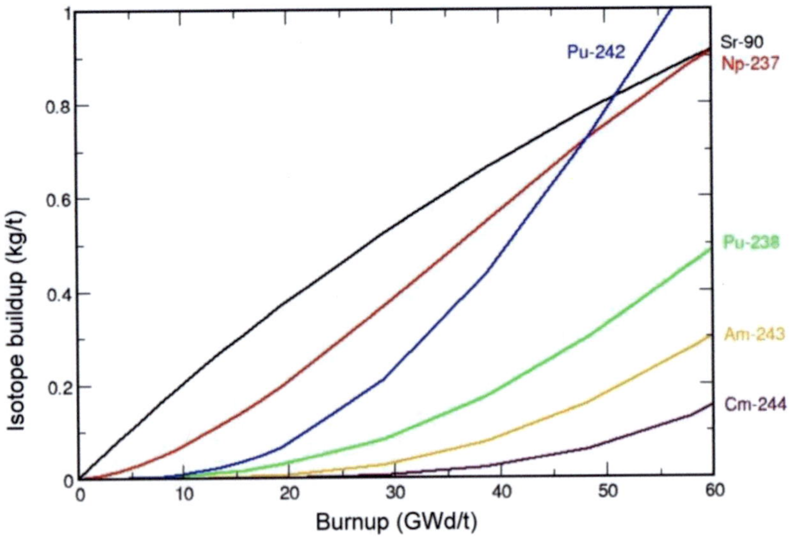

Figure 4.8b: Isotope concentrations for Pu-242, Sr-90, Np-237, Pu-238, Am-243, Cm-244 as a function of burnup [GWd/ton] of LWR fuel. Initial fuel enrichment: 5% U-235 [32].

4.7 The conversion and breeding process

4.7.1 Uranium-plutonium cycle

Neutron capture in the fertile isotopes U-238 or Pu-240 creates the new fissile isotopes Pu-239 and Pu-241 in the reactor fuel. This is depicted in Fig. 4.7 and Fig. 4.8a and 4.8b.

Neutron capture in U-238 creates U-239 being subject of β^--decay to Np-239 (decay half-life 23.5 min). Similarly, Np-239 decays to Pu-239 with a decay half life of 2.35 days. Neutron capture in Pu-239 results in Pu-240 and neutron capture in Pu-240 leads to Pu-241 which follows β^--decay to Am-241 with a half-life of 14.3 y. Finally neutron capture in Pu-241 results in Pu-242. Further neutron capture in americium leads to the curium isotopes. Also the neutron capture in U-235 and U-236 as well as the buildup of Np-237 and Pu-238 are shown there (Fig. 4.7).

4.7.2 Thorium-uranium cycle

Similarly, as depicted in Fig. 4.9, the thorium conversion chain leads to Th-233 after neutron capture in the Th-232. This Th-233 decays first to Pa-233 with a decay half- life of 23.4 min and then to the fissile isotope U-233 with a rather long decay half-life of 27 days. Subsequent neutron capture in uranium isotopes leads to U-234, U-235 and U-236 (see equations in Section 4.6).

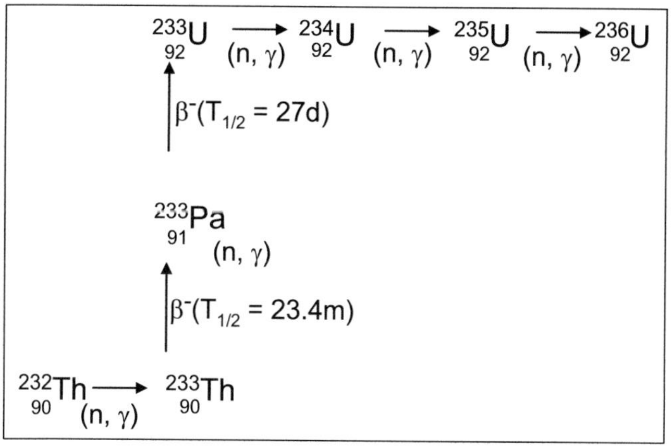

Figure 4.9: Thorium-uranium reaction chain.

4.7.3 Conversion and breeding process

The production of new fissile isotopes Pu-239 and Pu-241 in the U/Pu-cycle (Fig. 4.10) and U-233 in the Th/U-233 cycle is called conversion process. The conversion ratio CR is defined as the ratio of creation of new fissile isotopes to the rate of destruction of fissile isotopes. The conversion ratio CR depends on the fuel cycle and on the neutron energy spectrum in the reactor core.

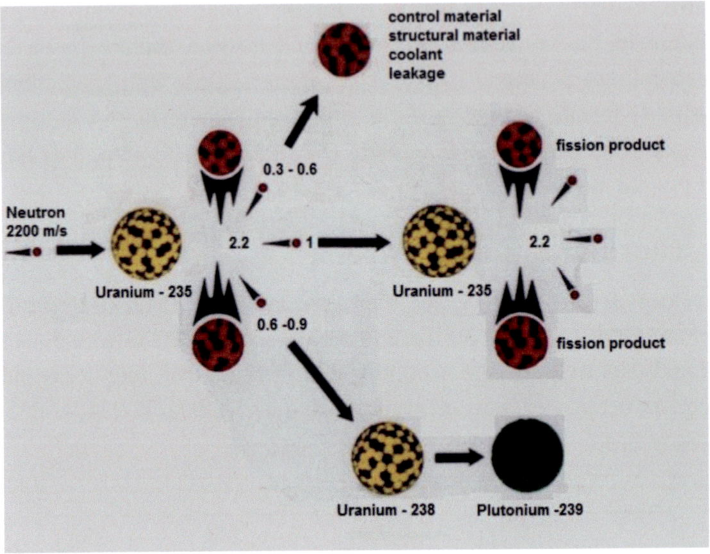

Figure 4.10: Conversion process of U/Pu fuel cycle.

It is determined by computer calculations in building the ratio of the sums of the total number of newly generated and destructed fissile isotopes in the reactor core between periodic refueling sequences. The conversion ratio for reactors with thermal neutron spectrum is CR <1. Only in the case of thermal breeders with U-233 it might slightly exceed unity. If the conversion ratio becomes >1 it is called breeding ratio BR. The conversion ratio CR can also be expressed by the relation

$$CR = \overline{\eta} - 1 - \overline{a} - \overline{\ell} + \overline{f}$$

where $\overline{\eta}$ is the neutron yield, i.e., the total number of fission neutrons generated per neutron absorbed, averaged over all fissile isotopes, the neutron energy spectrum and the whole reactor volume. The quantities $\overline{a}$, $\overline{\ell}$ and $\overline{f}$ are equivalent corresponding averages. $\overline{a}$ describes the absorption of neutrons in the coolant, structural and control materials.

$\overline{\ell}$ is the leakage from the reactor core. $\overline{f}$ describes the small contributions of fast neutron fissions in U-238, Pu-240 and Pu-242 or Th-232 (fast fission effect.). The largest contribution in the above relation for CR is the neutron yield $\overline{\eta}$. It is shown for different fissile isotopes belonging to different fuel cycle options in Fig. 4.11:

More refined calculation methods for the determination of CR or BR and other parameters like doubling time were described, e.g. in [9].

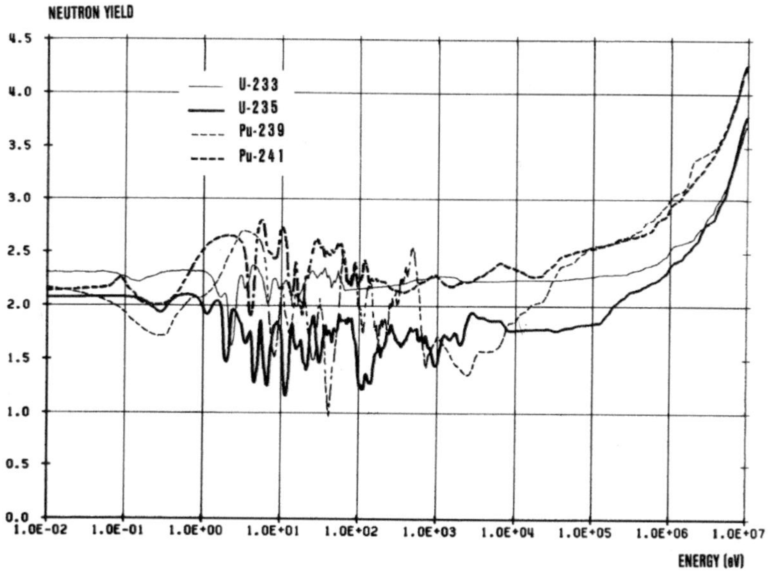

Figure 4.11: Neutron yield as a function of energy of incident neutron causing fission for U-233, U-235, Pu-239, Pu-241 [32].

In the range of thermal neutron energy spectra U-233 assumes the highest values of e.g. $\eta = 2.25$ for U-233, whereas U-235 has the lowest η-values. This is the reason for proposals of reactor designs with thermal neutron energy spectrum and U-233/Th-232 fuel.

In the neutron energy range >100 keV the two plutonium isotopes Pu-239 and Pu-241 assume the highest η values of $\eta \approx 2.6$-2.7. Therefore, in this neutron energy range a breeding ratio

$$BR > 1$$

is possible. Liquid metal cooled (lead or sodium cooled) reactors have a neutron spectrum in this range and are able to operate as breeder reactors. They are called (fast) breeder reactors (FBRs) due to their fast energy neutron spectrum. If particularly high BR values shall be achieved, e.g. in a rapidly expanding nuclear energy economy the neutrons escaping from the core can be captured in thick blanket regions with natural or depleted uranium.

Thermal energy spectrum reactors initially must start operation with the only available natural or enriched uranium. They can attain only a conversion ratio of 0.5 to 0.8 depending upon their absorption factor $\overline{a}$ in the coolant (light water, heavy water, graphite) and in their structural or absorber materials.

U-233 and Pu-239/Pu-241 are artificial man-made fissile materials which must be produced first in thermal or fast neutron spectrum reactors. Only then U-233/Th-233 or Pu-239/U-238 can be used in reactors which attain high conversion or breeding ratios. Table 4.5 shows the conversion or breeding ratios for the main important different reactor systems operated in the near future.

Table 4.5: Conversion/breeding ratios for different reactors.

Reactor type	Initial fuel	Fuel cycle	Conversion/ breeding ratio
PWR/BWR	3-5% U-235	U/Pu	0.55
PWR/BWR - Pu recycl.	5% Pu	U/Pu	0.7
PWR -U-233/Th	4% U-233	U/Th	0.8
FBR – Pu/U	15-25% Pu	U/Pu	1.2-1.4
FBR – U-233/Th	12-20% U-233	Th/U-233	1.03-1.15

4.8 Fuel utilization

The fuel utilization is defined as the fraction of original nuclear fuel that can be ultimately converted into fission energy. This includes also the conversion of fertile isotopes like U-238, Pu-240 or Th-232 into fissile isotopes. Some of the converted fissile nuclei can be extracted after unloading of the spent fuel by chemical reprocessing. This collected fissile fuel can be used for refabrication of new fuel elements which can be loaded again (recycled) in nuclear reactors.

This means that the determination of the fuel utilization must account for recycling of the fuel including possible losses of fuel during reprocessing and refabrication.

For the once-through fuel cycle the fuel utilization will be lowest. For Pu/U recycling, e.g. in LWRs, the fuel can be recycled only several times (Section 12). For FBRs multi-recycling of the fuel is possible. FBRs attain the highest fuel utilization factor. The fuel utilization factor depends on the enrichment of the fuel, the tails assay after enrichment, the conversion factor (reactor design and fuel cycle) and in case of recycling on the burnup of the fuel per cycle and fuel losses during reprocessing and refabrication.

The fuel utilization factor can be determined by detailed analysis following up the remaining fuel in a balance sheet for each operating cycle of the reactor core [31]. This leads to Fig. 4.12.

Fig. 4.12 shows the fuel utilization as a function of the conversion or breeding ratio. Nuclear reactors operating not in the fuel recycling mode, e.g. LWRs in the once-through fuel cycle with a conversion ratio of 0.5 to 0.6 attain only a fuel utilization factor of about 0.6%. Reactors operating in the recycling mode, e.g. LWRs with U/Pu recycling attain a conversion ratio of CR = 0.72 and a fuel utilization of almost 1%. This can be further improved somewhat by LWRs with tight lattice fuel elements and higher conversion ratios [34].

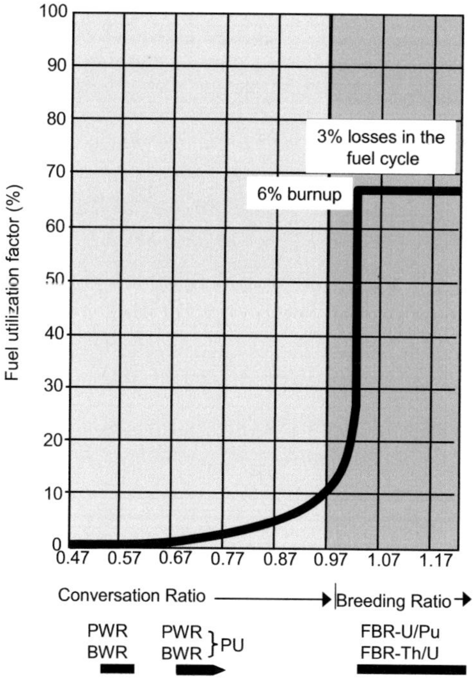

Figure 4.12: Fuel utilization [%] as a function of the conversion ratio for LWRs (PWRs or BWRs), plutonium recycling PWRs or BWRs and FBRs operating in the U/Pu-cycle or the Th/U-cycle [31].

The fuel utilization factor shows a sharp increase above conversion factors of CR = 0.9 and a step type increase for breeding ratios of about BR $\geq$ 1.03 (assumed fuel losses of 1.5% during reprocessing and 1.5% during refabrication). Above this limit the reactor system is able to make up for the fuel cycle losses by breeding sufficient fissile material. FBRs operating in the uranium/plutonium closed fuel cycle operate beyond this limit. The fuel utilization factor is then only dependent on the average fuel burnup (number of recycle steps or fuel loss during recycling). FBRs with a breeding ratio above, e.g. BR $\geq$ 1.03 can attain a fuel utilization factor of 60% which is about a factor of 100 higher than the fuel utilization factor of LWRs in the once-through cycle. Similar high fuel utilization factors can be attained by FBRs operating in the Th/U-233 cycle.

In thermal spectrum reactors the concentration of the fission products, especially Xe-135 and its precursor I-135 as well as Sm-149 must be accounted for because of their high absorption cross sections in the thermal neutron energy range. Their variation of concentration affects the criticality k_{eff} considerably after shut down of the reactor and may prevent restart of the reactor for a certain period of time. Due to the decay half-life of I-135 of 6.61 h the decay of this precursor may induce so-called Xenon oscillations in large power reactors during operation.

4.9 Radioactive Inventories in spent fuel

The amount of fission products and actinides as well as the radioactivity inventory in structural material is a function of the operating time or of the burnup in MWd/t_{HM} of the spent fuel. With the methods described in Section 4.5 and computer programs, e.g. KORIGEN [27] it is possible to calculate the amount of different substances in the spent fuel and their radioactivity. Table 6.4 lists the radioactivity of the most important fission products and actinides in 1 t_{HM} of spent fuel after a burnup of 60,000 MWd/t_{HM}. The radioactivity is measured in Curie [Ci] or Becquerel [Bq].

$$1 \text{ Curie} \mathrel{\hat{=}} 3.7 \times 10^{10} \text{ Bq}$$

1 Bq corresponds to one disintegration per second in a β-, α- or γ-decay of a nucleus.

Table 4.6: Radioactivity of fission products and actinides of spent PWR fuel after a burnup of 60 GWd/t_{HM}. The specific radioactivity is given for the time of unloading of the spent fuel and several years of cooling (Broeders [32]).

Fission products		Specific radioactivity (Ci/t_{HM})				
Isotope	half life	discharge	1 year	3 years	5 years	7 years
H 3	12.349 y	9.714E+02	9.184E+02	8.209E+02	7.337E+02	6.558E+02
KR 85	10.720 y	1.701E+04	1.595E+04	1.402E+04	1.232E+04	1.082E+04
SR 90	29.121 y	1.166E+05	1.139E+05	1.086E+05	1.035E+05	9.871E+04
Y 90	2.6667 d	1.171E+05	1.139E+05	1.086E+05	1.035E+05	9.873E+04
ZR 95	63.981 d	1.752E+06	3.354E+04	1.229E+01	4.501E-03	1.649E-06
NB 95	35.150 d	1.765E+06	7.548E+04	2.728E+01	9.992E-03	3.660E-06
RU106	1.0080 y	9.339E+05	4.696E+05	1.187E+05	3.002E+04	7.591E+03
RH106	29.900 s	1.040E+06	4.696E+05	1.187E+05	3.002E+04	7.591E+03
CS134	2.0619 y	4.163E+05	2.975E+05	1.519E+05	7.755E+04	3.960E+04
CS137	29.999 y	1.847E+05	1.805E+05	1.723E+05	1.645E+05	1.571E+05
BA137M	2.5517 m	1.752E+05	1.707E+05	1.630E+05	1.556E+05	1.486E+05
CE144	284.26 d	1.409E+06	5.784E+05	9.745E+04	1.642E+04	2.767E+03
PR144	17.283 m	1.423E+06	5.784E+05	9.745E+04	1.642E+04	2.767E+03
PM147	2.6235 y	2.038E+05	1.629E+05	9.605E+04	5.663E+04	3.339E+04
EU154	8.6001 y	1.473E+04	1.359E+04	1.157E+04	9.845E+03	8.380E+03

Actinides		Specific radioactivity (Ci/t_{HM})				
Isotope	half life	discharge	1 year	3 years	5 years	7 years
U 234	2.45E05 y	3.380E-02	5.823E-02	1.080E-01	1.573E-01	2.058E-01
U 235	7.04E08 y	1.220E-02	1.220E-02	1.220E-02	1.220E-02	1.220E-02
U 236	2.34E07 y	4.034E-01	4.034E-01	4.035E-01	4.035E-01	4.036E-01
U 238	4.5E09 y	3.067E-01	3.067E-01	3.067E-01	3.067E-01	3.067E-01
NP237	2.15E06 y	6.062E-01	6.183E-01	6.188E-01	6.197E-01	6.208E-01
NP239	2.3553 d	2.718E+07	6.636E+01	6.635E+01	6.634E+01	6.633E+01
PU238	87.744 y	8.287E+03	8.766E+03	8.752E+03	8.620E+03	8.485E+03
PU239	24064. y	3.748E+02	3.822E+02	3.821E+02	3.821E+02	3.821E+02
PU240	6537.3 y	6.997E+02	7.012E+02	7.039E+02	7.064E+02	7.087E+02
PU241	14.399 y	1.996E+05	1.902E+05	1.727E+05	1.569E+05	1.425E+05
PU242	3.87E05 y	4.637E+00	4.637E+00	4.637E+00	4.637E+00	4.637E+00
AM241	432.23 y	1.810E+02	4.928E+02	1.072E+03	1.596E+03	2.069E+03
AM243	7380.2 y	6.629E+01	6.636E+01	6.635E+01	6.634E+01	6.633E+01
CM242	163.19 d	1.206E+05	2.569E+04	1.155E+03	5.200E+01	2.398E+00
CM244	18.110 y	1.458E+04	1.406E+04	1.303E+04	1.207E+04	1.118E+04

4.10 Inherent Safety Characteristics of Converter and Breeder Reactor Cores

4.10.1 Reactivity and Non-Steady State Conditions [2,5,11,12,13,26]

It has been explained in Section 4.4 that $k_{eff} = 1$ corresponds to the steady state condition of the reactor core, in which case the production of fission neutrons is in a state of equilibrium with the number of neutrons absorbed and the number of neutrons escaping from the reactor core. For $k_{eff} \neq 1$, either the production or the loss term becomes dominant, i.e., the number of neutrons varies as a function of time. The neutron transport equation or, by way of approximation, the multigroup diffusion equation for calculation of the time dependent neutron flux must then be solved for the non-steady state case. However, in most cases it is a sufficiently good approximation to solve the so-called point kinetics equations in connection with equations describing the temperature field and its impacts on k_{eff}.

Axial movements of the absorber rods in the core or compaction of the core change the loss term for neutrons and influence k_{eff}. The relative change as a function of time of $k_{eff}(t)$ is called reactivity:

Reactivity

$$\rho(t) = \frac{k_{eff}(t)-1}{k_{eff}(0)} = \frac{\Delta k_{eff}(t)}{k_{eff}(0)} \tag{4.4}$$

The point kinetics equation describing the reactor power as a function of time can be derived from the time dependent multigroup diffusion equation and the time dependent equations for the precursors of the delayed neutrons under the assumption of separation into space and time functions. The point kinetics equations read

$$\frac{dP(t)}{dt} = \left[\frac{\rho(t)-\beta_{eff}}{l_{eff}}\right]\cdot P(t)+\lambda\cdot C(t) \tag{4.5}$$

$$\frac{dC(t)}{dt} = -\lambda\cdot C(t)+\frac{\beta_{eff}}{l_{eff}}\cdot P(t)$$

with the initial conditions of

$P(t = 0) = P_0$ (steady state reactor power)
$C(t = 0) = C_o$ (steady state concentration of the parent nuclei
(precursors, see Section 4.1) for all delayed neutrons
combined in one group).

In these equations, $P(t)$ is the reactor power, $C(t)$ describes the averaged concentration of parent nuclei (precursors) of the delayed neutrons, λ is the average decay constant of all parent nuclei of delayed neutrons, β_{eff}, the effective fraction of all delayed neutrons (integrated over all fissile isotopes and averaged over the reactor), l_{eff} the lifetime of the prompt neutrons in the reactor, i.e., the average time required for a prompt fission neutron to induce a new fission process. Values of β_{eff} and λ are calculated from those indicated in

Section 4.1 by suitably averaging the decay constants with the corresponding delayed neutron fractions of the usually used 6 groups of delayed neutrons. Values of l_{eff} are in the range of 10^{-3} to 10^{-5} s for thermal reactors and around 4 x 10^{-7} s for FBRs. The average decay constant for a one-group treatment is in the range of about $\lambda = 0.08\ s^{-1}$ for all fissile fuel isotopes.

For $\rho(t) > \beta$ the influence of the delayed neutrons can be neglected and the solution for P(t) can be simplified to

$$P(t) = \exp\left[\frac{\rho(t)}{l_{eff}} \bullet t\right] = \exp\left[\frac{k_{eff}(t)-1}{k_{eff}(0)\cdot l_{eff}} \bullet t\right] \quad (4.6)$$

The reactivity, $\rho(t)$, is composed of the superimposed or initiating reactivity, $\rho_i(t)$, which can be caused, e.g., by movements of absorber rods or fuel and the feedback reactivity, $\rho_f(t)$, which takes into account all repercussions of temperature or density changes in the reactor core,

$$\rho(t) = \rho_i(t) + \rho_f(t)$$

Movements of absorbers or density changes produce local changes in the macroscopic cross sections and the neutron flux in certain material zones in the reactor core and, accordingly, also in $k_{eff}(t)$ and $\rho_i(t)$ (superimposed or initiating perturbations). The resultant change as a function of time of the neutron field and the power level alters the temperatures in the reactor core. Temperature changes provoke changes in material densities (expansion and displacement) and microscopic cross sections by the Doppler broadening of resonances. Also the neutron flux spectrum can be shifted by changing the moderation of the neutrons. Moreover, the dimensions of the reactor core and its components are changed by thermal expansion. All these feedback reactivities, $\rho_f(t)$, resulting from changes in power and temperature together with external perturbation reactivities constitute a feedback circuit (Fig. 4.13).

For numerical treatment, the feedback reactivity, ρ_f, is split up into individual contributions by different temperature effects

$$d\rho_f = \mathrm{Sum}_{i=1}^{1} \frac{\partial k_{eff}}{\partial T_i}$$

where dT_i are the average changes in the temperatures of the fuel, moderator, coolant, structural or absorber materials. In LWRs, the coolant is also the moderator. In other types of reactors (HTGR), the moderator (graphite) is distinct from the coolant (gas).

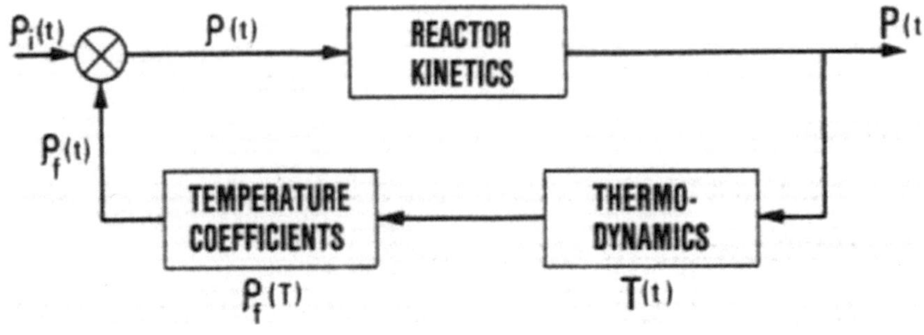

Figure 4.13: Schematic of the reactor dynamics feedback loop.

4.10.2 Temperature Reactivity Coefficients

4.10.2.1 Fuel Doppler Temperature Coefficient [11,12,13,25,26,32,33]

The fuel Doppler temperature coefficient is due to the fact that the neutron resonance cross sections depend on the temperature of the fuel and the relative velocities, respectively, of neutrons and atomic nuclei.

The resonance cross sections, $\sigma(E,T)$, for U-238, Th-232 and U-233, U-235, Pu-239, etc. show very pronounced peaks at certain neutron kinetic energies. An increase in fuel temperature, T_f, broadens this shape of the resonance curve which, in turn, results in a change in fine structure of the neutron flux spectrum in these ranges of resonance energy. The reaction rates are changed as a consequence. Above all, the resonance absorption for U-238 increases as a result of rising fuel temperatures, while the effect of a temperature change in the resonance cross sections of the fissile materials, U-235 and Pu-239, is so small that it can generally be neglected if the fuel enrichment is not extremely high. For these reasons, temperature increases in the fuel result in a negative temperature feedback effect (Doppler effect) brought about by the increase in neutron absorption in U-238. For Th-232, the effects are similar. The Doppler effect is somewhat less pronounced at very high fuel temperatures because adjacent resonances will overlap more and more. The resonance structure then is no longer as pronounced as at low temperatures, which leads to a reduction of the negative Doppler effect.

Due to the specific energy distribution of the neutron spectrum, the Doppler effect in thermal reactors follows

$$\frac{1}{k_{eff}} \frac{\partial k_D}{\partial T_f} = \frac{-1}{\sqrt{T_f}} \tag{4.7}$$

whereas in FBRs it follows the relation

$$\frac{1}{k_{eff}} \frac{\partial k_D}{\partial T_f} = -\frac{1}{T_F^X} \text{ with } \frac{1}{2} \leqq x \leqq 3/2 \tag{4.8}$$

The Doppler coefficient is always negative in power reactor cores because, given the relatively low enrichment in U-235 and Pu-239, respectively, the resonance absorption of U-238 will always dominate. Negleting the short time delay needed for heat transfer from PuO_2 to UO_2 grains it is an instantaneous negative feedback coefficient of reactivity, which immediately counteracts increases in power and temperature. (The Doppler feedback is virtually missing for highly enriched fuel.)

Besides the fuel Doppler coefficient, the fuel expansion coefficient also leads to a negative feedback coefficient of reactivity. Especially in fast breeder cores it may well attain an importance equal to the Doppler coefficient. However, the response times of these feedback effects are different.

4.10.2.2 Coefficients of Moderator or Coolant Temperatures [32]

The main contributions to the coefficients of moderator or coolant temperatures stem from changes in the densities of the moderator or coolant and from resultant shifts in the neutron spectrum and in the associated scattering transfer probabilities. Temperature rises decrease the density of the coolant and accordingly reduce the moderation of neutrons. The neutron spectrum is shifted towards higher energies. As a result of the lower moderator density and the correspondingly higher transparency to neutrons of the core it is also possible that far more neutrons escape from the reactor core and neutron losses due to leakage rate will increase.

For the present line of PWRs, the sum total of the individual contributions to changes in various energy ranges finally leads to a negative coefficient of the moderator temperature which, however, also depends on the concentration of boric acid dissolved in the coolant and the burnup condition of the reactor core. In large graphite moderated HTGRs containing U-233, the moderator temperature coefficient is usually positive. In small HTRs the moderator coefficient is negative.

Also in sodium cooled FBRs with core sizes in excess of about 100-150 MW(e), the coolant temperature coefficient is positive because the neutron spectrum is shifted towards higher energies as a consequence of the reduced moderation. The resultant increased contribution by the fast fission effect of U-238 as well as the higher η-values (see Fig. 4.11) add to the reactivity. These positive reactivity contributions cannot compensate all negative contributions coming from an increase in the leakage rate of neutrons escaping from the core (which is the dominating effect in small sodium cooled FBRs with power levels of less than approx. 100-150 MW(e)).

4.10.2.3 Structural Material Temperature Coefficient

Especially in FBRs, the structural material temperature coefficient also plays an important role. Increasing temperatures cause the core structure to expand radially and axially and, in this way, result both in indirect changes in material densities and in changes of size of the reactor core and, as a consequence, of neutron leakage. The structural material temperature coefficient must be determined by detailed analyses of all expansion and bowing effects for given core and fuel element structures also taking into account the core restraint (clamping) system. For FBRs, the structural material coefficient is also negative. This is accomplished by the specific design of the core support plate and the core restraint system.

For analysis of the control behavior of a reactor core and its behavior under accident conditions, the non-steady state neutron flux, power, temperature and all feedback reactivities must be considered in detail. Negative feedback reactivities or temperature coefficients always counteract increases in power and temperature. Positive coefficients of moderator or coolant can be tolerated as long as all the other temperature coefficients, above all the sufficiently fast prompt Doppler coefficient, are negative and larger in magnitude than the positive coefficients of moderator or coolant. Table 4.7 shows typical temperature coefficients of reactivity for various types of reactors.

Table 4.7: Typical temperature coefficients of reactivity for various reactor lines.

Temperature coefficient Δk/K	PWR	BWR	LMFBR
Moderator or coolant (fresh fuel)	-9×10^{-5}	-10×10^{-5}	$+5 \times 10^{-6}$
Doppler coefficient (500-2800 °C)	-2.7×10^{-5} to -1.7×10^{-5}	-2.5×10^{-5} to -1.3×10^{-5}	-1.1×10^{-5} to -2.8×10^{-6}

PWRs or BWRs have highly negative coolant or moderator temperature coefficients. Sodium cooled FBRs throughout their whole operating cycle have positive coolant temperature coefficients. Thermal reactors, such as PWRs and BWRs, have more negative Doppler coefficients than FBRs.

4.10.3 Reactor Control and Safety Analysis [2,9,18,27,28]

4.10.3.1 Reactivity Changes During Startup and Full Power Operation

As the reactor core is slowly being started up from zero power to full power the temperatures of coolant and core structure rise by several 100 °C. At the same time, the fuel temperature increases by more than 1000 °C. This causes a negative reactivity effect, which must be overcome by moving absorber (control) rods out of the reactor core. In LWRs, this reactivity span is in the range of several percent. In sodium cooled FBRs, it is somewhat smaller mainly because of the lower value of the negative Doppler coefficient.

The buildup of fission products and actinides as well as the burnup of fissionable isotopes leads to a reactivity loss of up to 12% in LWRs and about 3% in sodium cooled FBRs. Sufficient excess reactivity, i.e., keff.> 1, therefore must be provided in a core with fresh (non-irradiated) fuel and zero power at the beginning of an operating cycle. At this time, the excess reactivity is counterbalanced by the insertion into the core of such absorber materials (Section 4.4), which provide a sufficient reactivity span for reactor control. Due to the burnup effects as well as fission product buildup mentioned in Section 4.6, the negative reactivity must be reduced during the operating cycle. This is accomplished by several methods, e.g., withdrawing absorber rods, reducing the concentration of soluble poisons, such as boric acid, and by the diminished absorption effect of burnable poisons, such as gadolinium or erbium contained in fixed rods. The strongly absorbing isotopes of these elements suffer considerable depletion during reactor operation.

The fraction of excess reactivity for fissile isotope burnup and fission product buildup designed into the fresh core determines the length of operation of a core (operating cycle). The reactor core is shut down by moving into the core absorber rods with sufficient negative reactivity. In this case, the reactivity span from full power (high temperature) to zero power (low temperature) has to be overcome. In addition, the reactor core must be held subcritical, which means that it has to attain and maintain a keff well below 1.

References Section 4

[1] Weinberg, A.M., Wigner, E.P., The Physical Theory of Neutron Chain Reactors, University of Chicago Press, Chicago (1958).

[2] Glasstone, S., Edlund, M.C., The Elements of Nuclear Reactor Theory, D. Van Nostrand, Princeton, NJ (1952).

[3] Lamarsh, J.R., Introduction to Nuclear Reactor Theory, 2nd ed., Addison-Wesley, Reading, MA (1983).

[4] Duderstadt, J.J., Hamilton, L.J., Nuclear Reactor Analysis, Wiley, New York (1976).

[5] Henry, A.F., Nuclear-Reactor Analysis, MIT Press, Cambridge, MA (1975).

[6] Bell, G.I., Glasstone, S., Nuclear Reactor Theory, Van Nostrand Reinhold, New York (1970).

[7] Meghreblian, R.V., Holmes, D.K., Reactor Analysis, McGraw-Hill, New York (1960), pp. 160-267 and 626-747.

[8] Radkowsky, A., ed., Naval Reactors Physics Handbook, U.S. Atomic Energy Commission, Washington, DC (1964), Chap. 5.

[9] Ott, K. O. et al., Introductory nuclear reactor statics, American Nuclear Society, La Grange Park, Illinois, USA (1983).

[10] Michaudon, A.: Nuclear Fission and Neutron Induced Fission Cross Sections. Oxford: Pergamon Press (1981).

[11] Keepin, G.R., Physics of Nuclear Kinetics, Addison-Wesley, Reading, MA (1965).

[12] Ash, M., Nuclear Reactor Kinetics, McGraw-Hill, New York (1965).

[13] Hetrick, D.L., ed., Dynamics of nuclear systems, University of Arizona Press, Tucson, AZ (1972).

[14] Lewis, E.E., Miller, W.F., Computational methods of neutron transport, Wiley-Interscience, New York (1984); reprinted by American Nuclear Society, La Grange Park, IL (1993).

[15] Ronen, Y., ed., CRC Handbook of Nuclear Reactors Calculations I, CRC Press, Boca Raton, FL (1986).

[16] Alcouffe, R.E., et al., DANTSYS: A diffusion accelerated neutral particle transport code system, LA-12969-M, Los Alamos National Laboratory, USA (1995).

[17] Lawrence, R.D., The DIF3D nodal neutronics option for two- and three-dimensional diffusion theory calculations in hexagonal geometry, ANL-83-1, Argonne National Laboratory, USA (1983).

[18] Briesmeister, J.F., editor, MCNP – A general Monte Carlo N-Particle transport code, Version 4C, Technical Report, LA-13709-M, Los Alamos National Laboratory, USA (2000).

[19] Koning, A., et al., The JEFF-3.1 Nuclear Data Library, JEFF Report 21, NEA No. 6190, OECD/NEA, Paris (2006).

[20] Roussin, R.W., Young, P.G., McKnight, R., Current status of ENDF/B-VI, Proc. Int. Conf. Nuclear Data for Science and Technology, Gatlinburg, TN, Vol. 2, p. 692 (1994).

[21] Kikuchi, Y., JENDL-3, Revision 2: JENDL 3-2, Proc. Int. Conf. Nuclear Data for Science and Technology, Gatlinburg, TN, Vol. 2, p. 685 (1994).

[22] Askew, J. et al., A general description of the lattice code WIMS. J. British Nuclear Energy Society, Vol. 5, 564 (1966).

[23] Bondarenko, I.I. et al., Group constants for nuclear reactor calculations, Translation – Consultants Bureau Enterprice Inc., New York (1964).

[24] Oldekop, W.: Einführung in die Kernreaktor- und Kernkraftwerkstechnik, Teil I. München: Karl Thiemig (1975).

[25] Waltar, A., et al., Fast Breeder Reactors, Pergamon Press, New York (1981).

[26] Stacey, W., Nuclear reactor physics, John Wiley & Sons, New York (2007).

[27] Wiese, H.W., Fischer, U., KORIGEN - Ein Programm zur Bestimmung des nukle-aren Inventars von Reaktorbrennstoffen im Brennstoffkreislauf. Kernforschungs-zentrum Karlsruhe, KfK-3014 (1981).

[28] Haeck, W. et al., An optimum approach to Monte Carlo burnup, Nucl. Sci. Eng., 156, 180-196 (2007).

[29] Fission Product Nuclear Data (FPND) - 1977. Proc. Second Advisory Group Meeting on Fission Product Nuclear Data, Energy Centrum Netherlands, Petten, 5-9 September 1977. Vienna : International Atomic Energy Agency, IAEA-213 (1978).

[30] SCALE: A modular code system for performing standardized computer analyses for licensing evaluations, ORNL/TM-2005/39, Version 5, Vols I-III (2005).

[31] Heusener, G., Personal communication, KfK Karlsruhe (1980).

[32] Broeders, C.H.M., Personal communication, KIT Karlsruhe (2010).

[33] Poston, D.I. et al., Development of a fully-automated Monte Carlo burnup code MONTEBURNS, LA-UR-99-42 (1999).

[34] Broeders, C.H.M. et al., Neutronenphysikalische und thermodynamische Auslegung eines Referenzentwurfs für einen FDWR-Gleichgewichtskern, KFK-Nachrichten, Jahrg. 23, 1/91, p. 16 (1991).

[35] Dresner, L., Resonance absorption in nuclear reactors, Pergamon Press, New York (1960).

[36] Trellue, H.R., Safety and neutronics: a comparison of MOX vs UO_2 fuel, Progress in Nuclear Energy, 48, p. 135 (2006).

5 Nuclear reactors with a thermal neutron spectrum

5.1 Introduction and historical development

The probability to fission the isotope U-235 by neutrons is highest if the neutron velocity is relatively low, i.e. at so-called thermal energies (average 0.025 eV corresponding to 2200 m/s). The fission neutrons originate at relatively high average kinetic energies (≈2 MeV). Therefore the neutrons are slowed down in thermal nuclear power reactors by scattering processes in a moderator, e.g. light water, heavy water or graphite.

The first commercial nuclear power reactors were ordered in the USA in 1963. These were pressurized water reactors (PWRs) developed from the experience with submarine nuclear reactors. Boiling water reactors (BWRs) were also ordered after successful testing of the Experimental Boiling Water Reactor at Argonne National Laboratory (USA). Both reactor types needed low enriched uranium oxide fuel.

Russia developed PWRs around the same time. It built the first experimental reactor at Obninsk in 1954 and commercial size reactors (Novo-Voronezh) in the 1960. A first experimental BWR (Dimitrovgrad) was not pursued for development to a commercial size reactor.

In the begin the USA started with graphite moderated reactors cooled by water (Hanford). The UK and France started with MAGNOX reactors using graphite as a moderator and pressurized carbon dioxide as coolant gas. These MAGNOX reactors could use natural uranium as fuel.

Canada also started with natural uranium as fuel, but used heavy water as moderator and coolant. These CANDU reactors were developed and built as commercial size units with an electrical power from 200 to 700 MW(e).

MAGNOX and CANDU reactors initially used natural uranium fuel. Therefore the burnup of their fuel elements was relatively low (7 GWd/t). Their fuel is being enriched now to about 1.5-2% U-235 to attain a burnup between 12-15 GWd/t for commercial reasons. They are designed to load and unload their fuel continuously at full power. High temperature gas cooled reactors (HTRs) are being developed in Europe, USA, Japan and China.

PWRs and BWRs are operated in a batch loading and unloading scheme with shutdown of the reactor every 12-24 months for exchange of the spent fuel. For better economy the burn up of the fuel was steadily increased. PWRs and BWRs need enrichment up to 4.5 and 5% to reach a burnup of their fuel of 55-60 GWd/t.

5.2 European Pressurized Water Reactors (PWRs)

About 265 PWRs were operating in the world in 2008 (Section 2.1). These PWRs were manufactured and built by several manufacturers, in the USA, Russia, Europe and Japan. Their technical concept is very similar.

The 1600 MWe European Pressurized Water Reactor (EPR) designed, manufactured and built by AREVA as a so-called Generation-3 PWR will be described below. Fig. 5.1 explains the main design principles [1].

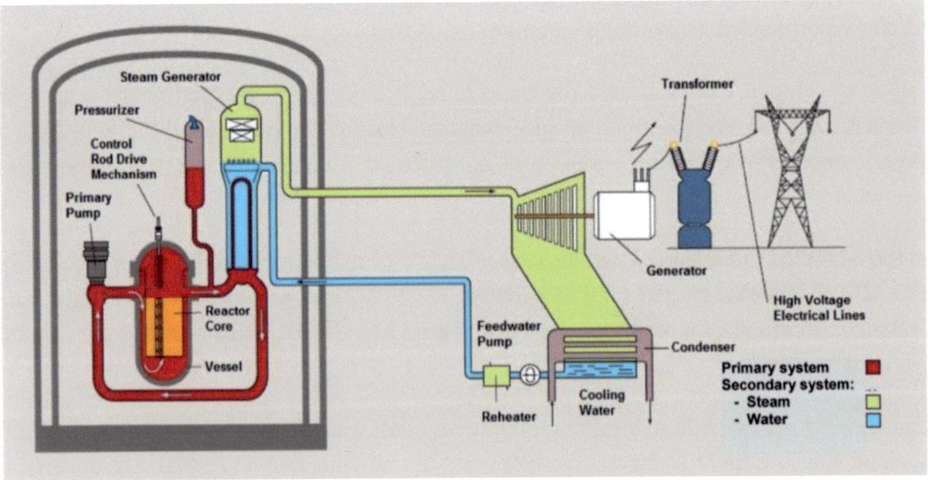

Figure 5.1: Main design principles of the European Pressured Water Reactor (AREVA).

The heat generated by nuclear fission in the reactor core is transferred from the fuel rods to the coolant in the primary coolant system. The power is controlled by absorber rods. The highly pressurized water (15.5 MPa) is circulated by the primary coolant pumps and heated in the core from 295 °C to 328 °C (Table 5.1). A pressurizer controls the primary pressure of the coolant. The primary coolant flows to four steam generators, where steam of 7.8 MPa and 293 °C is generated. This steam drives the turbine and generator. Behind the turbine the saturated steam is precipitated in the condenser and the condensate water is pumped back as feedwater into the steam generators. The waste heat is discharged from the condenser to the environment either to a river or through a cooling tower to the atmosphere. The thermal efficiency is about 35.5%.

5.2.1 Core with fuel elements and control elements

The EPR core consists of 241 fuel elements (Fig. 5.2 and Fig. 5.3). For the initial core these fuel elements are split into four groups of different enrichment in U-235. Different fuel enrichments shall adapt already from the beginning to the later equilibrium core. This equilibrium core will have fuel elements with different burnup which can be unloaded every one and a half to two years (batch loading and unloading). The fuel elements can reach a maximum burnup of about 70 GWd/t. Some fuel elements contain gadolinium in the uranium fuel as a burnable poison either mixed homogeneously with the fuel or arranged heterogeneously in special rods. In addition, the coolant water contains soluble boric acid as neutron absorber. Both gadolinium and boric acid compensate from the beginning and during the operation for the negative reactivity effects of the decrease on U-235 enrichment and build-up of fission products. They guarantee an operation cycle length of 1.5 to 2 years. The fuel assembly of square geometry consists of 17x17 rods, 265 of which are fuel rods. The fuel element has bottom and top pieces and 10 spacer grids distributed over the axial length of 4.2 m (Fig. 5.3). The spacer grids hold the distance between the fuel rods (lattice pitch). In addition, 24 guide thimbles support the whole structure of the fuel element. The guide thimbles are also used as guide tubes for moveable absorber rods or for the in-core instrumentation [2,3,7].

Table 5.1: Characteristic data of EPR reactor core (AREVA).

Reactor core	
Thermal power	4,500 MWth
Operating pressure	155 bar
Nominal inlet temperature	295.6 °C
Nominal outlet temperature	328.2 °C
Equivalent diameter	3,767 mm
Active fuel length	4,200 mm
Number of fuel assemblies	241
Number of fuel rods	63,865
Average linear heat rate	156.1 W/cm

The fresh (non irradiated) fuel rods contain a stack of low enriched uranium dioxide sintered pellets with a U-235 enrichment of about 5%. The fuel rod cladding consists of zircaloy-M5 with an outside diameter of 9.5 mm and a radial thickness of 0.57 mm. A gas plenum is provided axially at the upper end of the fuel rod where the fission gases can accumulate.

The core has a fast acting shutdown control system consisting of 89 rod cluster control assemblies (RCCA). Each RCCA contains 24 absorber rods which dive into the 24 guide thimbles of fuel assemblies. These absorber rods contain neutron absorbing materials like Ag, In, Cd or boron carbide pellets.

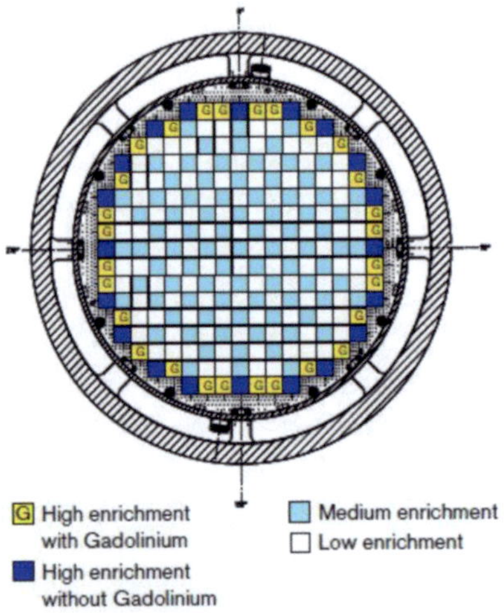

Figure 5.2: Nuclear reactor core of EPR with fuel elements (AREVA).

5.2.2 Reactor pressure vessel

The reactor pressure vessel (RPV) and its closure head are made of ferritic steel. The RPV contains the inlet and outlet water nozzles for the coolant flow and the upper penetrations for control rod drive mechanisms (CRDM) and instrumentation tubes. Its outside diameter is about 4.9 m and its height including the closure head is 12.7 m. Its cylindrical wall thickness is 250 mm. The bottom wall thickness is 145 mm, the closure head wall thickness is 230 mm. The RPV internal structures support the fuel assemblies of the core. The space between the polygonal core structure and the cylindrical core barrel is filled with a neutron reflector. This reduces the neutron leakage from the core and protects the reactor pressure vessel wall from too high neutron damage. The upper internal structures house the rod cluster of the control assembly guide tubes. They also maintain the fuel elements axially in their position.

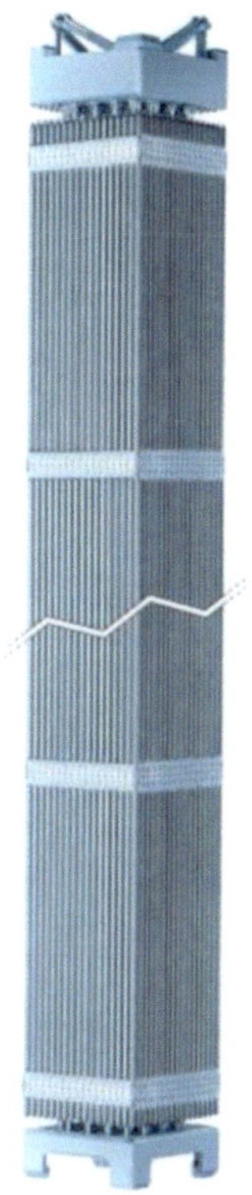

Figure 5.3: EPR 17x17 rods fuel element EPR (AREVA).

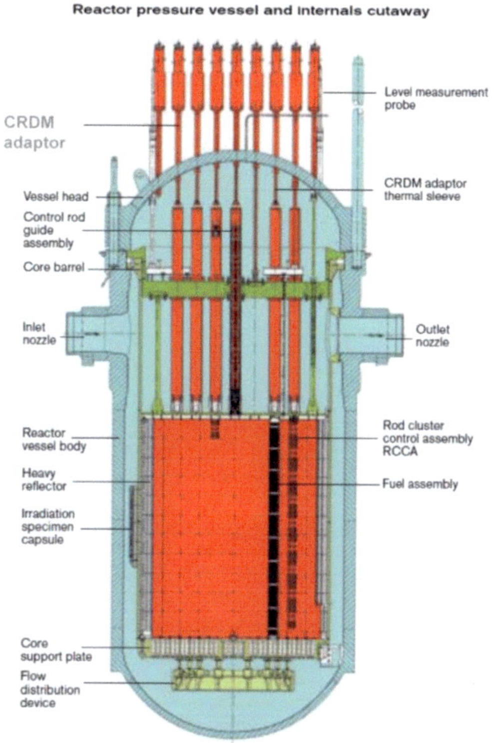

Figure 5.4: Cutaway of reactor pressure vessel of (AREVA).

5.2.3 Primary coolant pumps, pressurizer and piping

The primary coolant pumps provide the forced convection circulation of the coolant water which transports the heat from the reactor core to the steam generators (Fig. 5.1). The power needed for one primary pump is about 9 MWe. The coolant pipings have an inside diameter of 780 mm and a wall thickness of 70 mm.

The pressurizer maintains the pressure of the primary system of 155 bars within narrow limits. It is connected via a so-called surge line to the hot leg of one of the four primary circuits. The pressurizer (Fig. 5.1) is equipped with electrical heaters to raise the pressure and, with a water spray system, to lower the pressure. Relief and safety valves at the top of the pressurizer protect the primary system boundaries against overpressure. Additional motorized valves provide the operator with the possibility to rapidly depressurize the primary system in case of specific accident situations.

5.2.4 Steam generators

The four steam generators are vertical U-tube natural circulation heat exchangers (Fig. 5.5) equipped with an axial economizer. It consists of two parts:

- the lower part ensuring vaporization of the secondary feed water
- the upper part for drying the steam water mixture.

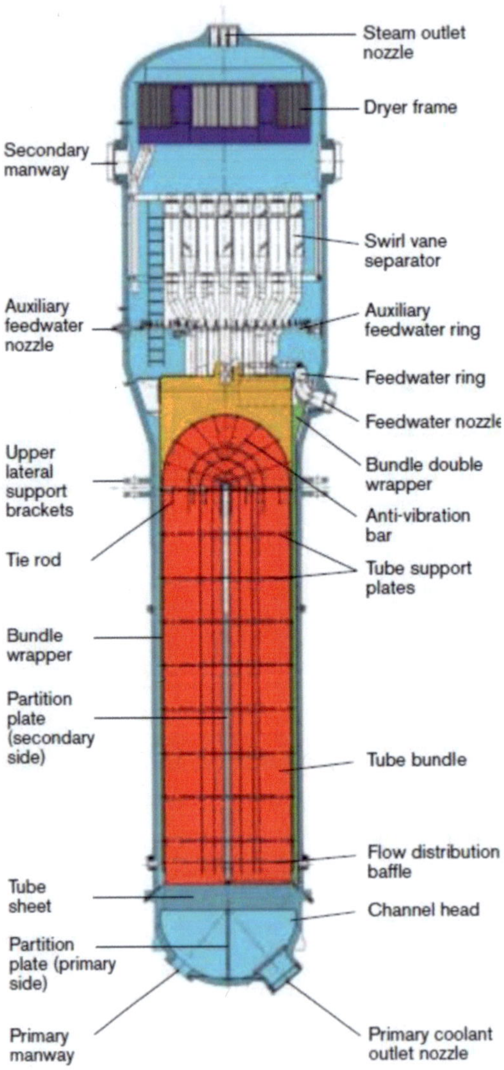

Figure 5.5: Cutaway of EPR steam generator (AREVA).

It produces saturated steam of 78 bars and 293 °C. The secondary feedwater is split between the cold and hot legs, which leads to an overall thermal efficiency of about 35.5%. The steam generator consists of 5980 tubes made of Inconel 690 with 19.05 mm diameter and 1.09 mm wall thickness. More data are given in Table 5.2.

Table 5.2: Characteristic design data of EPR steam generator (AREVA).

CHARACTERISTICS	DATA
Steam generators	
Number	4
Heat transfer surface per steam generator	7,960 m^2
Primary design pressure	176 bar
Primary design temperature	351 °C
Secondary design pressure	100 bar
Secondary design temperature	311 °C
Tube outer diameter/wall thickness	19.05 mm / 1.09 mm
Number of tubes	5,980
Triangular pitch	27.43 mm
Overall height	23 m
Materials	
• Tubes	Alloy 690 TT*
• Shell	18 MND 5
• Cladding tube sheet	Ni Cr Fe alloy
• Tube support plates	13% Cr improved stainless steel
Miscellaneous	
Total mass	500 t
Feedwater temperature	230 °C
Moisture carry – over	0.1%
Main steam flow at nominal conditions	2,554 kg/s
Main steam temperature	293 °C
Saturation pressure at nominal conditions	78 bar
Pressure at hot stand by	90 bar

* TT: Thermally treated

5.2.4.1 Chemical system and volume control system

The chemical system:

- ensures permanent monitoring and adjustment of the boron concentration in the coolant water
- enables adjustment of other chemical characteristics of the coolant water

The volume control system:

- provides the pressurizer spray water
- injects water in the primary pump seal system
- provides filling and draining of water during reactor shut down or power rise up conditions.

5.2.5 Safety injection and residual heat removal system

The safety injection system has four independent trains each injecting water at medium pressure (92 bars) into the primary system from water stored in accumulators. A low pressure injecting system pumps water into the primary system when the pressure will have been already decreased to low pressure. The residual heat removal system cools the reactor core when the reactor is shut down and the steam generators cannot provide efficient cooling, e.g. at lower than 120 °C. In addition it cools the spent fuel pool.

5.2.6 In-containment refueling water storage tank (IRWST)

The IRWST contains a large amount of borated water and collects water which is discharged inside the containment. It is located at the bottom of the containment.

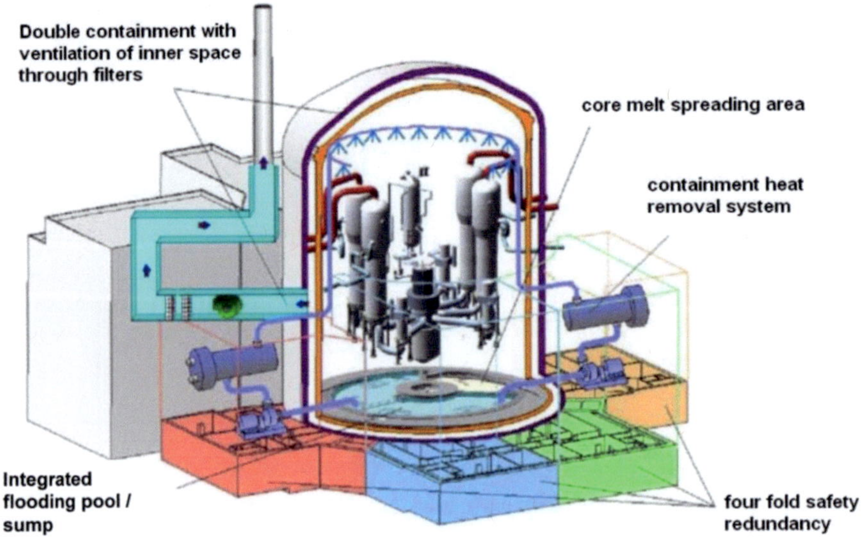

Figure 5.6: EPR double containment with RPV, cooling systems and molten core spreading area (AREVA).

5.2.7 Emergency feed water system (EFWS)

The EFWS ensures that water is supplied to the steam generators when all those systems are unavailable which supply the feedwater under normal conditions.

5.2.8 Emergency power supply systems (EPSSs)

The EPSSs ensure power supply in case of loss of external electrical power supply by electrical grids. These are four emergency diesel generators in a protected concrete building. In case these emergency diesel generators should fail (station black out) two additional generators (diesel or gas turbine) provide the necessary power. They are located in separate buildings.

5.2.9 EPR safety concept and containment system [4,5]

The EPR safety concept follows the optimization of all safety systems according to the results of a probabilistic safety analysis. This leads to an extremely low probability of occurrence for core melt down accidents. In addition the containment of the EPR is designed such that accidents are eliminated which could lead to large releases of radioactive materials. Relocation or evacuation outside of the immediate vicinity of the plant, limited sheltering or long term food ban would not be necessary in case of a core melt accident.

The EPR nuclear reactor system is, therefore, equipped with a strong double containment of prestressed concrete which can withstand the mechanical consequences of severe accidents. In addition it contains a molten core spreading area and cooling system below the reactor pressure vessel. Leak tightness of the containment and filter systems guarantee extremely low releases of radioactive materials even in case of severe core melt down accidents.

5.3 Russian Light Water Reactors

5.3.1 Main design characteristics [6]

Russian LWRs, called VVERs, are built at electrical power of 640 MW(e), 1000 MW(e) and 1500 MW(e). The design characteristics of the reactor pressure vessel and of the reactor core of these VVERs are similar to PWRs built in Europe (Section 5.2). However, the fuel elements have hexagonal shape, and the steam generators are arranged horizontally.

Table 5.2 shows the main design characteristics of the VVER reactors. Fig. 5.7 shows a cross section of VVER reactor pressure vessel with the core and the control rod drive mechanisms.

Table 5.3: Some design characteristics of Russian VVERs [6].

	VVER-640	VVER-1000	VVER-1500
Electrical Power [MWe]	640	1000	1500
Thermal Power [MWth]	1800	3000	4250
Pressure of primary water [MPa]	15.7	15.7	15.7
Pressure steam generator [MPa]	7.1	6.3	7.1
Average linear power [W/cm]	100	166	156
Outer diameter of fuel rod [mm]	9.1	9.1	9.1
Outer diameter of RPV [m]	4.54	4.54	5.3
Number of coolant loops	4	4	4

5.3.2 Safety concept of VVERs

The safety concept of modern VVERs is very similar to the EPR safety concept. It is based on active and passive emergency cooling systems. The VVERs are shut down by 121 control/shut down rods falling into the core by gravity. In addition, boric acid is injected into the primary coolant.

In case of a loss of coolant accident (LOCA) the pressure in the reactor pressure vessel drops and water must be inserted from four high pressure emergency hydro accumulators which are initially at a pressure of 5.9 MPa. These are followed by eight low pressure hydro accumulators at a pressure of 1.5 MPa. The water reservoir of these hydro accumulators is sufficient to cool the reactor core for at least 24 hours.

Special depressurization valves connect the hot and the cold legs of the loops with the spent fuel pool. They open passively at a certain pressure difference in case of a large coolant pipe break or long term loss of coolant. The coolant coming from the rupture of the coolant pipe is collected in the lower part of the containment forming the so called emergency pool. When the emergency pool level will have risen to a level between the cold leg and hot leg, the valves connecting the emergency pool and the spent fuel pool will open. From that moment on all water in the emergency and the spent fuel pool will be available for cooling the reactor core and the spent fuel. A molten core retention and cooling device is located underneath the reactor pressure vessel.

The reactor containment is a double containment. The inner containment contains measures (hydrogen igniters) to alleviate the consequences of hydrogen combustion. Radioactive materials leaking out of the inner containment into the space between the double containment are passed through filters. As a consequence similar safety standards are attained as described in Section 5.1 for EPR.

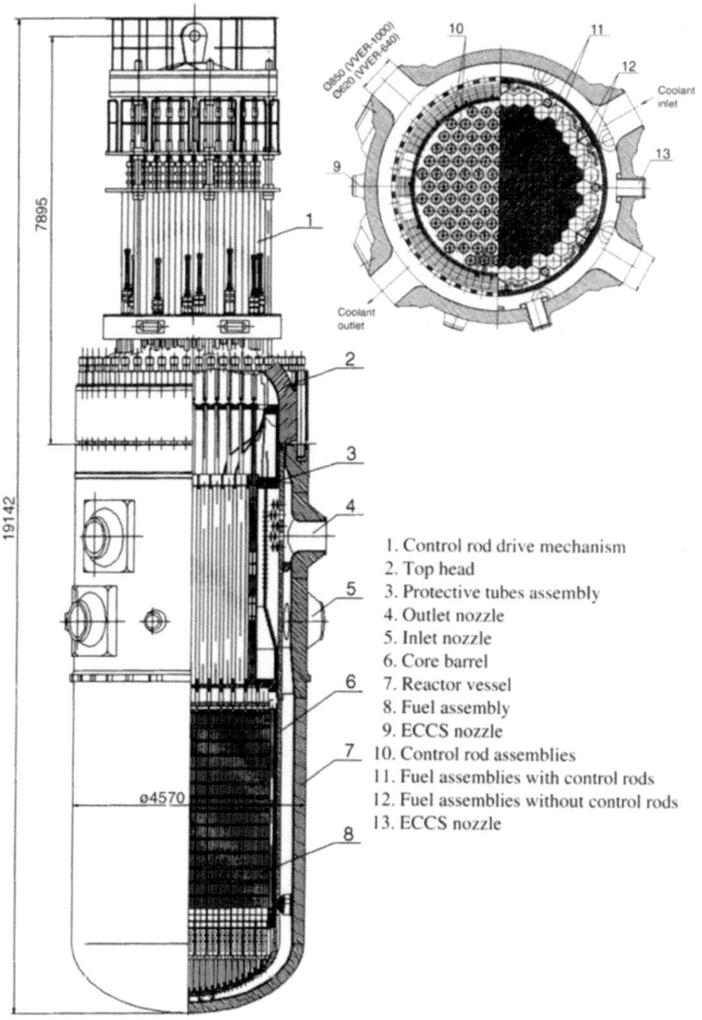

Figure 5.7: Cross section view of Russian VVER reactor pressure vessel with core and control rod drive mechanics [6].

5.4 Boiling Water Reactors (BWRs)

The development of commercial boiling water reactors (BWRs) started in the USA in 1956. The BWRs built today by a number of manufacturers in the USA, Europe and Japan are characterized by almost identical technical designs. This chapter will mainly deal with the 1250 MW(e) Generation-III+ SWR-1000 designed by AREVA in Europe. This modern BWR is characterized by passive safety systems. Its reactor core and fuel elements are very similar to other modern BWRs. Fig. 5.8 shows the main design principles of the SWR-1000.

The saturated steam produced in the reactor core flows from the reactor pressure vessel directly to the turbo-generator system and is pumped back from the condenser to the pressure vessel. The condenser is cooled by cooling water from a cooling tower or from a river.

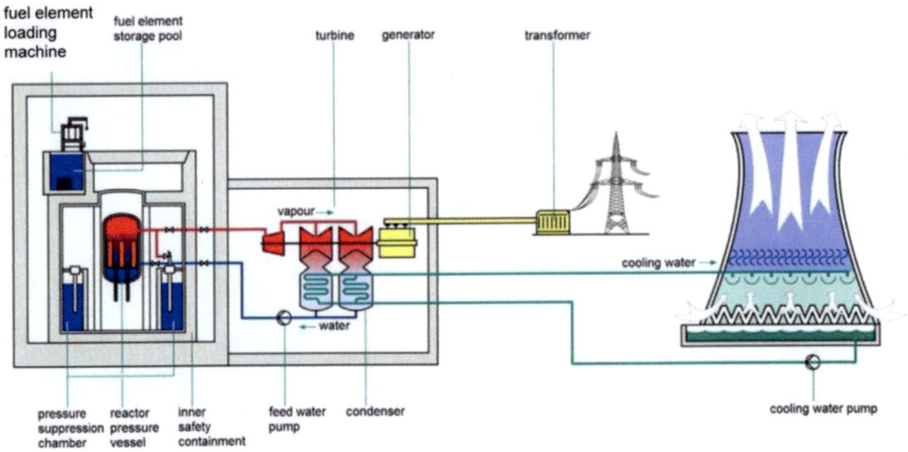

Figure 5.8: Main design principle of the boiling water reactor SWR-1000 (AREVA).

5.4.1 Core, Pressure Vessel and Cooling System

The reactor core consists of an array of 664 fuel elements about 3.0 m long [3,7,8,10]. The fuel element contains 128 fuel rods with outer diameters of 10.28 mm in a closed square box called ATRIUM-12 fuel elements. Fig. 5.9 shows as an example an ATRIUM-10 BWR fuel element having 8 fuel rods less than the ATRIUM-12 fuel element, but equal fuel rod design parameters. For moderation of the neutrons and cooling of the core, water flows through the core and is allowed to boil in the upper part of the core. 157 cruciform absorber rods, containing boron carbide as the absorber material, are installed in between a set of four fuel elements. The absorber rods are moved hydraulically into and out of the reactor core from below. The fuel rods have claddings of Zircaloy and contain UO_2 pellets with an average enrichment of about 5% U-235. The fuel is unloaded after a maximum burnup of 65,000 MWd(th)/t. Roughly one quarter of the fuel elements are unloaded, in a four batch reloading scheme after 18 months and replaced by fresh fuel elements. Fuel elements which have not attained their maximum burnup at that time are reshuffled in the core.

Some fuel rods contain gadolinium as burnable poison to compensate for the burnup of fissile material and the build-up of absorbing fission products in the fuel during reactor operation. An internal water channel of 4x4 cm in the ATRIUM-12 fuel element is designed for power flattening across the fuel element. The average power generation density in the core is 51 kW/l or 24.7 kW(th)/kg uranium. The water inlet temperature in the core is 220 °C; the outlet temperature is 289 °C, which corresponds to a saturation steam pressure of 7.5 MPa.

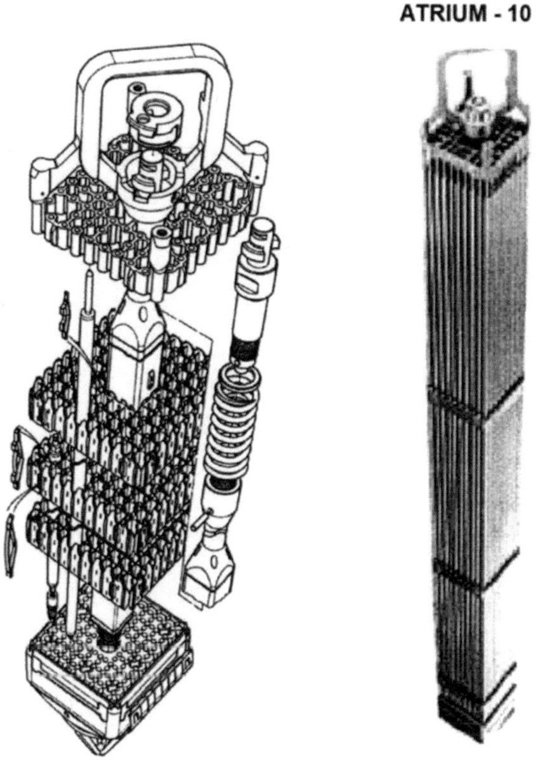

Figure 5.9: ATRIUM-10 boiling water reactor fuel element [7,9,10].

The steam is generated by water boiling in the reactor core. To provide sufficient core flow for ample heat transfer, BWRs employ internal jet pumps. The core with the absorber rods is contained in a steel pressure vessel of 23.4 m height and 7.1 m diameter (Fig. 5.10). Steam separators and steam driers are arranged above the core. The reactor vessel head can be removed for loading and unloading of the fuel elements. The reactor pressure vessel has a wall thickness of about 150 mm. It is made of 22NiMoCr37 steel, the inside being plated with austenitic stainless steel.

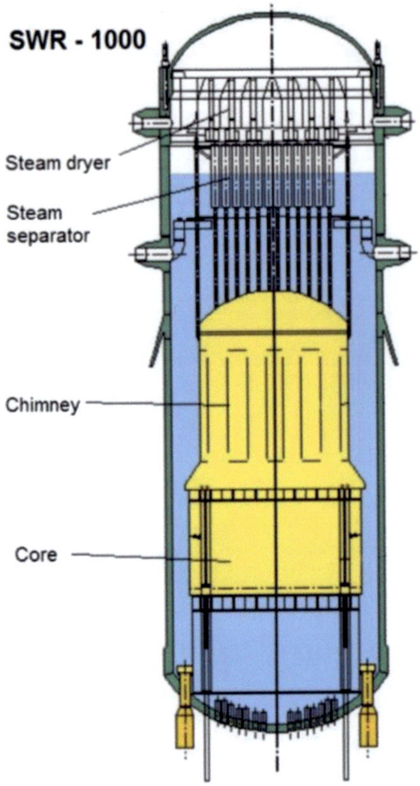

Figure 5.10: Reactor Pressure Vessel of SWR-1000 (AREVA).

The water circulation is driven by eight internal jet pumps in the reactor pressure vessel and through the core. The velocity of the circulating water influences the evaporation rate in the core and can be used for changing the reactor power. Reduction of water flow through the core will result in a higher evaporation rate and in a larger volume of bubble formation. Increasing the volume of steam in the core reduces the moderation of neutrons. As a consequence, the reactivity and the reactor power will be reduced. In this way, changes in the water flow can be used to control the reactor power without movement of control rods. Therefore, BWRs can automatically follow the load requirements of the turbine. The reactor power can be controlled by sensing pressure disturbances at the turbine, transmitting these signals to the recirculation flow control valve and regulating core flow.

In order to ensure high quality of the reactor feed water, all the feed water recirculated from the turbine condenser is pumped through filters (demineraliser units) and cleared of any corrosion products and other impurities.

5.4.2 The SWR-1000 inner containment system [8,9]

The inner containment system is a reinforced concrete containment with an inner steel liner. It is subdivided into a pressure suppression chamber and a drywell as well as four large hydraulically coupled core flooding pools (Fig. 5.11). The core flooding pools serve as a heat sink for passive heat removal from the reactor pressure vessel by emergency condensers and the pressure relief valves.

The reactor pressure vessel, the three main steam lines and the two feed water lines are located in the drywell. The core flooding pools contain four emergency cooling condensers for passive heat removal in accident situations. In the upper part of the inner containment the large shielding and storage pool is located, together with four containment cooling condensers. The large shielding pool is hydraulically connected to the fuel element storage pool.

The drywell also contains the core flooding lines for passive flooding of the reactor pressure vessel in case of accident situations and the passive pressure pulse transmitters for the initiation of safety functions. Finally, the drywell is equipped with two 100% capacity recirculation air cooling systems, the high pressure part of the cooling water cleaning system and the lines of the residual heat removal system. Table 5.4 indicates the number of different safety systems in the inner containment of the SWR-1000. The residual heat removal system pumps and the heat exchangers are installed underneath the pressure suppression chamber (Fig. 5.11). The whole inner containment is inertized by nitrogen to ensure fire protection and prevent hydrogen-oxygen chemical reactions (hydrogen deflagration or detonation) in case of a serious core melt down accident.

The three main steam lines and the two feed water lines connected to the reactor pressure vessel are equipped each with two containment isolation valves, one located inside and another one outside of inner containment penetrations (Fig. 5.11). These containment isolation valves can be closed in the case of any pipe rupture in the inner containment. This action isolates the reactor pressure vessel from the water turbine or condenser cycles.

The control rod drive and shut down rod system is acting from below the reactor pressure vessel. The pressure suppression chamber acts as a heat sink in the event of accident situations and provides water inventory for make up in the reactor pressure vessel via the residual heat removal system.

Table 5.4: Number of passive shutdown core flooding and residual heat removal system components.

ITEM	Number
Pressure suppression chamber	1
Vent pipe	16
Spring loaded pilot valve	8
Safety & Relief valve	8
Scram system	2 x 2
Core flooding pool	4
Emergency condenser	4
Passive pressure pulse transmitter	3 x 4
Pilot valve	15
Core flooding system	4
Shielding storage pool	1
Containment cooling condenser	4
Passive outflow reducer	4

5.4.3 Safety relief valve system

The safety relief valve system acts for short term removal of excess steam after a turbine trip and protects the reactor coolant pressure boundary against overpressure exceeding allowable limits. It prevents high pressure melt ejection in case of severe core melt down. Similarly it depressurizes in case of pipe rupture and in the event that the water level in the reactor pressure vessel falls below specified limits.

The safety relief valve system consists of 8 safety relief valves together with relief lines and steam quenchers. The latter are installed in the four core flooding pools. All relief lines lead to the core flooding pools but not into the pressure suppression pool as in earlier BWR designs. The safety relief valves are spring loaded valves or they act by solenoid pilot or diaphragm pilot valves via the 12 passive pressure pulse transmitters. No actuation by signals from the instrumentation and control system is required.

5.4.4 Emergency condensers

The emergency condensers are located in the four core flooding pools (Fig. 5.11). They also function as completely passive devices for residual heat removal from the reactor pressure vessel. They are actuated when the water level in the reactor pressure vessel drops below a certain limit. In this case the upper part of the emergency condensers is flushed with steam from the reactor pressure vessel which condenses and returns to the lower part of the pressure vessel. Passive flow reducers installed in the nozzles of the reactor pressure vessel direct the mass flow in the right direction.

5.4.5 Containment cooling condensers

Four containment cooling condensers are located in the part of the inner containment above the core flooding pools (Fig. 5.11). They remove passively the residual heat from the containment atmosphere to the shielding and storage pool. The tube bundles of these condensers are arranged at a slight angle to horizontal. In that way natural circulation of the water inside the tube bundles develops and transfers the heat from the inner containment to the shielding pool.

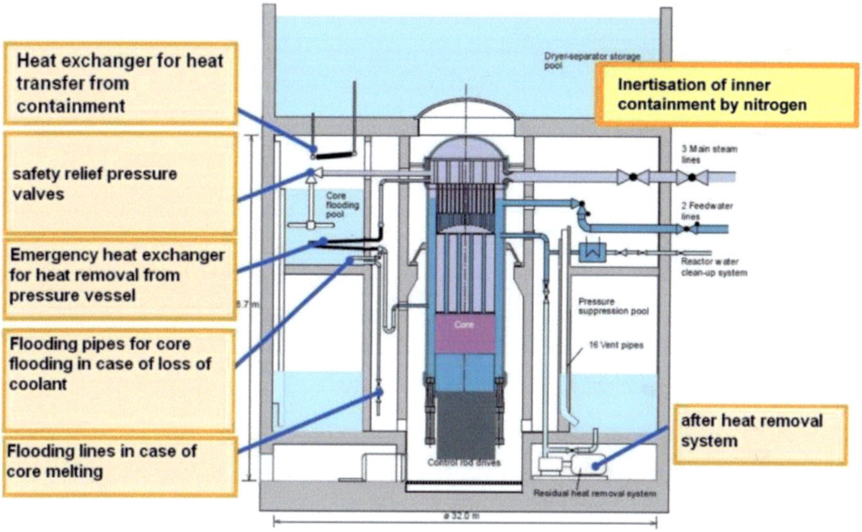

Figure 5.11: Containment and Internals (AREVA).

5.4.6 Passive Pressure Pulse Transmitter

The passive pressure pulse transmitters function without electrical power supply or actuation by the instrumentation and control system. They serve to initiate reactor scram, as well as containment isolation of the main steam lines and depressurization of the reactor pressure vessel.

The passive pressure pulse transmitters consist of a small heat exchanger which is connected to the reactor pressure vessel. When the water level in the reactor pressure vessel drops the primary side of the small heat exchanger fills with steam. This causes the water on the secondary side of the small heat exchanger to evaporate leading to a rapid pressure rise. This pressure rise triggers the safety function of the diaphragm pilot valves.

5.4.7 Residual Heat Removal and Active Core Flooding Systems

Two active low-pressure core flooding and heat removal systems ensure the cooling during shutdown conditions. They also remove the heat from the core flooding pools and from the pressure suppression pool in the event of a loss of coolant accident. In addition they also transfer water during refueling conditions.

5.4.8 Safety Shutdown Systems

If there are reactivity perturbations or losses of coolant flow, the reactor is shut down in a short time by rapid insertion of the absorber rods. This is achieved by two diverse shut down systems:

- an electrical motor driven operational shut down system
- a hydraulically acting fast shut down (SCRAM) system.

As a backup shutdown system, the SWR-1000 can poison the coolant (moderator) with a neutron absorbing boric acid and, in this way, also quench the nuclear reaction and shut down the reactor. This is a completely independent additional shutdown system.

5.4.9 Cooling after a severe core melt

The coolant flooding system can transport water from the core flooding pools to the lower area of the drywell. This water pool can cool the lower part of the reactor pressure vessel from the outside. In this way a core melt can be retained within the control rod guide structures of the lower part of the reactor pressure vessel. The residual heat of the core melt can be conducted through the lower steel structures to the water pool. This heat power can be transferred by steam to the containment cooling condensers.

5.4.10 Emergency power supply

Emergency power for the cooling systems can be supplied by an external electrical emergency power grid. In addition, diesel generators can take over in case the external electrical emergency power grid would fail.

5.4.11 SWR-1000 safety concept and containment system

The SWR-1000 safety concept follows the optimization of all safety systems according to the results of a probabilistic safety analysis. The incorporation of passive safety systems together with proven active safety systems, the application of fail-safe principles and the principles of redundancy and diversity provide an optimal overall safety design. This leads to extremely low probabilities of occurrence for core melt down accidents. In addition the safety systems and the containment system of the SWR-1000 are designed such that accidents which would lead to large releases of radioactive materials are eliminated. Relocation or evacuation outside the immediate vicinity of the plant, limited sheltering or long term food ban would not be necessary in case of a core melt accident.

The SWR-1000 is equipped with a strong double containment of prestressed concrete (Fig. 5.8) which can withstand the mechanical consequences of severe accidents. Its inner containment is inertized by nitrogen against hydrogen detonations. Leak tightness of the containments and filter system guarantee extremely low releases of radioactive materials even in case of severe core melt down accidents.

5.5 Other Types of Fission Reactors

A number of additional types of reactors, with other coolants and neutron moderators, exist on the market for nuclear electricity generation. Other reactor lines have been proposed as projects but not so far put into practice. They will be described very briefly below.

5.5.1 Pressurized Heavy Water Reactors

This line uses heavy water as the moderator and heavy or light water as the coolant. It was originally developed in Canada (CANDU reactor). Its version using heavy water as the neutron moderator and coolant can be run on natural uranium. The fuel elements are replaced continuously on-load. More recent developments, such as the Advanced CANDU Reactor (ACR), also use light water as the coolant and will be operated on 2% low-enriched uranium. This type is offered in unit sizes of up to 1000 MWe. CANDU reactors represent roughly ten percent of all nuclear reactors built and operated in the world (Section 2.1).

The Steam-generating Heavy Water Reactors (SGHWR) developed in the United Kingdom use light water in pressure tubes surrounded by heavy water as the moderator. The light water in the pressure tubes attains boiling temperature. Other heavy water reactors were developed in Germany and Japan. However, these lines are not pursued any further. The UO_2 fuel of heavy water reactors reaches a maximum burnup of 7-18 GWd/t. The spent fuel can be reprocessed. (Several types of research reactors moderated and cooled by heavy water have been built and are still operated world wide.)

5.5.2 Gas-cooled Reactors

Gas-cooled reactors use graphite as the neutron moderator and a gas (carbon dioxide or helium) as the coolant. The gas-cooled reactors developed first in the United States (Hanford), the United Kingdom, Russia and France allowed the use of natural uranium fuel. Accordingly, the attainable fuel burnup was only approx. 6-7 GWd/t. Further development of these reactors led to the Advanced Gas-cooled Reactors (AGRs) with a fuel enrichment of approx. 2-2.5% U-235 and a maximum burnup of approx. 18 GWd/t. The spent UO_2 fuel can be reprocessed.

In the United States, Germany and Japan the gas-cooled reactor line was advanced still further in an attempt to achieve high gas temperatures of 900-950 °C (Very-High-Temperature Gas-cooled Reactors, VHTGRs). These gas temperatures are to be employed technically for the generation of process heat. Besides graphite as the moderator, helium needs to be used as a coolant in these designs. Moreover, the UO_2 or ThO_2 fuel is used in small particles coated with pyrolytic carbon and silicon carbide. These fuel particles are embedded in a graphite matrix. The graphite matrix is surrounded by graphite as a neutron moderator. This results in two fuel element designs: so-called prismatic fuel elements and spherical fuel elements (spheres or pebbles). The prismatic designs are used in demonstration reactors in the United States (Fort St. Vrain), the UK (Dragon) and in Japan (HTTR). The spherical fuel elements are employed in so-called pebble bed reactors in Germany (AVR, THTR) and China (HTR-10). These fuel elements attain burnups of 100 GWd/t and more.

However, reprocessing them chemically is fraught with immense difficulties as the graphite must first be separated from the fuel particles, and the fuel particles with their layers of pyrolytic carbon and silicon carbide must be broken up. Only after these steps can the fuel be dissolved and reprocessed chemically.

High-temperature pebble bed reactors at present are planned mainly as small modular HTRs of 200-300 MWth. These modular HTRs are to be proposed to produce process heat of high temperatures in Europe, China, USA, Russia, and South Africa.

5.5.3 Molten Salt Thermal Breeder Reactor (MSBR)

The MSBR originally was developed at Oak Ridge, USA in the 1960s. It is fuelled with a homogeneous salt fluid containing both the fissile uranium and fertile thorium fuel. The fuel carrier is a mixture of fluorides of lithium, beryllium, and thorium. The fissile uranium is present as UF_6. Both U-235 and U-233 can be used and plutonium was used as well during the Molten Salt Reactor Experiment (MSRE) at Oak Ridge (USA). The fuel carrier salt is pumped through a graphite core structure. The heat produced by the core salt is transferred in a heat exchanger to a secondary coolant (molten salt). This molten salt in a steam generator transfers the heat to a water cycle to produce steam. The fission products and protactinium, Pa-233, are continuously removed chemically by a purification and on-line reprocessing system.

Although there are design proposals for this reactor line no plans for construction are known as yet.

5.5.4 Limitation to LWRs and LMFBRs

The studies and findings described above focus almost exclusively on LWRs. LWRs stand for eighty percent of the nuclear reactors currently in operation in the world and planned after 2010. LMFBRs, i.e. sodium-cooled (SFR) and lead-bismuth-eutecticum-cooled (LBE-FR) breeder reactors with fast neutrons, probably will supplement or replace LWRs on a large scale after 2050. The UO_2 and PuO_2/UO_2 fuels, respectively, of these LWRs and LMFBRs can be reprocessed chemically and recycled. The necessary facilities of the Pu/U fuel cycle, i.e. reprocessing and refabrication plants, are on stream already and will be expanded further in the near future.

The CANDU reactors and ACRs moderated with heavy water will not be analyzed any further in this context as the low burnup (7-18 GWd/t) of the fuel elements makes the plutonium produced not proliferation-proof (Section 9 to 14). Also, the line of modern gas-cooled reactors, such as the HTGR and modular HTR reactors, is not analyzed any further as the technical feasibility of reprocessing their fuels has not yet been demonstrated.

The focus on today's LWR and later LMFBR fuel cycles therefore covers most of the presently existing and later operating nuclear reactors. A sufficient number of analyses have been published about the possibility of future proliferation-proof fuel cycles. These proliferation-proof future fuel cycles also can be combined with transmutation and incineration of the minor actinides, neptunium, americium (Section 9 to 14).

References Section 5

[1] EPR – European Pressurized Water Reactor, the 1600 t MWe Reactor, www.areva.com

[2] Sengler, G. et al., EPR core design, Nucl. Eng. and Design, 187, Issue 1, 79-119 (1996).

[3] Güldner, R. et al., Contribution of advanced fuel technologies to improved nuclear power plant, The Uranium Institute 24th Annual Symposium, London (2009).

[4] Czech, J. et al., European pressurized water reactor: safety objectives and principles. Nucl. Eng. and Design, 187, Issue 1, 25-32 (1999).

[5] Weisshäupl, H.A., Severe accident mitigation concept of the EPR, Nucl. Eng. and Design, 187, Issue 1, 35-45 (1999).

[6] Lutkin, G., The Russian advanced VVER design, Nuclear News, June 2002.

[7] Strasser, A., et al., Fuel fabrication process handbook, Advanced Nuclear Technology International, Surahammer, Sweden (2005).

[8] SWR-1000: An advanced boiling water reactor with passive safety features, AREVA brochure (2003).

[9] Stosic, Z. et al., Boiling water reactor with innovative safety concept: The generation III+ SWR 1000, Nuclear Eng. and Design, 238, 1863-1901 (2008).

[10] Garner, N. et al., ATRIUM-10: ten years of operational experience. Nucl. Eng. International, Vol. 47, No. 572 (2002).

6 Fast Neutron Reactors (FRs)

Fast neutron reactors, with a fast neutron spectrum, operate in the U-238/Pu fuel cycle or in the Th-232/U-233 fuel cycle (Section 7). In case of the U-238/Pu fuel cycle plutonium is used from reprocessed spent fuel of e.g. LWRs. The core of FRs can be loaded with either metallic (Pu-U-Zr-alloy)-, oxide (PuO_2/UO_2)-, carbide (PuC/UC)- or nitride (PuN/UN)-fuel. Sodium was used in most cases as coolant so far, but also lead, lead-bismuth and helium gas were proposed.

6.1 Breeding process

As explained in Section 4.7 the relatively high η-value (neutron yield) of Pu-239 and Pu-241 leads to a breeding ratio >1 in the neutron energy spectrum range of about 200 keV. This neutron energy range can be achieved with either sodium, lead, lead-bismuth or helium gas. In the breeding process essentially U-238 is converted into Pu-239 which is fissioned. Fig. 6.1 shows the breeding process for a breeding ratio BR = 1.2.

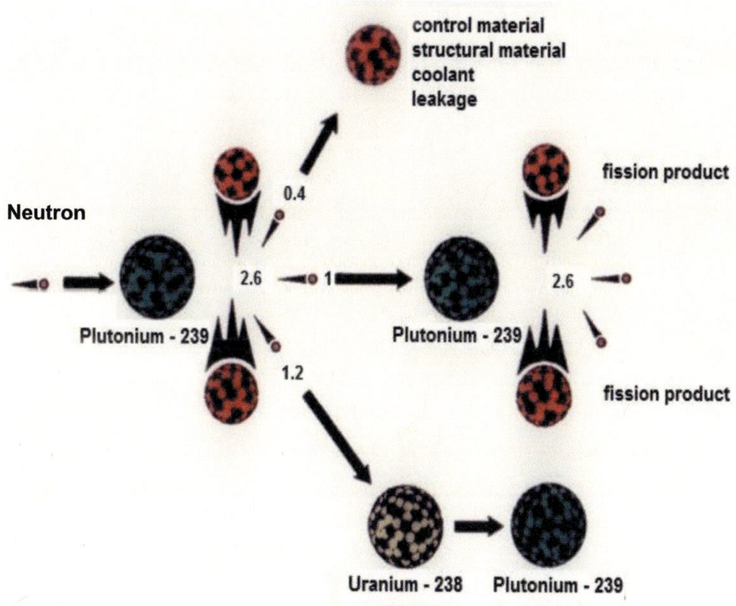

Figure 6.1: Breeding process (U-Pu-cycle).

As explained in Section 4.8, the breeding ratio BR >1 results in a fuel utilization of more than 60%. This can be compared with the 0.6% fuel utilization of a LWR. As the LWR has a natural uranium consumption of about 170 t per GW(e)y at a plant load factor of 0.85 (Section 2.7), the FR has a uranium consumption of a factor 100 lower, which is only 1.7 t per GW(e)y.

FRs are loaded initially with plutonium or U-233/U-235 coming from reprocessing of spent fuel of, e.g. LWRs. After this starting phase FRs need only depleted uranium (0.2% U-235 and U-238) or thorium. As natural uranium contains only 0.72% U-235 and 99.28% U-238 the consumption rates for FRs for depleted uranium and natural uranium are roughly the same.

In the U/Pu fuel cycle FRs are operated with mixed plutonium-uranium fuel. The uranium commonly will be depleted uranium (tails assay from enrichment plants or reprocessed uranium). The initially loaded plutonium is acting then like a catalyst and is permanently replaced in situ by converted Pu-239 along the nuclear reaction chain:

$$^{238}_{92}\mathrm{U} \xrightarrow{n,\gamma} {}^{239}_{92}\mathrm{U} \xrightarrow[23.5\ \mathrm{min}]{\beta^-} {}^{239}_{93}\mathrm{Np} \xrightarrow[2.35\ \mathrm{d}]{\beta^-} {}^{239}_{94}\mathrm{Pu}$$

Indirectly only U-238 is fissioned and converted into thermal energy.

A similar breeding process holds for the Th-232/U-233 fuel cycle:

$$^{232}_{90}\mathrm{Th} \xrightarrow{n,\gamma} {}^{233}_{90}\mathrm{Th} \xrightarrow[22.1\ \mathrm{min}]{\beta^-} {}^{233}_{91}\mathrm{Pa} \xrightarrow[27.1\ \mathrm{d}]{\beta^-} {}^{233}_{92}\mathrm{U}$$

Like the initially loaded plutonium also the U-233 must first be generated, e.g. in LWRs or HWRs to obtain the initial core inventory of a Th-232/U-233 fuelled FR.

Regarding the available uranium and thorium resources in the world (Section 2) the breeding process opens up an energy potential which can be good for many thousand years. The time ranges discussed in Section 2 regarding the natural uranium availability of reactors with a thermal neutron spectrum (converter reactors) can be multiplied roughly by a factor 100 based on the same nuclear energy production capacity. With the breeding process fission nuclear energy can provide energy on a time scale far beyond any presently conceivable planning interest. This is comparable with the energy potential that is hoped to be tapped by fusion reactors operating on the D-T cycle with lithium as the breeding material [1,2,3].

6.2 Development of FRs

The principle of breeding had been understood from the onset of development of nuclear fission reactors. Accordingly, FRs have been designed, constructed, and operated in the USA, the UK, and the USSR since the 1950s [3]. The first generation of FRs were built and operated with the aim of investigating fast neutron reactor physics, control stability and to demonstrate the selected technical solutions. Early small FRs like Clementine, EBR-I in the USA, BR1 and BR-2 in the USSR, Zephyr and ZEUS in the UK were followed by larger experimental reactors like EBR-II and EFFBR in the USA, DFR in the UK, BOR-60 in the USSR, Rapsodie in France, and the KNK-II test reactor in Germany. They were equipped with uranium or plutonium metal – or by PuO_2/UO_2-fuel and most of them were mainly cooled by sodium [3].

In the 1960s it was recognised that FRs needed a fuel allowing high burn-up in the range of 100 GWd/t for economical reasons. Therefore, mixed oxide plutonium-uranium (MOX) fuel was selected. In addition sodium was adopted exclusively as coolant. Helium gas and lead or lead-bismuth were also proposed as coolants.

Two principle design concepts have been adopted for sodium cooled fast reactors (SFRs). Both design concepts need an intermediate sodium coolant cycle. In the loop-type concept, the primary pumps and intermediate heat exchangers with non-radioactive sodium are located outside the reactor vessel. They are interconnected by coolant pipes. In the pool-type concept a larger reactor sodium filled tank houses the pumps and intermediate heat exchangers. Fig. 6.2 shows the scheme of a pool-type SFR. The heat produced in the fuel elements is transferred by the primary radioactive sodium to the intermediate non-radioactive sodium coolant circuit and to the steam generators. Sodium temperatures of 488 °C and steam at a pressure of 12.7 MPa are attained. This leads to a thermal efficiency of 42%.

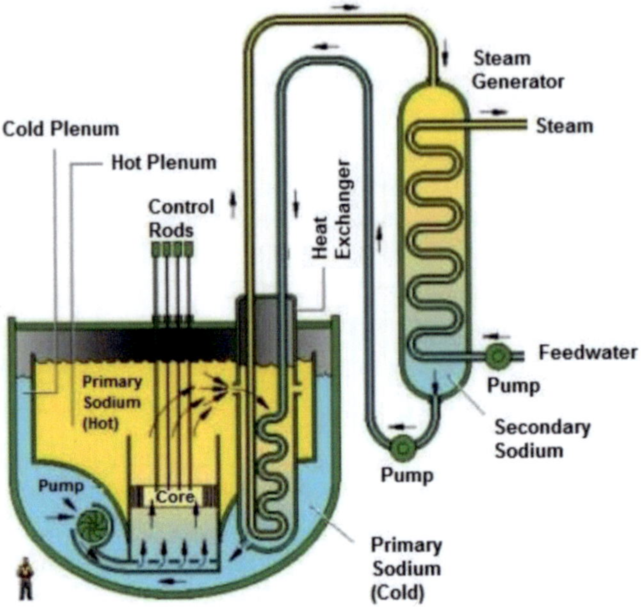

Figure 6.2: Design scheme of a pool type SFR (AREVA).

6.3 Sodium coolant properties [4-12]

Sodium has a melting point of 98 °C and a boiling point of 880 °C at atmospheric pressure. It has a high thermal conductivity of 66.1 W/cm °C and a specific heat capacity of 1.30 kJ/kg °C, at 527 °C. Due to its low neutron moderation capabilities the average neutron energy of the neutron energy spectrum is in the range of 200 keV. The excellent thermal properties allow a power density of 300-400 kW(th)/l in the core and a max. linear rod power of the fuel rods of 400-450 W/cm. SFRs have a relatively small reactor core volume and an enrichment of the MOX fuel of 15 to 25% fissile plutonium (Pu-239 and Pu-241). Due to relatively low microscopic fission cross sections of plutonium in the energy range of 100 keV the neutron flux in the core must be in the range of about $5x10^{15}$ n/(cm^2·s). This high neutron flux results in relatively high irradiation damage in the structural materials.

Neutron capture in sodium, i.e. in Na-23, leads to Na-24 which is radioactive and decays via β^-decay with a half-life of 15 hours.

6.4 Demonstration SFRs [4-13]

Three sodium cooled demonstration FRs were already taken in operation in the early 1970s. The Soviet BN-350 reached first criticality in 1972 and operated until 1999. The French Phénix was connected to the electrical grid in 1973 delivered full power until 2009 and the British PFR delivered electricity from in 1975 until 1994. For all three demonstration FRs, with a power output of 150-250 MWe, the original design characteristics were confirmed in terms of fast reactor core physics, control stability, safety engineering and sodium technology. Somewhat later, the FFTF (400 MWth) reached its first criticality in the USA. It served as a fuel and materials test facility until 1992.

First commercial size power SFRs, e.g. BN-600 [600 MW(e)] started operation in the USSR in 1980/82 and Superphénix [1200 MW(e)] began operation in France in 1985/86 and was shut down in 1998. Other SFR demonstration reactor projects, e.g. CRBR in the USA or SNR-300 in Germany were either stopped during the design phase or not taken into operation for political reasons.

Phénix in France operated over 35 years with good operational performance. The Russian BN-600 was still operating by 2010 with excellent operational performance over 20 years. This demonstrates the technical feasibility of sodium cooled FRs.

In Japan, the prototype demonstration plant MONJU (280 MW(e)) will not be taken into operation following a decision Japanese government end of 2016. Other commercial size SFRs were still under construction by 2010 (BN-800 with 800 MW(e) in Russia and PFBR with 500 MW(e) in India).

Fig. 6.3 shows the design details of the core fuel element of MONJU. The hexagonal fuel element contains 169 fuel rods. These fuel rods have an outer diameter of 6.5 mm, a length of about 2.8 m and a cladding thickness of 0.47 mm. Spiral wires around the fuel rod guarantee the proper spacing between the fuel rods. The fuel rod contains the PuO_2/UO_2 pellets of the core zone and the UO_2 pellets of the lower and upper axial blanket. The inner radial core zone is enriched by 16% in Pu-239 and Pu-241, whereas the outer radial core zone has an enrichment of 21%. Above the upper axial blanket a gas plenum zone is located where the gaseous fission products are collected [11].

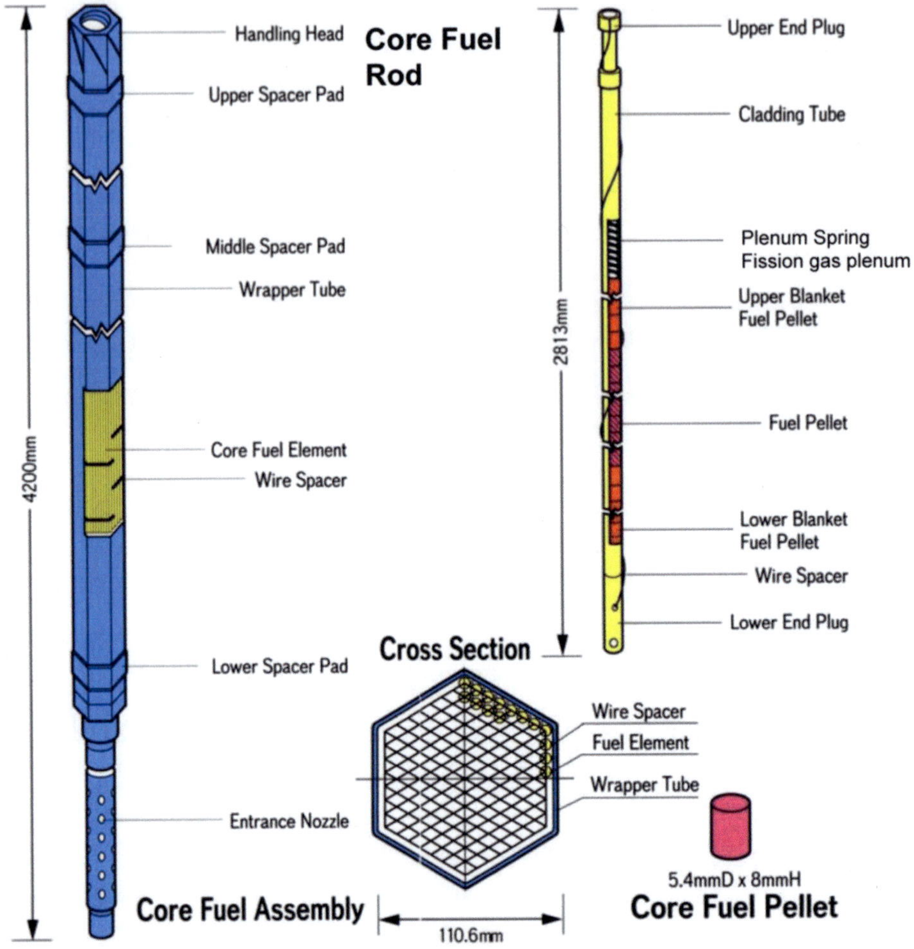

Figure 6.3: Hexagonal core fuel element of the MONJU demonstration FBR [12].

6.5 Large scale deployment of SFRs

There are several reasons for the delayed large scale commercial introduction of SFRs, despite the fact that they were developed in Europe, the USA, the USSR and Japan already since about 1950.

- The presently assured uranium resources are higher than originally prognosticated.
- The projection for nuclear energy needs during and after the oil crisis of the 1970s had been overestimated.

- LWRs are dominating the nuclear energy application now. They demonstrated excellent operational and economic performance. They will be built and operated over the next 50-60 years until the enriched uranium availability will decrease, because of shorter natural uranium availability.
- The next step will be plutonium recycling in the same type of LWRs. The necessary reprocessing and MOX fuel refabrication plants have been built already in France, the UK, Russia and Japan.
- SFRs – despite of their technical maturity – are not as economic yet as LWRs. The reasons are their higher technical sophistication (two sodium coolant circuits, the use of austenitic steels etc) and their higher fuel cycle costs.
- The use of plutonium in SFR cores and the possibility of breeding relatively pure plutonium in their blankets have lead to a stop of SFR development in the USA and Germany between 1980-90 as a consequence of their non-proliferation policy.

Only recently, scientific and technical solutions have been worked out to solve these non-proliferation problems. They will be explained in Sections 13 and 14.

As already explained in Section 2 a large scale deployment of SFRs and their fuel cycle can be expected around 2040-2050.

6.5.1 Commercial size SFRs

Only the near commercial size sodium cooled fast reactor BN-600 will be operating during the next years. In addition the commercial size BN-800 and the Indian PFBR (500 MW(e)) will go into operation in the near future. Also several small lead-bismuth cooled fast reactors were decided for construction in Russia based on experience with submarines.

Japan, France and Russia are developing SFRs up to 1500 MW(e) which shall become competitive with LWRs around 2050. SFRs must be developed together with their fuel cycle. It will take several decades to construct and operate several 1500 MW(e) size SFRs together with reprocessing and refabrication plants.

6.5.2 BN-600 in Russia [9-15]

The 600 MWe SFR BN-600 reached first criticality in 1980 and full power operation in 1982. It is a pool type SFR design with three secondary heat transfer loops and three steam generators. Each steam generator consists of eight sections for the evaporator, superheater and reheater. In case of failing tubes, one of these eight different sections can be isolated and be replaced while the reactor is operating on partial load. The core and the intermediate heat exchangers together with the centrifugal primary pumps are housed in a large sodium filled pool tank (Fig. 6.4). A large head shield plug closes the upper part of this pool tank. It contains excentrically rotating plugs for positioning the fuel element transfer machine exactly above a fuel element position for reshuffling or replacement procedures. The upper sodium surface in the shield tank is covered by argon gas. The pressure of this cover gas is slightly above atmospheric pressure.

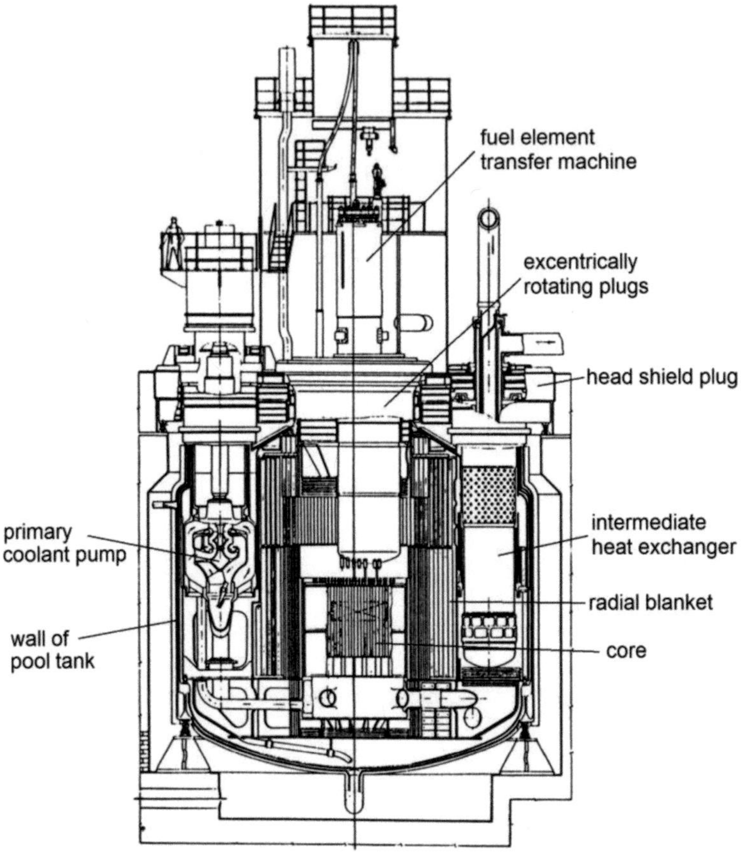

Figure 6.4: BN-600 pool type sodium cooled fast reactor [9].

The core contains 369 hexagonal fuel elements. The height of the core is 103 cm and its diameter is 206 cm. The radial blanket has a thickness of 47 cm. The upper and lower axial blankets are 30 cm thick each. The sodium flows with a velocity of about 5 m/s upwards through the core. The sodium core inlet temperature is 365 °C, its outlet temperature is 535 °C. BN-600 was originally fueled by enriched uranium oxide with 17%, 21% and 26% enrichment in U-235 in the three radial core zones. From 1998 on this enriched UO_2 fuel was partly replaced by vibro-compacted PuO_2/UO_2 mixed oxide fuel. A maximum fuel burn-up with this fuel of 110 GWd/t was attained.

BN-600 had already demonstrated excellent operational experience over 28 years by 2010. Early fuel rod failures and tube failures in steam generators could be overcome by developing better materials and by design improvements [14,15].

The excellent operational experience of the French demonstration reactor Phénix (250 MWe) and Superphenix (1250 MWe) as well as of the Russian BN-600 demonstrated already the technical feasibility of these pool type sodium cooled FRs.

Table 6.1: Characteristic design parameters of BN-600 (pool type LMFBR) [9,11].

		BN-600
Reactor Power		
Thermal	MW (th)	1400
Electrical net	%	600
Plant efficiency		41
Reactor Core		
Fuel		UO_2 and PuO_2/UO_2
Core outer diameter	cm	205
Core height	cm	103
Pu eq. enrichment		
Inner core zone	%	17
Middle/Outer core zone	%	21/26
Total breeding ratio		0,85-1,0
Pu eq. mass	tonne	2.6
Total UO_2/PuO, mass in core	tonne	12
Fuel rod outer diameter	mm	6.9
Length of fuel pin	mm	2445
Core power density		
Average	kW (th)/1	445
Maximum	kW (th)/1	603
Residence time of fuel	D	420
Max. fuel rod power	W/cm	480
Max. burnup	MWd (th)/tonne	110.000
Blankets		
Fuel		UO_2
Axial thickness	cm	30
Radial thickness	cm	47
Fertile rod outer diameter	cm	1,40
Fissile Fuel Bundles		
Number of bundles		369
Number of pins per bundle		127
Pin total length	m	2,4
Bundle total length	m	3,5
Cladding material		stainless steel
Cladding maximum rated temperature	°C	695

Fertile Fuel Bundles		
Number of bundles		362
Number of pins per bundle		37
Pin total length	m	1,8
Bundle total length	m	3,5
Cladding material		stainless steel
Control Bundles		
Main shutdown system:		
Number of bundles		14
Number of absorber elements per bundle		7/31/8
Pin length	m	1,1
Cladding material		stainless steel
Primary System		
Coolant		sodium
Primary Na mass	tonne	770
Rated flow	tonne/s	6
Core sodium inlet temperature	°C	365
Core sodium outlet temperature	°C	535
IHX sodium inlet temperature	°C	533
IHX sodium outlet temperature	°C	362
Secondary System		
Coolant		sodium
Secondary Na mass	tonne	830
Rated flow	tonne/s	6,1
SG sodium inlet temperature	°C	510
SG sodium outlet temperature	°C	315
IHX sodium inlet temperature	°C	315
IHX sodium outlet temperature	°C	510
Water-Steam System		
SG water inlet temperature	°C	240
Turbine steam inlet temperature	°C	502
Turbine steam inlet pressure	MPa	13,2

6.5.3 Commercial size SFR design [14-21]

Studies on commercial size FR designs with a power output of 1200-1500 MWe were performed in Europe, Russia and Japan since about 1990. The objective of such studies was to investigate the technical and economical feasibility of such large FRs and how they could be introduced into the already existing market of nuclear power reactors. One of these design proposals, representing a Japanese sodium cooled loop type fast reactor (JSFR) [16,17,18,19] will be described in this section. It is based on the construction experience of the loop type Japanese demonstration fast reactor MONJU.

JSFR has a thermal power of 3570 MWth and an electrical power output of 1500 MW(e). The reactor core rests on steel support structures and is housed in a sodium filled reactor tank. The free surface of the sodium is covered by argon gas at a pressure of slightly above 0.1 MPa. The reactor tank is covered by a thick shield cover plate with eccentrically rotating plugs. The guide structures of the control and absorber rods penetrate this shield cover plug from above. The fuel element loading and transfer machine is operating from above the rotating plugs after the control rod drive mechanisms will have been decoupled.

The primary radioactive sodium coolant enters the reactor tank with a temperature of 395 °C and flows from the lower entrance plenum upward through the core. It is heated up in the core to an outlet temperature of 550 °C and flows to the intermediate heat exchangers.

JSFR has only two cooling circuits. The primary pumps are integrated into the intermediate heat exchangers. The secondary non-radioactive sodium is pumped to two steam generators (SGs) where steam of at 19.2 MPa and 497 °C is produced. The thermal efficiency of the JSFR plant would be 42%.

In comparison to earlier loop type demonstration SFRs, e.g. MONJU, the 1500 MWe JSFR has much shorter sodium pipings outside of the pool tank. This is achieved by using high chromium steels and simplified geometric configurations with inverse L-shaped pipes. The reactor tank and all primary and secondary sodium piping are double walled to avoid sodium fire in case of leakage. The space in between double walled pipings is filled with nitrogen gas which can be heated. Electrical trace heating on sodium piping can be avoided by this design. The two steam generators are equipped with especially developed double walled tubes to prevent sodium-water interactions in case of failing steam generator tubes.

PuO_2/UO_2 mixed oxide is used as fuel in the core. The reactor core has two enrichment zones for radial power flattening. The inner core zone has an enrichment of 18.3% Pu-239/241, the outer zone 20.9% enrichment in Pu-239/241. This leads to a fissile plutonium core inventory of 8.5 t. The breeding ratio is 1.10. The cladding is made of vanadium oxide dispersed steel (ODS). It allows a neutron fluence of 5×10^{23} n/cm^2 equivalent to 250 dpa or a burnup of the core fuel up to 150 GWd/t over eight years of full operation. After an operation cycle period of about 26 months about one fourth of the core will be unloaded. The reactor core and radial blanket are surrounded by a core barrel which restrains the core in radial direction in order to fulfill earthquake design requirements. Also, the whole coolant circuit system is designed to obey Japanese aseismic design criteria [16,17,18,19,20,21].

JSFR has two diverse shutdown systems, one of which is designed with flexible joint absorber parts. This allows absorber insertion under robust restraint conditions in case of earthquakes. A third shutdown system is based on the thermomagnetic properties of ferromagnetic alloy in the control rod guide structures. The shutdown function is initiated passively when the sodium outlet temperature exceeds the Curie point of the holding magnets. This third shutdown system prevents sodium outlet temperatures of more than about 750 °C in case of severe accident situations for which a failure of the shut down systems is assumed. Thus, sodium boiling and failure of fuel rods in case of anticipated failure of the first two shutdown systems is avoided.

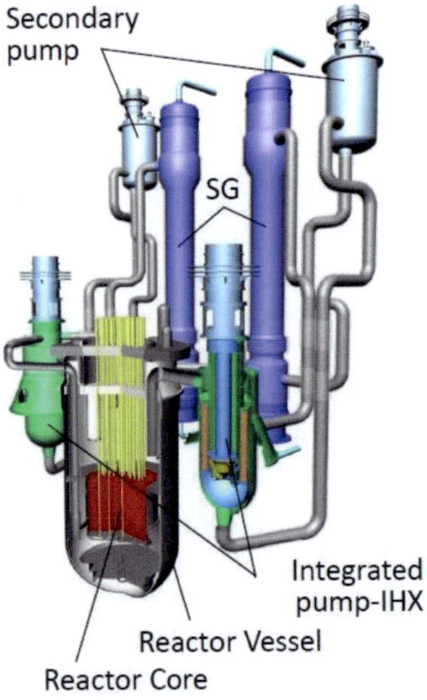

Figure 6.5: 1500 MWe sodium cooled JSFR design proposal (JAEA) [16,17,18,19].

Multilayered molten core debris tray structures are arranged underneath the reactor core support structures. These molten core debris tray structures shall retain molten core fuel, avoid recriticalities and cool the molten fuel.

Decay heat removal can be accomplished by natural convection of the sodium in the primary and secondary coolant circuits. Under accident conditions additional emergency decay heat removal systems start passively. They act on the basis of natural convection of the sodium with sodium-air coolers and dampers. No pumps, no pony motors and no air blowers are needed in such cases.

The inner reactor containment is a concrete containment with inside and outside steel cladding which can resist all mechanical and thermal loads in case of severe accidents. The surrounding outer containment must be designed against external loads, e.g. earthquakes, flooding. etc. As modern pressurized water reactors, e.g. EPR or SWR-1000, also future SFR containments must have extremely low leakage conditions and very efficient filter systems to avoid large radioactivity releases in case of severe accidents. Thus, evacuation or relocation of the population outside of the plant can be avoided even in case of severe accidents.

6.6 Lead-Bismuth cooled FRs

Based on experience with lead-bismuth cooled submarine reactors in Russia lead-bismuth cooled FRs were proposed first in Russia and later investigated also in Japan and in Europe.

6.6.1 Lead-bismuth coolant properties

Lead-bismuth (44.5% lead and 55.5% bismuth eutectic alloy (LBE)) has a melting point at 125 °C and a boiling point at 1670 °C. Its density at 400 °C is 10.24 kg/m^3. Its thermal heat conductivity at 400 °C is 13.7 W/(m °C) and its heat capacity is 0.146 kJ/(kg·K). Due to its low neutron moderation capabilities the average neutron energy of the neutron energy spectrum is in the range of about 200 keV. The excellent thermal properties allow a similarly high power density in the core as in case of SFRs. The corrosion properties of LBE require special cladding surface treatment and protection layers of steels [22,23,24,25,26]. The oxygen content must be controlled accurately. LBE does hardly react with oxygen or water and, therefore, simplifies the design of LBE cooled FRs.

6.6.2 Design proposals for Lead-bismuth FRs

The core of a LBE cooled reactor has hexagonal fuel elements. The coolant fraction in the subassembly design is only about 25% due to the thermal properties of LBE. Corrosion concerns lead to a LBE coolant velocity of about 2 m/s. The core fuel can be about 16% enriched UO_2 as in the small size Russian modular type SVBR-75/100 [29,30]. Also ThO_2/UO_2 mixed oxide fuel or PuN/UN mixed nitride were proposed as fuel in the Russian BREST-300 design [11] or in the Japanese LBE 550/750 MWe design [27,28]. With 16% U-235 enriched uranium fuel only a conversion ratio CR = 0.85 is attained. The SVBR has no radial blanket fuel elements.

The coolant circuits can be drastically simplified due to the low chemical affinity of LBE against oxygen and water. The steam generators can be directly integrated into the pool type tank. Fig. 6.6 shows the Russian small modular lead bismuth cooled SVBR-75/100

[29,30]. The core, the primary pumps and the steam generators are integrated into a pool tank. The reactor tank structures must be designed to withstand the weight of the LBE coolant.

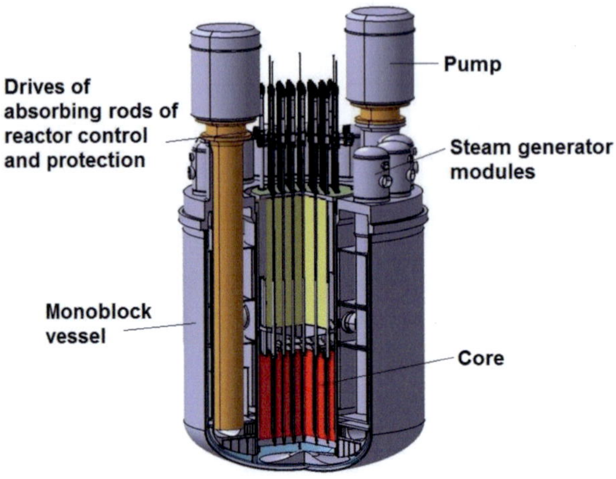

Figure 6.6: SVBR-75/100 reactor design [29,30].

Emergency cooling relies entirely on natural convection as described for the case of SFRs. In case of failing steam generators the afterheat can also be conducted radially through the wall of the reactor tank to an outside water tank. The SVBR-75/100 is shut-down by independent and diverse shut down systems. Due to the relatively high density of the lead-bismuth coolant all steel structures must have higher thickness than in SFRs. Special care must be given to an earthquake resistant design. For that reason design proposals for lead-bismuth cooled FRs are restricted to medium power size up to 500 or 750 MWe in Japan [16,27,28]. The small modular type SVBR-75/100 can be assembled to a cluster type plant with 8 or 16 SVBR-75/100 reactors with a total output of 800 MWe or 1600 MWe [29,30].

6.7 The Integral Fast Reactor (IFR)

The IFR is a sodium cooled pool type fast reactor using metallic uranium-plutonium-zirconium alloy (U-Pu-Zr)-fuel in combination with pyrometallurgical fuel reprocessing and remote injection casting fuel refabrication [31,32,37]. The reactor plant, the pyroprocessing plant and the metallic fuel refabrication plant are collocated at one site at Idaho National Laboratory. The optimization of reactivity temperature coefficients, e.g. Doppler coefficient, sodium expansion coefficient, structural expansion coefficients including control rod drive line expansion and natural convection flow of the coolant sodium results in an inherent control behavior of the reactor without reliance on control rod scram systems [31,32,33]. Off normal events with very low probability of occurence, e.g. loss of coolant flow, loss of heat sink or run-out of a control rod followed by failure of the shut down systems will lead only to a sodium coolant temperature rise up to about 600 °C which is about 200 °C below the boiling point of sodium. The decay heat of the core can be safely removed by natural convection flow.

IFR core designs were reported for 340 MW(e), 600 MW(e) and 1350 MW(e). Table 6.2 gives an impression of the core design and main design characteristics of a 340 MW(e) IFR design. The IFR was designed as an LMFBR [31,32,33] with core internal blanket fuel elements or as an FR burner reactor (Advanced burner reactor) for the incineration of the transuranium elements (plutonium, neptunium, americium, curium [34,35]).

Table 6.2: 900 MWth IFR Core Design Parameters for 340 MW(e).

Electric power MW(e)	340
Reactor outlet temperature (°C)	510
Reactor ΔT (°C)	135
Core Concept	Heterogeneous
Fuel residence time (cycles)	
Driver	4
Blanket*	4
Cycle length (full-power days)	292
Fuel material	
Driver	U-Pu-10% Zr
Blanket	U-10% Zr
Clad and duct material	HT-9
Active fuel height (cm)	
Driver	91
Blanket	112
Axial blanket thickness (cm)	0.0
Number of pins per assembly	
Driver	271
Blanket	169
Fuel pin diameter (cm)	
Driver	0.72
Blanket	1.0
Cladding thickness (cm)	0.056
Duct wall thickness (cm)	0.36

*Refers to internal and radial blanket

References Section 6

General, FBR plant description

[1] Häfele, W., et al., Fusion and Fast Breeder Reactors. International Institute for Applied Systems Analysis, RR-77-8, Laxenburg, Austria (1977).

[2] Kessler, G., Nuclear fission reactors, Springer Verlag, Wien (1983).

[3] Status of liquid metal cooled fast breeder reactors, Technical report series No. 246, IAEA, Vienna (1985).

[4] Vendryes, G., PHENIX - A Special Feature on the French Prototype Fast Breeder Reactor. Nuclear Engineering International 16, 557-580 (1971).

[5] Kazachkovskij, O.D., et al., The Present Status of the Fast Reactor Programme in the USSR. In: Nuclear Power and Its Fuel Cycle, Proc. Int, Conference, Salzburg, 2-13 May 1977, IAEA, Vienna (1977).

[6] Carle, R., SUPERPHENIX: First Commercial Plant of the Fast Breeder Line. J. British Nuclear Energy Society 14, 183-190 (1975).

[7] PFR - A Special Feature on the British Prototype Fast Breeder Reactor. Nuclear Engineering International 16, 629-650 (1971).

[8] Fast Flux Test Facility (FFTF) – A Special Survey. Nuclear Engineering International 17, 613-628 (1972).

[9] Leipunskii, A.I., et al.: A Nuclear Power Station with the BN-600 Reactor. Soviet Atomic Energy (Atomnaya Energiya) 25, 1216-1221 (1968).

[10] International Nuclear Fuel Cycle Evaluation, Fast Breeders. Report of INFCE Working Group 5. Vienna: International Atomic Energy Agency (1980).

[11] IAEA Fast reactor data base, IAEA-TECDOC-1531 (2006).

[12] Fuji-ie, Y., Adamov, E., New generations of nuclear reactors, Handbook of Nuclear Chemistry (2010).

[13] Waltar, A. et al., Fast Breeder Reactors, Pergamon Press, New York (1981).

[14] Saraev, O.M. et al., BN-600 power unit 15 year operating experience, Conceptual designs of advanced fast reactors, presented at the Technical Committee Meeting, Kalpakkam (India) October 3-6, IAEA-TECDOC-907, p. 27-31 (2005).

[15] Oghkanov, N.N. et al., 30 years of experience in operating the BN-600 sodium cooled fast reactor, Atomic Energy, Vol. 108, No. 4 (2010).

JSFR

[16] Kotake, S., et al., Feasibility study on commercialized fast reactor cycle systems, current status of FR systems design, Proc. of GLOBAL 2005, Tsukuba, Japan (2005).

[17] Ogura. M., et al., Conceptual design study of JSFR: overview and core concept. Proc. of Int. Conf. on Fast Reactors and Related Fuel Cycles: Challenges and Opportunities (FR09), Kyoto, Japan, December 7-11, 2009.

[18] Etoh, M., et al., Conceptual design study of JSFR – reactor system. Proc. of Int. Conf. on Fast Reactors and Related Fuel Cycles: Challenges and Opportunities (FR09), Kyoto, Japan, December 7-11, 2009.

[19] Ichikawa, K., et al., Conceptual design study of JSFR-reactor cooling system. Proc. of Int. Conf. on Fast Reactors and Related Fuel Cycles: Challenges and Opportunities (FR09), Kyoto, Japan, December 7-11, 2009.

[20] Ohyama, K., et al., Decay heat removal by natural circulation for JSFR. Proc. of Int. Conf. on Fast Reactors and Related Fuel Cycles: Challenges and Opportunities (FR09), Kyoto, Japan, December 7-11, 2009.

[21] Kurome, K., et al., Steam generator with straight double-walled tube. Proc. of Int. Conf. on Fast Reactors and Related Fuel Cycles: Challenges and Opportunities (FR09), Kyoto, Japan, December 7-11, 2009.

Lead-bismuth cooled FRs

[22] Subbotin, V., et al., Lead-bismuth cooled fast reactors in nuclear power of the future, Proceedings of the International Topical Meeting on Advanced Reactor Safety, ARS'94. Pittsburgh, pp. 524–529

[23] Tupper, R., et al., Polonium hazards associated with lead-bismuth used as reactor coolant, Int. Conf. on Fast Reactors and Related Fuel Cycles, Kyoto, Oct. 1991.

[24] Weisenburger, A., et al., Corrosion, corrosion barrier development and mechanical properties of steels foreseen as structural materials in liquid lead cooled nuclear systems, Proc. 17th Int. Conf. on Nucl. Eng. (ICONE 17), Brussels, July 2009.

[25] Müller, G., et al., Control of oxygen concentration in liquid lead and lead-bismuth, Journal of Nuclear Materials, 321, 256-262 (2003).

[26] Weisenburger, A., et al., Behaviour of chromium steels in liquid Pb-55.5. Bi with changing oxygen content and temperatures. Journal of Nuclear Materials 358, 69-76 (2006).

[27] Enume, Y., et al., Conceptual design of medium scale lead-bismuth cooled fast reactors, Genesa/ANP, Kyoto, Japan (2003).

[28] Hayafune, H., et al., Conceptual design study of Pb-Bi cooled fast reactor plant system in the feasibility study in Japan, Proceedings of GLOBAL 2005, Tsukuba, Japan, October 9-13, 2005.

[29] Poplavsky, V., et al., Core design and fuel cycle of advanced fast reactors with sodium coolant. Int. Conf. on Fast Reactors and Related Fuel Cycles, Kyoto (2009).

[30] Zrodnikov, A., et al. Innovative nuclear technology based on modular multi-purpose Lead-bismuth cooled fast reactors, Progress in Nuclear Energy, 50, 170-178 (2008).

Integral Fast Reactor

[31] Wade, D.C. et al., The integral fast reactor concept: physics of operation and safety, Nucl. Sci. and Eng., 100, p. 507 (1988).

[32] Wade, D.C. et al., The design rationale of the IFR, Progress in Nucl. Energy, 31, No. 1/2, p. 13 (1997).

[33] Wade, D.C. et al., The safety of the IFR, Progress in Nuclear Energy, 31, No. 1/2, p. 63 (1997).

[34] Kim, T.K. et al., Core design studies for a 1000 MW_{th} advanced burner reactor, Annals of Nucl. Energy, 36, p. 331 (2009).

[35] Dunn, F.E. et al., Preliminary safety evaluation of the advanced burner test reactor, ANL-AFCI-172 (2006).

[36] Wigeland, R. et al., Mitigation of severe accident consequences using inherent safety principles. Int. Conf. on Fast Reactors and Related Fuel Cycles, Kyoto (2009).

[37] Pahl, R.G. et al., Experimental studies on U-Pu-Zr fast reactor fuel pins in the Experimental Breeder Reactor II, Metallurgical Transactions A, Vol. 21A, 1863-1870 (1990).

7 The Nuclear Fuel Cycles

7.1 Storage of Spent Fuel Elements after Discharge

After discharge from the reactor core, the fuel elements are stored at the reactor site for a period of one or two years to allow for radioactivity decay and cooling. Spent fuel elements are then transported in spent fuel transport casks either to intermediate storage facilities or to storage pools at reprocessing plants. The intermediate storage facilities can also be located at the reactor site [1,2,3].

7.1.1 Transport of Spent Fuel Elements

Spent fuel elements are transported in special fuel transport casks, which weigh between 60 and some 120 t and have load capacities for up to about 12 t of spent fuel (Tab. 7.1). Fuel transport casks can be transported by special trucks on the road or on special rail cars. Also barge shipments on both inland waterways and oceans are made. The spent fuel elements are cooled within the casks either by air (dry casks) or water (wet casks). About 38 spent fuel elements are unloaded from a 1300 MWe PWR per year (Fig. 7.1). They can be transported, e.g. in Grmany in two CASTOR V/19 transport cask to an intermediate storage facility [4,7].

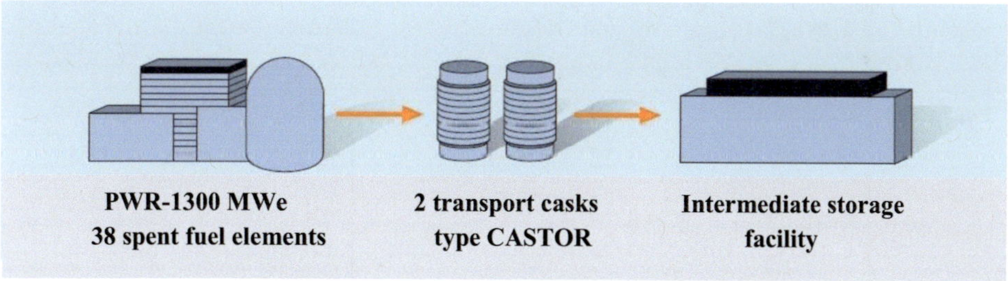

Figure 7.1: Transport of spent fuel elements from reactor to intermediate storage (VGB).

The transport casks contain the necessary shielding with steel, lead and water or borated water. They are cooled by natural airflow over fins on the outer surface or by forced air circulation. Spent fuel casks are designed to withstand severe accident conditions during shipment. Releases of radioactivity under such conditions must be rendered impossible. Therefore, the casks must be able to withstand such impacts as thermal tests (fire), drop tests under gravity, crash tests, and water immersion tests before being licensed for actual transport. Special international shipping regulations have been elaborated.

7.1.2 Intermediate Storage of Spent Fuel Elements

Spent fuel elements can be stored for intermediate periods of time in water pools (wet storage), air cooled casks (dry storage) or in special containers. For wet storage in intermediate storage pools or storage pools of reprocessing plants, the spent fuel elements are arranged in racks or baskets kept in water pools. The water serves as a heat transfer medium for the heat generated in the fuel elements and provides the necessary shielding of the fuel elements. It is maintained at a sufficiently high level to provide shielding during all fuel handling operations. The walls and floors of storage pools are made of reinforced concrete lined with stainless steel [6].

LWR spent fuel elements can be stored, if needed, in water pools for many decades. During this time period, the fuel elements will not experience appreciable water corrosion on their outer surfaces.

Dry storage of LWR spent fuel elements is feasible in air cooled storage casks made of cast iron. European cast iron spent fuel casks take up to 19 PWR or 52 BWR fuel elements (FE) (Table 7.1). They are equipped with outside cooling fins and can be stored in large intermediate storage buildings (Fig. 7.2). The storage building is cooled by air [5,7].

Table 7.1: Design characteristics of fuel element transport casks [4].

Country	Type	Number of fuel elements (FE)	Total weight (tonnes)	height/ diameter (m)
Germany	CASTOR V/19	19 PWR-FE	121	5.86/2.44
	CASTOR V/52	52 BWR-FE	123	5.45/2.44
	CASTOR IIa	9 PWR-FE	116	6.01/2.48
	CASTOR 440/89	84 WWER – 440 FE	116	4.08/2.66
France	TN 13/2	12 PWR-FE	105	5.60/2.5
Great Britain	Excellox 4	7-15 PWR-FE	91 t	5.6/2.2

Dry storage is also used for HWR and HTGR graphite fuel elements. Spherical graphite fuel elements of HTRs can be stored under dry conditions in gastight cans.

LMFBR fuel elements are kept first for some time in sodium cooled storage pools on the reactor site. For intermediate storage they are filled in cans, cooled either by sodium and then after cleaning be stored under water or only cooled by air or an inert gas (nitrogen). Before reprocessing, the sodium is removed from the fuel element surface by melting or steam cleaning in a hot inert gas atmosphere.

Dry storage is also used for HWR and HTGR graphite fuel elements. Spherical graphite fuel elements of HTRs can be stored under dry conditions in gastight cans.

LMFBR fuel elements are kept first for some time in sodium cooled storage pools on the reactor site. For intermediate storage they are filled in cans, cooled either by sodium and then after cleaning be stored under water or only cooled by air or an inert gas (nitrogen). Before reprocessing, the sodium is removed from the fuel element surface by melting or steam cleaning in a hot inert gas atmosphere.

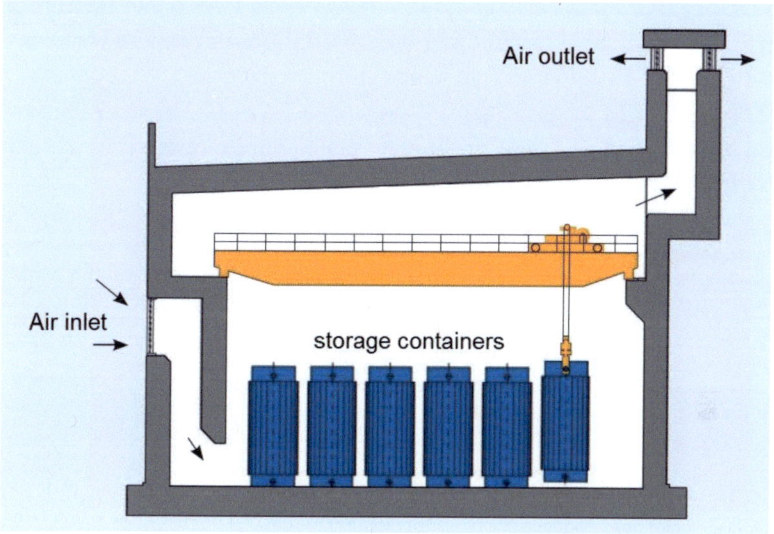

Figure 7.2: Storage of intermediate storage containers in air cooled storage building (VGB).

7.2 The Uranium-238/Plutonium Closed Fuel Cycle [8,9,10,11]

Uranium can be utilized more efficiently in a closed fuel cycle with reprocessing and recycling of the fissile and fertile material. This applies to fuel used in LWRs and LMFBRs. For LMFBRs the closed fuel cycle is imperative. Technical aspects of reprocessing and recycling (refabrication) in the uranium/plutonium fuel cycle will be described in the following sections.

7.2.1 Reprocessing of Spent UO_2 Fuel Elements

Spent fuel elements with irradiated UO_2 fuel and stainless steel or zircaloy claddings are transported to the reprocessing plant and stored there prior to chemical reprocessing. The steps of disassembly of such fuel elements, dissolution of the fuel as well as chemical separation are the same in principle for LWR and LMFBR fuel elements operated on UO_2 or PuO_2/UO_2 fuel.

7.2.1.1 Mass Inventories of Spent Fuel and Waste

The mass inventories, their heat generation and their potential of radiotoxicity constitute important parameters on which to base engineered safety measures [7].

Customarily, these data are based on 1 t of heavy metal (HM) fuel. In that case, roughly 1.14 t of UO_2 or UO_2/PuO_2 corresponds to 1 tHM. When loaded into the core, 1 tHM of fresh LWR fuel in an equilibrium cycle with 5% U-235 enrichment contains 50 kg of U-235 and 950 kg of U-238 (Fig. 7.3). When unloaded from the LWR core after a burnup of 60 GWd(th)/t, 1 tHM of spent fuel still contains 0.7% of U-235 and 91.96% of U-238, but 0.66% of U-236, some 1.05% of different plutonium isotopes, 5.5% of fission products (FPs), 0.13% of Np-237, americium, and curium.

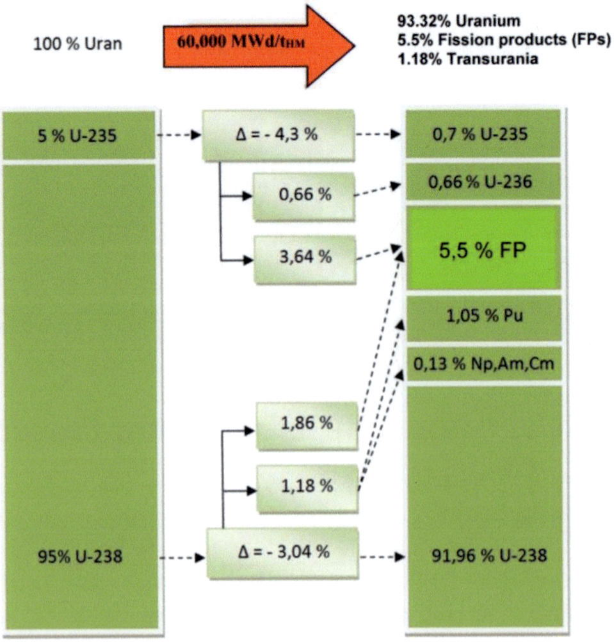

Figure 7.3: Contents (%) of fresh uranium fuel (U-235, U-238) and spent fuel after 60,000 MWd(th)/ton uranium (U-235, U-236, U-238), fission products, plutonium and minor actinides (Np, Am, Cm).

7.2.1.2 Decay of radioactivity of spent fuel

The decay of radioactivity of the spent fuel is initially dominated by the fission products. Fig. 7.4 shows the decaying radioactivity in Bq per tonne during the first 10 years after the spent fuel elements were unloaded from the reactor core. It also illustrates the time periods for fuel transport, reprocessing and waste treatment (vitrification of the high level waste concentrate).

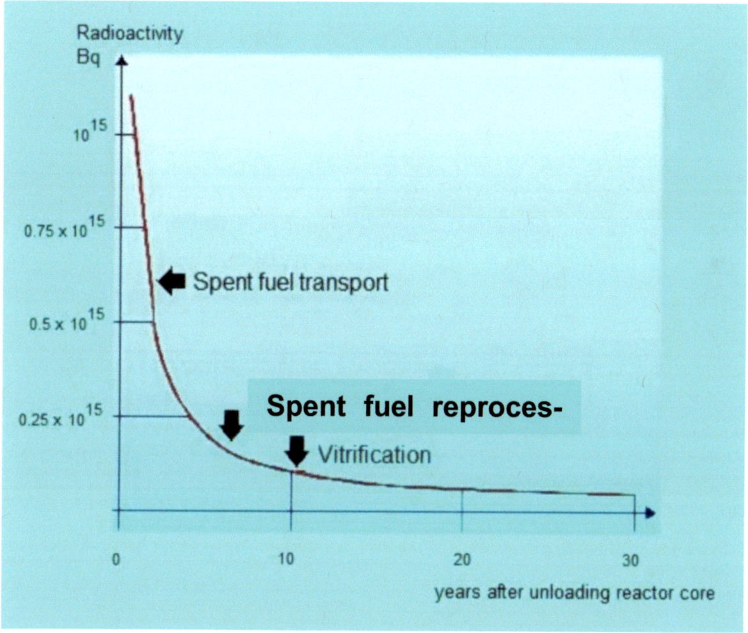

Figure 7.4: Radioactivity of 1 t of spent fuel as a function of time.

7.2.2 LWR Fuel Element Disassembly and Spent Fuel Dissolution

In a reprocessing plant (Fig. 7.5 shows the head end of such a plant) the storage pools are arranged close to the fuel element disassembly cells. The fuel elements are moved by means of a crane from the storage pool into the disassembly cell above it. In this cell, LWR fuel elements are cut up by large bundle shears. After the end parts have been removed from the fuel elements, the fuel rod bundles are chopped into pieces several cm long. The bundle shear is operated remotely and is designed so that it can also be repaired by remotely operating tools. The fuel element and fuel rod sections drop directly into a dissolver basket located in the dissolver cell underneath. The basket is filled with boiling nitric acid, which leaches the fuel out of the chopped fuel rod hulls. After leaching of the fuel, the remaining hulls with tiny particles of fuel and fuel element structural parts are dumped from the basket into a container, and the container is moved into the hull storage facility.

The fuel solution still contains small solid parts. This undissolved fraction of plutonium is about 1%. The undissolved solid particles are removed through coarse filters or by centrifuges.

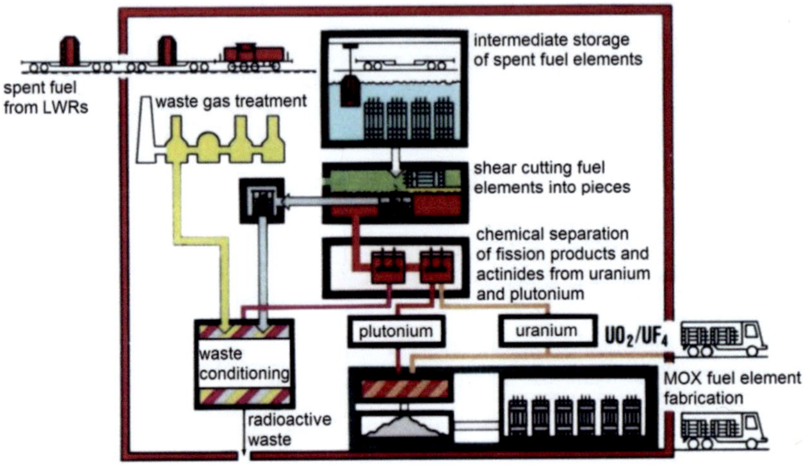

Figure 7.5: Head end with waste gas purification as well as reprocessing of spent fuel and waste conditioning.

7.2.3 Gas Cleaning and Retention of Gaseous Fission Products

During the processes of chopping and dissolution of the fuel, gaseous and volatile fission products are released. They must be removed together with water vapor, nitrous gases and nitrogen. This mixture of volatile fission products, vapors and gases must be treated in the waste gas cleaning system. Gaseous and volatile fission products are made up of the following components:

- Tritium produced by ternary fission and by (n,T)-reactions in light atomic nuclei.
- Carbon, C-14. is produced by an (n,α)-reaction from O-17 and by the (n,p)-reaction of N-14. In the gaseous effluent it appears as 14CO2.
- Krypton is generated as a gaseous fission product. Some 7% of the krypton fission products produced consist of Kr-85 isotopes.
- Xenon is another gaseous fission product. However, only traces of the Xe-133 isotope produced must be considered.

All the other fission product noble gases generated are either stable or have very short halflives.

I-129 and traces of I-131 are partially volatile isotopes initially found in dissolved fuel. Ru-106 may volatilize as ruthenium tetroxide evaporating from strong nitric acid solutions, but only some 10^{-4} fractions of Ru-106 enter into the gaseous effluent stream. In a similar way, small traces of such β-emitters as strontium or α-emitters as uranium and plutonium can penetrate into the gaseous effluent as aerosols. However, only some 10^{-4} to 10^{-6} fractions of the fuel inventory are carried into the gas stream as aerosols.

These gaseous effluents are first passed through a condenser. Afterwards, the nitrogen oxides are oxidized and washed out. The remaining aerosol fractions only amount to 10^{-6} to 10^{-8} times the inventory. Scrubbers and high-efficiency particulate aerosol (HEPA) filters are used next to remove the aerosols. Iodine is retained very efficiently in silver impregnated ($AgNO_3$) filter materials. Tritium as HTO contained in water vapor and ^{14}CO, are retained in molecular sieves. The removal of Kr-85 can be achieved by means of low temperature rectification. In the same process, the xenon noble gas can also be removed. The separated krypton can be stored in compressed gas cylinders. Alternatives may be the entrapment in zeolites (crystallized silicates).

7.2.4 Chemical Separation of Uranium and Plutonium (PUREX process) [8,9]

The most applied process to date is the PUREX process (plutonium and uranium recovery by extraction). The PUREX process uses tri-n-butyl phosphate (TBP), which may be diluted by kerosene or n-paraffin (hydrocarbon) solvents as organic solvents to extract uranium and plutonium. TBP is stable in nitric acid and can selectively extract tetravalent and hexavalent uranium and plutonium nitrate complexes. However, this selective extraction capability of TBP does not apply to trivalent plutonium nitrate complexes.

For extraction, the fuel solution acidified with nitric acid and containing uranium, plutonium, higher actinides and fission products is moved from the middle of column A (Fig. 7.6) in a liquid-liquid counter current extraction flow past the specifically lighter organic solvent (TBP in kerosene) rising from the bottom. In that process, the organic solvent extracts uranium and plutonium, while the fission products and actinides remain in the aqueous solution. The solution with nitric acid leaves the column at the bottom as high level aqueous waste (HLW). It contains the fission products and higher actinides. The aqueous waste is evaporated to recover the nitric acid. The remaining concentrate is further treated as high level waste concentrate (HLWC).

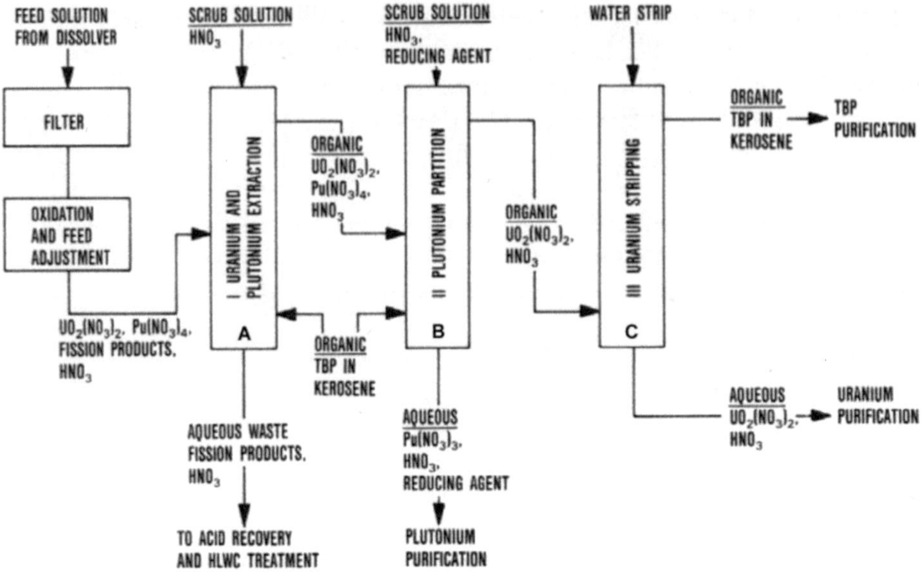

Figure 7.6: Simplified PUREX process flowsheet [9].

The rising organic solvent contains uranium and plutonium and small traces of fission products, which are removed by a nitric acid solution injected at the top of the column. The organic solvent leaves the columns at the top and is introduced into column B, where the tetravalent and hexavalent plutonium is reduced to trivalent plutonium by means of a reducing agent stream, e.g. U (IV) nitrate with hydrazine nitrate, hydroxylamine nitrate or Fe(II) sulfamate. The most elegant method developed uses electrolytic reduction within the extraction apparatus. This trivalent plutonium is soluble in organic TBP-kerosene and, as a consequence, is re-extracted into the aqueous phase, while hexavalent uranium remains in the organic TBP-kerosene phase. Small amounts of re-extracted uranium are extracted again by organic TBP-kerosene introduced at the bottom of the second column. The aqueous plutonium product stream leaves the second column at the bottom, while the organic uranium product stream leaves at the top and enters the bottom of the third column C, where it is met by a countercurrent stream of diluted nitric acid as an aqueous re-extraction solution flowing from the top. The uranium product stream with nitric acid then leaves column C at the bottom, while the organic solvent leaves at the top. After removal of organic decomposition products and fission products by washing, the organic solvent can be recycled into the system.

For sufficient decontamination of uranium and plutonium, the uranium and plutonium product streams are required to pass through two further decontamination cycles. The final products, after concentration and purification, are plutonium nitrate, $Pu(NO_3)_4$, and uranyl nitrate, $UO_2(NO_3)_2$. The plutonium nitrate and a part of the uranyl nitrate solution are mixed to form a so-called master mix which has already the plutonium enrichment which is needed for the PuO_2/UO_2 mixed oxide (MOX) fuel. The remaining uranylnitrate can be used as reprocessed uranium for later re-enrichment, or as blanket material in LMFBRs. The resulting waste streams must be treated separately.

7.2.5 Mass flows and radioactivities in a reprocessing facility [10]

A reprocessing facility with a throughput of 2 t_{HM}/d of spent LWR fuel operating 300 d per year has a yearly capacity of 600 t_{HM}/y spent fuel. Such a reprocessing plant could serve up to 30 GWe of LWRs.

In the first extraction cycle A such a reprocessing plant would produce about 1 m^3/d of high level waste concentrate (HLWC) with a radioactivity of about $2.2x10^{16}$ Bq/m^3, about 1 m^3/d of hulls and structural materials with a radioactivity of about $0.9x10^{15}$ Bq/m^3 and about 0.1 m^3/d of sludge and insoluble residuals with a radioactivity of about $2x10^{16}$ Bq/m^3.

In the second and third uranium and plutonium decontamination cycles B and C some 0.2 m^3/d of organic solvent is produced which contains some traces of uranium/plutonium. It contains about $2x10^{10}$ Bq/m^3 of radioactivity. In addition, about 3 m^3/d aqueous MLW of $2x10^{12}$ Bq/m^3 with traces of uranium/plutonium are produced. Krypton and Tritium enriched water are recycled and conditioned.

The 600 t_{HM}/y reprocessing plant extracts 20 kg plutonium and 1,98 tonnes of uranium per day. About 0.5% of the plutonium and uranium are found in the waste streams.

7.2.6 Reprocessing capacity for spent UO_2 fuel [11]

The PUREX process is entirely used in large scale spent UO_2 fuel reprocessing. The largest commercial reprocessing plants are operating in France (1700 t_{HM}/y) and the UK (1200 t_{HM}/y) whereas medium scale spent UO_2 fuel reprocessing facilities are also available in Russia (500 t_{HM}/y) [17] and Japan (800 t_{HM}/y) (Table 7.2). Smaller scale reprocessing facilities operate also in India and China. The total commercial UO_2 spent fuel reprocessing capacity in the world was somewhat more than 4000 (ts_{HM}/y) in 2010. In addition the UK still operates a 1500 (t_{HM}/y) reprocessing plant for MAGNOX type fuel.

It is interesting to note that – except for Japan – all other commercial spent UO_2 fuel reprocessing capacity is located in Nuclear Weapon States.

Table 7.2: Commercial reprocessing plant capacity for UO_2 spent LWR fuel in the World (IAEA) in 2010.

Country	Plant	Fuel type	Reprocessing capacity (t_{HM}/y)
France	Cap de la Hague	LWR	1700
United Kingdom	Sellafield Windscale	LWR AGR	1200 1500
Japan	Tokai-mura Rokkasho-mura	LWR LWR	90 800
Russia	Mayak	LWR	500
India	Tarapur Kalpakkam	CANDU FBR	100 100
China	Lanzhou	LWR	50

7.3 Conditioning of waste from LWR fuel reprocessing [12,13,14]

7.3.1 Storage and cooling of liquid high level waste concentrates (HLWC)

The HLWC from the first extraction cycle A of the reprocessing plant can be stored in stainless steel tanks and cooled by tube coils with circulating water such that the temperature remains <65 °C. The decay heat production of the HLWC is about 7 kW/m^3. The tanks are installed in so-called hot cells lined with steel plates. The tanks stand in a type of pans which can collect any leakages. The HLWC can be kept in these tanks for at least 30 years.

7.3.2 Solidification of the HLWC by vitrification

For solidification the HLWC is calcinated (expulsion of liquid) and decomposed in oxides at about 400 °C. Then it is mixed with glass frites (borosilicate glass) and molten together in a furnace by either induction or Joule heating. The molten borosilicate glass can be loaded with about 18% of fission products and about 0.4% actinides. The molten radioactive glass is then filled into stainless steel canisters with a diameter of 43 cm and an overall height of 134 cm. They contain about 0.16 m^3 or 400 kg radioactive borosilicate glass (Fig. 7.7). The decay heat of such a borosilicate container is about 2.5 kW at the time of vitrification and about 0.4 kW after 50 years of storage.

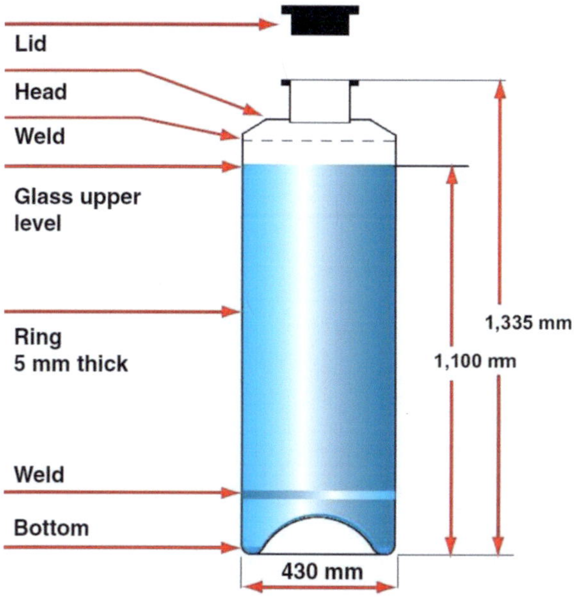

Figure 7.7: Borosilicate HLW container [13].

7.3.3 Conditioning of solid HLW from reprocessing plants

The hulls and structural pieces of the spent fuel elements together with some insoluble residues represent α-emitter contaminated long-lived MLW from the reprocessing plant. They are initially stored under water.

Conditioning is achieved by compaction by a factor of five. The structural parts are introduced in a strong metallic cylinder and compressed with a 250 MPa press to a metallic pancake. Several of these pancakes are filled in so called CSD-C containers (Fig. 7.8) which have the same outer dimensions (43 cm diameter, 134 cm height) as the containers with vitrified borosilicate glass.

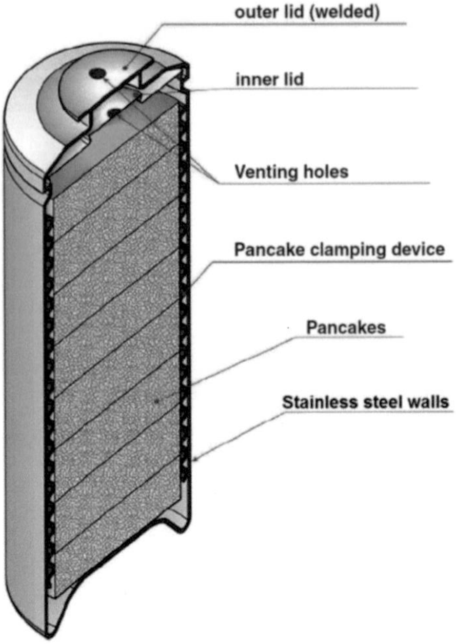

Figure 7.8: CSD-C container with compacted hulls and steel parts [13].

7.3.4 Conditioning of solid organic waste from reprocessing plants, refabrication plants and nuclear reactors

Solid wastes like α-emitter contaminated papers, plastics, ion exchange resins sludges etc. from reprocessing and fuel refabrication plants as well as nuclear reactors are incinerated by medium temperature pyrolysis systems (400 °C) and treated by calcination (900 °C). The resulting ashes contain more than 99% of the original radioactivity. They are mixed with cement based materials (paste of cement, mortar, concrete) and filled in containers of different size.

7.3.5 Conditioning of liquid organic MLW

Aqueous MLW solutions are concentrated by evaporation and then treated in a calcinator at 400 °C. The concentrates are mixed with cement based materials and filled in containers of different size. Another waste treatment technique is mixing the concentrate with hot bitumen. The product is again filled in drums.

7.3.6 Treatment of Krypton-85 and Tritium

Kr-85 with a half-life of 10.8 years can be forced into pressurized steel cylinders of 50 l volume. The Kr-85 cylinders are stored in shafts in a Krypton storage facility.

Tritium with a half-life of 12.3 years can be concentrated during the PUREX process. The water with high concentrations of tritium can be stored in tanks.

7.3.7 Transport and Storage of HLW and MLW

The transport and intermediate storage of HLW and MLW containers (Sections 7.3.2 to 7.3.4) is done in, e.g. CASTOR transport and storage containers. These CASTOR containers can be filled with 33% of glass containers and 67% of CSD-C containers. Two CASTOR containers are sufficient to take the conditioned waste produced by a 1.3 GWe LWR per year (Fig. 7.9). These CASTOR containers can be stored in intermediate storage facilities for several decades until the waste containers can be finally stored in a deep repository.

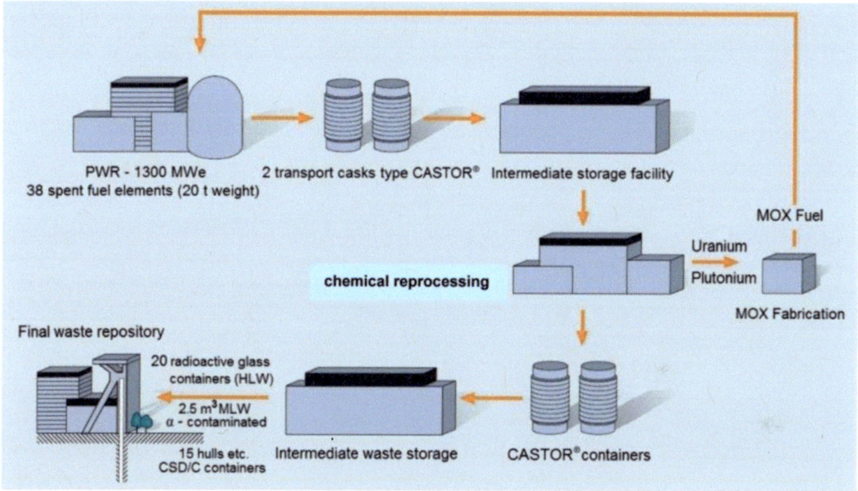

Figure 7.9: HLW radioactive glass containers and α-emitter contaminated decay heat producing MLW from a 1.3 GWe LWR plant generated per year (VGB).

7.4 Long Term Waste Disposal

7.4.1 Low level waste disposal without long-lived α-emitters

LLW without long-lived α-emitters are conditioned in bitumen or filled in concrete containers. It is then stored in concrete building structures which are covered by a concrete roof and many meters of sand. Below this concrete building, possible leaks can be detected. After about 300 years such a LLW repository can be released from surveillance, because the radioactivity will have essentially decayed. An example of such a non α-emitting waste containing LLW is located at Aube (France).

7.4.2 Repositories for low heat producing HLW/MLW

Such repositories for low heat producing HLW/MLW are in operation in granite type of geological formations about 500 m deep in Sweden and Finland. These repositories are accessible through a vertical shaft or a ramp type gallery. The MLW/LLW containers are stored in concrete structures and sealed with bentonite.

Another type of repository for low heat producing LLW/MLW shall be operated in Germany. It is a 800-1200 m deep former iron mine which is protected from water ingress by very thick layers of clay located above the LLW/MLW repository.

7.4.3 Repositories for HLW in deep geological formations

HLW containers and containers with heat producing MLW will be stored in 500-1000 m deep geological formations. These deep repositories are arranged like deep mines with a vertical shaft and horizontal galleries. The HLW containers can be stored in boreholes or arranged horizontally [15] (Fig. 7.10).

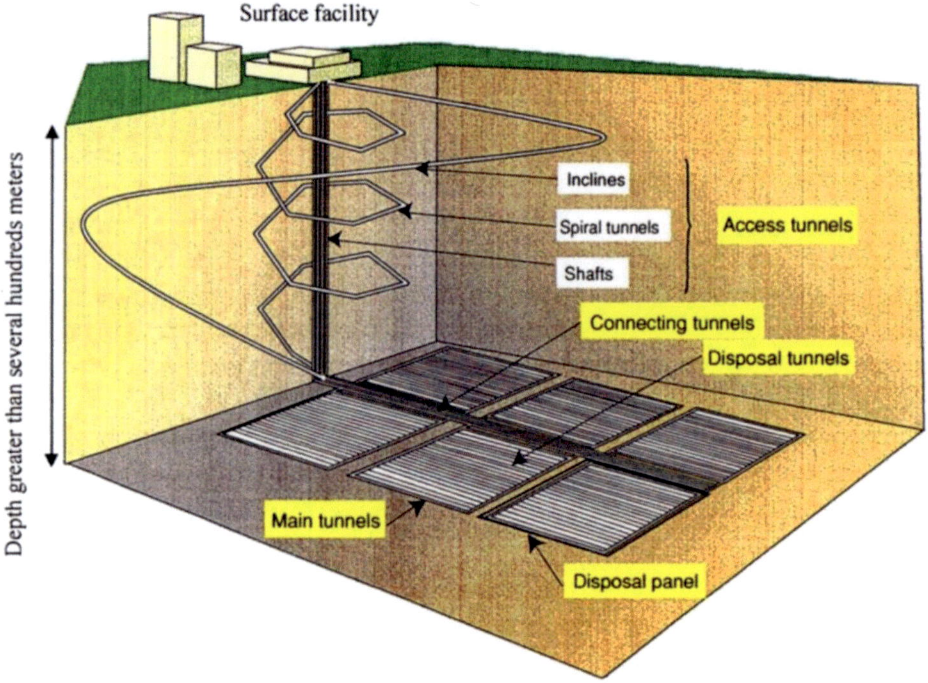

Figure 7.10: Deep repository for HLW and α-emitter contaminated decay heat producing waste [15].

Salt domes or thick layers of salt at a depth of several hundred meters are ideal for deep repositories. However, granite, gneiss, tuff and basalt formations as well as clays are also considered attractive. In 2010, there was no HLW repository in deep geological formations in operation yet. But many countries have test sites for such HLW repositories.

A technical multibarrier system surrounds the HLW glass in order to increase the long term safety of the vitrified waste against water ingress and dissolution of radioactive molecules out of the borosilicate glass. This technical multibarrier system consists of e.g. thick steel containers surrounded by copper or titanium layers. In addition a thick layer of bentonite and silica sand forms the outer layer. The heat of the HLW will be transferred by thermal conduction to the outer layers and to the geological formations. When the HLW canisters will have been placed in prepared holes by remote handling techniques the remaining void will be backfilled by bentonite and sand.

7.5 Direct Disposal of Spent Fuel

Direct disposal of spent fuel was favored in the United States from 1977 on in order to avoid reprocessing spent fuel and Pu recycling with MOX fuel following a strict non-proliferation policy. At the same time, this stopped the development of breeder technology with U/Pu fuel recycling.

This type of once-through fuel cycle will not be considered any further here for the following reasons:

- Direct disposal of spent fuel is technically feasible. Even while a repository is being filled, monitoring by IAEA inspectors is possible. However, once the repository has been closed, it cannot be monitored for many thousands of years. The repository would contain hundreds of tons of plutonium and neptunium in spent fuel. Historical experience of mankind speaks against successful long-term surveillance in this case.

The Yucca Mountain repository in the United States, which had been planned for direct disposal of spent fuel, for these very reasons had foreseen retrievability of the spent fuel elements after a specific period of time. As a result of a decision by the U.S. Department of Energy, the Yucca Mountain repository was given up again in 2010.

The direct disposal option would block the following technical possibilities for the future civil use of nuclear power:

- Plutonium multiple recycling with more than 99% incineration of the plutonium;
- Multiple recycling of neptunium and americium and their incineration to a level of more than 99%.
- Only incineration of Pu, Np, and Am allows the radiotoxicity level and the heat load of the waste to be reduced drastically.
- Breeder technology and the use of U-238 are possible only by multiple recycling of plutonium. This extends the supply of fuel for the civil use of nuclear power to many thousands of years.
- Only reprocessing and Pu recycling allow the production of proliferation-proof plutonium with a Pu-238 content of about 5%. As is shown in Sections 9 to 14, this is the only way to achieve a future long-term civil use of nuclear power with proliferation-proof plutonium.

7.6 Mixed Oxide Fuel Fabrication [16,17,18,19,20]

Plutonium recycling in thermal reactors, e.g., LWRs, requires the fabrication of mixed oxide U/Pu (MOX) pellets. The MOX-powder (master mix) coming from the reprocessing plant is blended together with UOX-powder to achieve the required fissile enrichment. The mixed powder is precompacted and granulated into a freely flowing powder. This is turned into cylindrical pellets which are first sintered at temperatures of 1000-1700 °C and then ground to the required dimension (Fig. 7.11). Finally, they are loaded into zircaloy or steel tubes, to which end caps are welded. These fuel rods are assembled to fuel assemblies.

A certain fraction of the mixed oxide pellets will be imperfectly fabricated and are rejected during inspection procedures. Such material and grinder fines, which are designated clean rejected oxide, can be recycled directly into the manufacturing process. However, a small fraction of pellets and powder are contaminated by corrosion products, etc.. These must be dissolved in a nitric acid/fluoride solution and, along with filtrates from wash-leach processes, be treated chemically as in a PUREX reprocessing step to recover the uranium and plutonium.

As MOX fuel must be reprocessed after having attained its design burnup, the fuel fabrication process must guarantee high solubility (> 99%) of the irradiated MOX fuel in nitric acid. Such high solubility is required to minimize the plutonium loss in the residues during chemical reprocessing.

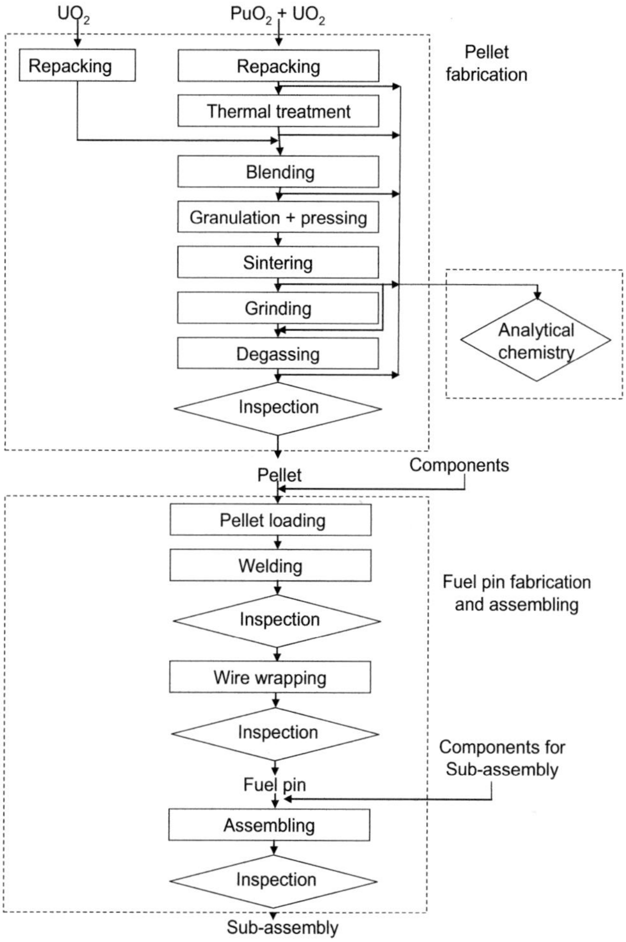

Figure 7.11: Process scheme for the fabrication of MOX elements.

Three other refabrication processes can be applied. The sol-gel process allows the direct fabrication of spherical MOX particles, which can be pressed and sintered into fuel pellets.

The AUPuC (ammonium (U,Pu) carbonate) refabrication process also allows the fabrication of relatively coarse grain MOX powder. This crystal powder is fabricated essentially free of Am-241 and then pressed and sintered into MOX fuel pellets.

A third MOX fabrication process is based on vibro compaction. The MOX fuel is broken into small fuel particles of about 0.1 - 0.5 mm size. These are filled into the cladding tube and compacted by vibro compaction.

A MOX fuel fabrication plant with an annual fabrication capacity of 150 tHM/y roughly corresponds to the plutonium mass flow produced by the 2 tHM/d or 600 tHM/y model reprocessing plant described in Section 7.2.6.

7.6.1 MOX fuel refabrication capacity in the world

In most cases MOX fuel refabrication plants are collocated with spent fuel reprocessing plants. This avoids long distance transportation of UO_2/PuO_2 powders.

The MOX fuel refabrication plants are – except in Japan – entirely located in Nuclear Weapon States. Table 7.3 shows the capacities of MOX refabrication plants. In Russia [21] and USA additional MOX plants are built for the conversion of weapons plutonium.

Table 7.3: Reprocessing plant capacity for spent MOX fuel in the World (IAEA).

Country	Plant	Fuel type	Reprocessing capacity (t_{HM}/y)
France	Marcoule	LWR-MOX	195
United Kingdom	Sellafield	LWR-MOX	120
Japan	Rokkasho-mura Tokai	LWR-MOX FBR-MOX	130 20
USA	Savannah River	LWR-MOX	70*
Russia	Zheleznogorsk Tomsk	FBR-MOX LWR-MOX	60 70*

*Weapons plutonium

7.7 The Uranium/Plutonium Fuel Cycle of Fast Breeder Reactors [22,23,24,25]

7.7.1 Ex-Core time Periods of FBR Spent Fuel

Like all recycling converter reactors and near-breeder reactors, FBRs must work in a closed fuel cycle. Their systems inventory, consisting of the core fuel inventory and the fuel inventory passed through reprocessing and refabrication should be as small as possible for economic reasons. At present, it is generally assumed that an ex-core time of two years is feasible for the FBR fuel cycle.

After unloading from the reactor core, the core elements and the radial blanket elements are first stored at the reactor site for some 180 days. Then they are transported to the reprocessing plant in transport casks, which can contain six to twelve fuel elements each. Shipping the fuel elements takes about thirty days. Another thirty days are assumed for intermediate storage and pre-treatment of the fuel elements prior to cutting and dissolution. Assuming a reprocessing plant with annual reprocessing capacities of about 260 tHM/y (in more detail 170 t of core and axial blanket fuel mixed with 90 t of radial blanket fuel), the total time required for all steps, from chopping the fuel pins to conversion to PuO_2 and UO_2 powder, is estimated to be forty days. Sixty days are assumed for intermediate storage of the PuO_2/UO_2 oxide powder, another thirty days for transfer to the fuel refabrication plant. The reprocessing plant and the fuel refabrication plant are assumed to be collocated at one site.

The associated UO_2/PuO_2 fuel refabrication plant will have an annual capacity of about 110 t of mixed UO_2/PuO_2 fuel for the core and an annual capacity of about 150 t for UO_2 blanket fuel (about 65 t for the axial blankets and about 85 t for the radial blanket). The UO_2/PuO_2 powder will be stored for about thirty days and then transferred in batches to the fabrication lines. The fabrication process takes about sixty days, and another thirty days are required for fuel element storage prior to shipment to the FBR power plant. Shipment requires some thirty days; another thirty days are assumed for storage on the reactor site before the fuel is loaded into the core for power generation.

Assuming another 180 days for unforeseen delays, which may arise from imperfect synchronization between the various fuel cycle operations, the total ex-core or fuel cycle time adds up to 730 days or two years.

7.7.2 Mass Flow in an FBR Fuel Cycle [22]

A model fuel cycle for reprocessing the UO_2/PuO_2 fuel discharged from 10 GW(e) FBRs roughly corresponds to a capacity of 1 tHM/d or, at 250 equivalent full power load days, an annual capacity of 250 tHM. Such a fuel cycle includes reprocessing and refabrication plants on an industrial scale.

From the assumed 10 GW(e) FBRs, an annual 170 tHM of uranium and plutonium in core fuel elements and 90 tHM of uranium and plutonium in radial blanket elements are discharged and transported to the reprocessing plant per year. These spent fuel and blanket elements contain 6.45 t of fission products. Of this fission product volume, some 5.8 t is contained in the core fuel elements and in the axial blankets, and some 0.65 t in the radial blanket elements. In addition to these quantities of fuel and fission products, there are approximately 200 kg of higher actinides (Np-237, Am-241, Am-242m, Am-243, Cm-242 and Cm-244).

In the reprocessing plant, the fission products and the actinides are separated and go into the HLWC concentrate. Some 227 tHM of uranium and about 22.1 tHM of plutonium are recovered, of which 1.59 tHM of plutonium can be diverted as a breeding gain to start new LMFBRs or to feed converter reactors. Roughly 0.1%, i.e., 20 kg of plutonium and some 200 kg of uranium, initially remain as high level solid or liquid wastes accumulating in the reprocessing and refabrication plant. In additional waste treatment steps, plutonium is also recovered from the waste so that ultimately only some 20 kg/a of plutonium will be lost to the HLW and MLW.

7.7.3 FBR Spent Fuel Reprocessing [23,25,26,27]

The PUREX process is also used for reprocessing of spent FBR fuel elements. However, technical modifications are required to take into account the specific characteristics of FBR fuel elements enriched in plutonium.

The end pieces of the fuel elements are cut off, and the fuel element wrapper is removed mechanically. The fuel rods are then separately cut into pieces 2.5 cm long by means of a shear. In this step, core fuel and blanket fuel are mixed. The fuel rod pieces fall into the dissolver, where the fuel is dissolved in hot nitric acid. The dissolver geometry must be carefully adapted to the higher plutonium enrichment of LMFBR fuel to avoid criticality.

The fuel solution coming from the dissolver is first clarified by centrifuging or filtration. Then the PUREX counter current solvent extraction process is applied. However, the contact times of the solvent and the nitric acid solution must be short to limit radiolysis of the solvent. For this purpose, pulsed columns or centrifugal contactors are used.

When decontaminating plutonium and uranium, the higher plutonium concentration must be taken into account to ensure that the plutonium fraction in the waste is kept as small as possible. The process for conversion of plutonium nitrate and uranyl nitrate into PuO_2 and UO_2 is the same as in an LWR reprocessing plant. Also, plutonium nitrate and uranyl nitrate can be directly mixed, co-converted and co-precipitated into mixed UO_2/PuO_2. The modifications of the PUREX process described above, the smaller dimensions of all tanks (higher plutonium enrichment, criticality) ultimately require the construction of special FBR reprocessing plants.

7.7.4 FBR Fuel Fabrication [17,20,29,26]

The same fabrication processes are applied for FBR fuel as for LWR (MOX) fuel (Section 7.6.1). The reprocessing plant and the FBR fuel refabrication plant are collocated at one site. After storage in a buffer store, the MOX fabrication process begins with mechanical blending of UO_2 and master mix UO_2/PuO_2 powders to establish the desired enrichment. Afterwards, the fabrication process proceeds with pressing, sintering, grinding and drying of sintered pellets. This is followed by assembling the pellets inserting the pellets into cladding tubes. Finally, the fuel pins are welded and assembled into FBR fuel elements.

FBR fuel pellets have somewhat smaller diameters than LWR pellets. The pellet fabrication and pin loading operations are carried out in glove boxes. To protect the workers against α-radiation and neutrons originating from various plutonium isotopes and their radioactive daughters (spontaneous fission and (α,n)-reactions with oxygen), shielding must be provided at the fabrication lines. Future plants are expected to be operated remotely to a large extent. Besides the above described fuel fabrication technology also the sol-gel precipitation technique, the vibro-compaction technique, and the AUPuC process are being employed.

7.7.5 Status of FBR Fuel Reprocessing and Refabrication

Several small test and pilot plants for LMFBR fuel reprocessing and MOX fuel fabrication in the UK and France, had been operated for almost twenty years. They were closed after PFR, Phenix and Superphenix had been shut down (Table 7.4). Japan has small scale reprocessing test facilities in operation at Tokai-mura for the fuel of JOYO and MONJU. It also has a MOX fuel fabrication plant with a capacity of 5-10 tHM/y in operation at Tokai-mura. France operated a MOX fuel fabrication plant of 20 tHM/y throughput at Cadarache. Russia will start a 60 tHM/y plant at Zheleznogorsk to provide the UO_2/PuO_2 fuel for BN-800 (Table 7.4).

Table 7.4: Status of FBR fuel reprocessing and refabrication.

	Country	Location	Capacity t_{HM}/a	Status
FBR fuel reprocessing	UK France Japan	Dounreay Marcoule Tokai-mura	5 5 5	closed closed in operation
FBR fuel refabrication	Japan Russia France	Tokai-mura Zheleznogorsk Cadarache	5-10 60 20	in operation planned closed

7.8 The Closed Nuclear U/Pu MOX Fuel Cycle for PWRs

In a closed fuel cycle the spent fuel, after its radioactivity has decayed to a certain level in temporary storage facilities, is shipped to a reprocessing plant for chemical reprocessing. After chemical reprocessing, the fissile plutonium as well as the residual uranium can be re-used to fabricate new fuel elements and be recycled in PWRs (Fig. 7.12). A small fraction of the fissile material (about 0.1%) goes into the radioactive waste during chemical reprocessing and refabrication of the fuel, where it is lost i.e. stored in a deep geological repository. Recycling improves the utilization of fuel (see Section 4.8) and, consequently, decreases the consumption of natural uranium.

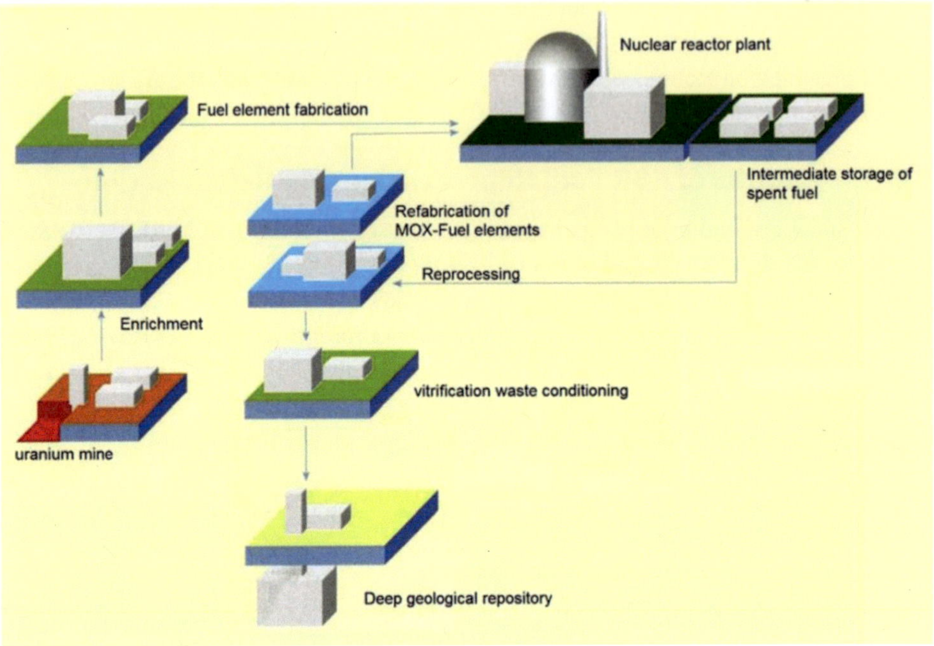

Figure 7.12: The closed nuclear U/Pu (MOX) fuel cycle for PWRs.

Fissile plutonium from several reactors may be collected and used exclusively in special fuel MOX recycle reactors. Also, every PWR can recycle its own plutonium generated in preceding operation cycles. In both cases this is then called self-generated recycling (SGR).

7.8.1 Plutonium Recycling as Plutonium Uranium Mixed Oxide (MOX) Fuel in the SGR mode

In the SGR mode (shown by Fig. 7.13), continuous mixing of recycled plutonium with "fresh" plutonium from spent LEU UOX fuel elements leads to an optimum high percentage of the fissionable plutonium isotopes Pu-239 and Pu-241 such that up to, e.g. three full recycles of the plutonium become possible. Thus, the safety parameters, e.g., the coolant void coefficient of full MOX core PWRs can be kept at tolerable values.

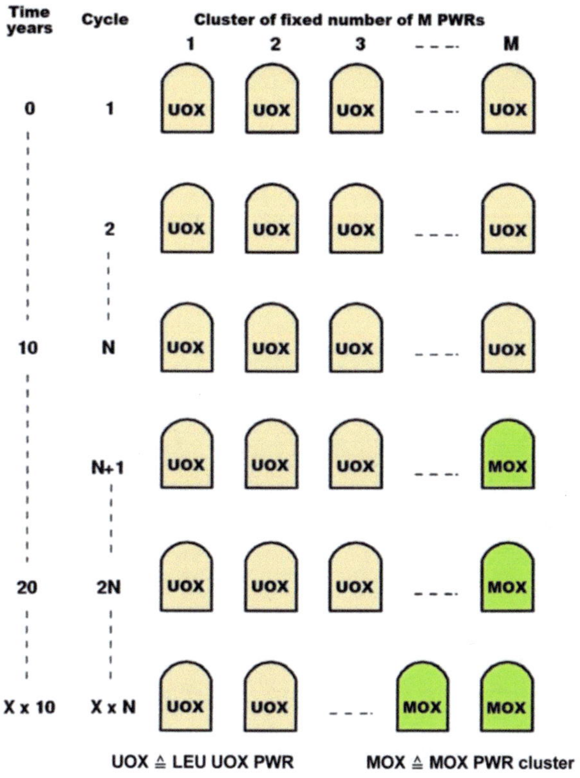

Figure 7.13: Scenario for plutonium multi-recycling in the SGR mode for a cluster of M of PWRs.

The plutonium bearing MOX fuel elements have the same structural design as LEU UOX fuel elements of PWRs. Each fuel rod of such a fuel element contains UO_2/PuO_2 mixed oxide (MOX) fuel pellets. The average fissile plutonium enrichment is chosen such that the same discharge burnup as in the LEU fuel elements can be achieved. If a certain fissile plutonium enrichment of, e.g., 6% shall not be exceeded for safety reasons, small amounts of U-235 are added to achieve the required criticality values.

Enrichment zoning within the fuel elements is necessary to avoid local power fissile peaks. This is achieved by giving the fuel rods at the periphery of the fuel subassemblies a lower Pu enrichment, while all remaining fuel rods of the inner part of the fuel element have a higher enrichment.

Also water filled rods can be arranged evenly distributed in the fuel element. In this case all MOX fuel rods can have all the same fissile enrichment.

Table 7.5: Plutonium isotopic compositions of PWR spent fuel after 50 MWd/t_{HM} at burnup and 10 years of intermediate storage. (This isotopic plutonium composition differs somewhat from the plutonium in Table 12.2 due to different cross section sets used. Also 1.9% americium must be added for 100%) [28].

Isotopic composition	Plutonium in wt% 10 years after unloading (50 GWd/t_{HM})
Pu-238	2.8
Pu-239	55.1
Pu-240	23.3
Pu-241	9.3
Pu-242	7.6

7.8.2 Plutonium incineration in PWRs during several recycling steps

Broeders [28] analyzed recycling of plutonium in PWRs applying the SGR mode. The plutonium from reprocessing of spent fuel element of several LEU MOX PWRs is collected until a first full MOX PWR can be started. This is explained in Fig. 7.13. The assumptions of Broeders [28] for this SGR recycling strategy were:

- 4.5% U-235 enrichment of the fuel element of the UOX PWRs
- 6% max. Pu_{fiss} enrichment for MOX fuel (if needed, low enriched U-235 is added to fullfil criticality conditions)
- the plutonium isotopic composition is shown by Table 7.5
- 6 burnup cycles within 10 years, i.e. 20 months per burnup cycle
- 7 years cooling time of the spent after unloading from the core and time for reprocessing
- 3 years time for refabrication of the MOX fuel including time for transport.

For a maximum average burnup of 50 GWd/t_{HM} and the above time periods for burnup, reprocessing, refabrication and transport it is appropriate to consider a cluster of M = 8 PWRs. Initially these PWRs are operated with LEU UOX fuel with 4.5% U-235 enrichment. Their spent fuel is reprocessed and MOX fuel elements with about 6% Pu_{fiss} enrichment are fabricated. These MOX fuel elements with an isotopic mixture M1 given in Table 7.5 are loaded into a first MOX PWR which can be started after 10 years (Table 7.6). The plutonium from the spent fuel elements of this full MOX-PWR is again mixed with plutonium coming from 7 operating UOX PWRs. This mixing procedure is then continued until after 20 years a new plutonium isotopic mixture M2 can be loaded to the first full MOX PWR.

Table 7.6: Scenario for multi recycling of plutonium in full MOX PWR cores in a pool of 8 PWRs adequate for a target burnup 50 GWd/t_{HM} [28].

Time		Reactors							
year	Cycle	1	2	3	4	5	6	7	8
0	1	U	U	U	U	U	U	U	U
	2	U	U	U	U	U	U	U	U
	-	U	U	U	U	U	U	U	U
10	6	U	U	U	U	U	U	U	U
	7	U	U	U	U	U	U	U	M1
	-	U	U	U	U	U	U	U	M1
20	12	U	U	U	U	U	U	U	M1
	13	U	U	U	U	U	U	U	M2
	-	U	U	U	U	U	U	U	M2
30	18	U	U	U	U	U	U	U	M2
	19	U	U	U	U	U	U	M3	M3
	-	U	U	U	U	U	U	M3	M3
40	24	U	U	U	U	U	U	M3	M3
	25	U	U	U	U	U	U	M4	M4
	-	U	U	U	U	U	U	M4	M4
50	30	U	U	U	U	U	U	M4	M4
	31	U	U	U	U	U	U	M5	M5
	-	U	U	U	U	U	U	M5	M5
60	36	U	U	U	U	U	U	M5	M5
	37	U	U	U	U	U	U	M6	M6
	-	U	U	U	U	U	U	M6	M6
70	36	U	U	U	U	U	U	M6	M6

U: PWR with UOX fuel; Mi: PWR with MOX fuel of generation i = 1....6

After 30 years a second full MOX PWR can be added. These two full MOX PWRs operate with plutonium isotopic mixture M3 between 30 and 40 years, with plutonium isotopic mixture M4 between 40 and 50 years, with plutonium isotopic mixture M5 between 50 and 60 years and with plutonium isotopic mixture M6 between 60 and 70 years. Table 7.7 shows the plutonium isotopic composition for the MOX fuels M1 to M6.

Table 7.7: Plutonium isotopic compositions of PWR spent fuel after 50 MWd/t_{HM} at burnup and 10 years of intermediate storage. (This isotopic plutonium composition differs somewhat from the plutonium in Table 12.2 due to different cross section sets used. Also 1.9% americium must be added for 100%) [28].

Plutonium Compositions (wt%)	Pu-238	Pu-239	Pu-240	Pu-241	Pu-242
M1	2.8	55.1	23.3	9.3	7.6
M2	3.5	49.4	26.2	10.0	9.4
M3	3.9	46.8	27.9	9.2	10.8
M4	4.3	43.1	28.9	9.9	12.3
M5	4.6	41.5	29.3	9.5	13.6
M6	4.8	40.4	29.6	9.1	14.1

This SGR mode plutonium recycling procedure can be terminated, e.g. after 30 or 70 years (Table 7.6) and the plutonium can be loaded into the cores of FRs having better neutronic characteristics for the incineration of plutonium compositions like M2 to M6. Broeders [28] noticed that plutonium isotopic compositions like M3 or M4 would lead to untolerable coolant temperature coefficients. This unacceptable feature could be counteracted by the selection of wider spacings of the MOX fuel rods in the MOX fuel assemblies.

As can be seen from Table 7.7 the percentage of Pu-238 increases steadily from 2.8% (M1) to 4.8% (M6) whereas Pu-239 decreases from 55.1% (M1) to 40.4% (M6). Also, the Pu-240 increases from 23.3% (M1) to 29.6% (M6), the Pu-241 from 7.6% (M1) to 9.1% (M6) and Pu-242 even from 7.6% (M1) to 14.7% (M6). These changes in plutonium compositions occur over 60-70 years of plutonium recycling in operating MOX PWRs. In order to fulfill the requirements that the fissile part of the plutonium does not exceed the 6% limit and that a certain initial criticality value must be attained also low enriched uranium must be added (Tab. 7.8).

Table 7.8: U-235 enrichment of uranium to be added to the different plutonium mixtures M1 through M6.

Uranium added to plutonium mixture	Enrichment wt%
M1	0.7
M2	1.5
M3	2.9
M4	2.5
M5	3.0
M6	3.3

7.8.3 Balance of plutonium inventories and incineration of plutonium

The balance of the plutonium inventories in the fuel cycle of the cluster of M = 8 PWRs reveals that a considerable part of the plutonium which is generated by the UOX PWRs is incinerated by the full MOX PWRs. Figure 7.14 shows the plutonium inventory of the cluster of M = 8 PWRs for two cases:

- The straight full line represents the direct spent fuel storage strategy (no reprocessing). The plutonium is accumulating and would have to be disposed in a deep geological repository.
- The SGR plutonium recycle strategy is represented by a polygon type set of straight dotted lines which are bending more and more, according to the incineration of the plutonium in the full MOX PWRs with plutonium MOX fuel isotopic mixture M1 to M5. The difference between the full line and the dotted lines represents the plutonium which is incinerated by the MOX PWR normalized to 1 GWe. The essential result is that over 60 to 80 years of SGR recycling about 50% of the plutonium generated by the UOX PWRs could be incinerated, i.e. utilized by the MOX PWRs. This is due to the fact that one full MOX PWR can incinerate about 420 kg Pu/GWe·y.

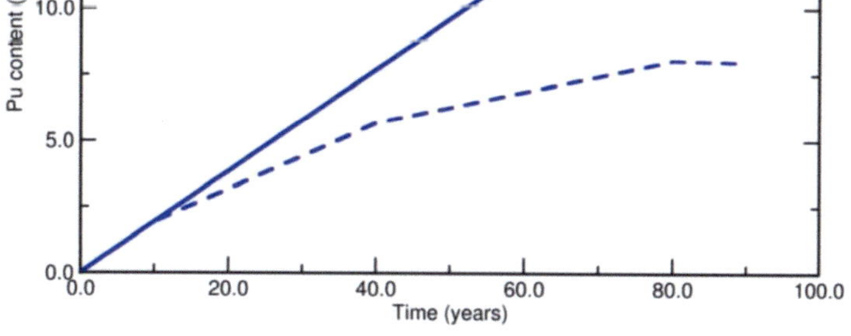

Figure 7.14: Balance of plutonium collected from a cluster of M = 8 UOX: PWRs with 10 GWe total power operating in the once through direct spent fuel disposal mode (full line) and inventory of plutonium in the cluster of M = 8 UOX and MOX PWRs with 10 GWe power (Pu content or inventory normalized to 1 GW(e)) [28].

A similar Pu-recycle scenario can be analyzed by replacing the full MOX PWRs by FR burners (Pu incinerators). This is described in Section 7.8.5. FR-burners can achieve a plutonium incineration rate of 570 kg Pu/GW□y [29,30].

7.8.4 Neptunium and Americium generation in the SGR plutonium recycle scenario

Figure 7.15 shows that the neptunium generation in the SGR plutonium recycle scenario is only slightly different between the UOX-PWR direct spent fuel disposal case and the UOX and MOX PWRs operating in the SGR mode. In the SGR mode more neptunium-237 is transmuted into Pu-238 by neutron capture.

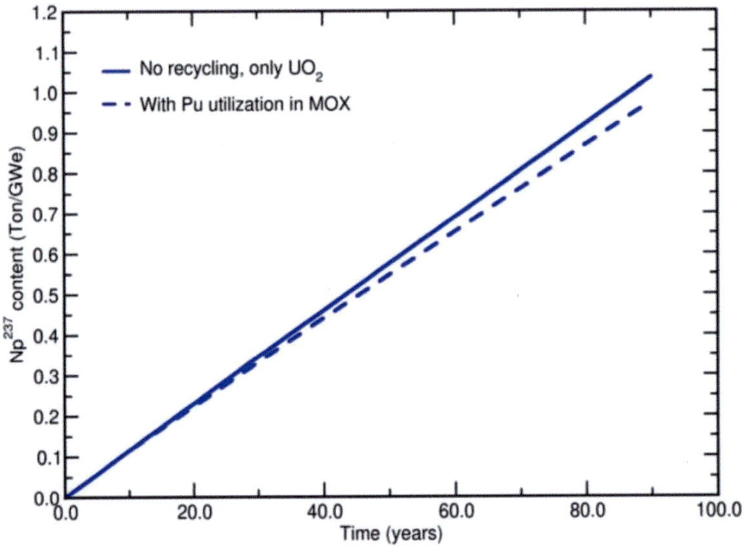

Figure 7.15: Amount of neptunium generated by a cluster of M = 8 UOX PWRs (full line). Amount of neptunium generated by a cluster of M = 8 UOX and MOX PWRs operating in the SGR-mode (dotted line). The amounts are normalized to 1 GWe [28].

The americium production via neutron capture in Pu-242 and beta-decay of Pu-242 is shown for the SGR plutonium recycling case and for the UOX direct spent fuel disposal case in Figure 7.16. In the SGR scenario considerably more (factor 2.5 after 50 years and a factor of 3.5 after 70 years) americium are produced.

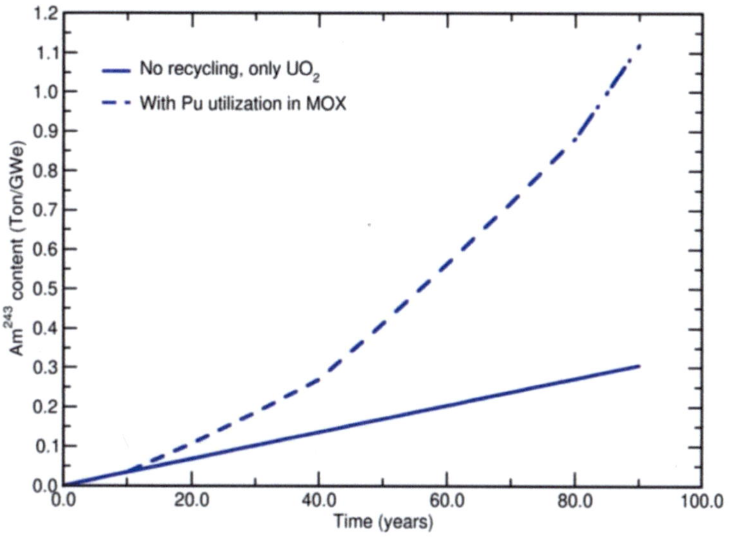

Figure 7.16: Amount of americium generated by a cluster of M = 8 UOX PWRs (full line). Amount of americium generated by a cluster of M = 8 UOX and MOX PWRs (dotted line) operating in the SGR mode. Amounts are normalized to 1 GWe [28].

7.8.5 Plutonium incineration in a MOX-PWR or FR burner or ADS strategy

If instead of the MOX-PWR strategy described in Section 7.8.1 and 7.8.2 FRs or accelerator driven systems (ADSs) are loaded with the plutonium of UOX PWRs similar results are obtained [29,30]. However, the plutonium incineration rates are higher. Table 7.9 shows the higher incineration rates of FR burners, e.g. so-called CAPRA-FRs and the plutonium incineration rates of ADSs compared to those of MOX PWRs.

Table 7.9: Incineration rates of different plutonium burner reactors.

Nuclear Reactor Type	MOX-PWR	FR-burner (CAPRA)	ADS
Incineration rate (kg/GW(e)·y)	420	570	700

Assuming these plutonium incineration rates similar analyses as described in Sections 7.8.1 and 7.8.2 can be performed. This leads to Figure 7.17 which shows the plutonium inventories in the fuel cycle of a cluster of M = 8 UOX-PWRs operating in symbiosis with either MOX-PWRs or FR burners (CAPRA type) or ADSs. The inventories are normalized in tonnes per GW(e). The straight line represents the once through fuel cycle with direct spent fuel storage in the deep geological repository. The plutonium is accumulating as a function of time following an almost straight line in Figure 7.17. (Only the decay of Pu-241 to Am-241 and Pu-242 to Am-243 results in small deviations from the straight line).

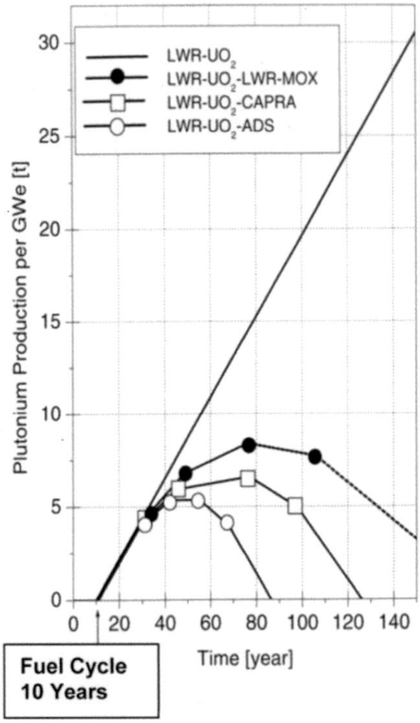

Figure 7.17: Balance of plutonium (normalized to 1 GW(e)) from a cluster of M=8 UOX-PWRs with 10 GW(e) total power operating in either the once through direct spent fuel disposal mode (straight line). The polygon type lines show the pluto-nium inventory (normalized to 1 GW(e) for the cases of UOX-PWRs operating in symbiosis with either MOX-PWRs or FR-burners (CAPRA-type) or ADSs.

The SGR plutonium recycle strategy is represented by polygon type straight lines where each new line represents different incineration rates and the introduction of an additional MOX-PWR or FR-burner (CAPRA-type) or ADS. The difference in tonnes of plutonium between the full line (once through cycle) and the lines for the UOX PWR strategy with MOX-PWRs or FR-burners (CAPRA) or ADS represents the plutonium inventory which is incinerated by these recycling burner reactors. As to be expected the FR-burners and even more the ADSs incinerate the plutonium produced by the UOX-PWRs more efficiently and faster than MOX-PWRs.

For the MOX-PWR some lines are shown as dotted lines from about 75 years on, because the coolant temperature coefficient as an important safety related reactivity coefficient could become intolerable (Section 7.8.2). This is not the case for UOX-PWRs operating in symbiosis with ADS or with FR-burners (CAPRA-type). Both show a potential for incineration of all plutonium produced by the UOX-PWRs in a time frame of about 125 years (FR burners) or about 85 year (ADSs).

The results of Figs. 7.15 and 7.18 are theoretical examples following a certain strategy. In reality first the UOX-PWR in symbiosis with the MOX-PWR are started. Later FR-burners will follow when FRs will be deployed on large scale. This strategy might perhaps be followed by the introduction of several ADSs.

7.9 Chemical separation (partitioning) of minor actinides

The PUREX process for the separation of uranium and plutonium from the fission products and minor actinides was described in Section 7.2. The separation efficiency for uranium and plutonium achieved in the large reprocessing plants La Hague (France) and Sellafield (UK) is approximately 99.9% [31,32,33].

7.9.1 Joint chemical separations of plutonium and neptunium

With slight modifications of the PUREX process neptunium and plutonium can also be separated together from uranium on the one side and fission products and the remaining actinides on the other side. This was already demonstrated in the large reprocessing plant La Hague and by the Japan Nuclear Fuel Cycle Development Institute. The separation efficiencies achieved for neptunium were 99% [31,34]. It is also possible to separate the neptunium from the liquid HLW which is still stored at reprocessing plants [31,33].

7.9.2 Aqueous chemical separation of americium and curium

The chemical separation of americium and curium is more difficult, since some of the fission products (lanthanides) and americium as well as curium exist in the threevalent state [32]. Therefore, americium, curium and the lanthanides must first be separated together. Several chemical processes were developed for this separation step:

- the DIDPA process by the Japan Atomic Energy Research Institute (JAERI) [35]
- the TRUEX process by the Argonne National Laboratory (ANL) in USA [36]
- the TRPO process by the Tsinghua University in Beijing (China) [37]
- the DIAMEX process by the French Commissariat à l'Energie Atomique (CEA) [34,38].

These processes allow separation factors of 99.9%. Some of them were already tested on pilot facility scale, e.g. the DIAMEX process in the ATALANTE facility at Marcoule in France [39].

7.9.3 Chemical separation of Americium/Curium from the Lanthanides

Special chemical separation processes were developed for the separation of americium/curium from the lanthanides. These are:

- the DIDPA process with the element agent DTPA of JAERI [35,40]
- the TRPO process with the chemical agent CYANEX 301 by the Tshinghua University in Beijing (China) [41]
- the SANEX process with the chemical agent BTP by the Karlsruhe Research Center (Germany) and the Institute of Transurania (European Commission).

Separation efficiencies of 99.9% were achieved for these separation processes. The SANEX process was also tested in the pilot scale test facility ATALANTE in Marcoule, France [34,42,43].

7.9.4 Chemical separation of Americium from Curium

Finally, the chemical separation processes TRPO in China as well as EXAm and LUCA in Europe were developed to separate americium from curium [41,44,45]. Again 99.9% separation efficiencies were achieved, e.g. with the TRPO process. Figure 7.18 shows the sequence of aqueous chemical separation processes as developed in France for the partitioning of uranium, plutonium, neptunium, americium and curium.

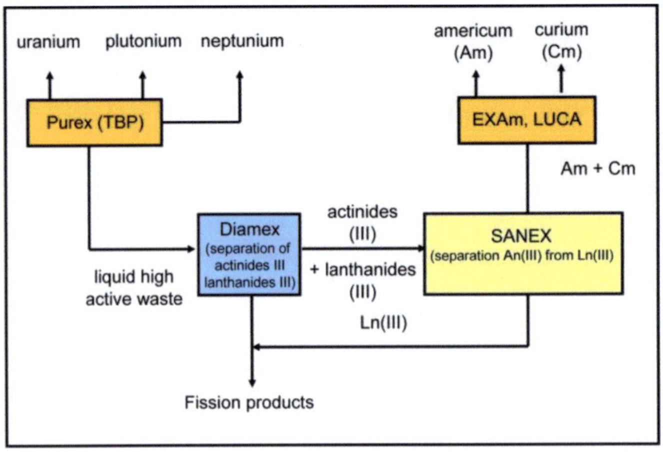

Figure 7.18: Partitioning processes for uranium, plutonium, neptunium, americium and curium in Europe [34].

7.9.5 Pyro-metallurgical methods for the separation of Uranium, Plutonium and Minor Actinides

Besides the above aqueous chemical separation methods also pyrometallurgical separation methods were developed in USA, Russia, Europe and Japan.

7.9.5.1 The Integral Fast Reactor Pyroprocessing Process

The Argonne National Laboratory (ANL) in USA develops the pyrometallurgical separation process in combination with the Integral Fast Reactor (IFR) development (Section 6.7).

It is based on experience gained with metallic U-Pu-Zr fuel of the EBR-II experimental program at Idaho, USA [46,47]. Pyroprocessing technologies aim also at a separation factor of >99.9%. However, the plutonium and minor actinides (neptunium, americium, curium) in combination with about 30% uranium remain together. The pyroprocessing method is a batch mode process, contrary to the aqueous partitioning processes which operate in a continuous mode [48,49].

In pyroprocessing the chopped U-Pu-Zr fuel rod pieces are placed in steel baskets which are placed into an electrorefiner vessel. The bottom of this vessel is covered by liquid cadmium (melting point 321 °C). This cadmium layer is covered by a thick layer of a eutectic mixture of lithium chloride and potassium chloride (melting point 350 °C). The electrorefiner is operated in a temperature of 500 °C.

The perforated steel basket with pieces of fuel rods is acting as the anode. The actinides from the fuel are transported as ions from the anode to two kinds of cathodes. A solid cathode, where most of the uranium is collected and a liquid cadmium cathode where the remaining uranium together with plutonium, neptunium, americium and curium are collected. Finally, this mixture of uranium, plutonium and minor actinides is recovered in a vacuum furnace [49,50].

The metal ingots from processing in the vacuum furnace are sent to an injection casting station for refabrication of metallic fuel rods (Section 7.9.6.1).

A similar pyroprocessing approach as at ANL (USA) was developed in Russia for the UO_2/PuO_2 fuel of FRs BN-600 and BN-800 (Section 6). In this Russian DOVITA process [51,52] the UO_2/PuO_2 MOX fuel is converted into chlorides and separated in a melt of NaCl-KCl at 650 °C.

7.9.6 Fuel refabrication for incineration of actinides

At present the refabrication technologies for actinide fuel are still based on existing technologies for MOX fuel as described in Section 7.6. The minor actinides are mixed as oxide powders into the MOX powders. The following actinide fuels can be fabricated [56]:

$$(U\ Np)O_2 \quad (U\ Pu\ Am)O_2 \quad (U\ Pu\ Np)O_2$$

However, the neutron radiation caused by americium and curium requires heavy shielding. Therefore, dust-free aqueous fabrication process lines SOL-GEL techniques for the fabrication of micro spheres are applied. The micro spheres can be pressed and sintered to fuel pellets or vibrocompacted into fuel rod claddings. In addition actinide nitric solutions can be infiltrated into porous pellets or micro spheres [52,53,54,55,56].

7.9.6.1 Metallic fuel

Argonne National Laboratory (ANL) developed the injection casting refabrication method in connection with the IFR pyroprocessing of metallic U-Pu-Zr fuel. The fuel ingots from processing furnace are induction heated under vacuum and homogenized to the required enrichments. Then the injection casting system is pressurized and the molten fuel is injected into tubes (molds) which are rapidly cooled. The tubes are removed, the fuel rods cut to length and inserted into cladding tubes together with a small amount of sodium which acts as bonding between the fuel rod and the inner wall of the cladding. End caps are welded to the fuel rods, the fuel rods are assembled to fuel elements [48,49,50].

7.9.7 Intermediate storage of Curium [63]

The application of pyrochemistry does not allow to separate curium from the other actinides. Therefore, uranium, plutonium, neptunium, americium and curium do remain together for recycling of the transurania in IFRs.

The application of aqueous reprocessing technologies does allow the partitioning of all minor actinides individually (Sections 7.9.1 to 7.9.4). Curium causes high neutron radiation and heat loads during refabrication and reprocessing. Therefore, it is proposed to chemically separate curium and store it for about 200 years. This allows the isotopes Cm-243 (halflife 29 y) and Cm-244 (halflife 18 y) to decay into Pu-239 and Pu-240. (The isotope Cm-242 (hallife 163 d) decays already during reactor operation and subsequent cooling of the spent fuel). The small amounts of Cm-245 (halftime 8500 y) does remain and can be recycled later together with plutonium.

The curium nitride solutions can be infiltrated into porous beads which are subsequently calcinated and sintered. These beads can be filled into containers and stored for about 200 years, before they will be reprocessed again to separate Pu-239 and Pu-240. This treatment of the curium isotopes is necessary for aqueous chemical separation and refabrication of the actinide containing fuel in order to avoid refabrication of Cm-243 and Cm-244 fuel elements.

If curium together with the other actinides would be recycled in LWRs californium-252 will generated producing very high neutron radiation. As Cf-252 has a halflife of about 2 years such spent fuel would have to be intermediately stored until the Cf-252 will have decayed.

7.9.8 Incineration of minor actinides in nuclear reactors [58,59,60,61,62]

The minor actinides can be incinerated in LWR or FR cores or blankets. Neutron absorption reactions lead to transmutation or fission of the minor actinides.

Neutronic analyses for destruction rates of neptunium and americium have been performed. The following destruction rates were obtained if neptunium is mixed homogeneously to the fuel of PWRs or FRs. The americium containing fuel must be arranged at the periphery of the cores of either PWRs or FRs. The reason being not to change the safety coefficients to intolerable values. Table 7.10 shows typical destruction rates for neptunium and americium.

Table 7.10: Destruction rate for neptunium and americium in MOX PWRs or MOX LMFBRs [59].

	PWR 1.3 GW(e) Destruction rate kg/GW(e)·y	LMFBR 1.5 GW(e) Destruction rate kg/GW(e)·y
Neptunium	85	78
Americium	39	110

If the actinides uranium, plutonium, neptunium, americium, curium, are kept together in metallic U-Pu-An-Zr fuel the destruction rates for transurania is also about 230 kg/GWe·y [60,61,62].

References Section 7

[1] International Nuclear Fuel Cycle Evaluation, Spent Fuel Management. Report of INFCE Working Group 6. Vienna: International Atomic Energy Agency (1980).

[2] The Safe Transport of Radioactive Materials. Special Issue. IAEA Bulletin 21 (6), 2-75 (1979).

[3] Shipments of Nuclear Fuel and Waste: Are They Really Safe? Washington: US Department of Energy, DOE/EV-0004 (1977).

[4] Brennelementbehälter, http://de.wikipedia.org/wiki/Brennelementbeh%C3%A4lter (2010).

[5] Brennelement-Zwischenlager Ahaus. Hannover: Deutsche Gesellschaft für Wiederaufarbeitung von Kernbrennstoffen (DWK) (1979).

[6] Spent Fuel Storage Factbook: Facts Booklet. Washington: US Department of Energy, DOE/NE-005 (1980).

[7] Entsorgung von Kernkraftwerken, Arbeitskreis Abfallmanagement des VGB Power Tech. www.vgb.org

[8] Baumgärtner, F., Chemie der nuklearen Entsorgung. München: Karl Thiemig (1978).

[9] Benedict, M., et al., Nuclear Chemical Engineering. New York: McGraw-Hill (1981).

[10] Bericht über das in der Bundesrepublik Deutschland geplante Entsorgungszentrum für ausgediente Brennelemente aus Kernkraftwerken. Hannover: Deutsche Gesellschaft für Wiederaufarbeitung von Kernbrennstoffen (DWK) (1977).

[11] International Nuclear Fuel Cycle Evaluation, Reprocessing, Plutonium Handling, Recycle. Report of INFCE Working Group 4. Vienna: International Atomic Energy Agency (1980).

[12] Radioactive waste management, http://www.world-nuclear.org/info/info04.html

[13] Nuclear Waste Conditioning, Nuclear Energy Division Monograph, Commissariat à l'énergie atomique, Gif-sur-Yvette CEDEX, France (2009).

[14] International Nuclear Fuel Cycle Evaluation, Waste Management and Disposal. Report of INFCE Working Group 7. Vienna: International Atomic Energy Agency (1980).

[15] H12: Project to establish the scientific and technical basis for HLW disposal in Japan, Report JNC TN 1410, 2000-001, JNC, Tokai-mura (2000).

[16] Final Generic Environmental Statement on the Use of Recycle Plutonium in Mixed Oxide Fuel in Light Water Cooled Reactors (GESMO). Washington: US Nuclear Regulatory Commission, NUREG-002 (1976).

[17] Lakey, L.T., A Safety Analysis for the Nuclear Fuel Recovery and Recycling Center. Nuclear Technology 43, 213-221 (1979).

[18] Leblanc, J.M. et al., Chemical Aspects of Mixed Oxide Fuel Production. Radiochimica Acta 25, 149-152 (1978).

[19] Herbig, R., et al., Vibrocompated fuel for the Liquid Metal Reactor BOR-60, J. Nucl. Materials, 204, 93-101 (1993).

[20] Krellmann, J., Plutonium processing at the Siemens Hanau fuel fabrication plant, Nucl. Technology, 102, 18-28 (1993).

[21] Nuclear power in Russia, http://www.world-nuclear.org./info/inf45.html

[22] Status of Liquid Metal Cooled Fast Breeder Reactors, Technical Report Series 246, IAEA (1985).

[23] Bishop, J.F., et al., Fast Reactor Fuel Design and Development. In: Nuclear Power and Its Fuel Cycle, Proc. Int. Conference, Salzburg, 2-13 May 1977, Vol. 3, 377-391. Vienna: International Atomic Energy Agency (1977).

[24] International Nuclear Fuel Cycle Evaluation, Fast Breeder Reactors. Report of INFCE Working Group 5. Vienna: International Atomic Energy Agency (1980).

[25] auteron, J., Technologie du retraitement des combustibles des réacteurs rapides. In: Nuclear Power and Its Fuel Cycle, Proc. Int. Conference, Salzburg, 2-13 May 1977, Vol. 3, 633-645. Vienna: International Atomic Energy Agency (1977).

[26] Auchapt, P., et al., The French R&D Programme for Fast Reactor Fuel Reprocessing. In: Fast Reactor Fuel Reprocessing, Proc. Symposium, Dounreay, 15-18 May 1979, pp. 51-59. London: Society of Chemical Industry (1980).

[27] Allardice, R.H. et al., Fast Reactor Fuel Reprocessing in the United Kingdom. In: Nucle¬ar Power and Its Fuel Cycle, Proc. Int. Conference, Salzburg, 2-13 May 1977, Vol. 3, 615-630. Vienna : International Atomic Energy Agency (1977).

[28] Broeders, C.H.M., Investigations related to the build-up of transurania in pressurized water reactors, FZKA 5784, Forschungszentrum Karlsruhe (1996).

[29] Kessler, G., Requirements for nuclear energy in the 21st century, Progress in Nuclear Energy, 40, 3-4, 309-325 (2002).

[30] Kessler, G., et al., Wohin mit dem deutschen Plutonium?, atw 44. Jg., Heft 3, 156-164 (1999).

[31] Gompper, K., Zur Abtrennung langlebiger Nuklide, in: Radioaktivität und Kernenergie, Forschungszentrum Karlsruhe, Karlsruhe (2001).

[32] Clefs, Le future du retraitement, une synergie des procédés améliorés et d'approches nouvelles, p. 39, No. 33, CEA-France (1996).

[33] Emin, J.L. et al., AREVA NC experience of industrial scale MOX treatment in UP2-800, Proc. of GLOBAL 2009, Paris (2009).

[34] Warin, D., Minor actinide partitioning, 1st ACSEPT Int. Workshop, Lisbon (2010).

[35] Morita, Y. et al., Diisodecylphosphoric Acid, DIDPA, as an Extractant for Trans-uranium Elements; Int. Conf. on Evaluation of Emerging Nuclear Fuel Cycle Systems (Global 95), Vol. 2, p. 1163, Versailles, France (1995).

[36] Schulz, W.W., The TRUEX Process and the management of Liquid TRU-Waste, Sepa-ration Science and Technology, 23, pp. 1355-1372 (1988).

[37] Zhu, Y., The Removal of Actinides from High Level Radioactive Waste by TRPO Ex-traction. The Extraction of Americium and some Lanthanides, Chinese J. Nucl. Sci. Eng., Vol. 9, pp. 141-150 (1989).

[38] Madic, C. et al., High Level Liquid Waste Partitioning by Means of Completely In-cinerable Extractants, Report EUR-18038 (1998).

[39] Poinssot, Ch. et al., The CEA Atalante facility: a laboratory for fuel cycle R&D, Proc. of Global 2009, Paris (2009).

[40] Kubota, M., Development of the Four Group Partitioning Process at JAERI, in: Proc. of the 5th Int. Information Exchange Meeting, Mol, Belgium, 1998, p. 131, OECD (1999).

[41] Zhu, Y., The Separation of Americium from Light Lanthanides by CYANEX 301 Ex-traction, Radiochimica Acta, Vol. 68, pp. 95-98 (1995).

[42] Vitorge, P., Complexation de Lanthanides et d'Actinides Trivalents par la TRIPYRIDYL-TRIAZINE, Applications en Extraction Liquide-Liquide, CEA-R-5270 (1984).

[43] Kolarik, Z., et al., Selective extraction of Am(III) over Eu(III) by 2,6-ditriazolyl and 2,6-ditriazinylpyridines, Solvent Extr. and Ion Exch. 17, no. 1, pp. 23-32 (1999).

[44] Welden, A. et al., 1-cycle SANEX process development studies performed at Forschungszentrum Jülich, 1st ACSEPT Int. Workshop, Lisbon (2010).

[45] Modolo, G. et al., Selective separation of americium III from curium III, californium III and lanthanides III by the LUCA process, Proc. of the GLOBAL 2009, Paris (2009).

[46] Chang, Y.I., The integral fast reactor, Nucl. Techn., 88, 129 (1989).

[47] Stevenson, C.E., The EBR-II fuel cycle story, American Nuclear Society, LaGrange Park, Illinois (1987).

[48] Burris, L.R. et al., The application of electrorefining for recovery and purification of fuel discharged from the Integral Fast Reactor, AIChE Symposium Series No. 254, 83, p. 135 (1987).

[49] Laidler, J.J. et al., Development of pyroprocessing technology, Progress in Nuclear Energy, Vol. 31, No. 1/2, p. 131 (1997).

[50] Todd, T.A., The US fuel cycle research and development program: Separations research and development, First ACSEPT Int. Workshop, Lisbon (2009).

[51] Bychkov, A.V. et al., Pyrochemical reprocessing of irradiated FBR MOX fuel III. Experiment on high burn-up-fuel of the BOR-60-reactor. Int. Conf. on Future Nuclear Systems (Global 97), Yokohama (Japan), Vol. 2, p. 12 (1997).

[52] Bychkov, A.V. et al., RIAR experimental base development concept: Multi-purpose pyrochemical complex for experimental justification of innovative closed fuel cycle technologies, Proc. of GLOBAL 2009, Paris (2009).

[53] Fernandez, A. et al., Fuels and targets for incineration and transmutation of actinides: the ITU programme. ATALANTE 2000, Int. Conf.: The Nuclear Fuel Cycle on the Back-End of the Fuel Cycle for the 21st Century, Avignon (2000).

[54] Herbig, R. et al., Vibrocompacted fuel for the Liquid Metal Reactor BOR-60, J. Nucl. Materials, 204, 93 (1993).

[55] Pouchon, M.A., Conversion process: Internal gelation and the sphere-pac concept. 1st ACSEPT Int. Workshop, Lisbon (2010).

[56] Deptula, A. et al., Synthesis of uranium oxides by complex SOL-GEL processes (CSGP), 1st ACSEPT Int. Workshop, Lisbon (2010).

[57] Konings, R.J.M. et al., Transmutation of actinides in inert-matrix fuels: fabrication studies and modelling of fuel behaviour, Journal of Nuclear Materials, 274, p. 84 (1999).

[58] Tommasi, J. et al., Long-lived waste transmutation in reactors, Nucl. Techn., 111, p. 133 (1995).

[59] Messaoudi, N. et al., Fast burner reactor devoted to minor actinde incineration, Nucl. Techn., 137, 84 (2002).

[60] Status and assessment report on actinide and fission product partitioning and transmutation, OECD/NEA (1999). http://www.nea.fr/html/pt/pubdocs.htm

[61] Physics and safety of transmutation systems, A status report. OECD/NEA report No. 6090 (2006).

[62] Salvatores, M., Nuclear fuel cycle strategies including partitioning and transmutation, Nucl. Eng. and Design, 325, p. 805 (2005).

[63] Pillon, S. et al., Aspects of fabrication of curium-based fuels and targets, Journal of Nucl. Materials, 320, pp. 32-40 (2003).

8 The IAEA Safeguards System

INFCIRC/153 concentrates on the surveillance of nuclear material: "The objective of safeguards is the timely detection of diversion of significant quantities of nuclear material from peaceful nuclear activities to the manufacture of nuclear weapons or of other nuclear explosive devices or for purposes unknown, and deterrence of such diversion by the risk of early detection." (INFCIRC/153, Article 28) [1].

For practical purposes, a factor much more important than deterrence to the countries involved is the finding by IAEA that no diversion has taken place, the so-called assurance of non-diversion. Often this is the condition for a state being supplied with nuclear material, representing a general confidence building measure.

In order to document that no nuclear material was diverted, states bind themselves to build up national material accountancy systems, in which nuclear materials inventories are continuously recorded by accountancy and measurements. Inventories and inventory changes are reported to IAEA in a system of records and reports.

For this purpose, the signatory country to the NPT [2] first of all reports to IAEA design information about all nuclear facilities on its territory. Next, the material balance areas and the types of balancing and reporting are defined in facility attachments for each plant. IAEA verifies the information provided by plant operators by inspections and independent measurements in a defined maximum scope. If there are discrepancies, attempts are made to settle them. The system is to prevent and discover, respectively, diversions of nuclear materials by a country. Every state is obliged to establish a physical protection system so as to prevent diversion of nuclear material at a sub-national level. Recommendations for the physical protection of nuclear material are given by IAEA in document INFCIRC/225 Rev. 4 [3].

8.1 Material Balance Measurements

Nuclear materials often come in forms that cannot be counted or measured exactly, e.g. powders, liquids, metal pieces, scraps etc.. On the other hand the fissile inventories of fuel rods and fuel assemblies can only be measured with certain accuracy, i.e. there exists some unavoidable measurements accuracy.

For information to be generated about the nuclear material inventory in a plant, the initial inventory and the quantities coming in and going out must be known. This requires stocktaking of the real inventory and continuous measurement of all incoming and out going nuclear materials streams. From time to time new inventories will have to be determined. If a new inventory value is within the error limits of material balance measurements, it is stated that all material is still present. However, such statement implies an uncertainty, which will be explained in more detail below. The material balance is described by

$$I_2 = \Sigma M_i - \Sigma M_o + I_1 + MUF$$

with: I_1 = initial inventory
I_2 = inventory measured after time t
ΣM_i = sum of material inputs after time t
ΣM_o = sum of material outputs after time t
MUF = *m*aterial *u*naccounted *f*or.

MUF includes measuring errors, process losses and, if applicable, also diverted material [1,4]. If MUF exceeds the measuring error, the IAEA inspector can state a loss of material. However, such diversion may have been due only to a particularly large statistical variation and no actual losses may have occurred. In that case, the IAEA inspector's statement would be a so-called "false alarm". On the other hand, despite a loss of material, statistical variations may have made MUF so small as to leave the loss undiscovered.

The relations connecting detection probability, rate of false alarms, detectable quantity and measuring accuracy are explained in Fig. 8.1. If a quantity M_o is present, the probability P(M), of obtaining the quantity M in a measurement is given by

$$P(M) = \frac{1}{\sqrt{2\pi \cdot \sigma}} \cdot e^{\frac{(M-M_o)^2}{2\pi \cdot \sigma^2}}$$

with

σ = standard deviation, if a Gaussian distribution is assumed for the measurement errors.

With 90% probability, M will be in the range of $M_o \pm 1.65\sigma$. If M<M', with $M' = M_o - 1.65\sigma$, an alarm is initiated in the example given, i.e. a discrepancy between the operator's data and the measurement is reported. In five percent of the cases, this may be a false alarm due to statistical variations.

If there is a loss of material such that actually only a quantity M_1 is available, with $M_o - M_1 = 3.3\sigma$, a loss is discovered with 95% probability. The detection probability is $1-ß = 95\%$, $ß = 5\%$. Losses of $M_o - M \geqq 3.3\sigma$ are discovered with $p \geqq 95\%$ probability at a false alarm rate of 5%. If the false alarm rate is reduced to 0.14% with a detection probability of 95%, detectable losses are $M_0 - M \geqq 4.65\sigma$ (Figure 8.1). This likewise applies to an improvement in detection probability.

The criteria to be met by safeguards systems thus are defined by

(1) the missing amount to be discovered (significant quantity),
(2) the detection time (timely detection),
(3) the detection probability for a significant amount,
(4) the permissible rate of false alarms.

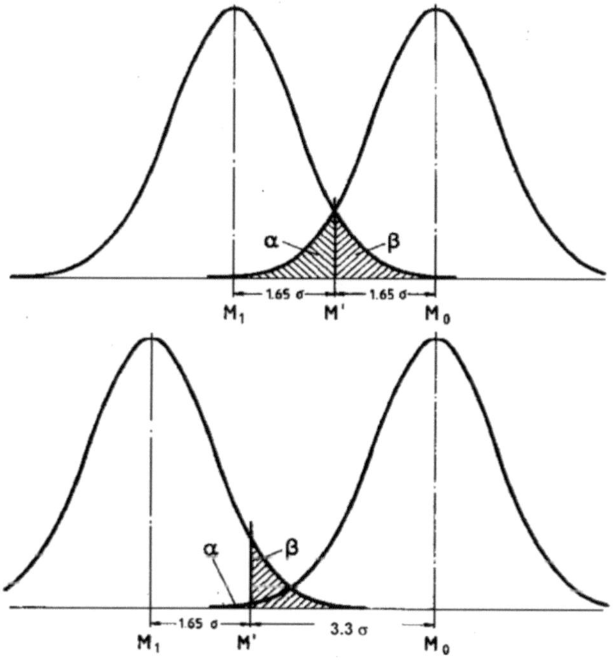

Figure 8.1: Relations between detection probability, rate of false alarms, detectable quantity and measuring accuracy.

8.1.1 Significant quantities of fissile materials and timely detection

The "significant quantity" and "timely detection" concepts have been quantified in the course of the implementation of IAEA safeguards agreements. The Standing Advisory Group on Safeguards Implementation (SAGSI) to IAEA has confirmed values for "significant quantities," which are indicated in Table 8.1. They are derived from so-called threshold amounts of special fissile material, which are defined as approximate quantities needed for nuclear explosive devices.

Table 8.1: Quantities of nuclear material of safeguards significance (IAEA).

Material	Quantity of safeguards significance	Threshold amounts	
"Direct use" material			
Pu (<80% Pu-238)	8 kg	Pu (>95% Pu-239)	8 kg
U-233	8 kg	U-233	8 kg
Uranium (≧20% U-235)	25 kg	U(>90-95% U-235)	25 kg
(Plus rules for mixtures where appropriate)			
"Indirect use" material			
Uranium (<20% U-235)	75 kg		
Natural uranium	10 t		
Depleted uranium	20 t		
Thorium	20 t		
(Plus rules for mixtures where appropriate)			

The IAEA has taken a very conservative approach and considers all plutonium capable of being used for nuclear weapons. The only exception is plutonium with ≥80% Pu-238 being not considered to be useable for nuclear weapons due to the large heat generation by the Pu-238 alpha decay.

Hence the IAEA does not distinguish between reactor-grade or weapon-grade plutonium in its safeguards efforts. However, the IAEA does care about uranium isotopic composition focusing on fissile U-233 and U-235. Highly enriched uranium (HEU) is uranium with ≥20% U-235 or ≥12% U-233 [4].

The "detection time" should correspond, in orders of magnitude, to the "conversion time" for fissile material. The "conversion time" is defined as the minimum time required converting different chemical forms of nuclear material to the metallic components of a nuclear explosive device. The "detection time" is defined as the maximum time which may elapse between a diversion and its detection by IAEA safeguards. Table 8.2 indicates the conversion times. IAEA uses these values as guidelines while additional practical experience will be acquired in implementing safeguards in different fuel cycle facilities.

Table 8.2: Estimated material conversion times (IAEA).

Material classi-fication	Beginning material form	End process form	Estimated conversion time	IAEA timely detection
1	Pu, HEU, or U-233 metal	Finished plutonium or uranium metal components	Order of days (7-10)	one month
2	PuO_2, $Pu(NO_3)_4$ or other compounds. HEU or U-233 oxide or other pure compounds. MOX or other **non-irradiated** pure mixtures of Pu or U (U-233 + U-235) >20%. PU, HEU and/or U-233 in scrap or other miscellaneous impure compounds.	Finished plutonium or uranium metal components	Order of weeks (1-3)	one month
3	Pu, HEU or U-233 in **irradiated** fuels	Finished plutonium or uranium metal components	Order of months (1-3)	3 months
4	U containing <20% U-235 and U-233, thorium		Order of one year	one year

With these preliminary guidelines on "timely detection" of diversions of "significant quantities" of nuclear material, IAEA strives for a safeguards system, which has a high probability of meeting these goals. There is a growing tendency, however, not to base the credibility of a safeguards system primarily on the degree to which these goals are met but to include many other factors which confirm compliance of a state with the NPT requirements.

The a priori probability of detection sought is usually 90% or higher and most often 95%. Since a statement of diversion represents a grave incident, the rate of false alarms must be very low.

In addition to the "abrupt diversion" case, to which the "detection time" applies, a case must also be considered of somebody constantly trying to divert small quantities so that, e.g., after one year, a "significant amount" will have accumulated (protracted diversion). Discovering such action imposes the most stringent requirements on the accuracy of the measurements.

In order to ensure that really all material has been considered in a material balance, containment and surveillance measures are adopted. That is to say, containers and fuel elements are sealed and transport processes are surveyed optically. Seals can also be used to reduce the measuring expenditure in inventory taking. Since sealed containers with undamaged seals still contain their material, no new quantitative assay is necessary.

8.1.2 Methods of Safeguards Techniques

Implementing the safeguards concept on the basis of measurements and material balances requires the existence of suitable measuring techniques. As a consequence, a worldwide effort was initiated around 1968 to develop methods of measurement, measuring equipment, seals and cameras. It was also tried, in a parallel effort, to cast into more precise terms the basic theoretical principles of the system, verify its applicability to specific plants, improve the mathematical methods of data analysis, and develop systems for handling the large amounts of data.

Some reactors require additional safeguards measures. For instance, pebble bed reactors do not permit fuel element identification, so active neutron interrogation is used for nuclear materials assays. Similarly, in fast breeders, fuel elements will be under opaque coolants like sodium or lead-bismuth and thus not accessible for direct identification. Passive neutron counting during fuel element transportation in the reactor area and sealing of the core fuel elements are proposed here. Ultrasonic viewing to confirm seals of the fuel elements is required.

8.1.3 Material Balance Areas (MBAs)

For safeguarding such facilities as enrichment plants, fuel fabrication plants, nuclear reactors and reprocessing plants, where the nuclear material is present in an unsealed form, the concept of material balance areas is applied.

8.1.3.1 Material Balance Area for a Light Water Reactor

The Light Water Reactor (LWR) (Figure 8.2), unlike more complicated reprocessing or enrichment plants, has only one MBA. A LWR will have the following key measuring points (KMPs):

- arrival (receipt) of fresh low enriched fuel
- loading of fresh fuel to the core
- discharge of spent fuel from the core
- transfer to spent fuel pool
- transfer from spent fuel pool to dry storage or reprocessing.

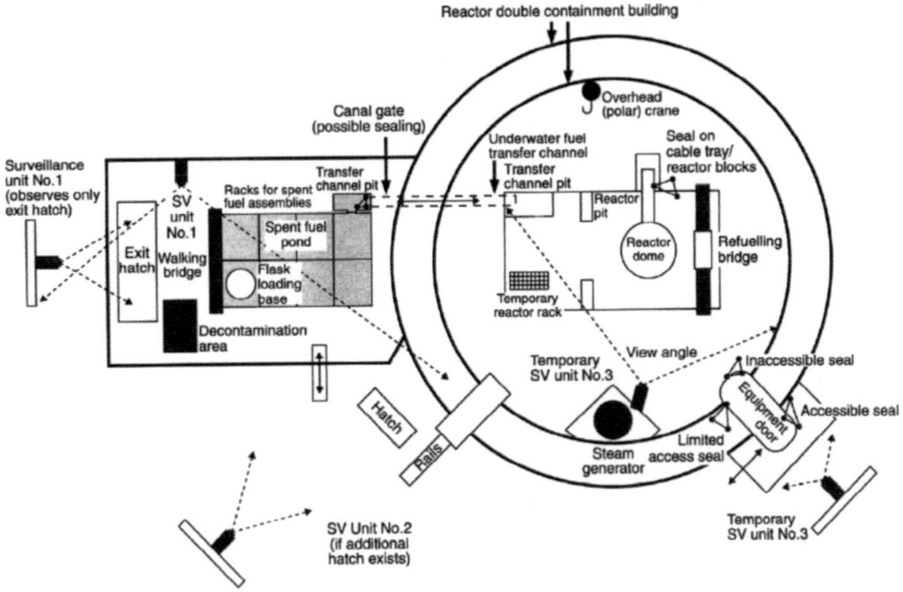

Figure 8.2: Material balance area of a Light Water Reactor.

8.1.3.2 Material Balance Areas for a Reprocessing Plant

Reprocessing facilities are e.g. typically divided into three material balance areas, which include the spent fuel storage area, the process area, and the product storage area (Figure 8.3). In the spent fuel storage area (MBA1), the spent fuel elements arriving from the nuclear reactor plant can be verified. However, their fissile material content can be determined by non-destructive techniques only to an accuracy of several percent. A better quantitative analysis is possible in the second material balance area (MBA2), which includes the whole chemical process area. Finally, MBA3 contains the storage area for the purified end product. For MBAs 1 and 3, the statement that no material has been diverted can be based on containment and surveillance measures, such as seals and cameras. For MBA2, the difference between output and input has to be measured and compared with the inventory change. This requires inventory taking in the whole process area, which includes a washout of the process equipment; this is time consuming and expensive. Thus, only one or two inventory takings per annum are acceptable from the plant operator's point of view. This may conflict with the timeliness of detection.

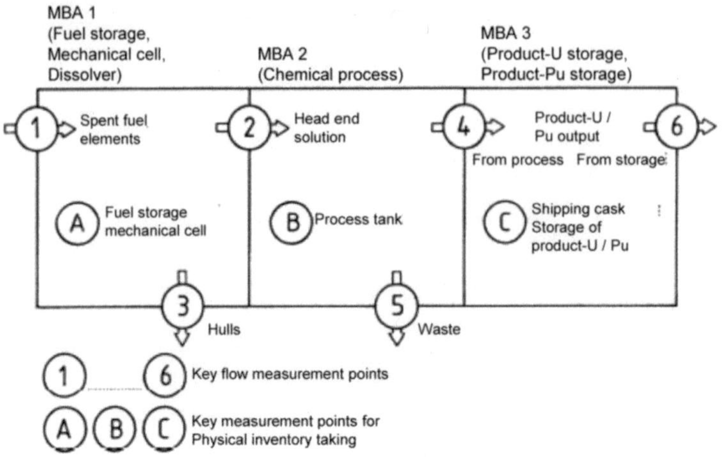

Figure 8.3: Material balance area and key measuring points, e.g. in a reprocessing facility.

No problem exists in small pilot size plants, where inventory changes remain below one significant amount of nuclear material. Typically, also the MUF accumulated during one year is less than a significant amount (Table 8.1).

This is different in large commercial size reprocessing plants. Let us assume, for illustration, a plant with an annual throughput of 800 tonnes of heavy metal, of which roughly 0.9% is plutonium. The plutonium throughput then would be 7.2 t per annum. The largest error in the material flow measurement is attributable to the input accountancy tank, where the best values achievable are about $2\ \sigma = \pm 1\%$, including sampling, chemical analysis, and volumetric assay (Section 8.4). This leads to an accumulated 3σ-error during one year of ±246 kg of plutonium, disregarding other sources of error. Getting the accuracy down to below 8 kg of plutonium would require taking inventories thirty times a year, which is economically unacceptable. A significant improvement in input accountability measurement does not seem to be feasible in the near term and, moreover, would not solve the problem with respect to the many other sources of error [5,6,7,8].

Similar problems may be encountered in large fuel fabrication plants with plutonium bearing fuel, or in large enrichment plants. Advanced safeguards concepts must therefore be applied.

8.1.4 Advanced Safeguards Approaches

8.1.4.1 Near-Real Time Accountancy and Extended Containment/Surveillance Systems

Two different approaches are applied to solve the problems outlined above: near-real time accountancy (NRTA) and containment surveillance (C/S) systems. In NRTA, inventory taking at the expense of a complete halt of the process is replaced by an essentially continuous determination of the plant inventory from readings of the process instruments, such as level indicators of vessels, densitometers or flow meters, combined with periodic analyses of samples from the vessels. Although these measurements may not be as accurate as those normally used for material balance purposes, they do solve the problem of abrupt diversion and they greatly facilitate the detection of protracted diversion.

In addition, the large number of measurements conducted throughout a year allows trends to be discovered, which are well below the measuring error, provided that sophisticated statistical methods are applied. This helps in detecting protracted diversion. Problems are posed by the fairly large inspection effort required, the limited ability of the inspector to verify the origins or the samples received for chemical analysis, and the detailed insight the inspector obtains into plant operation (problem of intrusion) [5,6,7,8,9].

A totally different approach is the so-called "extended containment and surveillance concept" where the inspection effort is concentrated on measurements and surveillance at the periphery of a facility, which has the character of a containment. In the pure form of this concept, the statement by the inspector that no nuclear material has been diverted is no longer based on a material balance and on verification of the presence of the material, but on the fact that no diversion of nuclear material across the periphery of the facility has been detected.

This system requires no overly accurate input accountability measurements, since the material in the main process stream can be sealed and traced as it leaves the plant. Only the waste streams, which contain about 1% of the nuclear material, have to be assayed. This greatly relaxes the accuracy requirements for the measurements. On the other hand, all inputs and outputs to the plant, including the inactive ones, must be controlled. Extensive uses of cameras and doorway monitors are characteristic of this approach. If the monitor system of a passage fails, even for a short time, the inspector is unable to assess whether nuclear material has been diverted, and how much. In addition, the inspector must ensure that there are no clandestine exits that could have been used to divert nuclear material. The system has two advantages: It can be used in parallel, as a national physical protection system to detect theft on a subnational basis, and it fulfils the requirement of non-intrusiveness. However, some combination with materials accountancy is said to be needed, where the C/S-system covers the abrupt diversion aspect.

Even if advanced safeguards systems succeed in meeting the requirement of timely detection of significant amounts of nuclear material, there still remains political concern about large scale nuclear fuel cycle plants in NNWSs (see Sections 8.4, 8.7 and 8.8). There is the possibility of a state suddenly quitting the NPT and immediately possessing large amounts of nuclear material for weapons use.

Various measures have been proposed to reduce this risk, including international plutonium storage -or proliferation– proof nuclear fuel cycles [10,29,32] (Sections 10 to 14).

8.1.5 Safeguards Measurement Technologies

The main measurement technologies used for safeguards surveillance are destructive analysis (DA) and non-destructive analysis (NDA). DA refers to chemistry, mass spectrometry etc. which can be done only in specific laboratories. These methods usually achieve the highest accuracy. NDA measures entire items which can no more be destructed for assay. These methods are applied for process control, fuel element assay etc.. The typical NDA methods are gamma-ray measurements or neutron assay methods.

8.1.5.1 Destructive analysis (DA)

DA refers, e.g. to weight and volume measurements, chemical analysis and mass spectrometry. It requires a specific laboratory to receive and analyze the samples of nuclear materials [11,12]:

- Gravimetry consists of very precise weighing of plutonium or uranium which has been oxidized to PuO_2 or U_3O_8. Non-volatile impurities are determined spectrographically and the gravimetrical results are corrected accordingly. The precision of such measurements can have a relative standard deviation of ±0.05%.
- Chemical analysis applies reduction-oxidation titration procedures. Chemical tritration measures electrical properties of solutions containing a compound that undergoes a chemical reaction, while precisely measured amounts of another chemical are added. The IAEA does such chemical analysis in their Seibersdorf (Austria) laboratories with a precision of ±0.02% for uranium and ±0.04% for plutonium.
- Mass spectrometry measures the mass of ionized particles passing through a magnetic field. It allows the determination of the isotopic composition of uranium and plutonium samples. The U-235 enrichment can be analyzed to a precision of ±0.014% and the Pu-239 enrichment to ±0.02%.
- Laser induced break down spectroscopy (LIBS) focuses a pulsed laser on a small sample of material. This creates a plasma and the resulting excited atoms and ions emit light at very precise wavelengths. This light is transmitted through an optical fiber to a grating spectrometer. The accuracy can be ±0.1% for uranium or plutonium.

- Alpha spectrometry and isotope-dilution alpha spectrometry measures the radioactive alpha particles emitted by plutonium. It is possible to accurately determine the total plutonium concentration in dissolver solutions if the plutonium isotopic composition is known from other measurements. A well known Pu-242 spike is added to the plutonium sample and this is then deposited on a filament. The ratio of the Pu isotopes to the spike allows calculating the plutonium concentration with an accuracy of ±0.2%.
- K-edge absorption densitometry measures the absorption of X-rays generated by Co-57 and Se-75 whose energies are close to the point at which plutonium absorbs X-rays most strongly (e.g. at around 110 to 120 keV). It is applied for plutonium concentrations in input solutions and process solutions (inline measurements). An accuracy of ±0.4% can be attained.
- X-ray fluorescence measures well characterized emissions from various elements when they are stimulated by X-rays. This measurement technique has µg-detection limits and measures the amounts of each element present (but not individual isotopes). It is primarily used as online instrumentation for process solutions for accurate determination of the Pu/U ratio. It is often used in combination with K-edge measurements to determine Pu/U ratio in low concentration solutions, e.g. Pu-dissolver solutions.
- Spectrometry determines the plutonium concentration of a solution by measuring light transmitted through it at a wavelength which is absorbed by plutonium. This method is widely used for process control and material accountancy at all stages of the process.

Verification by IAEA consists in taking and analyzing samples independently in order to verify the data furnished by an operator. Often this is done by random sampling. Weight measurements (e.g., in fuel rods) can be checked in principle by having the inspector bring along a standard only known to him and having it weighed. Volumetric assays can be conducted by adding to the solution a "spike" of the inspector and subsequently assaying for the concentration.

The IAEA inspector can also collect small material samples and ship them to the IAEA-Seibersdorf analytical laboratory for precise measurement. As an alternative he also can send samples to laboratories located in IAEA member states which have been certified by the IAEA.

8.1.5.2 NDA techniques measuring spontaneous or stimulated radiation.

NDA techniques have been developed to allow inspectors to determine the contents of nuclear material in a final product, e.g. fuel elements. Another application of NDA techniques is in containers with scrap and waste, which must be included in a material balance [11,12].

The radiation emitted by the nuclear material is used for measurement. Only neutral particles, i.e., gamma-rays, neutrons or anti-neutrinos, have the necessary penetrating capability. The easiest way is to use the characteristic radiation of the material investigated. Figure 8.4 shows, as an example, the gamma-ray signature, measured by gamma-ray detectors for plutonium isotopes, neptunium, americium and tin (Sb-125).

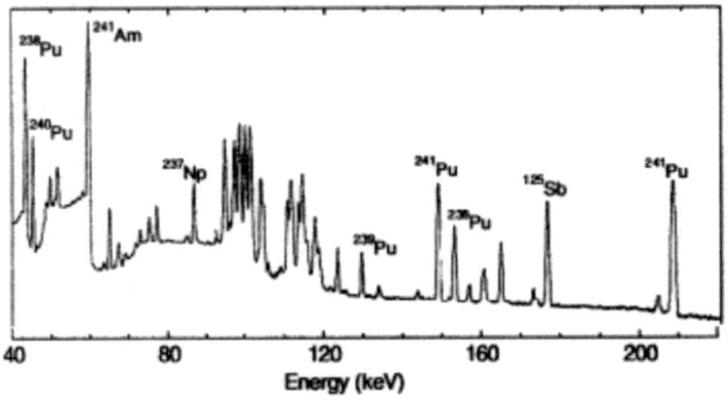

Figure 8.4: High resolution gamma spectra of Sb, Np-, Pu- and Am-fraction [13].

Table 8.3 shows the major gamma-ray signatures for uranium, plutonium and americium isotopes. The most important one for U-235 is the 185.7 keV gamma ray signature. Using high resolution Ge-detectors or CdZnTe- and LaBr3-detectors it is possible to measure the complete isotopic composition of uranium (U-232, U-233, U-234, U-235, U-236 and U-238). The uranium isotope U-235 can be determined with accuracy better than ±2% and with special efforts even with ±0.2%.

Similar measurement capabilities are possible for the plutonium isotopes using the gamma-ray signatures shown by Table 8.3. However, high resolution Ge- or CdTe-detectors cooled by liquid nitrogen to 77 K must be used. In addition, computer programmes must be applied [11]. The most important gamma-ray signatures for Pu-239 are at 129.28 and 413.69 keV. Pu-242 is essentially an emitter of α–particles. Figure 8.5 shows a multichannel analyzer system with Ge-detector used by IAEA.

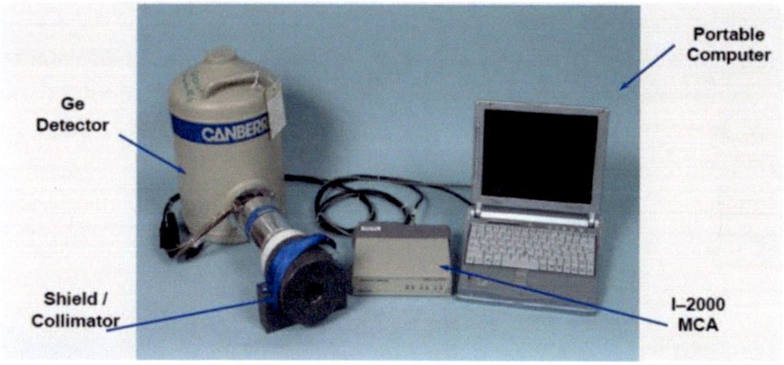

Figure 8.5: Multi-channel analyzer with Ge-detector used by IAEA [30,31,34].

Table 8.3: Major gamma-ray signatures for different isotopes [11,14,15,16,17].

Isotope	Energy (keV)	Intensity (1/g·s)
U-232	57.8	1.6×10^9
U-233	54.6 317.2 261.4	6.4×10^4 8.3×10^4 10^5
U-234	53.2	2.8×10^5
U-235	185.72	4.3×10^4
U-236	49.2	1.9×10^3
U-238	1001.10	10^7
Np-237	312.2	10^7
Pu-238	43.5 99.9 152.77 766.40	2.5×10^8 4.6×10^7 6.5×10^6 1.5×10^5
Pu-239	129.28 375.00 413.69	1.4×10^5 3.6×10^4 3.4×10^4
Pu-240	45.2 100.4 160.3	3.8×10^6 3.4×10^4 3.5×10^4
Pu-241	148.60 207.98	7.5×10^6 2.0×10^7
Am-241	59.54	4.6×10^{10}

A problem is encountered when strong gamma ray attenuation is present. Table 8.4 shows material thicknesses in cm in which the gamma-rays mentioned above are attenuated to e^{-1}. A number of sophisticated techniques have been developed to determine the gamma-ray attenuation of homogeneous or nearly homogeneous materials. Nevertheless, the main areas of application of passive gamma counting are uranium and plutonium enrichment, liquid samples of plutonium-uranium solutions in reprocessing plants and low density waste.

Table 8.4: Gamma-ray attenuation in various materials to e^{-1} [7].

Gamma ray energies	186 keV	414 keV
Materials	Attenuation distance (cm)	
Low density waste	7.7	≈105
H_2O	7.1	9.6
Al	3.0	4.07
Fe	0.84	1.4
UO_2	0.065	0.40

8.1.5.3 Neutron assay

Fast neutrons have a much higher penetration capability than gamma-rays. They are emitted during spontaneous fission of many heavy nuclei. Table 8.5 lists neutron emission rates by spontaneous fission and (α,n)-processes in oxygen (UOX or MOX fuel) of the most important isotopes. If the isotopic composition of a material is known, the amount of that material in a sample can be determined by neutron counting, provided the system has been calibrated. Neutron assay and gamma-ray measurements can, therefore, be complementary, because the determination of the isotopic composition can be obtained from gamma-ray measurements.

The neutron yield of (α,n)-processes (by reaction with light elements, e.g. beryllium, boron, oxygen, fluorine) depends on many details and therefore is not well suited as a signature for plutonium or uranium. To eliminate the (α,n)-contribution, use is made of the fact that spontaneous fission neutrons are mostly emitted in pairs, whereas only one neutron is emitted in the (α,n)-reaction. Thus, a coincidence logic is applied, which only counts the neutron pairs (spontaneous fission neutrons) and suppresses the (α,n)-neutrons. Figure 8.6 shows a neutron coincidence counter used by IAEA.

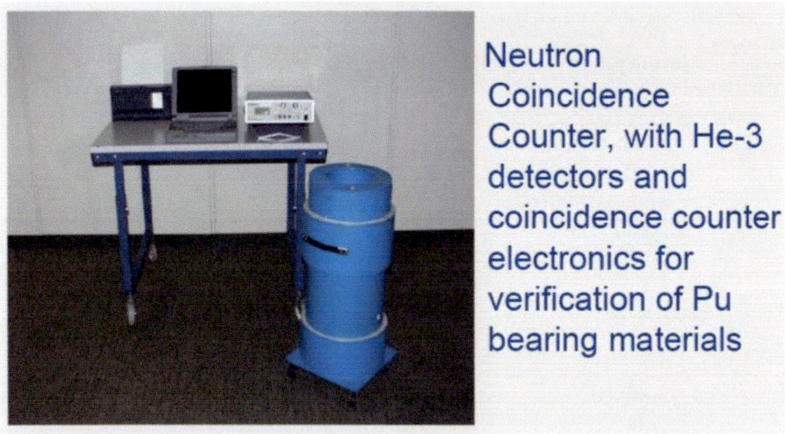

Figure 8.6: Neutron coincidence counter used by IAEA [30,31,34].

Neutron coincidence counting is widely used for assaying of plutonium contaminated waste. Its main drawback is the fact that primarily the isotopes Pu-240 or Pu-242 are measured (highest spontaneous fission neutron emission rate). The uncertainty in isotopic composition limits the achievable accuracy of plutonium assays. For uranium or neptunium, the spontaneous fission neutron emission rate is too low. In addition, even small amounts of curium render the method inapplicable, since the spontaneous fission neutron emission rate of curium is four orders of magnitude higher than that of plutonium.

If no "passive" assay technique is applicable, a so-called "active interrogation" has to be applied. For this method, fissions are induced in the material of interest by an external neutron source and the neutrons or gamma-rays released in the fission process are detected. It is then necessary to discriminate between the induced fission neutrons and the source neutrons. Several methods are available:

(1) Discrimination by energy. When the energy of the source neutrons is below the threshold of the detector, only the induced fission neutrons are counted.
(2) Discrimination by time. Delayed neutrons are counted in the time interval when the source is switched off.
(3) Discrimination by multiplicity. The high multiplicity of gamma and neutron emissions in the fission process is used in a manifold coincidence measurement.

Table 8.5: Neutron emission rates for spontaneous fission of isotopes [11, 14-24].

Isotope	Halflife $T_{1/2}$(a) spont.fiss.	$\overline{\nu}$ spont. fiss.	Neutron emission rates	
			(n/g·s) spont.fiss.	(n/g·s) (α,n)reaction
U-232	$<6.8\cdot10^{15}$	1.71	1.3·	$2.5\cdot10^{4}$
U-233	$>3\cdot10^{17}$	1.76	$<1.9\cdot10^{-4}$	11
U-234	$1.5\cdot10^{16}$	1.81	$6.8\cdot10^{-3}$	7
U-235	$1.8\cdot10^{17}$	1.88	$5.9\cdot10^{-4}$	$2.5\cdot10^{-3}$
U-236	$2.5\cdot10^{16}$	1.91	$4.2\cdot10^{-2}$	$7\cdot10^{-2}$
U-238	$9.86\cdot10^{15}$	2.00	$1.1\cdot10^{-2}$	$3.9\cdot10^{-4}$
Np-237	10^{18}	2.15	$1.1\cdot10^{-4}$	0.8
Pu-238	$5.0\cdot10^{10}$	2.21	$2.59\cdot10^{3}$	$2.0\cdot10^{4}$
Pu-239	$5.5\cdot10^{15}$	2.30	$2.18\cdot10^{-2}$	72
Pu-240	$1.32\cdot10^{11}$	2.15	$1.02\cdot10^{3}$	265
Pu-241	$6\cdot10^{16}$	2.25	$5\cdot10^{-2}$	
Pu-242	$7.0\cdot10^{10}$	2.14	$1.72\cdot10^{3}$	4.4
Am-241	$1.15\cdot10^{14}$	2.45	1.2	$4\cdot10^{3}$
Cm-242	$6.6\cdot10^{6}$	2.51	$2.08\cdot10^{7}$	$3.9\cdot10^{6}$
Cm-244	$1.27\cdot10^{7}$	2.68	$1.15\cdot10^{7}$	$9.5\cdot10^{4}$

A large number of such active interrogation techniques and instruments have been proposed and developed [11]: An oscillating Cf source with delayed neutron counting was developed as an active interrogation for waste drums containing more than 1 mg of U-235 or transuranic waste (differential die-away technique). For the latter also a pulsed 14 MeV neutron generator can be used. Detection limits of a few mg to a few 10 mg of Pu-239 or U-235 are attained which satisfy the existing waste disposal criteria [24]. Also a combined thermal-epithermal neutron interrogation system was developed for better matrix penetrability of samples in waste assay [25].

8.1.5.4 Calorimetry

Calorimetry [11,26] is used as the most accurate NDA method with measurement accuracies of ±0.5 to ±1%. Calorimetry takes advantage of the heat produced by isotopes like Pu-238, reactor-grade plutonium, Am-241 or Tritium (Table 8.6). However, the calorimetric measurements must be combined with an isotopic analysis (mass-spectrometry or gamma-ray measurements). Calorimetric measurements are time consuming, typically several hours up to one day. They can only be made in specific laboratories [26]. More recent calorimetry developments are based on solid state sensors.

Table 8.6: Specific thermal power in mW/g for different isotopes or mixtures [26].

Isotope	Specific thermal power (mW/g)
U-232	680
Pu-238	5708
Tritium	324
Am-241	114
weapon-grade Pu	3.25
Pu (30 GWd/t)* reactor-grade	12.1
Pu (50 GWd/t)* reactor-grade	19.0
Pu (60 GWd/t)* reactor-grade	24.6
Pu (70 GWd/t)* reactor-grade	31.47
U-233	0.28
Np-237	0.032
HEU 93% U-235	$2x10^{-3}$

*burnup of spent fuel

8.1.6 Containment and surveillance methods [27,34]

In the majority of facilities to be inspected, the nuclear material is found in the form of separate items only. Item counting, serial number identification and surveillance by visual inspection or cameras are adequate safeguards measures in such cases. This has greatly facilitated the implementation of safeguards.

8.1.6.1 Ultrasonic seals for fuel elements

A promising technique is the identification of randomly distributed discontinuities in a matrix by ultrasonic wave propagation and echo registration. The characteristics of these echoes vary with each variation of the discontinuity position in the matrix. Such a matrix forms part of a cap seal e.g. for LWR fuel bundles. It has to be destroyed when the seal would be removed. The seal may contain a fuel element identification known only to the inspector.

8.1.6.2 Electronic seals

Electronic seals may be applied to spent fuel transport and storage casks, concrete lids of power reactors, doors and gates of storage areas, piping, and containers with special nuclear material.

The sealing function is realized by using a fibre-optic cable (FOC) or, alternatively, an electrical wire. The FOC or wire is looped through the locking mechanism of the container or site to be sealed. The two ends are attached to the seal body. The fibre-optic concept has a light source (i.e., optical transmitter) and a light sensor (i.e., optical receiver) with the light being transmitted through an external FOC of practically arbitrary length. In the wire concept the electrical current is monitored as well as the resistance of the wire. Both concepts are designed for multiple connection and disconnection of the FOC or wire, i.e., "closing and opening of the seal".

The VAriable COding Sealing System (VACOSS) consists of a seal body containing the electronic circuitry and battery, a fibre optic cable, and an interface box to provide communication between the seal and the reader.

The Integrable Re-usable Electronic Seal (IRES) can be used with both a FOC and an electrical wire, while the seal detects whether it is being used with a FOC or a wire. The FOC is a multimode cable and the light source is a light emitting diode (LED) emitting random frames (8 bits) every 500 μs in the infrared range. Communication with the seal takes place via a serial interface or wireless link (radio module).

The Electronic Optical Sealing System (EOSS) is used with a FOC only. The FOC is a single mode fibre, and the light source is a laser. The open/closed status of the FOC is monitored by transmitting and receiving short light pulses at certain time intervals. Communication with the seal takes place via a serial interface.

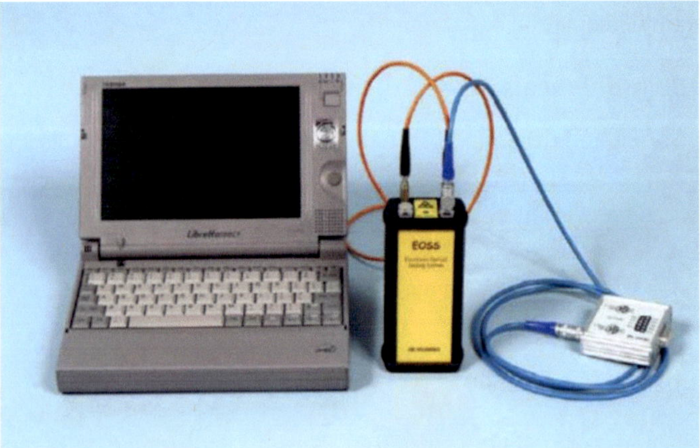

Figure 8.7: Electronic optical sealing system with fibre-optic cable [27,30,34].

8.1.6.3 Laser technology based safeguards survey [34]

Laser technologies are becoming increasingly important, providing new and novel verification and detection tools for current and future safeguards activities. Three-dimensional (3D) laser imaging in nuclear facilities is applied for design information verification (DIV) to confirm the absence of undeclared structural changes. Also, scanning a seal's unique microscopic surface structure is used as an inherent "fingerprint" which provides increased assurance against seal-counterfeiting.

8.1.6.4 Remote monitoring [34]

Remote monitoring transmits NDA measurements and signals of seals, pictures or videos to the remote monitoring data processing centers of the IAEA. This provides a full view of all data at any time and leads to reduced inspection frequencies. The data transmission can be performed either by radio signals, via the INTERNET or via satellites.

8.1.6.5 Environmental sampling [28]

Tunable diode laser spectroscopy (TDLS) can be used to detect parts per billion concentrations of hydrogen fluoride (HF) or UF_6 which is often associated with some forms of uranium conversion and enrichment processes. Swipes can be taken by 10x10 cm^2 cotton samples which are analyzed in specific laboratories.

8.1.6.6 Satellite imagery

Satellite imagery can be used as a powerful tool for non-proliferation verification. Publically available sattelite imagery has attained a resolution of 44 cm for black and white images and 2.4 m for multispectral application. Some imaging sattelites can acquire images from the microwave (radar) region or from the infrared bands.

Satellite imagery showed already in a number of cases to be very effective in detecting, evaluating and monitoring clandestine activities and nuclear facilities [41].

8.1.7 Anti-Neutrino measurement

8.1.7.1 Introduction

The fission of heavy nuclei (uranium, plutonium and other minor actinides) in nuclear reactors produces fission products. These fission products usually decay to more stable nuclei via beta-decay and emit aniti-neutrinos. An operating 1 GWe nuclear reactor emits anti-neutrinos on the order of 10^{21} per s. The distribution of the fission products is slightly different for the fissile isotopes U-235, U-238, Pu-239, Pu-241. It also depends on the velocity of the incident neutrons (fast or thermal) [35-39]. The released energy per fission, the emitted number of anti-neutrinos and their energy depend also on the fissile isotope that undergoes fission. Table 8.7 shows the characteristic data for anti-neutrinos originating from the isotopes U-235, U-238, Pu-239 and Pu-241.

Table 8.7: Characteristic data for anti-neutrinos originating from fission of different heavy nuclei [37].

	U-235	U-238	Pu-239	Pu-241
Released energy per fission [MeV]	201.7	205.0	210.0	212.4
Mean energy of anti-neutrinos [MeV]	1.46	1.56	1.32	1.44
Number of anti-neutrinos per fission	5.58	6.69	5.09	5.89

The theoretical anti-neutrino energy spectra are shown for U-235 and Pu-239 in Figure 8.8. They differ by about 50% [35,36,37].

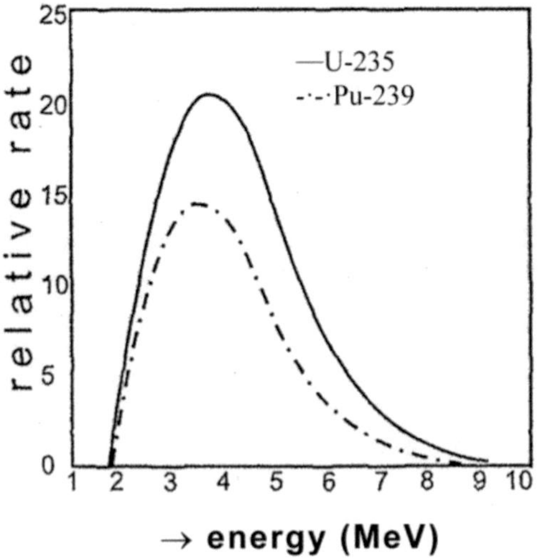

Figure 8.8: Theoretical anti-neutrino energy spectra for U-235 and Pu-239 [35].

At constant power of a fission reactor, e.g. LWR, the measured anti-neutrino flux and the shape of the anti-neutrino energy spectra are determined by the fissile fuel composition. As a function of operation time of the nuclear fission reactor the isotope U-235 decreases and the isotope Pu-239 (and other plutonium isotopes) slightly increases. Estimates of the power of the nuclear fission reactors by measuring the anti-neutrino flux, therefore, requires the knowledge of the fissile fuel composition as a function of operation time [37].

8.1.7.2 Anti-neutrino Detection

The principle employed for the detection of anti-neutrinos is the inverse beta decay process

$$\overline{\nu}_e + p \rightarrow e^+ + n$$

where the proton target is the nucleus of a hydrogen atom of a liquid scintillator. The positron produces a prompt signal of light measured by photomultiplier tubes optically coupled with the liquid target. The neutron is tagged by its capture on Gd nuclei dissolved in the liquid. This event occurs after a thermalization of the neutrons or a few ten μs after the prompt signal due to the positron energy loss. The neutron capture in Gd produces an emission of gamma rays with a total energy of about 8 MeV. Therefore, the signature of anti-neutrino interaction is provided by a delayed coincidence between these two events [35-39]. Figure 8.9 explains the detection principle for anti-neutrinos.

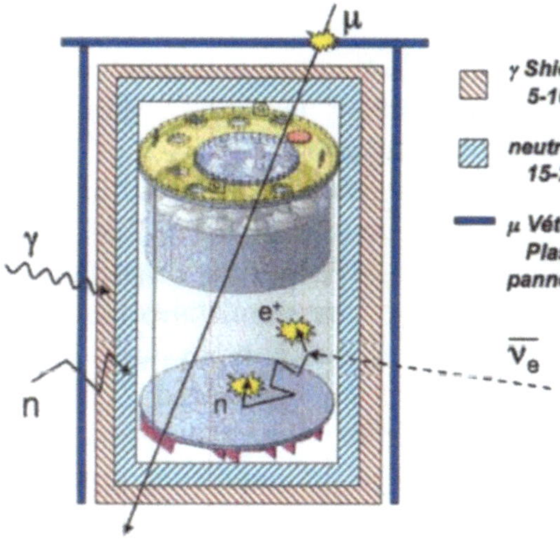

Figure 8.9: Detection principle for anti-neutrinos [37].

Cosmic rays and their associated secondary particles induce sources of background radiation. The muon véto shield tags the transit of these particles in or near the detector. In addition shielding of gamma rays and external neutrons is necessary.

As explained above the fuel composition (burnup of U-235 and generation of plutonium) determine the rate of anti-neutrinos and the energy spectrum of the anti-neutrinos emitted. Therefore, sophisticated computer programs must predict the spatial fuel composition (uranium, plutonium, minor actinides) and the fission product composition. These computer programs also allow calculating the anti-neutrino flux and the accompanying energy spectra.

As measurements show, only the total (superposition from all fissionable isotopes) of the rate and the energy spectra of the anti-neutrinos, both measurements and calculations must be compared [36,37].

8.1.7.3 Application of anti-neutrino detectors for safeguards

Measurements at several reactors in Russia, USA, Europe and Japan demonstrated that anti-neutrino detectors located at a certain distance, e.g. 100 m, from the nuclear fission reactors allow an independent measurement of the power level and of shutdown periods. In combination with sophisticated computer programs also the amounts of uranium and plutonium can be measured.

IAEA inspectors are not using this detection method yet. But it certainly represents an important possibility for independent safeguards measurements for the future [38,39,40]. All shut down events of the nuclear reactor can be clearly detected.

8.1.8 Unattended monitoring systems

Continuous automatic monitoring of nuclear facilities and transmission of the recorded data to the IAEA headquarters is an important tool of international safeguards. Such so-called unattended monitoring systems (UMS) were installed by IAEA already in 44 nuclear facilities in 22 countries by 2004.

UMS replace periodic visits of IAEA inspectors to nuclear facilities and allow uninterrupted facility operation. The UMS have the following basic characteristics:

- they are permanently installed and monitor automatically and continuously without the need of human interaction
- they use sensors for radiation, pressure, temperature, flow, vibration, electromagnetic fields etc.
- all components, e.g. cameras, sensors, computers etc., are enclosed on site in tamper-indicating enclosures to ensure authenticity.
- The IAEA must be able to independently verify its conclusions regarding the nuclear material in a facility. Therefore, data recording enclosures are designed such that the IAEA can detect any unauthorized tampering.

The software used is only available to the IAEA. Digital data follow 128 bit encryption algorithms. Data encryption is used both on- and offsite for all data. Power supplier are designed to be uninterruptable. The IAEA uses both virtual private networks and wireless solutions for data transmission to their headquarters.

8.1.9 Safeguards Application to the Different Parts of the Nuclear Fuel Cycle

In the following subsections the aforementioned safeguards methods are applied to the different parts of the nuclear fuel cycle. These are

- uranium enrichment plants
- nuclear reactors and fuel storage facilities
- fuel reprocessing and fuel refabrication plants.

References Section 8.1

[1] The Structure and Content of Agreements between the Agency and States Required in Connection with the Treaty on the Non-Proliferation of Nuclear Weapons. Vienna: International Atomic Energy Agency, INFCIRC/153 Corrected (1972).

[2] The Treaty on the Non-Proliferation of Nuclear Weapons; London, Moscow, Washington, 1 July 1968. Vienna: International Atomic Energy Agency, INFCIRC/140 (1970).

[3] IAEA (International Atomic Energy Agency). The Physical Protection of Nuclear Materials and Nuclear Facilities. IAEA-INFCIRC/225/Rev. 4 (Corrected), IAEA, Vienna (1999).

[4] Boyer, B., et al., International safeguards inspection: An inside look at the power, in: Doyle, J.: Nuclear Safeguards, Security and Nonproliferation, Butterworth-Heinemann, Elsevier (2008).

[5] Küchle, M., Plutonium inventory in storage tanks and process equipment and systematic errors in measurements. in: Gupta, D., Overview report of International Workshop on near Real Time Accountancy, KfK 3515, Karlsruhe (1983).

[6] Kluth, M., et al., Konzept zur Verfahrenstechnischen Auslegung einer 1000 Jahres-tonnen PUREX-Referenzanlage mit Basisdaten für eine Spaltstoffflusskontrolle, KfK 3204, Karlsruhe (1981).

[7] Kessler, G., Nuclear fission reactors, Springer, Wien (1983).

[8] Gupta, D., Investigations on detection sensitivity of the NRTA method for different size reprocessing facilities, KfK 4007, Karlsruhe (1985).

[9] Avenhaus, R., Material accountability, theory, verification, applications, Volume 2, The Wiley IIASA International Series on Applied Systems Analysis, Wiley, Chulchester, England (1977).

[10] Avenhaus, R., et al., Considerations on the large scale deployment of the nuclear fuel cycle, RR-75-36, IIASA, Laxenburg, Austria (1975).

[11] Reilly, D. et al., Nuclear material measurements technologies, in: Doyle, J.: Nuclear Safeguards, Security and Nonproliferation, Butterworth-Heinemann, Elsevier (2008).

[12] Appendix A: Safeguarding reprocessing facilities. http://www.princeton.edu/~ota/disk1/1995/9530/953007.pdf

[13] Morgenstern, A., et al., Analysis of Np-237 in spent fuel solutions, Radiochim. Acta, 90, 389-392 (2002).

[14] Keepin, G.B., State of the art technology for measurement and verification of nuclear materials, in: Tsipis, K. et al., Arms control verification: the technologies that make it possible, Pergamon-Brassey's, Washington (1986).

[15] Holden, N.E., et al., Spontaneous fission half-lives for ground-state nuclides, Pure Appl. Chem., 72, 8, 1525 (2000).

[16] Kocherov, N., et al., Handbook of Nuclear Data for Safeguards, International Atomic Energy Agency Nuclear Data Section (Dec. 1998).

[17] Abhold, M., et al., Field detection of nuclear materials, in: Doyle, J.: Nuclear Safeguards, Security and Nonproliferation, Butterworth-Heinemann, Elsevier (2008).

[18] Manero, F., Konshin, V., Status of energy independent $\overline{\nu}$ -values for the heavy isotopes (Z>90) from thermal to 15 MeV and of the $\overline{\nu}$ -values for spontaneous fission, INDC(NDS)-34/G (1972).

[19] Segrè, E., Spontaneous fission, Phys., Rev. 86, 1 (1952).

[20] Roberts, J., Spontaneous-fission decay constant of ^{238}U, Phys. Rev., 174, 1482 (1968).

[21] Nuclear Data Sheets, Vol., 4, No. 6, Elsevier (1970).

[22] Gold, R., Spontaneous fission decay constant of Am-241, Phys. Rev., C1, 738 (1970).

[23] Barton, D. et al., The spontaneous fission half-life of ^{244}Cm, J. Inorg. Nucl. Chem., 32, 769 (1970).

[24] Nicholas, K., et al., Capability and limitation study of the DDT-PAN Waste Assay Instrument, LA-12237-MS (1992).

[25] Melton, R., et al., Development of advanced matrix correction techniques for active interrogation of waste drums using the CTEN instrument, LA-UR-97-399 (1997).

[26] Bracken, D., et al., Application guide to calorimetry, LA-13867-M (2002).

[27] Autrusson, B., Electronic safeguards seals, Esarda Bulletin, Nr. 37 (2007).

[28] Bush, W. et al., Model safeguards approach for gas centrifuge enrichment plants, Proc. of an Int. Safeguards Symposium, Vienna, October 16-20, 2006; IAEA-CN-148/98 (2005).

[29] Rineiski, A., Kessler, G., Proliferation-resistant fuel cycle options avoiding neptunium for thermal and fast reactors, Nucl. Eng. Design, 240, 500-510 (2010).

[30] Zendel, M., IAEA, safeguards equipment http://www-pub.iaea.org/MTCD/Meetings/PDFplus/2007/cn1073/Presentations/4A.4%20Pres_%20Zendel%20-%20IAEA%20Safeguards%20Equipment.pdf

[31] Zendel, M. et al., IAEA safeguards equipment http://www.bnl.gov/ISPO/BNLWorkshop07/Presentations/Zendel_Moeslinger.pdf

[32] Broeders, C.H.M., Kessler, G., Fuel cycle options for the production and utilization of denatured plutonium, Nucl. Sci. Eng., 156, 1-23 (2007).

[33] IAEA, Multilateral approaches to the nuclear fuel cycle: expert group report submitted to the director general of the IAEA, INFCIRC/640 (2005).

[34] Zendel, M., IAEA safeguards equipment, Int. Journal of Nuclear Energy Science and Technology, Vol. 4, No. 1 (2008).

[35] Bernstein, A., et al., Monitoring reactors with cubic meter scale antineutrino detectors, in: R. Avenhaus et al.: Verifying treaty compliance: limiting weapons of mass destruction and monitoring, Kyoto Protocol Provisions, Springer, Berlin (2006).

[36] Yermia, F., et al., The Nucifer experiment: Reactor monitoring with antineutrinos for nonproliferation purpose, Proceedings of Global 2009, Paris (2009).

[37] Fallot, M., et al., Nuclear reactor simulations for unveiling diversion scenarios: capabilities of the antineutrino probe, Proceedings of Global 2009, Paris (2009).

[38] Achkar, B., et al., Comparison of antineutrino reactor spectrum models with the Bugey-3 measurements, Physics Letters B, 374, 243-248 (1996).

[39] Cribier, M., Neutrinos and nonproliferation in Europe, Neutrino Science and Geophysics, Honolulu (2005).

[40] Nieto, M., et al., Detection of antineutrinos for non-proliferation, LA-UR-03-4500, Rev. (2008).

[41] Pabian, F., Commercial satellite imagery: another tool in the nonproliferation verification and monitoring toolkit, in: Doyle, J., Nuclear Safeguards, Security and Nonproliferation, Butterworth-Heinemann, Elsevier (2008).

[42] Schanfein, M., IAEA unattended monitoring systems, in: Doyle, J., Nuclear Safeguards, Security and Nonproliferation, Butterworth and Heinemann, Elsevier (2008).

8.2 Safeguards concept of uranium enrichment plants

8.2.1 Introduction

For uranium enrichment two main technologies are presently used in civil nuclear reactor technology. These are the gas diffusion and gas centrifuge technology. Both enrich the gaseous UF_6 molecules of U-235 starting from natural uranium which is found in the composition of 0.72% U-235 and 99.28% U-238 (the small fraction of 0.0055% U-234 is neglected here) in uranium ores. Gas centrifuge technology has been proven to be much more economic than the gas diffusion technology, originally developed and used in the USA and France. For economical reasons, these nations will switch over to centrifuge technology in the near future. Laser enrichment technology still has to be proven to become as economic as centrifuge technology (see Section 3).

A typical large scale gas centrifuge enrichment plant is characterized [1,14] by

- a separative work unit (t SWU) capacity of 2900 t SWU/y
- a feed capacity of 5000 t of U_{nat} in 590 cylinder containers (48 inches diameter)
- a tails capacity of 4500 t of depleted uranium (0.3% U-235) in 540 cylinder containers (48 inches diameter)
- a product capacity of 500 t of low enriched uranium, e.g. 5% U-235, in 290 cylinder containers (30 inches diameter).

Centrifuge enrichment plants consist of the feed and take off areas, blending stations, a large assembly of centrifuges per cascade and several cascade halls (containing cascades connected in parallel) (see Figure 8.10).

Figure 8.10: Centrifuge cascade hall of a gas centrifuge enrichment plant (URANIT).

As the operation of gas diffusion plants will be terminated in the future, only the safeguards concepts for centrifuge plants will be discussed [2,3].

Material Balance Areas (MBA) are the storage area for the feed product as well as the storage area for the final enriched product and the tails product. In addition there is the process area or centrifuge cascade hall. The conversion time of HEU into weapons useable material is set by IAEA to the order of weeks. Hence the time goal for detection is one month. IAEA inspectors, therefore, have full access to the cascade hall for monthly inspections. They must verify that no UF_6 is being produced beyond the declared enrichment levels e.g. high enriched uranium (HEU). A physical inventory verification (PIV) is performed every year to verify the entire stock of nuclear material on site. Design information verifications (DIVs) are carried out initially on the new plant and repeated every year. They must confirm that the cascades are built according to the design information documents which were originally submitted. Limited frequency unannounced access (LFUA) inspections shall verify that the cascade design has not been modified. Complementary access visits are required by the additional protocol to the NPT. These are unannounced visits several times per year.

Due to the chemical and physical properties of UF_6 [4] and due to design peculiarities of a centrifuge enrichment plant being equipped with gas centrifuges connected to many pipings, ducts, elbows, filters etc., special care must be given to so-called fissile material holdup. Holdup refers to fissile uranium which can deposit in the equipment or in transfer pipes. Holdup is difficult to measure. It can amount to kilograms of nuclear material and can thus limit the accuracy of the material balance. The deposit formation occurs due to the chemical properties of the UF_6. When UF_6 comes in contact with water vapor it reacts to hydrogen fluoride UF and uranylfluoride UO_2F_2. This leads to corrosion and to solid uranium compounds which deposit on surfaces. In addition, the alpha particle radiation of the uranium molecules leads to decomposition of UF_6 into HF and solid uranium compounds which also deposit on surfaces. (UF_6 is the only gaseous chemical component of uranium.)

The IAEA inspectors [1,4] must verify that a possible diverter

- does not produce higher enriched uranium by reintroducing higher enriched UF_6 into the feed of the cascades
- does not reconfigure the cascade system
- does not remove significant quantities (SQ) of highly enriched uranium by false reporting or by overstating the amount of nuclear material in the holdup of the plant.

8.2.2 Inspection techniques [5,6,7,9]

8.2.2.1 Weight verification

The UF_6 gas cylinder containers are weighed either with the IAEA inspector weigh cells or the weigh scales of the plant operator (Figure 8.11). The transportable load cell based system of IAEA senses the weight of the suspended product cylinder with an accuracy of better than ±1 kg. This information is used for the annual nuclear material balance check.

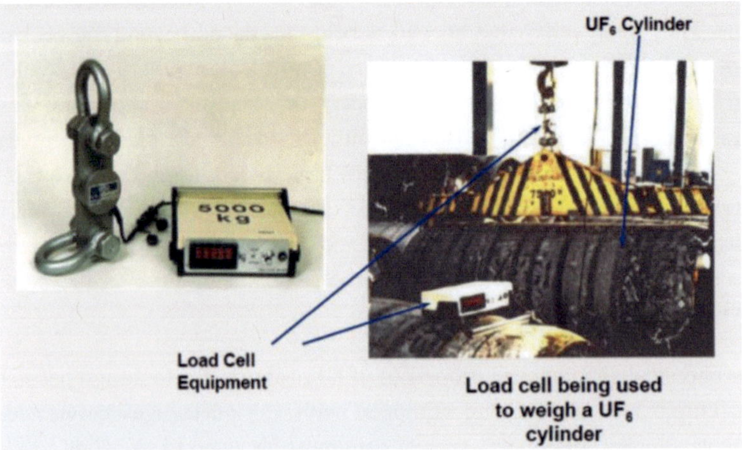

Figure 8.11: UF_6 cylinder containers and IAEA weigh cell (IAEA).

8.2.2.2 Destructive (DA) and non-destructive (NDA) Assay

For DA measurements gas mass spectrometers (section 8.6.1) are applied that can have high precision of even better than ±0.014% [8]. They are taken either from the UF_6 gas cylindrical containers or from the gas flow in the plant.

As NDA measurement techniques the gamma-ray techniques and to some extent also neutron assay techniques are applied [9] as described in Sections 8.1.5.2 and 8.1.5.3.

For gamma-ray measurements with a fixed detector sample geometry in the enrichment plant and for samples that allow to measure the 186 keV U-235 gamma-rays the count rates are directly proportional to the enrichment of the UF_6 [9]. Another enrichment measurement technique is the gas phase enrichment monitor. Samples are taken from plant sampling points in the feed, final product or tails lines. The sample cylinder is then connected to the monitor. For the measurement the attenuation of the 59.5 keV gamma-rays of an Am-241 source and the 185.7 keV gamma-rays of U-235 are compared. Measurement accuracies are in the range of ±1% [9].

There is also experience in using thermoluminiscent dosimeters (TLD) for certain equipments of the enrichment plant. Gamma-rays have advantages over neutrons in measuring holdups, because they can be collimated allowing the locations and distribution of deposits to be defined. Sodium iodine and bismuth germanate detectors can be applied. In addition, Peltier cooled cadmium tellurium detectors are in use as portable gamma-ray detectors.

The IAEA uses the online enrichment measurements CEM (continuous enrichment monitoring) and CHEMO (cascade header enrichment monitor) for safeguards verification. These are passive gamma-ray measurements combined with X-ray fluorescence detection to measure the enrichment of the UF_6 gas in outgoing piping from the centrifuge cascade hall [4,10]. This allows a Yes or No answer, whether enrichment ≥20% U-235 in the UF_6 gas is exceeded with a false alarm probability of 0.001. The CHEMO system can transmit a message daily to the IAEA confirming that the measurements are within the bounds of declaration.

8.2.2.3 Holdup of nuclear materials within the enrichment plant

The Generalized Geometry Holdup (GGH) assay method was developed to simplify the analysis of holdup measurements [5,6]. This method requires a portable spectroscopic system and a calibrated detector. As a final result the specific isotope mass for an area deposit is obtained.

Portable neutron monitors use four He-3 tubes surrounded by polyethylene as moderator. They are designed for either single counting or neutron coincidence counting.

A neutron holdup assay method for uranium enrichment facilities [7] uses Monte Carlo techniques modeling the centrifuge cascade hall. In addition a calibration method is used which relates the average neutron count rate at certain positions in the hall to the mass of the uranium holdup. The average neutron count rate is measured by portable detectors. This neutron assay method avoids the attenuation problems caused by structures etc. of the gamma-ray measurements.

The precision of gamma-ray or neutron holdup measurements is usually of the order of a few percent. However, the accuracy or the system error is very difficult to determine because it is difficult to know the true mass and location of nuclear material holdup in the complex facility [5]. Therefore, IAEA inspectors rarely rely on the measurement of holdups.

8.2.3 Unattended safeguard systems for enrichment plants

An unattended safeguard system was proposed by Pickrell et al. [11]. It is based on the fact that natural uranium is always accompanied by very small amounts (0.0055%) of U-234 which is an emitter of alpha particles. The latter react with the fluorine in UF_6 in an (α,n) reaction generating neutrons. U-234 is enriched by gas centrifuges with a higher enrichment factor than U-235. Therefore, the rate of neutrons generated by (α,n)-reactions increases considerably if U-235 is enriched over a certain limit. These neutrons can be detected by neutron monitors arranged in matrix form above the centrifuges. The neutron signals can be evaluated by IAEA.

A real time safeguards system for centrifuge enrichment plants was proposed by Delbeke et al. [15].

8.2.4 Containment and surveillance

As containment surveillance methods the IAEA applies three-dimensional laser imaging for design information verification (DIV) with the aim to confirm whether undeclared structural changes have been made.

Cameras as well as passive seals (metal and paper seals) or active seals: fiber optic seals (VACOSS, EOSS) and ultrasonic seals are important devices of the IAEA verification system (Section 8.1.6.2).

8.2.5 Environmental sampling

Tunable diode laser spectroscopy (TDLS) as well as light detection and ranging (LIDAR) are applied for the detection of parts-per-billion concentrations of HF (hydrogen fluoride to detect undeclared activities of uranium conversion or enrichment plants.

Within the enrichment plants swipes are taken many times a year at many locations of the enrichment plants. These swipes are sent to the Seibersdorf Laboratories of the IAEA or to licensed Network Analytical Laboratories for further analysis.

The use of swipes is based on the assumption that every component of the enrichment plant will release small amounts of UF_6 to the environment. Although these releases are extremely small, they are detectable and their analysis provides an indication (U-234, U-235, U-236, U-238 isotopic amounts) of the enrichment of the material that has been processed in the plant [12].

Environmental samples are collected by swiping selected areas of the enrichment plant with 10 x 10 cm^2 cotton cloth form sampling kits prepared in ultra clean conditions. These are sent to the laboratories and analyzed by using Thermal Ionisation Mass Spectrometry or Secondary Ion Mass Spectrometry.

The time required for particles to be released and settled on surfaces is estimated to be one to six weeks. A routine detection time of three months, therefore, appears achievable. This detection time is an important deterrent to a possible diverter [11,13].

References Section 8.2

[1] Bush, W. et al., Model safeguards approach for gas centrifuge enrichment plants, Proc. of an Int. Safeguards Symposium, Vienna, October 16-20, 2006; IAEA-CN-148/98 (2005).

[2] Steinebach, E., et al., Safeguards at the Dutch and German uranium enrichment plants of URENCO – development and experience -, 3rd Int. Conf. on Facility Operations – Safeguards Interface, San Diego, USA (1987).

[3] Friend, P., Urenco's views on international safeguards inspection, 8th Int. Conf. on Facility Operations – Safeguards Interface, March 30 – April 4, 2008, Portland, USA, 2008.

[4] Sharikov, D., Verification challenges for safeguarding uranium enrichment plants, ESARDA bulletin, No. 37 (2007).

[5] Reilly, D., Measurement of nuclear material process holdup, in: Doyle, J.: Nuclear safeguards, security and nonproliferation, Butterworth and Heinemann, Elsevier, New York (2008).

[6] Russo, P.A. et al., Evaluation of the integrated holdup measurement system with the M^3CA for assay of uranium and plutonium holdup, LA-13387-MS (1999).

[7] Beddingfield, D. et al., A new approach to holdup measurement in uranium enrichment facilities, LA-UR-00-2534 (2000).

[8] Reilly, D., et al., Nuclear material measurement technologies, in: Doyle, J., Nuclear Safeguards, Security and Nonproliferation, Butterworth and Heinemann, Elsevier, New York (2008).

[9] Keepin, G.B., State of the art technology for measurement and verification of nuclear materials, in: Tsipis, K., Arms control verification, the technologies that make it possible, Pergamon-Brassey's, New York (1986).

[10] Panasyuk, A. et al., Tripartite enrichment project: safeguards at enrichment plants equipped with Russian centrifuges, IAEA-SM-367/8/02, 2001

[11] Pickrell, M. et al., New types of unattended systems for enrichment plant safeguards, 2006. See: http://www.bnl.gov/ISPO/BNLWorkshop07/Presentations/Pickrell.ppt

[12] Bush, W., et al., IAEA experience with environmental sampling at gas centrifuge enrichment plants in the European Union, IAEA-SM-367/10/04.

[13] Cooley, J., et al., Experience with environmental scope sampling in a newly built gas centrifuge plant, INMM Annual Meeting (2000).

[14] Laughter, D., Profile of world uranium enrichment programs – 2007, ORNL/TM-2007/193 (2007).

[15] Delbecke, J., et al., The real time mass evaluation system as a tool for detection of undeclared cascade operation at gas centrifuge enrichment plants, 8th Int. Conf. on Facility Operations – Safeguards, Interface, Portland, Oregon (2008).

8.3 Safeguards for Light Water Reactors (LWRs) and spent fuel pools

8.3.1 Light Water Reactors and fresh fuel elements

Light Water Reactors (LWRs) contain two types of nuclear fuel. Low enriched, e.g. 5% U-235 and 95% U-238 fresh fuel, and irradiated fuel with additional amounts of fission products, plutonium and minor actinides (neptunium, americium and curium). When the irradiated fuel elements are unloaded from the reactor core operating in multi-batch mode after typically 12 to 18 months full power, the spent fuel contains about 0.8% U-235, 0.9% plutonium and 0.1% minor actinides.

The irradiated or spent fuel is considered direct-use material, since it contains plutonium (significant quantity 8 kg). The conversion time for converting the plutonium of the irradiated fuel into nuclear weapons useable metallic plutonium is considered to be 1 to

3 months. The IAEA goal for timely detection is assumed to be 3 months. Hence the IAEA inspector is required to visit the LWR plant on a 3 months basis [1,2,3] (see Section 8.1.1).

When the core will be refueled and spent fuel is unloaded the IAEA inspector will be responsible for verifying the new fresh fuel elements loaded into the core, the remaining core fuel elements with partially irradiated fuel and the unloaded spent fuel elements.

The material balance areas (MBA) for LWRs are shown in Section 8. 1.3.1 (Fig. 8.1.2).

The inspector relies on containment and surveillance methods (cameras and seals) and visual inspection as well as NDA methods. The fresh fuel elements can be measured by observing the 185.7 keV gamma rays of U-235 with the help of mini-multichannel analyzers (MMCA) connected to a CdZnTe-detector (Figure 8.12).

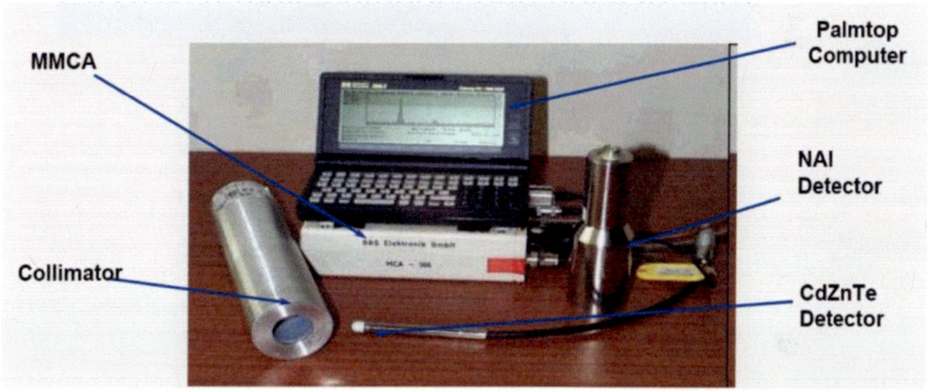

Figure 8.12: Mini-multi-channel analyzer (MMCA) with detectors for fresh fuel assay [4,5]

The safeguards objective of the IAEA is to verify that the operator of the reactor plant does not divert fissile material either in abrupt or protracted diversions, e.g. by replacing single irradiated rods or the irradiated fuel element by dummies and diverts the fuel elements with fissile fuel.

8.3.2 Safeguards surveillance of spent fuel elements

In spent fuel elements the gamma-ray signatures will be masked by the intense radiation emitted by fission products in irradiated or spent fuel elements. More than several hundred fission products are created in significant amounts during fission of uranium, plutonium or minor actinides. Some of these fission products emit delayed neutrons during their further decay. Only about 10 of these isotopes can be directly measured due to their characteristic gamma-radiation. Minor actinides which are built up after neutron absorption in plutonium or americium isotopes also emit spontaneous fission neutrons. Alpha-decay of these isotopes can lead to neutron emission after (α,n) reactions with light elements, e.g. oxygen, carbon etc.. For refueling LWRs will be shut down. The pressure vessel is opened and spent fuel elements having attained their specified burnup, e.g. 60 GWd/t, will be unloaded from the core and transferred under water into the intermediate fuel element storage pool. The spent fuel elements are usually replaced by fresh fuel elements. All fuel elements have seals or tags. The loading and unloading process is accompanied by video surveillance. The simplest verification measure is visual inspection by underwater cameras. Within the spent fuel storage pool Cerenkov viewing detectors (Figure 8.13) which amplify the Cerenkov light produced by spent fuel elements can be used to qualitatively verify the burnup and cooling time of the spent fuel elements.

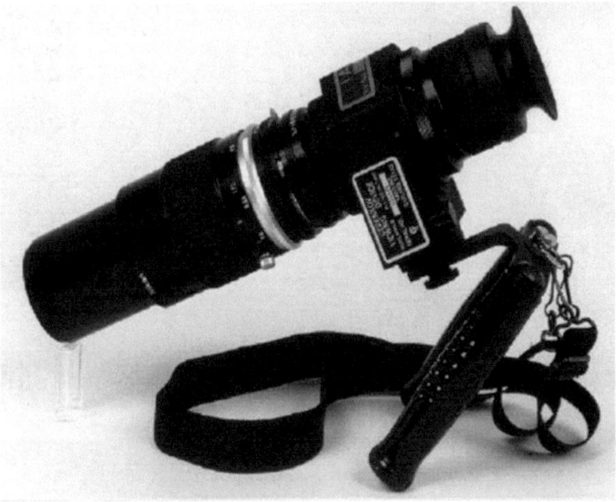

Figure 8.13: Cerenkov viewing detector [4,5].

For cooling times longer than about one year the total gamma-ray and neutron radiation is proportional to the burnup of the spent fuel elements [1,2]. This gamma-ray and neutron radiation can be measured by ionization and fission chambers. For this procedure the spent fuel elements are partially lifted from this storage position in the so called intermediate spent fuel storage pool. This measurement can be performed with the FORK detector (Figure 8.14) by IAEA inspectors [3,4,5,7]. Total neutron counting measurements have some advantages over total gamma-ray measurements, as the gamma-rays are attenuated by the fuel. However, total neutron counting is not precise enough to identify single missing fuel rods in an irradiated fuel element. Nevertheless they can identify misdeclared irradiated fuel subassemblies.

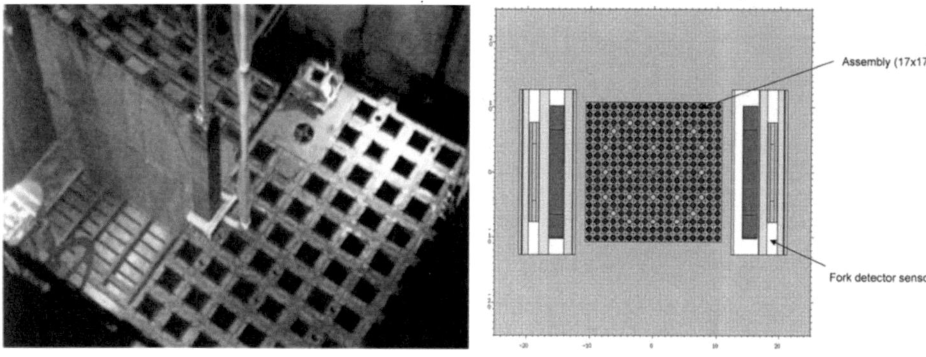

Figure 8.14: FORK detector and partially lifted fuel element.

8.3.3 Gamma-ray spectroscopy

High-resolution gamma-ray spectroscopy can be used if full knowledge of the fuel assembly geometry is provided and attenuation factors of the gamma-rays by the fuel are known. The spent fuel burnup can be obtained from the measurement of the activity of Cs-134 to Cs-137 [2].

The IAEA uses CdZnTe detectors and a multichannel analyzer to measure gamma-rays from fission products like Cs-134, Cs-137, Pr-144, Eu-154 and others [4,5].

8.3.4 Active and passive neutron interrogation methods

A method for the non-destructive assay of spent LWR fuel assemblies based on combined active and passive neutron counting was reported by Wuerz et al. [6]. The method allows the determination of the burnup, the total fissile content, the original enrichment of the spent fuel element as well as of the type of fuel (uranium or mixed oxide (MOX) fuel). The method was originally developed for criticality control at the front end of a reprocessing plant. It can also be used for fuel storage facilities.

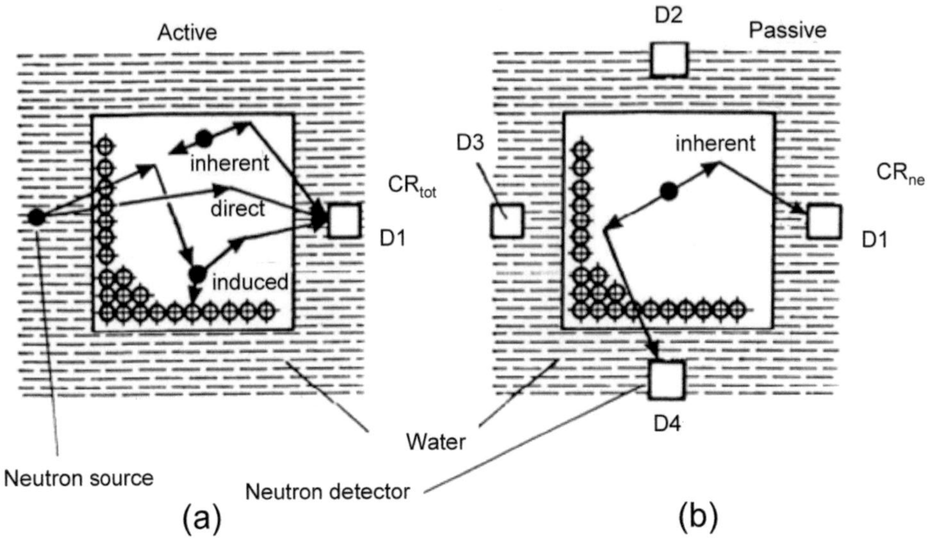

Figure 8.15: Principle of the method of active and passive neutron interrogation [6].

Measurements were undertaken in spent LWR fuel storage pools of a reprocessing plant and LWR power stations [6,7]. Without prior knowledge of any fuel assembly data, the burnup of uranium fuel assemblies can be determined with an uncertainty of ±1200 MWd/t. The initial enrichment of uranium fuel assemblies can be measured with an accuracy of ±5%. Using these data and accuracies, the total plutonium content can be determined from isotopic correlations with an accuracy of better than ±0.3 kg/t for pressurized water reactor (PWR) and ±0.5 kg/t for boiling water reactor (BWR) fuel assemblies.

The principle of the method is shown by Figure 8.15. The assay is performed under water. A Cf-252 neutron source is positioned next to the fuel assembly surface. A neutron detector Dl to D4 at the opposite side of the assembly is used for the measurement of thermalized neutrons. The total number CR_{tot} (total count rate) of neutrons measured consists of

- direct neutrons penetrating the assembly without being captured or scattered
- fission induced neutrons
- inherent spontaneous fission neutrons from the spent fuel CR_{ne}.

For the passive assay (Figure 8.16) the neutron source is removed and up to four neutron detectors measure the inherent spontaneous fission neutrons from the spent fuel. Both measurements contribute to the final experimental results.

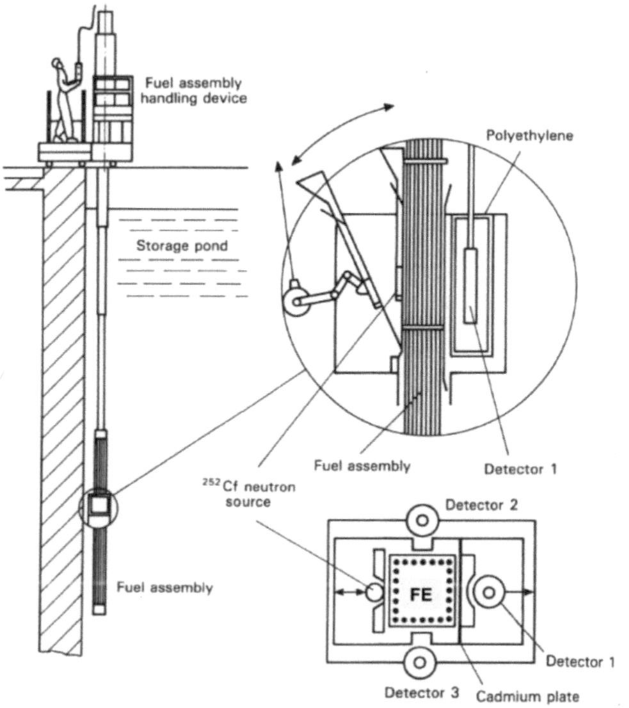

Figure 8.16: Arrangement of LWR fuel element and Cf-252 neutron source.

8.3.5 Advanced antineutrino measurements

Antineutrino detectors were developed in several countries between 1995 and 2010 and tested at nuclear reactors. They can measure the power level of a reactor plant from a distance of about 100 m away from the plant. The data could be processed and transmitted to the IAEA data center (see Section 8.1.7).

References Section 8.3

[1] Boyer, B., et al., International safeguards inspection; in: Doyle, J.: Nuclear safeguards, security and nonproliferation, Butterworth and Heinemann, Elsevier, New York (2008).

[2] Abhold, M., Irradiated fuel measurements, in: Doyl, J.: Nuclear safeguards, security and nonproliferation, Butterworth and Heinemann, Elsevier, New York (2008).

[3] Rinard, M., et al., Safeguarding LWR spent fuel with the FORK detector, LA-11096-MS (1988).

[4] Zendel, M., IAEA, safeguards equipment
http://www-pub.iaea.org/MTCD/Meetings/PDFplus/2007/cn1073/Presentations/4A.4%20Pres_%20Zendel%20-%20IAEA%20Safeguards%20Equipment.pdf

[5] Zendel, M. et al., IAEA safeguards equipment
http://www.bnl.gov/ISPO/BNLWorkshop07/Presentations/Zendel_Moeslinger.pdf

[6] Wuerz, H., A nondestructive method for Light Water Reactor fuel assembly identification, Nucl. Technology, 90, 191-204, 1990.

[7] Murphy, A., Calculating FORK detector response from spent fuel inventories using Monte Carlo techniques, ORNL/TM-2004/310, ORNL (2005).

8.4 Safeguards survey of large scale reprocessing plants

The safeguards approach of large scale reprocessing plants, e.g. the 800 t/y throughput of spent fuel plant at Rokkasho, Japan, follows systematic design information and verification (DIV) during all phases of construction, commissioning and operation. In addition it comprises installed, unattended radiation and spent fuel solution measurements as well as monitoring systems along with a number of inspector attended measurement systems. These independent or authenticated data will be transmitted over a network to a data recording center of the IAEA for evaluation. Near real-time accountancy (NRTA) uses short period sequential analysis which can be combined with plutonium/uranium solution monitoring data to provide higher assurance in the verification of fissile material (Sections 8.1.4 and 8.7). Containment and surveillance measures are combined with the NRTA concept.

Figure 8.17 explains the flow diagram of nuclear material in a large scale reprocessing plant. In the following sections the flow of fissile material through the reprocessing plant along with safeguards measurement techniques is explained. Section 8.4.9 deals with a specific safeguards concept proposed by IAEA for the Rokkasho reprocessing plant.

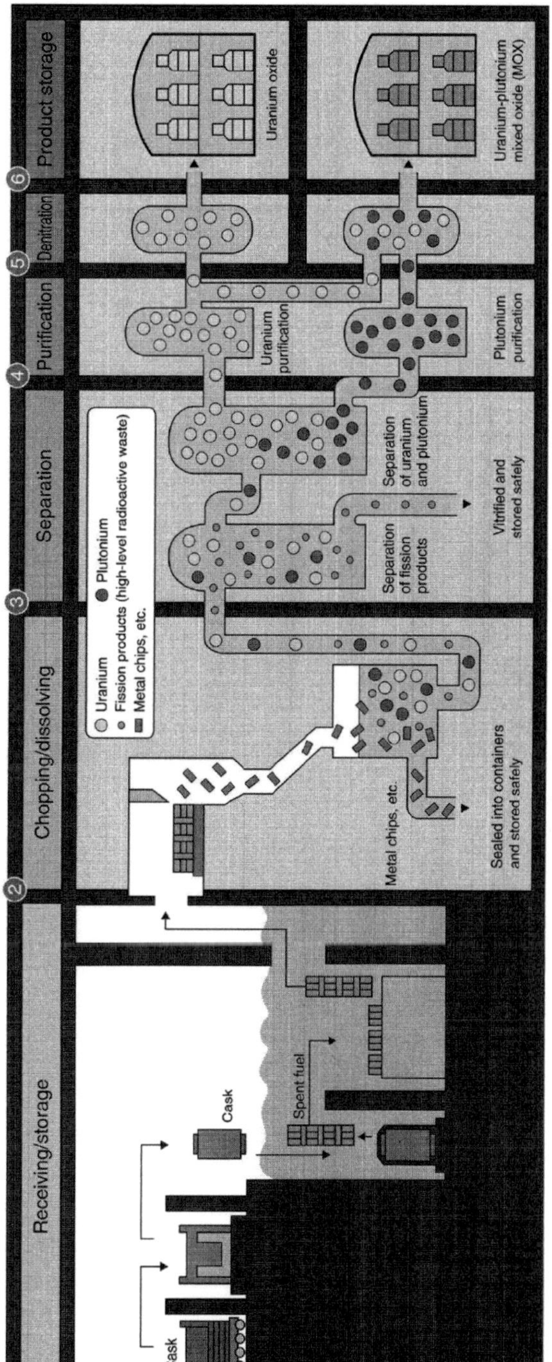

Figure 8.17: Flow diagram of nuclear material in a large scale reprocessing plant (Japan Nuclear Fuel Limited, 2008).

8.4.1 Spent fuel storage pool

Seals are attached to the shielded transport cask and to the spent fuel elements. In the spent fuel element storage pools active and passive neutron assay methods can be applied as described in Section 8.1.6. This allows confirmation of the burnup and the uranium/ plutonium content of each fuel element.

Systems of cameras with video recording and radiation detector systems are implemented in all areas, where fuel is moved or handled (spent fuel storage pool and fuel element chopping area).

The chopped pieces of fuel fall into the dissolver where the fuel is dissolved in boiling nitric acid. The remaining hulls and structural pieces of fuel elements are washed and become radioactive waste. The solution of dissolved fuel and fission products is clarified in a centrifuge to remove not dissolved products (sludge). The solutions are then collected in an accountancy tank where the uranium and the plutonium content is measured. All this equipment is located behind heavy shielding and not directly accessible

8.4.2 Safeguards measurement for the accountancy tank

For the fuel solution in the accountancy tank both DA and NDA measurement techniques are applied as already described in Section 8.1.5. Liquid samples from the accountability tank are pneumatically transferred to the measuring station. The X-ray fluorescence and the K-edge densitometry are applied to measure the total plutonium and uranium concentrations. In addition, volume and weight measurements of fuel solution samples or of the accountancy tank itself constitute important information.

For the X-ray fluorescence method an X-ray source, e.g. Co-57 (122 and 136 keV) irradiates the sample and ionizes electrons of the K-shell of the uranium or plutonium atoms. They emit so called K-shell X-rays which can be detected by a Ge-detector. The Co-57 X-ray source must be arranged such that the Ge-detector remains shielded [5].

K-edge densitometry works with two X-ray sources, one below and one above the K-edge of e.g. plutonium. Such X-ray sources are Co-57 with 122 keV and Se-75 with 121 keV (see section 8.1.5). The accuracy of measurement with both methods (X-ray fluorescence and K-edge densitometry) is between 0.2 and 0.6% [7]. Isotope dilution gamma-ray spectrometry (section 8.1.5) can also be used to measure dissolver solutions with uranium and plutonium. This method allows determining both the uranium and plutonium concentrations and the isotopic compositions with an accuracy of about 0.4% [7].

8.4.3 Separation of fission products and actinides

The high radiation of the fission products and the actinides require remotely controlled chemical processes. From the accountancy tank the fuel and fission product solution flows to the solvent extraction part of the plant. There, the fission products and actinides are separated first. After going through several purification cycles part of the uranium and plutonium solutions are mixed in a certain ratio (master mix). The remaining uranium nitrates are separately treated. Uranium and plutonium nitrates undergo a co-denitration and heat treatment process to be converted to UOX and MOX powders.

8.4.4 Near real time accountancy

Near real time accountancy (NRTA) is applied to give timely information on the fissile material balance in the plant and to analyze possible trends (Section 8.1.4). The movement of solutions is followed through the process to confirm transfers into and out of material balance areas (MBAs). The basic principle of NRTA is to monitor the in-process inventory of plutonium frequently (daily or weekly) applying a combination of direct measurements from in-process instruments, off-line analysis, and indirect data from computer simulations of the chemical processes. Most tanks and many process vessels are amenable to measurements or estimation of their inventories. One obvious advantage is that the throughput over a short interval, e.g. weeks, is significantly smaller than over an entire year. This means that the effects of some of the overall measurement inaccuracies are significantly reduced. Another advantage is that many more measurements are taken which improves the timely detectability.

8.4.5 Safeguards measurements in product storage areas

The MOX powder can be assayed by DA methods with an accuracy of ±0.15 to 0.2% [2,7]. The MOX powders are filled in cans, sealed and weighed with high precision. Several cans are put together in canisters which are again sealed. The canisters undergo additional measurements by neutron coincidence detectors and gamma detectors to determine the amount of uranium and plutonium and their isotopic compositions (section 8.1.5.3). The accuracy achieved is ±0.2 to 2% [5]. They are then transferred to the intermediate product storage areas, where also the remaining uranium dioxide products are stored.

8.4.6 Waste streams

There are two radioactive waste streams which contain fissile materials (Figure 8.17)

- the hulls and structural pieces of the fuel element with small amount of fissile material on their surfaces
- the high active waste containing the fission products and the minor actinides.
- The safeguards verification of those two waste products will be discussed in Sections 8.4.9.1 and 8.5.
- The low active and medium active waste streams do not contain fissile material.

8.4.7 Containment and Surveillance (C/S)

C/S technologies (seals, cameras and radiation detectors) are used extensively in the spent fuel storage pool, in the fuel element chopping area (head end) and in the final product (MOX- und UOX-powders) storage areas. The objective is to ensure that no fuel assemblies or fuel rods are removed without declaration.

8.4.8 IAEA resident inspectors

In addition to the above safeguards measurements for the material balance and containment and surveillance measures the IAEA can have resident inspectors at the reprocessing plant. They have full access to all facilities, to the operational staff and to measurement systems during continuous operation. An IAEA on site laboratory can also take samples for destructive analysis.

8.4.9 Material balance areas for a large scale spent fuel reprocessing plant

The safeguards concept of the Rokkasho reprocessing plant is described by Figure 8.18: The spent fuel reprocessing facility is divided into five Material Balance Areas (MBAs). These MBAs are subdivided into Key Measurement Points (KMPs) with the objective of timely detection as well as Physical Inventory Verification (PIV). Flow key measurement points (arrows in circles) have also been identified for all nuclear material streams or routes which cross MBA boundaries. In addition to flows that cross MBA boundaries, other strategic points are defined for verification of flows within the MBAs (points in circles).

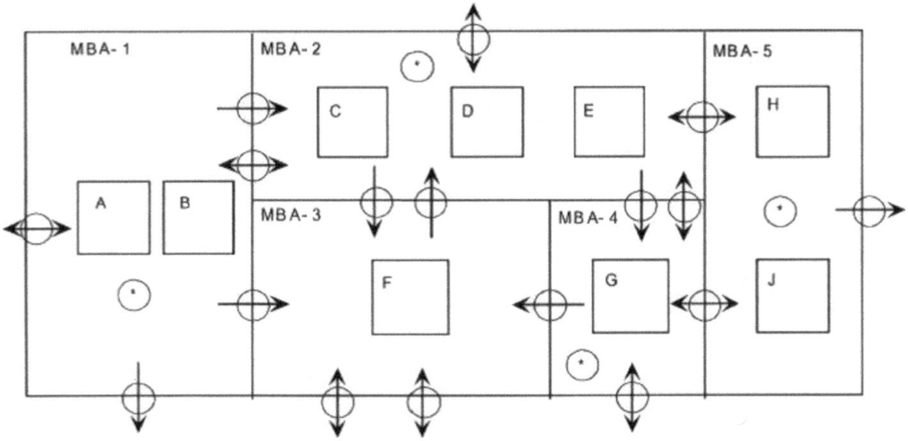

MBA-1:	Spent fuel receipt and storage area, Head-end area	MBA-3:	Waste treatment and storage area
MBA-2:	Main process area (including U conversion and laboratories	MBA-4:	MOX conversion area
		MBA-5:	Product storage area
KMP A:	Spent fuel receipt and Storage area		
KMP B:	Head-End area	KMP E:	Nuclear material in the U conversion area
KMP C:	Nuclear material in main process area	KMP F:	Nuclear material in the waste treatment and storage area
KMP D:	Nuclear material in the analytical laboratory	KMP G:	Nuclear material in the MOX conversion area
		KMP H:	UO3 product material in the storage area
		KMP J:	MOX product material in the storage area

Figure 8.18: Material balance areas (MBAs) and key measurement points (KMPs) in a large scale spent fuel reprocessing area.

8.4.9.1 MBA 1 – Spent Fuel Receipt and Storage, and Head-End Areas

Spent fuel assemblies received into the facility, are unloaded from the transport cask and transferred into the storage ponds where they are verified using the integrated spent fuel verification system. This system consists of time synchronized cameras and radiation detectors. Batches of dissolved spent fuel, for transfer to the main chemical process, will be verified for volume using a spent fuel solution measurement and monitoring system. Uranium and plutonium will be verified by sampling using an automatic sampling system or analysis by Hybrid K-edge Densitometry (Section 8.1.5.1 and [1,3,7]).

The flow of spent fuel assemblies from the storage pond transfer channel to the chopping machine and the movement of the hulls/end-pieces drums will be monitored using the integrated head-end verification system consisting again of a number of camera/radiation detectors mounted in the cell walls, with additional camera units installed in the shear cell.

Solid waste in the form of leached hulls and fuel end-pieces will be verified indirectly by determining the Cm-244 using the hulls monitoring system and then relating this to the U:Pu:Cm-244 ratios in the spent fuel dissolver solutions.

8.4.9.2 MBA 2 - Main Process Area

Verification of flow within the MBA

Assurance that the process flows and facility operations are as declared will be achieved through solution monitoring. This includes not only sensors for temperature and for pressures to determine solution levels and density, but also neutron sensors on the extraction systems. In addition, random samples will be taken during the month and analyzed. Short period (5-15 days), sequential evaluations of the material unaccounted for (MUF) will be performed using NRTA methods.

Verification of inventory

Inventory verification will be carried out on a monthly basis during which the volume of all vessels will be verified and samples will be taken according to a random sampling plan for destructive analyses. Un-measurable inventories will be estimated using established process design algorithms. The MUF will be evaluated on an interim basis using Near-Real-Time-Accountancy (NRTA) methods.

Plutonium will be analyzed using a spectrophotometric method and Uranium is analyzed using isotope dilution mass spectroscopy. High Active Liquid Waste (HALW) batches which are shipped to Waste Treatment will be verified for volume and sample taking. High Active Solid Wastes (hulls and end-pieces) that are shipped to the Waste Storage Area will be verified using the waste crate assay system, which is based on passive neutron counting to measure the plutonium content (see Sections 8.4.7.1, 8.4.7.3 and 8.5).

Verification of physical inventory

A physical inventory verification (PIV) will be carried out once per year, during which the clean-out status and remaining solutions will be verified.

8.4.9.3 MBA 3 - Waste Treatment and Storage Area

Canisters of vitrified HAW are verified using a vitrified canister assay system. This system determines the Cm-244 content of the canister from neutron emission and uses the ratio of Pu:U:Cm-244 established by sample taking.

Physical inventory verification carried out once per year, during which the declared clean-out and inventory in the liquid waste treatment area will be verified.

8.4.9.4 MBA 4 – MOX Conversion Area

Verification of inventory changes

The MOX powder canisters will be verified prior to transfer to storage using a plutonium canister assay system. This system is based on high level neutron coincidence counting and high resolution gamma spectroscopy. A camera is recording the canister identification seals. In addition, samples of MOX powder will be taken for destructive analyses.

8.4.9.5 MBA 5 – Product Storage Area

Verification of inventory changes

Receipt of MOX canisters is verified using a series of neutron detectors and surveillance cameras. The neutron detectors combined with surveillance are located so as to detect the passage and direction of travel of a filled or empty MOX canister as it moves to the MOX storage room.

Verification of physical inventory

Physical verification of uranium and plutonium in storage is carried out once per year.

References Section 8.4

[1] Picket, S., Case study: safeguards implementation at the Rokkasho reprocessing plant, in: Doyle, J.: Nuclear Safeguards, Security and Nonproliferation, Butterworth and Heinemann, Elsevier (2008).

[2] Appendix A: Safeguarding reprocessing facilities. http://www.princeton.edu/~ota/disk1/1995/9530/953007.pdf

[3] Johnson, J., et al., Development of safeguards approach for Rokkasho reprocessing plant, IAEA-SM-367/8/01.

[4] Zendel, M., Experiences and trends for safeguarding plutonium mixed oxide (MOX) fuel fabrication plants, Journal of Nuclear Material Management, Vol.1, p.26-32, Feb. 1993.

[5] Reilly, D., et al., Nuclear material measurements technologies, in: Doyle, J.: Nuclear Safeguards, Security and Nonproliferation, Butterworth and Heinemann, Elsevier (2008).

[6] Gutmacher, R., Measurement uncertainty estimates for reprocessing facilities, LA-11839-MS, DE90 017853.

[7] Mayer, K., et al., Sample analysis methods for accountancy and verification. A compendium of applied analytical methods. Esarda Bulletin, No. 31 (1999).

8.5 Nondestructive assay of residual fuel on leached hulls and dissolver sludge

In reprocessing plants the chopped pieces of spent fuel elements fall into a basket of the dissolver tank. Boiling nitric acid dissolves the fuel and the hulls of the chopped fuel rods (Section 8.4.9.1). Pieces of structural material of the fuel elements remain in the basket. They are washed with nitric acid, but some particles of the fuel remain on the surface of the hulls. In addition there is small sludge of undissolved material which also contains small amounts of fuel. The amount of hulls and sludge are about 0.2 m^3/t of spent fuel [3]. They become high level solid waste.

Several nondestructive assay methods apply passive and active neutron counting [1,2,3]. A method which applies both active and passive neutron counting was described in [3]. For the active neutron counting a stationary Cf-252 neutron source is used. The method relies on the same principle which was already mentioned in Section 8.3.2 (Fig. 8.3.3) for spent fuel element assay in spent fuel storage pools.

The measurement system is schematically shown in Figure 8.19. The dissolver basket with the hulls is put into a water tank which is surrounded by a polyethylene (PE) reflector. There the stationary neutron source and on the opposite side 3 boron-10 lined neutron detectors are located.

The neutron counting is performed in 2 steps – with and without – Cf-252 neutron source. The minimum detectable amount of residual fuel is 30 g plutonium/m^3 for hulls and 90 g plutonium/m^3 for sludge.

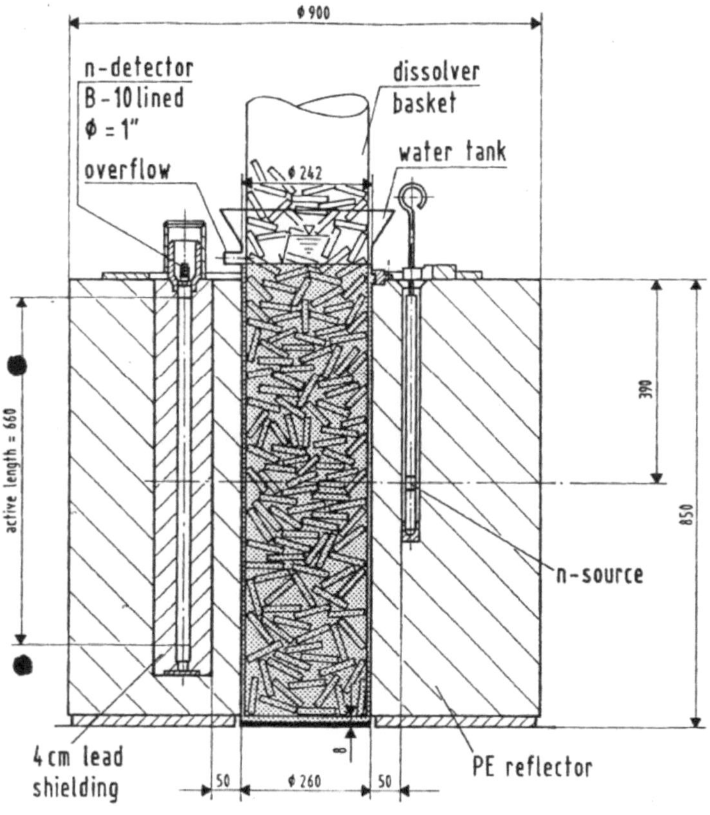

Figure 8.19: Arrangement for measurement of hulls in the dissolver basket [3].

References Section 8.5

[1] Reilly, D., Measurement of nuclear material process holdup, in Doyle, J.: Nuclear safeguards, security and nonproliferation, Butterworth and Heinemann, Elsevier, New York (2008).

[2] Pickrell, M. et al., New types of unattended systems for enrichment plant safeguards, (2006). See: http://www.bnl.gov/ISPO/BNLWorkshop07/Presentations/Pickrell.ppt

[3] Wuerz, H., Nondestructive determination of residual fuel on bunched hulls and dissolver sludges from LWR fuel reprocessing, Nucl. Eng. Design, 118, 123-131 (1990).

[4] Niura, N., et al., The use of curium neutrons to verify plutonium in spent fuel and reprocessing wastes, LA-12774-MS (1994).

8.6 MOX fuel fabrication process

In a large scale MOX fuel fabrication plant the MOX-powder (master mix) coming from the reprocessing plant (Section 8.4) is blended together with uranium oxide (UOX)-powder to achieve the required fissile enrichment. The mixed oxide powder is then mixed with a binder and pressed into green pellets. These are sintered in a sintering furnace at temperatures of 1000 and 1700 °C and then ground to the required dimensions. Finally they are loaded into zircaloy or steel tubes, to which end caps are welded. A high number (169 to 256) of fuel rods are then assembled to a fuel element [1]. Fig. 7.11 in Section 7.6 shows the MOX fuel fabrication process schematically.

8.6.1 MOX fuel fabrication plant

Modern MOX fuel fabrication plants are designed to produce MOX-fuel assemblies remotely in a fully automated process (Figure 8.20). MOX powder is transported in canisters from the reprocessing plant to the MOX fuel fabrication plant. There, the powder is transferred to the process glove boxes by an automated transfer machine. Due to the remote operation and the high radiation field in process areas, the MOX fuel storage areas are not normally accessible [1,3].

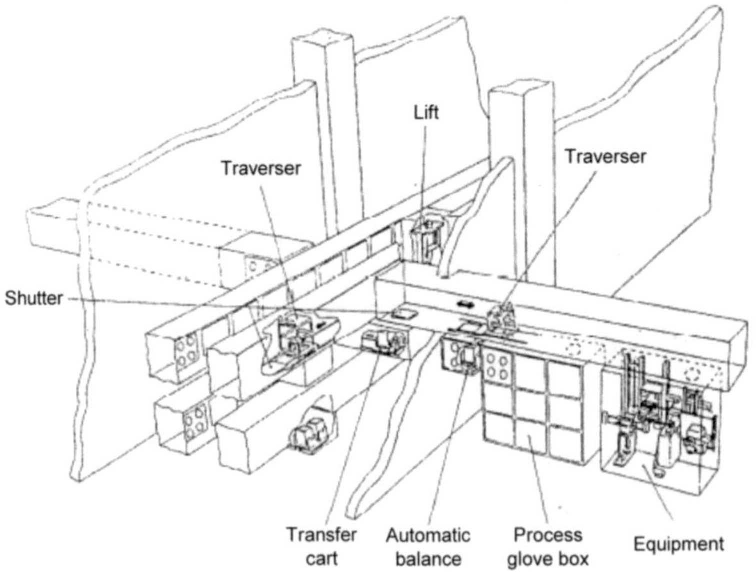

Figure 8.20: Glove box and intermediate storage area. These are connected by automatic transfer machines. During maintenance periods, the nuclear material can be removed from the glove box and placed in the storage area to reduce personnel exposure.

8.6.2 Safeguards approach

According to INFCIRC-153 the significant quantity (SQ) for MOX fuel is 8 kg (Table 8.1 Section 8.1.1). The timely detection period is one month. The entire MOX fabrication plant is included in one single Material Balance Area. The safeguards objective of IAEA is to detect both abrupt or protracted diversion and losses of nuclear materials across a material balance period of 1 month.

Key activities of the IAEA inspectors are:

- verification of transfers in and out of the plant
- interim inventory verification to confirm on a monthly basis that 8 kg of plutonium or more are not missing from the plant
- yearly physical inventory verification (PIV) to close the material balance period
- application of containment/surveillance measures

Nuclear material entering the plant in canisters contains seals which are verified by cameras. The oxide fuel powders are weighed before mixing and analyzed by destructive analysis (Section 8.1.5). The remote processing of the MOX powders and pellets takes place in glove boxes. These glove boxes are connected to intermediate storage areas. The material is transferred for processing in dedicated transfer containers from the intermediate storage to a glove box. When entering the glove box the container is weighed by a weigh cell. After processing in the glove box the fuel is transferred back to intermediate storage. The container is weighed again upon leaving the glove box [2].

The automated remote processing of the MOX fuel as well as the advancement of measurement capabilities and data processing allow the possibility of unattended mode verification by IAEA. The underlying measurement principle for the unattended mode verification is the measurement of the Pu-240 effective content by High Level Neutron Coincidence Counters (HLNC). These HLNCs stay continuously in the counting mode and automatically record the results which are transmitted to the data processing system. The HLNCs must be carefully calibrated [4]. In addition High Resolution Gamma Spectrometer (HRGs) and mass spectrometry are applied to determine the isotopic composition of the plutonium and confirm the effective Pu-240 content [1,2,3].

Hold-up of nuclear material in the process glove boxes is measured by a transportable glove box accountancy system. Near real time accountancy (NRTA) is applied in connection with all above described measurement and data processing systems. This allows the detection of possible falsifications of items in the inventory with virtual certainty within a sufficiently long time period [2]. The total process inventory is presented in accountable form and is verified at monthly intervals [2,3].

References Section 8.6

[1] Nakano, H., et al., Automation improves safety at Japan's Tokai MOX plant, Nucl. Eng. International, Dec 1989, p. 27-29.

[2] Sellinschegg, D., et al., Application of near real time accountancy (NRTA) at an automated mixed oxide (MOX) fuel fabrication plant, J. of Nucl. Materials Management (Feb. 1993).

[3] Zendel, M., Experiences and trends in safeguarding plutonium mixed oxide (MOX) fuel fabrication plant, J. of Nucl. Material Management, p.26-32, Feb 1993.

[4] Reilly, D., et al., Nuclear material measurement technologies, in Doyle, J., Nuclear Safeguards, Security and Nonproliferation, Butterworth and Heinemann, Elsevier (2008).

[5] Gmelin, W., Safeguards measures and efforts in conceptual fabrication plants for uranium and plutonium containing fuel elements, KfK 910, Kernforschungszentrum Karlsruhe (1969).

8.7 Assessment of Criticism of Safeguards for Large Scale Reprocessing Plants

8.7.1 Introduction

As referred to in Sections 8.1 and 8.4, the IAEA safeguards concept for large-scale reprocessing plants is based on a combination of

- materials accountancy,
- near real-time accountancy (NRTA),
- containment/surveillance measures (C/S).

This three-stage combined safeguards concept has been the subject of critical discussion again and again over the past few decades. One case in point referred to in that criticism was the Japanese reprocessing plant of Rokkasho-mura with its planned annual throughput of 800 tons of spent fuel. This reprocessing plant is the first commercial facility for reprocessing spent fuel to be built and operated in a non-nuclear-weapon state.

The two big commercial spent fuel reprocessing plants of La Hague (1700 t/y) and Sellafield (1200 t/y) are located in the two nuclear weapon states of France and the United Kingdom, respectively. They are not criticized by the same groups discussing critically the Rokkasho-mura plant [2,3]. France and the United Kingdtom are not required to place their reprocessing facilities under full IAEA safeguards [1]. Only parts of these large scale reprocessing plants are inspected by IAEA.

From the outset, the Rokkasho 800 t/y reprocessing plant had been planned in the presence of IAEA inspectors (Design Information Verification). The safeguards concept of the plant was defined jointly by the operator and IAEA after many years of public scientific discussion. So-called IAEA Resident Inspectors are permanently present. On site there is an IAEA laboratory facility for destructive analyses, which avoids time consuming shipment of radioactive samples from Rokkasho-mura to the IAEA laboratories at Seibersdorf.

8.7.2 Basis of Criticism

The basis of criticism can be summarized as follows [2,3]:

- INFCIR-153 requires all reactor-grade plutonium with a Pu-238 isotopic content ≤ 80% to be treated exactly like weapon-grade plutonium.
- The significant quantity of plutonium is 8 kg (Table 8.1). The time span for timely detection by IAEA inspectors is one month (Table 8.2). The detection time is the time span between diversion and detection by IAEA. Critics [1,2] assume a period shorter than a month, i.e. 7-10 days. The probability of detection recommended by IAEA-SAGSI is assumed to be 90-95% with a false-alarm rate of 5%.
- Exact compliance with these criteria also constitutes the basis of the U.S. Nuclear Non-Proliferation Act (NNPA) of 1978. In the opinion of critics, failure to observe the above criteria would not constitute a basis of effective safeguards control and, consequently, imply a high proliferation risk.
- At the first stage, "Materials Accountancy," of the IAEA safeguards concept it must be taken into account that measurement errors may occur in material balance accountancy of the nuclear materials in place (Section 8.1.1). As long as the measurement errors of the quantities of materials to be accounted for are relatively small compared to the significant quantity, SQ = 8 kg of plutonium, any diversion of nuclear materials can clearly be detected. However, there will be a problem when the errors in measurement of larger quantities of material greatly exceed a significant quantity, SQ, or even assume a multiple of an SQ. In that case, it will no longer be possible to distinguish clearly between a potential diversion of nuclear material and the measurement error produced within the established probabilities of 95% for detection or 5% for false alarm [1]. This is the case, for instance, when the overall measurement error is roughly 1%, and the quantity of plutonium to be accounted for over one year is approximately 7.2 tons contained in 800 tons of spent UOX fuel (0.9% plutonium in spent UOX fuel). This is in line with the design data of the Japanese large-scale reprocessing plant of Rokkasho.

- The Rokkasho plant annually, i.e. in 200 days of operation, processes 800 tons of spent UOX fuel elements with a mean concentration of 0.9% plutonium. This is 7.2 tons of reactor-grade plutonium per year. The measurement error in overall accountancy corresponds to 1%, i.e., the absolute value of the error in material unaccounted for, σ, (MUF) is 72 kg of plutonium or nine significant quantities, SQ. The 95% probability of detection and the 5% probability of false alarm correspond to 3 σ (MUF) = 246 kg of reactor-grade plutonium or more than 30 significant quantities.

  One obvious solution to the problem would be to reduce the period of one year to, e.g., one month or one week. However, this is not possible in practice as the plant must be subjected to a complete washout for every material balance accountancy operation. Only one or two inventory balance accountancy procedures in a year make practical sense.

  As a way out of this dilemma, the method of Near Real Time Accountancy (NRTA) and increased Containment/Surveillance (C/S) was developed.

8.7.3 Material Accountancy, Near Real Time Accountancy and Containment and Surveillance

NRTA means more frequent material balance accountancy without plant shutdown or washout. Continuous measurements by the process instrumentation are used for this purpose (Section 8.1.4.1). Both are available to the IAEA inspectors at any time.

At 200 days of full-load operation, the daily throughput of the Rokkasho 800 t/y reprocessing plant is 4 t/d of spent UOX fuel. At 0.9% plutonium concentration this leads to a plutonium throughput of 252 kg per week. The maximum measurement accuracy of 1%, for 95% probability of detection and 5% false alarm rate, results in a minimum diversion quantity of 3 σ (MUF) = 7.6 kg. This greatly reduces the risk of abrupt diversion. However, critics doubt [1,2] that the risk of protracted diversion can be diminished. This assertion is supported by an analysis by the U.S. Office of Technology Assessment [3], among other references, which states that IAEA first would have to substantiate its hopes in the NRTA method.

Employing more powerful containment/surveillance (C/S) methods as exemplified by seals, cameras etc. (Section 8.1.6) is considered by critics [1] as an important additional measure offsetting the deficits of the NRTA method referred to above. However, the absence of a logical way of combining the C/S method and the NRTA method into one quantifiable concept is criticized. Still, this is not a valid argument in itself.

IAEA together with the Japanese operator of the Rokkasho plant took into account all arguments put forward by the critics as outlined in the concept described in Section 8.4. This also applies to the MOX fuel refabrication plant built in Rokkasho-mura. Future operation of the reprocessing and refabrication plants will have to demonstrate the effectiveness of this three-stage combined safeguards concept of IAEA.

In addition, critics express concern that any NNWS acceding to the Non-Proliferation Treaty and running a reprocessing plant under IAEA safeguards may opt out of this NPT. It would then be able to use the available plutonium for building nuclear weapons right away.

Given the continued existence of the U.S. Non-Proliferation Act of 1978 and the non-proliferation policy accompanying it in the United States, there is a lot of mistrust of any reprocessing or MOX fuel refabrication activity in NNWSs.

References Section 8.7

[1] Report of the LASCAR-Forum: Large scale reprocessing plant safeguards, Appendix A, STI/PUB/922, IAEA, Vienna (1992).

[2] Miller, M., Are IAEA Safeguards on Plutonium Bulk-Handling Facilities Effective? Nuclear Control Institute, Washington, D.C. (1990).

[3] Barnaby, F., Planning for Failure – International Nuclear Safeguards and the Rokkasho-mura Reprocessing Plant, Oxford Research Group, Greenpeace International (2002).

8.8 Countercriticism and potential solution by proliferation-proof civil nuclear fuel cycles

The chief cause of existing mistrust is adherence to the ultraconservative provisions in INFCE-153 with the requirement that any kind of reactor-grade plutonium, except for plutonium with a Pu-238 isotopic content of ≥80%, be treated like weapon-grade plutonium. This requirement, that any reactor-grade plutonium be treated like weapon-grade plutonium, is not supported by scientific technical findings published in the open literature (Sections 9-13).

INFCE-153 dates from 1971. At that time, plutonium existed only in spent fuel elements from CANDU and gas-graphite reactors with <7 GWd/t burnup. That plutonium was far closer to the definitions of weapon-grade plutonium than today's reactor-grade plutonium in spent fuel elements with a burnup of 50-60 GWd/t from LWRs. These decisive differences are discussed in detail in Sections 9-10.

The difficulties of the NRTA concept emphasized by the critics are due very much also to this equating of reactor-grade plutonium with weapon-grade plutonium.

The differences between reactor-grade plutonium and weapon-grade plutonium are elaborated in Sections 9 and 10 below. A proposal is made to modify reactor-grade plutonium by changing the fuel in such a way that the concentration of the Pu-238 isotope is raised to roughly 5-6% or more. In this way, this is changed into denatured or proliferation-proof plutonium which cannot be used any more for making nuclear explosive devices (NEDs). Unlike some definitions in the literature and by IAEA, proliferation resistance as used in this publication does not mean an impediment to building nuclear weapons, but indicates that raising the Pu-238 isotopic concentration to more than, e.g., 5-6% makes NEDs technically unfeasible. The heat produced by such reactor-grade plutonium would cause the ambient chemical explosives to melt or initiate a chemical self-detonation of the explosive.

This is not meant to cast any doubt on the current IAEA safeguards system but remove the mistrust against reprocessing and plutonium recycling, and reduce it to a scientific basis.

Sections 9 and 10 show that this proliferation-proof reactor-grade plutonium can no longer be used for making nuclear weapons. Consequently, the risk anticipated to arise from the measurement error of 1%, which cannot be underrun in large-scale reprocessing and refabrication plants, is no longer relevant. There is no reason to fear either abrupt or protracted diversion. Opting out of the NPT and subsequently making use of the proliferation-proof reactor-grade plutonium to build nuclear weapons would make no sense. The reactor-grade plutonium then in possession of the government could not be used to make nuclear weapons.

A specific attempt to produce weapon-grade plutonium by shutting the nuclear reactor down and unloading it at an early point in time (after some weeks) and reprocessing such low-irradiated LEU UOX fuel elements or special U-238 elements can be detected by IAEA safeguards, e.g. antineutrino detectors (Section 8.3.5)combined with electronic data transmission to the IAEA headquarters.(Section 8.1.8).In this way very short reactor operation can be detected by IAEA and it can be verified,whether weapon-grade plutonium production is attempted by a State

If, however, the reactor is fuelled with MOX fuel from proliferation-proof reactor-grade plutonium, this too cannot be used any more for nuclear weapons at any point in time.

If an NNWS does not accede to the NPT and builds and runs on its own all nuclear plants required for making weapon-grade plutonium, it can be prevented from doing so only by diplomatic measures or other deterrents.

8.9 Proliferation-resistant or proliferation-proof?

It is known from pre-ignition theory [1] that the spontaneous fission neutron source produced by the even plutonium isotopes Pu-238, Pu-240 and Pu-242 decreases the attainable nuclear explosive yield of plutonium based nuclear explosive device (NEDs) (Section 9). Such NEDs become less attractive for military purposes. Therefore, a certain proliferation resistance is attributed to such plutonium isotopic compositions having higher contents of Pu-238, Pu-240 and Pu-242.

Plutonium resistance can also be attained by other means, e.g.

- denaturing of U-235 or U-233 by suitably large contents of U-238 increases the critical mass of such uranium isotopic compositions, e.g. to 800 kg for 20% U-235 in U-238 which would make such NEDs technically unfeasible. Therefore, enrichments of <20% U-235 in U-238 or <12% U-233 in U-238 are considered non-usable for NEDs [2,3].
- high spontaneous fission neutron radiation and gamma radiation does impede the fabrication process of NEDs
- plutonium with an isotopic content of >80% Pu-238 is considered non-usable for NEDs by IAEA (Section 8.1.1).
- reactor-grade plutonium from spent LWR fuel with a burnup of 60 GWd/t has such a high spontaneous fission neutron rate that pre-ignition would allow only so-called minimum nuclear explosive fizzle yields. In addition the relatively high alpha decay heat power would lead to intolerable temperatures in the implosion lenses of the NED (Section 9 and 10).

Theoretical concepts have been developed to define the degree of proliferation resistance. Attractiveness levels and safeguards categories were defined [3] and the multi-attribute utility analysis approach was proposed [4]. The figure of merit concept (FOM) in combination with different safgeguards categories was applied [3,5,6].

However, the results of such approaches are pre-determined by the definition of the attractiveness level or the definition of a figure of merit (FOM). Two examples shall explain the problem:

- the definition of the figure of merit is based on the formula [5,6]

$$\mathrm{FOM} = 1-\log\left(M\left[\frac{1}{800}+\frac{h}{4500}\right]+\frac{M}{50}\left[\frac{D}{500}\right]^{1/\log 2}\right)$$

M is the bare critical mass of the fissile metal product in kg
h is the heat content in W/kg
D is the dose rate of a fissile material sphere evaluated at 1 m from the surface in rad/(h·kg)

- This formula contains – among the two other criteria for critical mass and the radiation limit – a term for a heat power criterion which is based on the requirement that only plutonium with more than 80% Pu-238 isotopic content produces enough alpha-particle decay heat that it is non-useable for NEDs. The number 4500 is obviously based on a critical mass of 9.86 kg plutonium with 80% Pu-238 and an alpha-particle heat production of 570 W/kg Pu-238, i.e. 9.86x0.8x570 = 4500.
- The thermal analysis in Section 10 will show, however, that a near-critical metallic sphere of plutonium is already fully molten with only about 20-25% Pu-238 content. If this sphere of plutonium metal is surrounded by a reflector and by the implosion lenses the high explosives in the implosion lenses will melt or start chemical self-explosion already at a much lower Pu-238 content of about 5-7%. This means that the criterion of 80% Pu-238 in plutonium is unreasonably high.
- the definition of attractiveness factors based on the spontaneous fission neutron source only deals exclusively with the pre-ignition problem and neglects the alpha-particle heat production of the isotopes Pu-238 and Pu-240. Pellaud [7] proposed a criterion of 30% Pu-240 in the plutonium for sufficient proliferation resistance. As will be shown in Section 9 such a criterion would still lead to nuclear explosion yields around 0.5 kt TNT. The alpha-particle heat power in such a NED is so low that such NEDs would be technically feasible and the proliferation resistance is low.

Therefore, the above theoretical approaches are not applied in the following Sections 9 to 14. Only the theoretical methods of reactor physics and thermal analyses as well as material characteristics such as melting points or temperatures for chemical self-ignition are applied to a detailed analysis of so-called hypothetical nuclear explosive devices (HNEDs).

This allows limits for the Pu-238 isotopic content of reactor-grade plutonium to be determined above which reactor grade-plutonium becomes non-useable for NEDs. Such plutonium is considered **proliferation-proof** in the following Sections 9 to 14.

References Section 8.9

[1] Mark, J.C., Explosive properties of reactor-grade plutonium, Sci. Global Secur., 4, p. 100 (1993).

[2] Forsberg, C.W. et al., Definition of weapons-usable uranium-233, ORNL/TM-13517 (1998).

[3] National Academy of Sciences, The management and deposition of excess weapons plutonium, Committee on International Security and Arms Control (CISAC), National Academy Press, Washington, D.C. (1994).

[4] Charlton, W.S. et al., Proliferation resistance assessment methodology for nuclear fuel cycles, Nucl. Techn., 157, 143 (2007).

[5] Bathke, C.G. et al., The attractiveness of materials in advanced fuel cycles for various proliferation and theft scenarios, Proc. of Global 2009, Paris (2009).

[6] King, W.E. et al., The application of a figure of merit for nuclear explosive utility as a metric for material attractiveness in a nuclear material theft scenario, Nucl. Eng. and Design, 240, 3699 (2010).

[7] Pellaud, B., Proliferation aspects of plutonium recycling, J. Nucl. Material Management, 31 (1), p. 30 (2002).

9 Reactor-grade plutonium as a proliferation problem

9.1 Introduction

As outlined already in Section 1 reactor-grade plutonium presently is considered the main proliferation problem. After the International Fuel Cycle Evaluation Program [1], the USA decided around 1980-1982 to give up chemical reprocessing of spent LWR fuel and refabrication of plutonium/uranium mixed oxide (MOX) fuel as well as the use of MOX fuel in nuclear reactors. It had become apparent by then that the amount of civil plutonium produced in LWRs in the USA and elsewhere had exceeded the amount of military weapons plutonium in the NWSs (Sections 1.3 and 1.4). At the same time, leading US scientific organisations and US authors, e.g. the National Academy of Sciences [2], the American Nuclear Society [4], and Garwin [3] stated that reactor-grade plutonium could be used for nuclear weapons. These statements were repeated frequently. They are summarized in Fig. 9.1:

Unfortunately, plutonium of the quality produced in current nuclear power reactors, following separation from spent fuel and purification, can be used to make nuclear explosives, using technology comparable to that of the earliest plutonium weapons. While weapons made from this plutonium would have much less reliable and considerably lower explosive yields than similar weapons made from weapon-grade plutonium, they remain nevertheless highly dangerous nuclear explosives.

Figure 9.1: Statement regarding the misuse of reactor-grade plutonium [2,3,4].

Also the fast reactor (FR) programs were given up in several countries in the following decade. Declared US national nuclear waste policy was the direct disposal of spent fuel. Germany, among other nations adopted the same waste management policy for its nuclear fuel cycle in 2005. Other countries, such as France or Japan, did not follow this line. More recent discussions in the USA envisage retrievability of conditioned spent fuel from geological disposal sites to provide for a later turnaround of the fuel cycle policy of US-DOE [5]. Also, new research programs including recycling strategies for the nuclear fuel cycle (use of reprocessing and fuel refabrication) were initiated [5,6]. A number of publications, e.g. by DeVolpi [8,27], Pellaud [9], and DeVolpi et al. [10] either expressed doubts about the validity of the statement quoted above in Fig. 9.1 or complained that it was not scientifically sound, let alone justified [11]. Therefore, Kessler et al. [12], attempted to find a scientific basis for these statements and the issues it entailed. They analyzed the potential nuclear explosive energy and the thermodynamic characteristics of so-called hypothetical nuclear explosive devices (HNEDs) based on reactor-grade plutonium. There was found to be a limit for reactor-grade plutonium containing a certain percentage of Pu-238 at which such devices would become technically unfeasible [13]. The reason is that the heat produced by Pu-238 would cause the chemical high explosives around the fissile material either to melt or to ignite spontaneously in a chemical reaction. Denatured reactor-grade plutonium of this type could be produced in various fuel cycle options investigated [14]. They require closed fuel cycles with reprocessing and refabrication of the fuel. All plutonium can be burnt except for unavoidable losses in chemical reprocessing and refabrication, which must be disposed of together with the fission products [15]. These new fuel cycle strategies would allow a proliferation-proof denatured reactor-grade plutonium fuel cycle [14]. Even the minor actinides could be incinerated provided the necessary chemical separation processes were applied on a technical scale [16,17,18]. The minor actinides could also be taken care of by recycling in different reactors, e.g., LWRs, fast-spectrum reactors (FRs), or accelerator driven systems (ADSs) [15]. However, this raises also questions of possible misuse of neptunium and americium [23,24,25]. The main findings of the analyses will be outlined below in the following Sections.

9.2 Nuclear characteristic data of plutonium which are important for the assessment of the plutonium proliferation problem

For nuclear explosive devices (NEDs) the metallic form of the fissile materials, e.g. plutonium or uranium with its highest density is used. This leads to a neutron spectrum with an average neutron energy of about 1 MeV. In this energy range (Fig. 9.2) all plutonium isotopes (Pu-238, Pu-239, Pu-240, Pu-241, Pu-242) have relatively high microscopic fission cross sections [19]. Also Am-241 has a relatively high microscopic fission cross section. The spontaneous fission neutron rates of the plutonium isotopes Pu-238, Pu-239, Pu-240, Pu-241 and Pu-242 were already given in Table 8.5 (Section 8.1.5.3). These spontaneous fission neutrons are extremely important for the so-called pre-ignition problem of plutonium based NEDs and for the achievable nuclear explosive yield, as will be explained in Sections 11 and 12. The spontaneous fission neutron rates for the different plutonium isotopes are collected in Table 9.1. The critical masses for $k_{eff} = 1$ can be calculated by the numerical solution of the Boltzmann neutron transport equation applying the respective group cross section sets (Sections 4.3 and 4.4). The critical masses are listed in Table 9.1 [19].

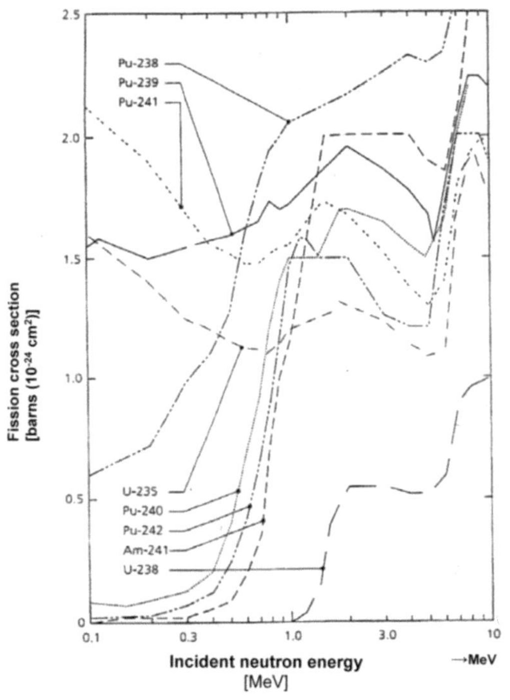

Figure 9.2: Fission neutron cross sections of the principal plutonium and uranium isotopes (and americium-241, decay product of Pu-241) as a function of neutron kinetic energy [19].

The heat produced by α-decay of the different plutonium isotopes Pu-238, Pu-239, Pu-240, Pu-241 and Pu-242 is also listed in Table 9.1. This alpha-particle heat produced in reactor plutonium can make so-called HNEDs technically unfeasible (Section 10). Pu-238 is a strong α-emitter with a half-life of 87.7 a. It produces 570 W/kg alpha-particle heat power. A reflector of natural uranium or beryllium etc. decreases these critical masses even somewhat further. For weapons plutonium with a high Pu-239 isotopic content (≈94%) the critical mass (with reflector) is in the range of 7.2 to 7.5 kg, whereas for reactor-grade plutonium the critical masses (with reflector) are in the range of 10 to 13 kg (Section 9.3).

The relatively low critical mass of the different plutonium isotopes (Table 9.1) is often the basis of conclusions that a NEDs – albeit perhaps of unpredictable yield because of the high spontaneous fission neutron source – could be designed using <u>any</u> isotopic composition of plutonium. However, this conclusion neglects the very important fact that a Pu-238 content of several percent makes such NEDs or better HNEDs technically unfeasible (Section 10).

Table 9.1: Characteristic data of different Pu-isotopes [8,19].

	Pu-238	Pu-239	Pu-240	Pu-241	Pu-242
α-heat power [W/kg]	570	1.9	6.8	3.3	0.15
spontaneous fission neutron source [n/g s]	2600	0.02	910	0.05	1700
Bare critical mass [kg] (Plutonium metal)	8.2	10.0	33.6	12.4	70.2

9.3 Isotopic compositions of weapons plutonium and reactor-grade plutonium

Table 9.2 shows the isotopic composition and characteristic data, e.g. critical mass, α-particle heat power and spontaneous fission neutron source of super-grade weapons plutonium and of weapons plutonium (as defined by US-DOE and US-NRC). In addition, it presents the characteristic data for plutonium which can be generated after very short irradiation time (1.250 MWd(th), e.g. in a heavy water moderated CANDU reactor. In AGRs or LWRs the plutonium isotopic composition would be similar for such short irradiation times of about 2 months [21]. Super-grade weapons plutonium can only be generated within even shorter irradiation times. While the LWRs would have to be shut down each time for unloading of such short irradiated fuel elements, the CANDUs and AGRs with continuous fuel element loading and unloading capabilities would be better suited for such operations.

Table 9.2 shows also an assumed isotopic composition of the US-test of 1962 reported by the US-DOE in 1977 and confirmed in 1994 [7] with plutonium originating from AGR spent fuel of the UK. The assumed isotopic content of 12% Pu-240 was estimated by Pellaud [9] and discussed by Carlson et al. [20].

Table 9.3 presents the plutonium isotopic compositions and characteristic data, e.g. critical mass, alpha-particle heat power and spontaneous fission neutron source for AGR and CANDU spent fuel after 5000 MWd/t and 7.500 MWd/t burnup. These reactors allow only low burn-up, since they are fueled with either natural uranium or only 1.5% U-235 enrichment of their fuel. Isotopic compositions of plutonium from MAGNOX and CANDU spent fuel are very similar.

Table 9.3 shows these characteristic data also for plutonium produced in the blankets of FBRs. This FBR blanket plutonium is very similar to the weapons plutonium category shown in Table 9.2.

Table 9.2: Isotopic composition (weight fractions), subcritical mass (reflector of 5 cm uranium metal), alpha-particle heat power and spontaneous fission neutron source of metallic weapon-grade plutonium and plutonium of low burnup fuel from CANDUs or AGRs.

Type	Plutonium Isotope	Pu-238	Pu-239	Pu-240	Pu-241	Pu-242	Subcritical mass [kg] keff= 0.98	alpha- particle heat power [W]	spontaneous fission neutron source [n/s]
A-1	Super-grade weapons Pu [2,4]	~0	>0.97	<0.03	~0	~0	7.22	14.8	$1.97 \cdot 10^5$
A-2	Weapon-grade Pu (DOE, NRC) [2,4]	0.00012	0,938	0,058	0,0035	0,00022	7,35	16,6	$3.93 \cdot 10^5$
A-3	HWR (CANDU) 1250 MWd/te	~0	0,969	0,029	~0	~0	7,20	14,7	$1{,}9 \cdot 10^5$
A-4	US-test 1962 UK-plutonium [9]	~0	0,88	0,12	~0	~0	7,4	18,4	$8{,}08 \cdot 10^5$

Table 9.3: Isotopic composition (weight fractions) as well as sub-critical mass (reflector of 5 cm uranium), alpha-particle heat power and spontaneous fission neutron source of plutonium separated from MAGNOX and CANDU or FBR blanket spent fuel (Chebeskov [21]).

Type	Reactor type	Burn-up, GWd/t	Plutonium isotopic composition					Subcritical mass [kg] $k_{eff} = 0.98$	alpha-particle heat [W]	Spont. neutron source [n/s]
			Pu-238	Pu-239	Pu-240	Pu-241	Pu-242			
B-1	MAGNOX	5.000	~0	0,685	0,250	0,053	0,012	9,14	29,05	$2{,}27 \cdot 10^6$
B-2	CANDU	7.500	~0	0,66	0,265	0,055	0,015	9,35	30,4	$2{,}49 \cdot 10^6$
B-3	FBR blanket	—	0,0002	0,956	0,042	0,002	0,00003	7,30	16,2	$2{,}83 \cdot 10^5$

Table 9.4 shows the plutonium isotopic composition of LWR spent fuel as a function of burnup. The average burnup of LWR spent fuel was about 33,000 MWd/t around 1990. This burnup was increased for commercial reasons up to about 55,000 MWd/t until 2010. The future objective for LWR fuel is a burnup of 60,000-70,000 MWd/t for LWRs.

Table 9.4 shows that the Pu-238 content increases steadily with burnup to about 5% for a burnup of 72 GWd/t. Due to the relatively short half life for α-decay of 87.7 y for Pu-238 this isotopic concentration decreases slightly (about 7% of its initial value) during 10 a after unloading of the spent LWR fuel from the reactor core. The even shorter halflife for β-decay of Pu-241 of $T_{1/2}$ = 14.35 a leads to a decrease of about 64% of the concentration for Pu-241 after 10 y from unloading the LWR spent fuel elements from the reactor core. This decrease of the concentration of Pu-241 leads to buildup of Am-241. This would require frequent chemical separation of the americium from the reactor-grade plutonium.

Table 9.4: Isotopic composition (weight fractions) as well as subcritical mass (reflector of 5 cm uranium), α-decay heat power and spontaneous fission neutron source of plutonium separated from LWR spent fuel versus its burn-up and for a decay time of 1 year (Chebeskov [21]).

Type	Spent fuel burnup [GWd/t]	Pu-238	Pu-239	Pu-240	Pu-241	Pu-242	Subcritical mass [kg] k_{eff} = 0.98	alpha-particle power heat [W]	Spont. neutron source [n/s]
C-1	30	0.016	0.565	0.238	0.128	0.054	9.24	112	$2.92 \cdot 10^6$
C-2	50	0.029	0.533	0.233	0.139	0.066	9.84	187	$4.15 \cdot 10^6$
C-3	60	0.038	0.518	0.231	0.142	0.072	9.85	243	$4.24 \cdot 10^6$
C-4	72	0.050	0.502	0.226	0.145	0.078	9.88	311	$4.61 \cdot 10^6$

The alpha-particle decay heat of 112 (W) for C1 is corrected against the value of 144 W in [13] because the alpha-particle heat of Pu-242 was too high.

Table 9.5 presents another plutonium isotopic composition evolving after three times recycling of the plutonium in MOX fuelled LWRs [22]. This plutonium isotopic composition was also used for the calculation of the potential explosive yield calculations [12] which will be discussed in Section 9.12. This plutonium (called D-1 in Table 9.5) is considered proliferation-proof [13,23].

Table 9.5: Isotopic composition (weight fraction) as well as subcritical mass (reflector 5 cm uranium), α-decay heat power and spontaneous fission neutron source for plutonium composition Pu(1) from [13,22].

Type	Pu(1)	Pu-238	Pu-239	Pu-240	Pu-241	Pu-242	Subcritical mass [kg] $k_{eff} = 0.98$	alpha-particle power heat [W]	Spont. fission neutron source [n/s]
D-1	multi-recycling in MOX-LWR	0.055	0.341	0.311	0.106	0.187	12.91	445	$9.6 \cdot 10^6$

In Table 9.6a and 9.6b additional reactor-grade-plutonium isotopic compositions and characteristic data are given, e.g. critical mass, α-particle heat power and spontaneous fission neutron source. They are all considered proliferation-proof as shown in [13,23,25]. This will be explained and discussed in Section 10. These plutonium isotopic compositions belong to future fuel cycle options which are specifically proposed in Section 12.4 to generate proliferation-proof reactor-grade-plutonium [14,23,25].

Table 9.6a: Different initial fuel compositions of PWR cores for the generation of denatured proliferation resistant plutonium [14].

Fuel type	pitch/ diameter P/D	Fuel composition	Th wt %	U wt %		Plutonium wt %		MA wt %
				Total	Fissile Fraction	Total	Fissile Fraction	
E-1	1.44	Re-enriched recycled U (RRU)	0	100	5.52	0	0	0
E-2	1.34	RRU U + Pu	0	93.9	5.00	6.1	64.5	0
E-3	1.41	RRU U + Pu + MA	0	92.5	5.00	6.5	64.5	1.0
E-4	1.59	Enriched U + Th + Pu+MA	52.6	35.1	6.00	10.7	64.5	1.6

RRU (Re-enriched recycled uranium), MA (minor actinides)

Table 9.6 b: Isotopic composition (weight fraction), spherical masses (reflector: uranium), α-decay heat power and spontaneous fission neutron source of proliferation-proof plutonium from different fuel types (Table 9.6a) as a function of burnup in [14,23].

Fuel type	Burnup [GWd/t]	Plutonium composition [%]					Subcritical mass [kg] (k_{eff} = 0.98)	Alpha-particle heat power [kW]	Spont. neutron source [n/s]
		Pu-238	Pu-239	Pu-240	Pu-241	Pu-242			
E-1	50	0.095	0.505	0.217	0.132	0.051	9.48	0.536	$5.04 \cdot 10^6$
	58	0.114	0.461	0.225	0.134	0.066	9.78	0.659	$6.0 \cdot 10^6$
E-2	58	0.058	0.41	0.263	0.178	0.091	10.61	0.380	$5.78 \cdot 10^6$
	66	0.068	0.398	0.258	0.18	0.096	10.64	0.441	$6.11 \cdot 10^6$
E-3	49	0.096	0.393	0.272	0.157	0.082	10.58	0.611	$6.73 \cdot 10^6$
	58	0.107	0.374	0.268	0.161	0.090	10.67	0.684	$7.2 \cdot 10^6$
E-4	49	0.089	0.337	0.329	0.150	0.095	10.49	0.567	$7.24 \cdot 10^6$
	58	0.10	0.306	0.333	0.158	0.103	10.59	0.640	$7.82 \cdot 10^6$

Tables 9.2 to 9.6.b show the wide variety of plutonium isotopic compositions generated by the application of civil nuclear energy. They have similar critical k_{eff} = 1 or subcritical (k_{eff} = 0.98) masses between 7 and 13 kg. However, the α-particle heat power varies between 14 (W) and 680 (W) for a subcritical (k_{eff} = 0.98) NED or HNED. The spontaneous fission neutron sources of such NEDs or HNEDs vary between about $2x10^5$ n/s and $7.8x10^6$ n/s.

The IAEA takes a very conservative approach and considers any kind of plutonium composition capable of being used for nuclear weapons (only exemption is plutonium with ≥80% Pu-238). The IAEA does define 8 kg plutonium as quantity of safeguards significance (Table 8.1, Section 8.1.1) and does not distinguish between weapon-grade plutonium and reactor-grade plutonium. This has lead to critical discussions [8,9,10,11,23,27]. It will be seen in Section 9.12 and 10 that the critique is justified.

Distinction should be made between:

- weapon-type plutonium (Table 9.2 cases A-1 to A-3 and Table 9.3 case B-3)
- reactor-grade plutonium from spent fuel with low burnup (cases B-1 and B-2 in Table 9.3)
- present LWR reactor-grade plutonium from spent fuel with a burnup 30,000 to 60,000 MWd/t (cases C-1, C-2 and C-3 in Table 9.4)
- proliferation-proof plutonium with a Pu-238 isotopic content of more than 5 to 6% (case C-4 in Table 9.4, case D-1 in Table 9.5, cases E-1, E-2, E-3, and E-4 in Table 9.6b).

9.4 The potential nuclear explosive yield of reactor-grade plutonium

9.4.1 Introduction and scientific approach

The scientific assessment of the question whether reactor-grade plutonium can be used for nuclear weapons must include the investigation and discussion of the following:

- a neutronic evaluation of the super prompt critical explosion of a subcritical sphere of reactor-grade plutonium initiated by an implosion type compaction process (Section 9.10 to 9.12)
- a thermal analysis of the implosion type design of a hypothetical nuclear explosive device (HNED) in case of considerable alpha-particle heat power produced by the reactor-grade plutonium including realistic dimensions and material characteristics (thermal conductivity, melting points etc.) (Section 10).
- an evaluation of the technical difficulties involved in constructing an implosion type nuclear explosive device with reactor-grade plutonium (Section 10.14.4).

The statements given by US scientific organizations shown in Fig. 9.1 seem to deal only with the neutronic evaluation. The problems resulting from the thermal and technical analyses were obviously considered as a minor problem in this statement. While this is correct for weapon-grade plutonium, it is not correct for reactor-grade plutonium from LWR spent fuel with higher burnup or even for reactor-grade plutonium with a Pu-238 isotopic content of e.g. about 5 to 6%. This latter plutonium can be considered proliferation-proof. Section 9.3 presented four fuel cycle options (see Table 9.6a and 9.6b) which lead to sufficiently high Pu-238 percentages in the spent fuel. Fuel types E-1 contains RRU in the fresh fuel elements and would lead to denatured proliferation-proof plutonium with more than 9% Pu-238 after a burnup of about 50,000 MWd/t. Fuel type E-2 contains plutonium separated from present spent LEU-UOX LWR fuel with a burnup of 50,000 MWd/te and additional RRU. It would lead to about 6% Pu-238 in the denatured proliferation-proof plutonium after a burnup of 60,000 MWd/t. Fuel type E-3 contains the plutonium and RRU as fuel type E-2 and in addition about 1% minor actinides. It would produce denatured proliferation-proof plutonium with about 11% Pu-238 after a burnup of 60,000 MWd/te. In fuel type E-4 some of the uranium in fuel type D-1 would be replaced by thorium. This again would lead to denatured proliferation-proof plutonium with 11% Pu-238 and about 3% U-233 (denatured in U-238) after a burnup of 60,000 MWd/te. All these Pu-238 percentages would lead to high enough alpha-particle heat power between 0.38 and more than 0.68 kW in the HNEDs. All these high Pu-238 percentages could, of course, be adjusted to obtain lower percentages of Pu-238 by adapting the initial fuel compositions.

9.4.1.1 Future proliferation-proof fuel cycles

In a future transition phase of nuclear energy all the existing present plutonium – after reprocessing – could be converted in into denatured, proliferation-proof plutonium within about five to six years of irradiation time in LWRs of NWSs. RRU could be enriched in already existing enrichment centers. Existing reprocessing centers could be extended by additional reprocessing centers preferably in NWSs. The conversion of present reactor-grade plutonium is only a question of reprocessing and refabrication capacity, since refabrication of the fuel and burnup of, e.g. the fuel type E-1, E-2, E-3 and E-4 and reprocessing of the spent fuel needs only about 12 years.

Denatured proliferation-proof plutonium would need only similar safeguards measures by the IAEA as presently required for LEU-UOX LWR fuel or <12% U-233 denatured uranium. This denatured proliferation-proof reactor-grade plutonium could then be incinerated in a civil denatured proliferation-proof fuel cycle in NNWSs.

Only for reactor-grade plutonium with an alpha-particle heat power below the above discussed limits and plutonium from fuel elements of research reactors or plutonium from defective fuel elements (which must be unloaded before reaching their maximum burnup) present IAEA safeguards measures for plutonium must remain.

9.4.1.2 Minor actinides as a proliferation problem

An additional proliferation problem arises for the minor actinides neptunium and americium. Reactor americium can be considered proliferation-proof as it appears in all fuel cycle strategies as a mixture of the three isotopes Am-241, Am-242m and Am-243. Also the isotope Am-241 which originates from the decay of Pu-241 can be considered as proliferation-proof. This was shown by Kessler [24] (Section 11).

The minor actinide neptunium is nuclear weapons usable and cannot be denatured. Therefore, all proposals to keep the minor actinides together with the plutonium in fuel type like, e.g. fuel type E-3 or E-4 are not yet a solution for the proliferation problem. Neptunium can be separated chemically like reactor-grade plutonium. This should be done only during the transition phase preferably in reprocessing centers of NWSs. In the subsequent phase of a civil denatured proliferation-proof fuel cycle neptunium must be avoided. A proposal for a fuel strategy which avoids neptunium, while denatured proliferation-proof plutonium and americium can be incinerated is discussed in Section 13. The reprocessing centers in the civil denatured proliferation resistant fuel cycle could also be multilateral reprocessing centers as recommended by IAEA [26].

9.4.2 Earlier analysis of the potential nuclear explosive yield

No detailed model for the analysis of the potential nuclear explosive yield of reactor-grade plutonium had been published until 2009. Only relatively simple parametric models were given by Seifritz [28], deVolpi [8], and by C. Mark [19].

In a new attempt to find a scientific basis for the statements quoted above and for issues raised, the theory and calculation procedures developed for reactor safety analysis of early plutonium metal-fueled fast reactors were used by Kessler et al. [12]. These calculation models from the early stages of reactor safety studies, the so-called reactor disassembly analysis (Bethe et al. [29], Jackson et al. [30], Smith et al. [31]), are very similar to the explosion analysis of an HNED, based on reactor-grade plutonium.

9.5 Design principle and geometrical dimensions of early NEDs

One part of the statement of Fig. 9.1 claims that reactors

> ".....can be used to make nuclear explosives, using technology comparable to that of the earliest nuclear weapons,.....".

It is, therefore, important to understand the design principles of the earliest NEDs. It is known from Los Alamos Primer [32] or historical books, e.g. by R. Rhodes [33], about the development of an atomic bomb that the so-called implosion method must be applied. The reasons are the high spontaneous fission neutron rates of some of plutonium isotopes, e.g. Pu-240. Weapons-grade plutonium must consist essentially of the isotope Pu-239 and only a very small amount (one to several %) of the isotope Pu-240. As can be seen from Figs. 4.8a and 4.8b (Section 4) this is only possible for fuel of extremely low burnup.

Information on the dimensions of a nuclear explosive device is very rare in the open scientific literature. According to R. Rhodes [33] the first nuclear fission bomb with weapon-grade plutonium (Fat Man) functioned along the implosion method and had the following design shown in Fig. 9.3.

- in the center of the NED was the super-grade weapons plutonium sphere of 6.2 kg Pu-metal (delta phase plutonium with a density of 15.8 g/cm^3 and 0.8% gallium. The bare sphere consisted of 2 hemispheres (4.5 cm radius) with a small 2.5 cm hole in the center for the initiator (neutron source). The plutonium isotopic composition was 99.1% Pu-239 and 0.9% Pu-240 [74].
- this Pu-sphere was surrounded by a 13.5 cm thick reflector or tamper of natural uranium metal (density 19.8 g/cm^3) weighing 120 kg [74].

- Around this reflector followed an aluminum layer of 11 cm thickness for dampening and smoothening the Rayleigh Taylor instabilities generated by small imperfections of the spherical shock waves.
- the aluminum layer was surrounded by 32 explosive lenses (20 hexagonal and 12 pentagonal). They had been molded and machined with high precision such that they fitted exactly together to a hollow sphere (Fig. 9.3). Each explosive lense had 2 electrical detonators. All detonators were fired simultaneously with a very small time jitter of about 10^{-6} sec.
- Around the explosive lenses was an aluminum shell, which was housed in an outer steel shell.
- The initiator (neutron source) consisted of 92% Po^{210}/Be and was inherent i.e. produced no neutrons up to the moment when the spherical shock wave from the chemical detonation of the implosion lenses destroyed the foils between the Po^{210} and the Be and mixed both materials together. This suddenly generated about $9.5 \cdot 10^7$ n/s. This had to happen only when the Pu was compressed at its maximum i.e. the shock wave had reached the center.
- All dimensions added up to an outer diameter of about 150 cm. Its total weight was 4.6 t. The nuclear explosive yield was 22 kt (TNT) [33,74].
- The sophisticated art of the implosion method consists in the geometry of the implosion lenses and the choice of the explosives. "Fat Man" had 2 different explosives: Baratol with a "relatively slow" detonation velocity (5 km/s) and Composition B with a "relatively fast" detonation velocity (8 km/s). When the explosive system is fired by the electrical detonators at the outer surface the detonation waves spread out spherically around each initial detonation point (Fig. 9.3). The detonation waves will show interferences with each other, but not have the form to be able to compress the U_{nat}-Pu-sphere. Therefore when the detonation wave reaches and penetrates into the slow explosive its propagation (detonation) velocity becomes smaller, whereas around it the detonation waves in the fast explosive overtake those in the slow explosive and assume a detonation front (convex to concave), which becomes more and more spherical with its center identical to the center of the plutonium-sphere. Oscillations (Raleigh-Taylor instabilities) whose amplitude would increase during propagation to the center are dampened and smoothened by the Al-hollow shell.
- The active chemical explosive lense system of "Fat Man" was 47 cm thick and weighed 2.5 tonnes [74].

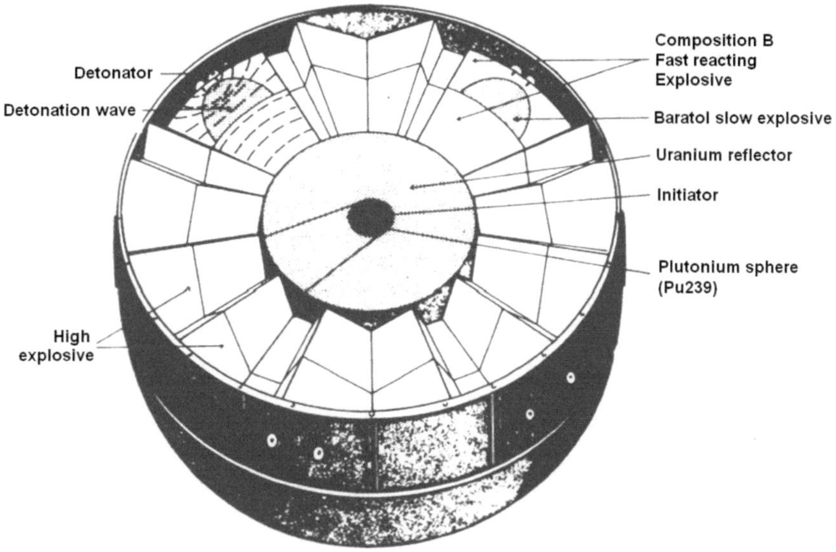

Figure 9.3: Design of the first plutonium NED (Fat Man) [33].

Hollow spherical shells of plutonium metal are more efficient for the implosion process than solid plutonium spheres (Seifritz [28] and Leuthäuser [34]). However, single hollow shells were not realized in the early phase of NED designs, because of fears that the hydrodynamic stability would not be sufficient. This stability problem was solved by putting a smaller solid Pu-sphere into a hollow Pu-sphere with some open space in between [28,35]. Therefore, part of the plutonium was arranged in the center levitated by wires (very thin but strong enough) and the remaining part was arranged as a hollow spherical shell. The implosion shock wave is accelerating the outer spherical shell parts and impacting then on the inner solid sphere. This is claimed to lead to higher compression and higher explosion yields [28,33,34]. It will be shown in Section 9.12 that these designs are not important for the analysis of HNEDs with reactor-grade plutonium and early pre-ignition. But they will be analyzed in the thermal analysis of HNEDs (Section 10).

In the USA and in Russia [36] the gun-type design was soon given up in favor of the implosion principle (hollow shell) also for U-235 as fissile material. Composite U-235 and Pu-designs (Pu as a solid inner sphere and U-235 as a hollow spherical shell) together with newly developed external impulse neutron initiators lead to a decrease in dimensions and weight and increased the yield by a factor 1.5 to 1.7. Obviously many nuclear weapons of the weapons countries followed this composite design.

9.6 Scientific basis for the discussion of the potential nuclear energy of reactor-grade plutonium

In an effort to find a scientific basis for the statements quoted in Fig. 9.1 and the issues raised about the potential yield of reactor-grade plutonium based HNEDs, Kessler et al. [12] described a model for calculating the power burst initiated by a strong positive reactivity insertion. This will be caused by compression of a plutonium metal sphere surrounded by a natural-uranium metal reflector. A verification of the calculation procedures and materials characteristics data employed, especially the equation of state, is achieved by recalculation of the results of a nuclear explosion published in part by Sandmeier [37].

The theoretical models of plasma physics will have to be applied here, as in reactor safety analysis the equation of state data of the fissile materials are needed only for a temperature range less than 7000 K. After all, calculations of nuclear explosion yields indicate that temperatures of millions of K and pressures of 10^3-10^4 TPa are reached. The theoretical model chosen, the materials data and the equation of state data employed are seen to reproduce the published results of Sandmeier [37] sufficiently well. In the later Section 9.9 a confirmation of the so-called Serber relation is also shown [28,32].

On this basis, the differences are discussed between nuclear explosive devices (NEDs) and so-called HNEDs (hypothetical nuclear explosive devices) for which reactor-grade plutonium shall be used.

Spherical critical arrangements with fast spectra based on uranium metal or plutonium metal, e.g. GODIVA or JEZEBEL, had been built and operated for studies of reactor physics and reactor kinetics (Paxton [38] and Wimett et al. [39]). Later, such safety analyses were also conducted for the early metal-fuelled experimental fast reactors, like EBR-I and EFFBR (McCarthy [40] and Nicholson [41]). Originally, these tools had been developed for calculating spherical geometries.

In such power burst experiments and safety analyses of spherical critical arrangements, the neutron density or power rises extremely fast after transition to the prompt critical regime. Unless the power rising exponentially is not turned around in due time by a sufficiently strong, inherent negative Doppler coefficient, the rapidly accumulating internal energy causes melting of the fuel, fuel vapor pressure buildup, and fast disintegration of the device under high internal pressures.

The high internal pressures drive the whole NED or HNED apart in an explosion expanding rapidly. The theoretical treatment of this so-called disassembly mechanism in the spherical HNEDs to be investigated is identical to the approach applied to the early spherical fast spectrum criticals or experimental fast reactors (McCarthy [40] and Nicholson [41]).

Although the calculation models of so-called unprotected reactivity accidents (no shutdown by absorber rods) in reactor safety and explosion yield of HNEDs are similar, many physics parameters differ greatly:

- the high density of metal fuel and the absence of a coolant or moderator make for a very hard neutron spectrum with an average neutron kinetic energy around 1 MeV which is above the resonance energy range. In addition, the small content of U-238 and Pu-240 is mainly responsible for the absence of a Doppler coefficient in HNEDs.
- the neutron lifetime in HNEDs is at least one order of magnitude shorter than in FR cores.
- the extremely high reactivity ramps and the absence of a Doppler coefficient raise the power or power density many orders of magnitude above the levels in reactor cores. Internal energies accumulate so fast that the internal pressure rises to approx. 10^4 TPa and the HNED explodes, releasing large amounts of energy.

All HNEDs investigated are spherical, containing a solid spherical or a hollow spherical metal core of reactor-grade plutonium (Fig. 9.4). From the very beginning of nuclear weapon development it had been obvious that – because of the strong internal source of spontaneous fission neutrons – only implosion in a spherical geometry would be successful if plutonium was used as a fissile material (Mark [19]). It will be shown in Section 12 that the reactor grade plutonium due to its higher spontaneous fission neutron rate always leads to very early pre-ignition of the nuclear power burst.

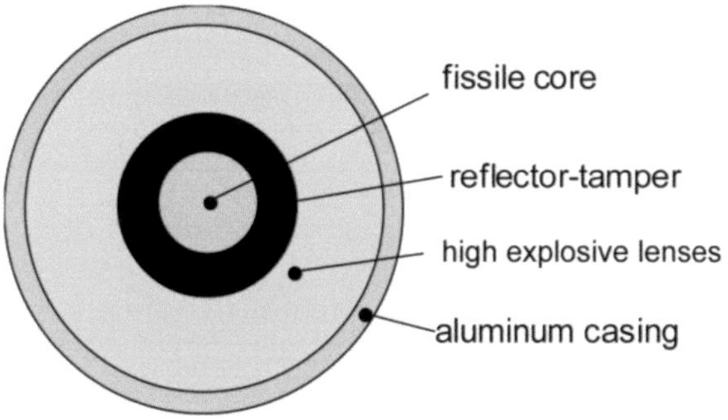

Figure 9.4: Geometric arrangement for studies of HNEDs with - plutonium.

9.7 Equations describing superprompt Critical power excursion and explosion

The compaction of reflector and the plutonium sphere by the spherical shock waves generated in high explosive lenses causes an increase in criticality $k_{eff}(t)$. After start of the neutron chain reactions, the neutron density or the neutron flux increase exponentially.

Under the assumption of spherical symmetry, the plutonium sphere and reflector can be divided into spherical shells described by materials characteristics, e.g. material density, nuclear cross sections, and distributions of neutron density, power, temperature etc. (Fig. 9.5).

Only the superprompt critical regime has to be considered as delayed neutron regime is passed extremely fast by strong positive reactivity insertions caused, e.g., by strong shock compression. Delayed neutrons do not play a role,decay times of the delayed neutron precursors being too long (Section 4.10).

Neutron density or neutron flux and power start at very low levels determined either by a neutron source suddenly generating neutrons or by inherent spontaneous fission neutrons causing pre-ignition.

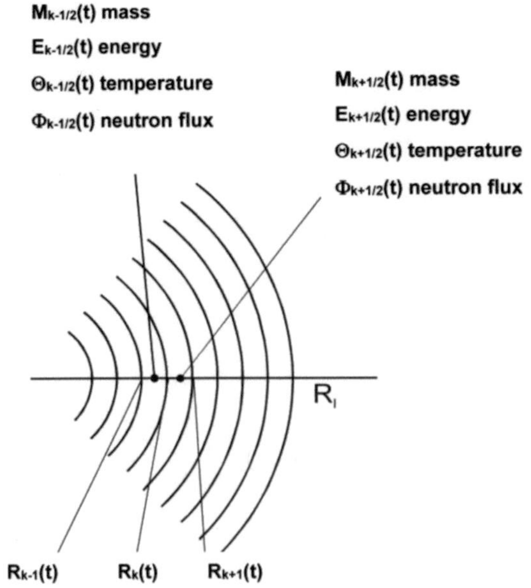

Figure 9.5: Spherical shells of the HNED for the numerical analysis.

Calculation starts from steady state. S_N-neutron transport calculation [42,43,44] for the spherical HNED leads to the initial k_{eff} at t = 0 and to the radial neutron flux and radial power distribution. The nuclear cross-sections for the S_N-calculation are provided from an appropriate multigroup cross section set [45]. Initial S_N-calculation furnishes the k_{eff} or $k_{eff} - 1 = \Delta k$ for the uncompressed state of the plutonium core-reflector system. Additional Monte Carlo [46] or perturbation calculations [43,47] lead to the effective neutron life time, ℓ_{eff} [s].

The power density evolution as a function of space and time L(r,t) in the superprompt critical regime is determined by these equations:

$$L(r,t) = L(t) \cdot w(r,t) \tag{9.1a}$$

with

$$L(t) = L(0) \cdot e^{\int_0^t \alpha(t')dt'} \tag{9.1b}$$

$$\text{where } \alpha(t) = \frac{\Delta k(t)}{\ell_{eff}(t)}, \tag{9.1c}$$

α(t) is called "Rossi alpha",

α(t) describes the increase in Δk(t) caused by external compression of the HNED, as well as the decrease of Δk(t) due to density variations within the HNED and compression or expansion effects during the explosion,

L(r,t) power density as a function of space and time within the HNED,

L(r,0) initial power density as a function of space at t = 0,

w(r,t) is equivalent to the normalized fission rate as a function of space and time as determined by S_N-calculations,

w(r,0) is the normalized fission rate as a function of space as determined by the initial S_N-calculation at t = 0.

As the power density, *L*(r,t), increases rapidly within very short times, no allowance needs to be made for heat transfer by conduction or radiation.

The energy density, Q(r,t), is given by

$$Q\,(\mathrm{r},\mathrm{t}) = \int_0^t L(\mathrm{r},\mathrm{t}')\,\mathrm{dt}' \tag{9.2}$$

the fissile material temperature, by

$$\theta(\mathrm{r,t}) = \frac{Q(\mathrm{r,t})}{\rho(\mathrm{r,t})\; \mathrm{c_p}(\theta)} \tag{9.3}$$

with

ρ (r,t)	materials density as a function of space and time,
$c_p(\theta)$	specific heat as a function of temperature.

The pressure, p(r,t), follows from the equation of state, generally written as

$$\mathrm{p(r,t)} = \mathrm{f}\left[\theta(\mathrm{r,t}),\rho(\mathrm{r,t})\right] \tag{9.4}$$

The spatial gradients of pressure furnish the acceleration of each mass point during disassembly as obtained by applying the law of inertia [29,40,41]. When a unit mass of material is displaced from its initial location, r, to a position, r + u (r,t), with the displacement, u(r,t), then

$$\frac{\partial^2 u(\mathrm{r,t})}{\partial \mathrm{t}^2} = -\,\frac{1}{\rho(\mathrm{r,t})}\cdot\frac{\partial \mathrm{p(r,t)}}{\partial \mathrm{r}} \tag{9.5}$$

Integration of acceleration over time leads to the velocity of each mass point. Integration of that velocity over time results in the displacement, u(r,t), of each mass point during disassembly.

9.7.1 Numerical Solution of the Coupled System of Equations

For numerical solution of the coupled system of Eqs. (9.1) through (9.5) the spherical HNED plutonium core and the natural-uranium reflector system are divided into a number of spherical shells (Fig. 9.6) with an inner radius, R_k, and outer radius, R_{k+1}. Each spherical shell is characterized by constant materials characteristics and constant values of these variables: mass, M_k; power density, P_k; internal energy density, E_k; temperature, θ_k; neutron flux, ϕ_k; density, ρ_k; pressure, p_k in a moving Lagrangian coordinate system (Harlow et al. [48]).

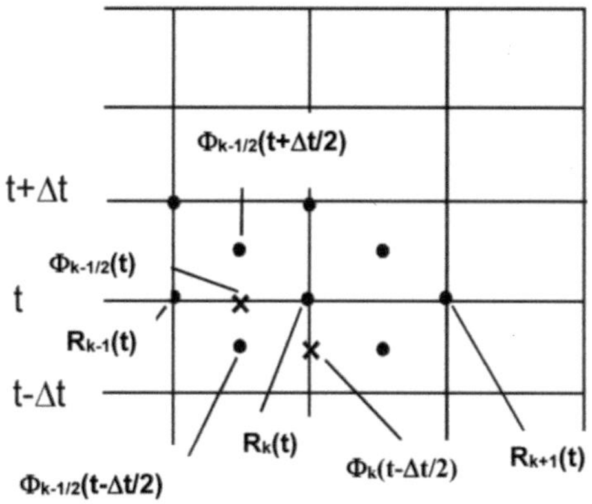

Figure 9.6: Mesh for space and time.

The time coordinate is subdivided into time increments, Δt (Fig. 9.6). For each time interval, Δt, Rossi alpha is considered constant while the total power or power density varies over Δt as

$$\sim\exp(\alpha\cdot\Delta t) \tag{9.6}$$

The total power over the time interval, Δt, is distributed over all spherical shells according to the fission rate distribution, w(r,t), calculated previously.

The density is determined from the original mass (conservation of mass) of each spherical shell and its new volume (calculated from the new boundaries) at time t. The new pressure is determined via the equation-of-state relation (Eq. (9.4)) from the energy, Q(r,t), (Eq. (9.2)) or temperature, θ(r,t), (Eq. (9.3)) and the new materials density, ρ(r,t), at time t. The spatial pressure gradients are used to calculate the average acceleration (Eq. (9.5)) of the mass points and, hence, the new velocity at time t. Further integration over the velocities leads to the new radial positions of each spherical shell and its boundaries. The numerical solution is performed in the Lagrangian coordinate system [48]. The so-called viscous-pressure approach [49] is used for hydrodynamic calculations in the presence of a steep shock front. If temperatures, densities and displacements change during thermodynamic and hydrodynamic calculations so as to exceed certain numerical limits, a new S_N-calculation is initiated to determine a new w(r,t) space function. Especially during the explosion process this S_N-calculation, because of the drastic expansion and the decrease of materials densities, leads to new, lower $k_{eff}(t)$ values and, hence, to negative $\Delta k_{expl}(t)$ or $\alpha_{expl}(t)$ values. These are added to the positive Δk_{comp} or α_{comp} values of the compression phase and lead to a new total $\alpha_{tot}(t)$,

$$\alpha_{tot}(t) = \alpha_{comp}(t) + \alpha_{expl}(t) \quad (9.7)$$

In a first step, this consecutive numerical procedure leads to rapid quasi-exponential power rise and, once sufficiently high energy density has been reached, to a fast increase in temperatures, materials densities, pressures, and to acceleration at almost all mass points. As the explosion progresses, more and more negative values of α_{expl} are generated until $\alpha_{tot}(t_p)$ becomes zero, at t_p (peak power)

$$\alpha_{tot}(t_p) = \alpha_{comp}(t_p) + \alpha_{expl}(t_p) = 0 \quad (9.7a)$$

This mitigates the quasi-exponential increase in power until peak power is reached at $\alpha_{tot}(t) = 0$.

Now the total power or power density, and also pressures, temperatures, and densities decrease rapidly while the whole HNED explodes. From a certain point of time on the release of nuclear energy as given by the total power integrated over time reaches an asymptotic value referred to as the explosion yield of the HNED.

9.8 Recalculation of the Sandmeier Case [12]

The only rather detailed analysis of a nuclear explosion ever published was part of a study of electromagnetic pulse analysis [37]. This recalculation of the "Sandmeier case" is an attempt to verify the theoretical concept of Section 9.7 and confirm the materials characteristics assumed, especially the equation of state used. An analytical model by Seifritz [51] was employed in the same way.

As not all input data and intermediate results were published at the time [37], additional guesses and assumptions had to be made about the geometry and critical dimensions, initial power and materials data. It is assumed that the publication of Sandmeier [37] describes an implosion-type device based on weapon-grade plutonium.

9.8.1 Dimensions for initial conditions

An implosion-type device based on weapon-grade plutonium (Table 9.7) with $k_{eff} = 0.98$ would have a radius of 4.806 [cm], if reflected by 5 [cm] of natural uranium metal. The $k_{eff} = 0.98$ is chosen for two considerations:

- safety reasons require a certain amount of subcriticality.
- k_{eff} should not be too low, for shock compression should lead to sufficient supercriticality

A $k_{eff} = 0.98$ was used by DeVolpi [8] and Kessler [12,13] for the same reasons.

If the above spherical implosion device (4.806 cm radius and 5 cm thick reflector) would be compressed into a smaller solid sphere of twice the original density (compression ratio, $cr = \rho/\rho_o = 2$, with ρ_o = uncompressed density, ρ compressed density) its radius would be 2.81 cm and the density would increase from 15.8 g/cm^3 to 31.69 g/cm^3. The compression to cr = 2 would lead to supercriticality of $k_{eff} = 1.4685$.

Table 9.7: Isotopic composition in % of weapon-grade plutonium [2,19].

Weapon-grade plutonium* [2,19]	
Pu-238	0.01 %
Pu-239	93.80 %
Pu-240	5.80 %
Pu-241	0.13 %
Pu-242	0.12 %
Impurities	0.14 %

*neutron background 52x10^3 n/kg·s, heat generation 2.5x10^{-3} kW/kg

9.8.2 Neutron Lifetime, ℓ_{eff}

The lifetime of the weapons-grade plutonium sphere with a 5 cm U_{nat} reflector is 1.18x10^{-8} s. This is similar to the 10^{-8} s assumed by Mark [19]. If compressed to cr =2 the lifetime would be about 20% smaller.

9.8.3 Rossi alpha

These supercriticality and lifetime results allow the Rossi alpha (Eq. (9.1c)) to be calculated. In the "Sandmeier case," [37] the published Rossi alpha is a constant value of α = 132x10^6 s^{-1} up to the start of expansion (explosion) of the plutonium sphere. This means that a compression ratio much higher than the cr = 2 must have been employed. Mark [19] mentions $\Delta k_{max} = 1$ or $k_{eff,max} = 2$, which would correspond to a homogeneous compression ratio of 4 or 5 (especially in regions around the center) of the weapon plutonium sphere with a uranium reflector (see Section 9.10.6). For example: with ℓ_{eff} = 0.8x10^{-8} s accounting roughly for a compression ratio cr = 4 to 5, and $\Delta k_{max} = 1$, this would lead approximately to α = 132x10^6 s^{-1}, as assumed by Sandmeier et al. [37].

9.8.4 Initial Power and Temperature at t = 0

In the Sandmeier case [37], the chain reaction is started by a neutron source right after full compaction of the weapon plutonium sphere and reflector system. At a time t = 30.32x10^{-8} s, the number of neutrons is 2.35x10^{17}. This is consistent with a neutron source of 10^8 n/s at t = 0 or 1 neutron per 10^{-8} s, as reported also in Rhodes [33]. It is easy to recalculate the Sandmeier value of 2.35×10^{17} neutrons by starting with 1 neutron and applying $\alpha = 1.32 \times 10^{6}\ s^{-1}$.

However, the initial power at t = 0 is required to solve Eqs. (9.1) to (9.5). For this reason, inhomogeneous neutron transport calculations were necessary with a homogeneously distributed internal neutron source of 10^{8} n/s. The results for the total fission power over the HNED at t = 0 is:

$$L(0) = 6 \times 10^{-3}\ [W]$$

In addition, an initial temperature of, e.g. $\theta(0) = 300$ K is chosen. The alpha heat power of 18 W, for the whole weapon plutonium sphere only causes a negligible temperature rise in the device. The alpha-particle heat power of the U_{nat} reflector is neglected.

9.8.5 Materials Data and Equation of State

9.8.5.1 Uranium Metal

The density of uranium metal is 18,900 kg/m^3, its thermal conductivity is 34 W/cm K. The specific heat of uranium metal is c_p = 0.142 kJ/kg K, the melting point is 1405 K, and the boiling point is 4093 K, the heat of fusion is 160 kJ/kg, and the heat of vaporization is 1752 kJ/kg [52].

9.8.5.2 Plutonium

Plutonium metal has five allotropic phases in its solid state (Table 9.8) [50,52].

Table 9.8: Allotropic phases and specific heat of plutonium metal [50,52].

Phase	Temperature [K]	Density [kg/m^3]	Specific heat c_p [kJ/kg]
α	340	19,700	0.155
ß	463	17,700	0.148
γ	543	17,140	0.155
δ	600	15,800	0.157
ε	773	16,500	0.147
liquid	923 933 973	16,600	0.173 0.175 0.175

melting point: 913 [K]
boiling point: 3528 [K]
heat of vaporization: 1409 [kJ/kg]

All publications in fast reactor safety research dealing with equation-of-state data of uranium and plutonium only cover the temperature range below 7000 K, which is a mere 3472 K above the boiling point [53,54,55]. Nuclear explosion yield calculations, however, range up to temperatures of millions of K and pressures to the order of 10^3 to 10^4 TPa. These cases require the use of refined models of kinetic gas theory [56] comprising ionization, radiation and quantum theoretical models. Specific internal energies of $5x10^5$ to 10^{15} J/kg at various particle densities are attained. The pressure[1], p (10^{-1} TPa), and the specific internal energy[1], E (10^8 J/kg), can then be described by this relation [28,34]:

$$p = (\gamma-1) \cdot \rho \cdot E \qquad (9.8)$$

p = pressure in 10^{-1} TPa $\qquad$ γ = polytropic exponent, c_p/c_v

ρ = density in 10^{-3} kg/m^3 $\qquad$ E = specific internal energy in 10^8 J/kg

In this case, the specific internal energy, E, can be determined from

$$dE = dQ - p \cdot dv \qquad (9.9)$$

where dQ is the increase in energy during the power excursion (see Eq. (9.2)),

dE = increase in specific internal energy, dv = differential change in $1/\rho = v$.

With these formulae no calculation of temperature but solely that of the change in specific internal energy, dE, is required.

The polytropic exponent, γ, is determined from plasma physics [28,56]. Figure 9.7 shows the polytropic exponent, γ, for specific internal energies between 10^5 and 10^{15} J/kg and particle densities of $4.78x10^{28}$ particles/m^3 down to 10^{23} particles/m^3 for uranium [34,57].

For the purposes of this study, the polytropic exponents, γ, are identical for uranium and plutonium. For high specific internal energies above 10^{14} [J/kg] of the ionized gas (plasma) of uranium or plutonium the polytropic exponent reaches an asymptotic value of $\gamma = 4/3$.

[1] The dimensions of specific internal energy, E [10^8 J/kg], pressure [10^{-1} TPa], and density [10^{-3} kg/m^3] correspond to Eq. (6) (Duderstadt et al., 1982 [56]).

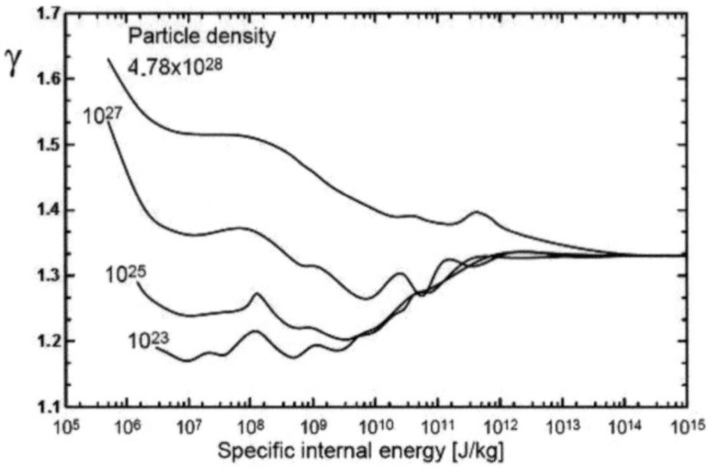

Figure 9.7: $\gamma = c_p/c_v$ as a function of the specific internal energy E (J/kg) and particle density per (m^3) uranium [34,51].

For densities higher than $\rho_o = 18{,}900$ kg/m^3 or particle densities higher than 4.78×10^{28} m^{-3} and specific internal energies above 5.4×10^5 J/kg, the following formula (regression curve) is used [51]:

$$\gamma = \frac{4}{3} + \frac{4.025}{E^{0.1771}} \left[1 + 0.0876 \left(\frac{\rho}{\rho_o} - 1 \right) \right] \tag{9.10}$$

E in J/kg.

The data for uranium and plutonium indicated above are employed in this procedure:

- The specific internal energy at the boiling point of plutonium and uranium is roughly 0.54×10^6 J/kg.
- Up to the boiling point
  a specific heat: 0.142 kJ/kg K for uranium,
  a specific heat: 0.159 kJ/kg K for plutonium
  is used.

For the pressure up to the boiling point

$$p = \alpha \cdot \rho + \beta \cdot \theta + \tau \tag{9.11}$$

with

$\alpha = 0.0072$ $\beta = 278.46$ $\tau = -0.3946$ for Pu-metal

$\alpha = 0.00619$ $\beta = 278.46$ $\tau = -0.4055$ for U-metal

is used. Here the following dimensions are valid:

p = pressure in 10^{-1} TPa ρ = density in 10^{-3} kg/m^3 θ = temperature keV.

The relation for p was suggested by Stratton [58]. It results in positive pressures only when an outside pressure of, e.g. 0.1 TPa (pressure of the chemical high explosive acting on the outside surface of the uranium reflector) is exceeded.

- Above the boiling point, Eqs. (9.8) through (9.10) and Fig. 9.7 are appplied as described above.

9.8.6 Results of Recalculation of the "Sandmeier Example" [12]

Figure 9.8 shows the relative radial fission rate or relative power density as a function of the radius for t = 0. Due to the lower enrichment (0.7% U-235) in the U_{nat} reflector, the fission rate is at least one order of magnitude lower than in the weapon plutonium sphere.

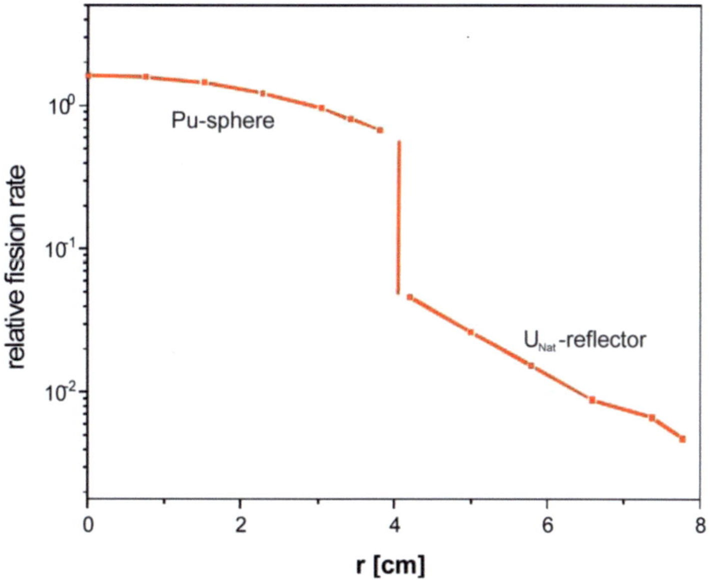

Figure 9.8: Sandmeier example, weapons-grade plutonium, relative fission rate as a function of radius at t = 0.

In the first phase, the power rapidly increases from its initial level as an exponential function of time, as only the constant Rossi alpha of $132 \times 10^6\ s^{-1}$ and no negative expansion reactivity is acting. At 0.309×10^{-6} s the boiling point of plutonium metal is reached. From then on, the internal pressures increase.

Figure 9.9 shows the Rossi alpha value as a function of time in 10^{-6} s and the expansion of the outer radius, R(t), of the weapon plutonium sphere. After 0.39×10^{-6} s, the weapon-grade plutonium sphere starts to expand more and more rapidly. This causes a negative feedback reactivity, or Δk_{expl}.

The curves for R(t) and $\alpha(t)$ can be considered roughly inversely proportional, which is one of the essential approximate assumptions underlying the derivation of the Serber relation [32]. Initially, Rossi alpha $\alpha(t)$ remains constant. The exponential power rise produces only sufficient nuclear energy in the time period up to approximately 0.39×10^{-6} s. At that time, the specific energy will be high enough for high internal pressures to be generated and the weapons-grade plutonium sphere to start expanding. This causes a decrease of $\alpha(t)$ (Fig. 9.9).

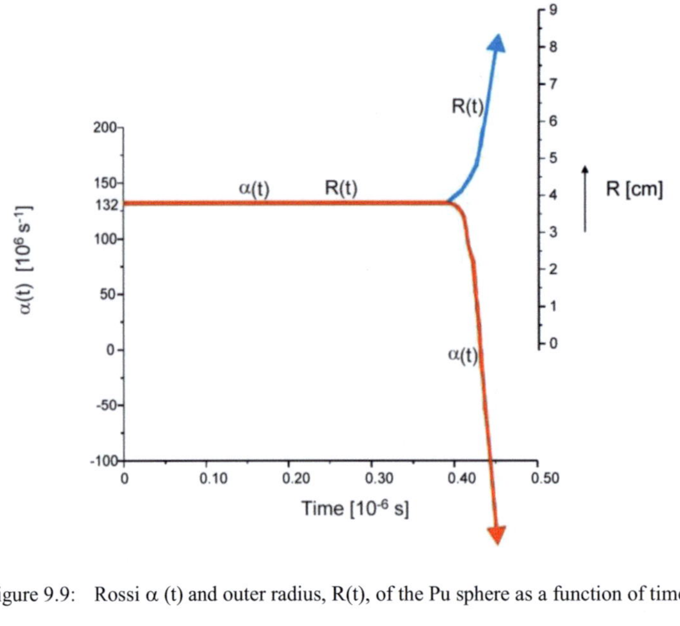

Figure 9.9: Rossi α (t) and outer radius, R(t), of the Pu sphere as a function of time.

Figure 9.10 shows the total power in the weapons-grade plutonium sphere and the U_{nat} reflector as a function of time. When α_{tot} becomes zero, the power as a function of time reaches a maximum and then decreases rapidly. Only the time between 0.36×10^{-6} s and

$0.45x10^{-6}$ s is shown (Fig. 9.10) when major essential contributions to the explosion energy release occur. Figure 9.10 also shows the partial contributions to the release of explosion energy integrated over power and time for each time interval of $0.01x10^{-6}$ s.

The values in kt of TNT equivalent are based on the

energy equivalence of $4.187x10^{12}$ J≡ A 1 kt TNT [50].

The integral over the intervals between $0.36x10^{-6}$ s and $0.45x10^{-6}$ s adds up to 24.7 kt of TNT (equivalent).

Figure 9.11 shows the nuclear energy released or the nuclear energy yield as a function of time in 10^{-6} s.

$$\text{Energy release: } 1.036x10^9x10^5 \text{ J x } \frac{10^{-12}}{4.187}\left[\frac{\text{kt}}{\text{J}}\right] = 24.7 \text{ kT TNT}$$

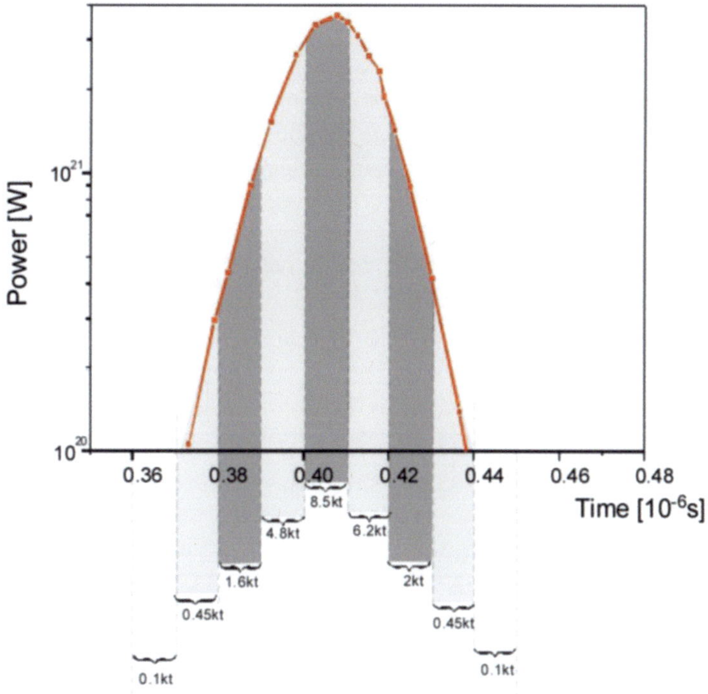

Figure 9.10: Power in W as a function of time. Only the time range between $0.36x10^{-6}$ s and $0.45x10^{-6}$ s is shown, where essential released energy occurs, indicated in kt.

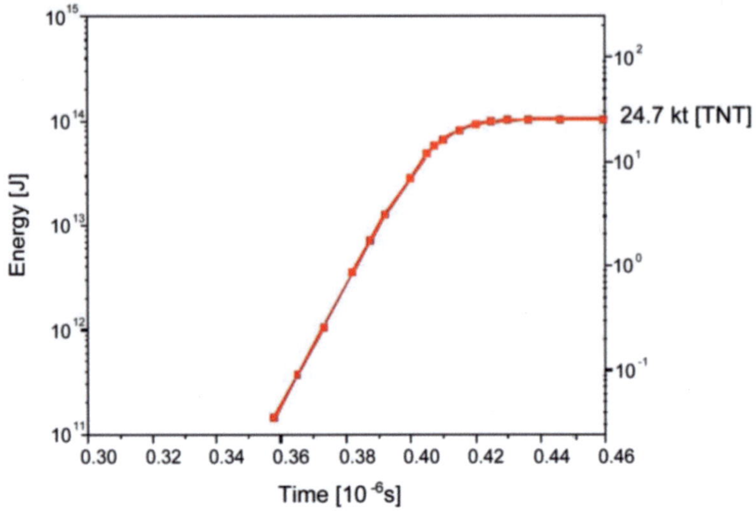

Figure 9.11: Energy release as function of time 10^{-6} s.

9.8.6.1 Comparison with the Results Published by Sandmeier

The increase in internal pressure up to the TPa range, begins at 0.386×10^{-6} s in Sandmeier's publication [37] and at 0.39×10^{-6} s in the above calculation. Peak power is reached at 0.44×10^{-6} s versus 0.41×10^{-6} s and the total nuclear explosion energy or nuclear explosion yield is 24.2 kt of TNT (equivalent) [37] as against 24.7 kt of TNT (equivalent) [12].

These results allow the conclusion to be drawn that the theoretical model described in Section 9.7 and the materials data and the equations of state in Section 9.8.5 employed are able to describe a nuclear explosion sufficiently well.

9.9 Verification of Serber's Relation, $Y: \frac{\Delta k_{max}^3}{l_{eff}^2}$

The so-called Serber relation [32], was used in earlier publications by Locke et al. [57], DeVolpi [8], Seifritz [28], and Mark [19]. According to this relation the energy yield of an exploding nuclear device is proportional to Δk_{max}^3 (maximum reactivity introduced by spherical compression) and inversely proportional to l_{eff}^2 (effective neutron lifetime). This is based on the assumption of roughly inverse proportionality between Rossi alpha, $\alpha(t)$, and the expanding outer radius of the plutonium sphere, R(t), in the expansion phase of the nuclear explosion (see Fig. 9.9).

Serbers relation can be verified by performing several calculations for plutonium spheres compressed to different states. Calculations were run for three different compression states with three different values of Rossi alpha: $\alpha = 132\text{x}10^6\ \text{s}^{-1}$ (which is equal to the α used in the "Sandmeier example" [37]), $\alpha = 76.2\text{x}10^6\ \text{s}^{-1}$ and $34.6\text{x}10^6\ \text{s}^{-1}$.

As can be seen from Fig. 9.12, the Serber relation,

$$Y : \frac{\Delta k_{max}^3}{l_{eff}^2} = \alpha_{max}^2 \cdot \Delta k_{max} \tag{9.12}$$

is roughly met by the calculations according to the physics model (Section 9.7). Deviations in relative yield between the Serber relation and results of the above model calculations can be of the order of 30%.

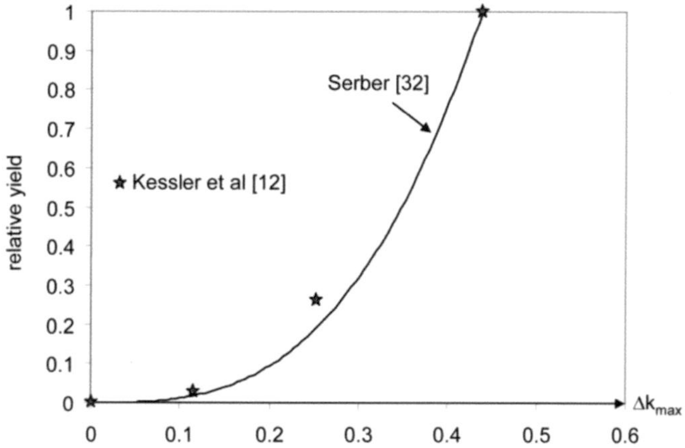

Figure 9.12: Comparison of Serber relation [32] and calculations by Kessler et al. [12].

9.10 Calculation of explosion yields of HNEDs with reactor-grade plutonium

A constant Rossi alpha of $\alpha = 132\text{x}10^6\ \text{s}^{-1}$ was set for recalculation of the "Sandmeier example" [37]. No compression or reactivity calculations were necessary since this Rossi alpha value was assumed to be based on the shock wave already having reached the center and the **weapons-grade plutonium** metal sphere being fully compressed.

Calculations of the explosion yields of HNEDs with **reactor-grade plutonium** require the density changes during the compression phase caused by the shock waves to be calculated and the reactivity increase, Δk(t), or the Rossi alpha, α(t), from the onset of compression to be determined. This is necessary because -ignition (buildup of persistent fission chains) can start immediately after prompt criticality has been reached (see Section 9.11). Also, the shock wave will be stopped when the internal pressure caused by the rapid energy increase, at a certain point, r, within the reactor plutonium sphere, will exceed the pressure of the shock wave moving inward.

9.10.1 Initial conditions for shock compression by outer chemical high-explosive lenses

The geometric shape and the dimensions of chemical high-explosive lenses, and the choice of high explosives, in practice must be such as to produce an almost perfectly convex spherical shock wave impinging upon the outside surface of the U_{nat} reflector [19]. Since a perfectly convex spherical shock wave must be assured for the calculation the resulting nuclear energy yield will constitute an upper bound of the nuclear energy yield attainable in practice.The geometry and the composition of spherical high-explosive lenses are classified. Any effort to model the spherical high-explosive lenses thus is impossible.

9.10.1.1 Shock compression in planar geometry

Zeldovich et al. [59], Johansson et al. [60], Leuthäuser [34], and Seifritz [28] explained that the pressure of the shock wave on the outer surface of the reflector layer in a planar geometry can be determined by having the Hugoniot curve of the uranium layer (U_{nat} reflector) intersect the pressure curve of the detonating chemical high explosive (Chapman-Jouguet point). The Hugoniot curve for uranium was determined by Johansson et al. [60]. These data were used to produce the characteristic results [34] for the shock wave (pressure p, velocity u, temperature T) on the outside of an idealized planar uranium reflector listed in Table 9.9.

Comparing the detonation pressures and shock velocities for the three high explosives in Table 9.9 with data of other high explosives (Dobratz [61], Gibbs et al. [62], Mader et al. [63]) leads to the conclusion that the Chapman-Jouguet point data for a planar geometry are very similar for most other chemical high explosives.

Table 9.9: Characteristic shock wave data at the Chapman-Jouguet point for various explosives [34].

	Composition B	RDX	HMX
p [TPa]	0.052	0.059	0.068
u [m/s]	750	830	920
T [K]	540	660	770

9.10.1.2 Shock compression in a spherical geometry

The physics of spherical shock waves is much more complicated than for planar geometry. The convergent flow in a spherical geometry leads to higher velocities and pressures as the shock wave travels outside in, i.e. towards smaller radii. Bushmann et al. [64] point out that maximum particle velocities of up to 8000 or 9000 m/s and pressures of a few 0.1 TPa can be achieved only in a spherical convergent flow. However, these pressures and high particle velocities can only be attained at very small locations close to the center of a spherical geometry. Experiments of this kind are very difficult technically because of the requirement of accurate initiation of the detonation over the outer surface and the need for advanced measurement techniques. Although the detonation (shock) velocity is around 8000 to 9000 m/s for most explosives such propulsion velocities are practically unattainable by finite-size explosion charges. These high velocities cannot be achieved with thick impactors because the time for chemical interaction of the explosion products is too short [64].

9.10.1.3 Maximum velocities and pressures achieved

Hypervelocity launchers allow particle velocities of up to 11 km/s and 15.8 km/s to be achieved (Kinslow [65] and Chabildas et al. [66] Table 9.10). However, such flyer plate experiments are not representative of the case described here, as they are only possible with small samples of less than 10^{-3} kg.

The highest pressures ever recorded in measurements by Al'tshuler (planar geometry) of underground nuclear explosions were 3.4 TPa and a compression ratio of 3.4 (Al'tshuler et al. [67]).

Table 9.10: Summary of high velocity impact experiments [65,66].

Type of experiment	highest velocity [km/s]	impact weight [g]
Two stage light gas gun	15.8	0.21
	11.3	0.44
	4.7	2500
Electromagnetic gun	6	0.024
	9.5	0.01
Explosively driven guns	8	7.4
	5.8	102
High explosives and	16.5	0.08
shaped charge accelerators	5	0.18

9.10.1.4 Amplification of shock wave pressure in spherical chemical high-explosive lenses

The chemical high-explosive lenses surrounding the fissile part of a HNED have two functions:

- Turning the concave spherical shock wave at the ignition points on the outside surface into a convex concentric wave acting on the uranium reflector of the HNED (see Fig. 9.3) (Rhodes [33]).
- To increase the pressure as a function of the decreasing radius (convergent flow).

Shock wave similarity calculations for 25-40 cm thick spherical hollow shells of high explosive materials [12] revealed that the shock wave pressure of the spherically convergent flow can be magnified up to 0.11 TPa for 25 cm thick implosion lenses. Thicker than 25 cm spherical shells of high explosive material are not technically feasible if reactor-grade plutonium with more than 2.1% Pu-238 or with more than 0.120 kW of the particle-heat power of the HNED is used [12,23].

On the basis of these results two levels of technology were defined for the pressure acting on the outer surface of the reflector [12]:

low technology	0.06 – 0.08 TPa,
very high technology	0.08 – 0.11 TPa.

9.10.2 Hydrodynamic shock compression

The simplified Navier-Stokes equations for the conservation of mass, momentum, and energy must be solved for spherical geometry to describe hydrodynamic shock compression. As very high pressures and temperatures are involved, usually the simplified plasma hydrodynamic equations for the electrons and ions are solved. Deriving those differential equations in time and space is best described by Zel'dovich et al. [59] and Duderstadt et al. [56]. For numerical calculations, these hydrodynamic differential equations are solved in a Lagrange coordinate system [48,56].

Kessler et al. [12] used the similarity between the implosion of a solid or hollow spherical pellet after external exposure to laser beams or ion beams in inertial confinement fusion research [56] and the actual problem of imploding a solid or hollow Pu-sphere by shock waves initiated by chemical high-explosive lenses. The problem is adapted by setting an outside boundary condition of a convex perfectly spherical shock pressure exerted on the outer reflector surface.

9.10.3 Equation of state (EOS) data for compression of Pu and U metal

Data published by Johansson et al. [60] and Benedict et al. [68] for the range up to 0.06 TPa were used by Kessler et al. [12]. Johansson et al. [60] also describe an experimental relation between the shock velocity, D, and the particle velocity, u_p, for uranium metal,

$$D = c_o + S \cdot u_p,$$

where D is the shock velocity [10^3 m/s], c_o and S are constants, u_p is the particle velocity [10^3 m/s].

Particle velocities and shock velocities will be shown in Section 9.10.4 below. The relation presented in Fig. 9.13 was derived (Baumung [69]) by applying the Hugoniot equations for the densities, pressures, and velocities on both sides of the shock front where the ratio, ρ_o/ρ (ρ_o – initial density, ρ – compressed density), is given as a function of pressure, p, in 0.1 TPa

$c_o = 2.6 \times 10^3$ m/s	$S = 1.45$	for U metal,
$c_o = 2.51 \times 10^3$ m/s	$S = 1.3$	for Pu metal.

For the pressure range of >0.05 TPa, also data derived from Kirzhnits [70] and Höbel et al. [71] on the basis of KATACO calculations are shown in Fig. 9.13. They are applicable especially in the 0.1 TPa range.

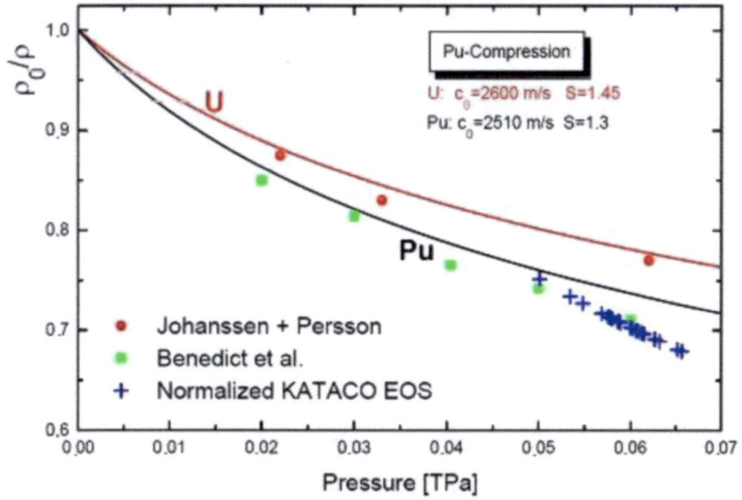

Figure 9.13: Equation of state or compression data for plutonium and uranium metal.

9.10.4 Calculations of Hydrodynamic Shock Compression in HNEDs

Results of shock compression calculations of HNEDs will be presented below. In these calculations, a concentric shock pressure of 0.11 TPa is assumed to act on the outside of the reflector.

The plutonium sphere and the natural-uranium reflector system are divided into 40 spherical shells (the plutonium sphere into 20 spherical shells, natural uranium reflector into 20 spherical shells). As can be seen from Fig. 9.14, radial shock compression follows the familiar pattern of theoretical prediction (Duderstadt et al. [56]). The shock front travels from the outside radius to the center of the reactor-grade plutonium sphere roughly in 27×10^{-6} s (Fig. 9.14). It would then be reflected, and a relaxation wave would travel back through the system of the reactor-grade plutonium sphere and the reflector. The onset of bending of each line shows the arrival of the shock wave at the radii of the different shells. The slope of the lines after arrival of the shock wave represents particle velocity.

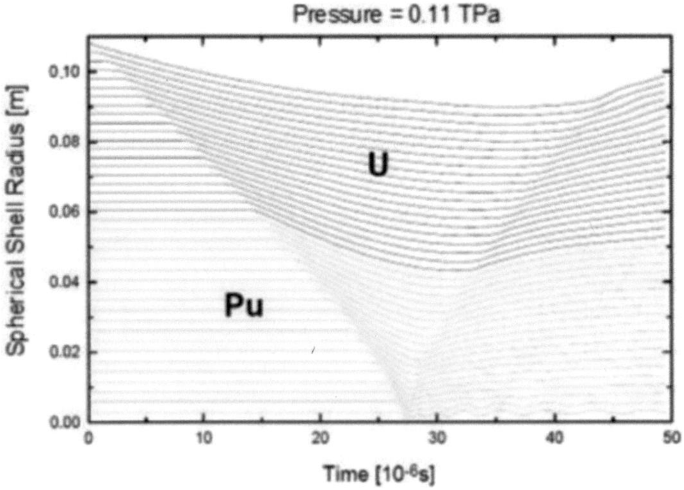

Figure 9.14: Spherical shell radius as a function of time [10^{-6} s] in a solid plutonium sphere and a natural uranium reflector.

Figure 9.15 presents the pressure in each of the 20 spherical shells of the uranium reflector and each of the 20 spherical shells of the reactor-grade plutonium sphere. Pressures in the uranium reflector slowly rise to some 0.15 TPa and then drastically surge to approximately 0.4 TPa and more when the convergent shock wave approaches the center of the reactor-grade plutonium sphere. (Theoretical prediction would have a singularity at $r = 0$).

Figure 9.15 shows density, ρ, as a function of time for the different spherical shells as the shock front progresses from the outside boundary of the uranium reflector to the center of the plutonium sphere. The density, or compression ratio, is not constant over space and time. The uranium metal is compressed in the uranium reflector from 18,900 kg/m^3 to some 32,000 kg/m^3 and becomes the higher the closer the shock wave approaches the center. Density increases twofold and threefold and more in the very small region around the center of the plutonium metal sphere.

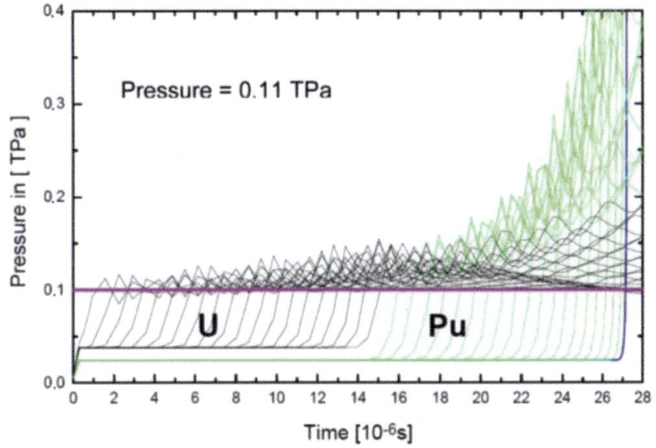

Figure 9.15: Pressures in each spherical shell as a function of time (10^{-6} s) in the solid plutonium sphere with a natural-uranium reflector. The outside pressure at the reflector boundary is a constant 0.11 TPa.

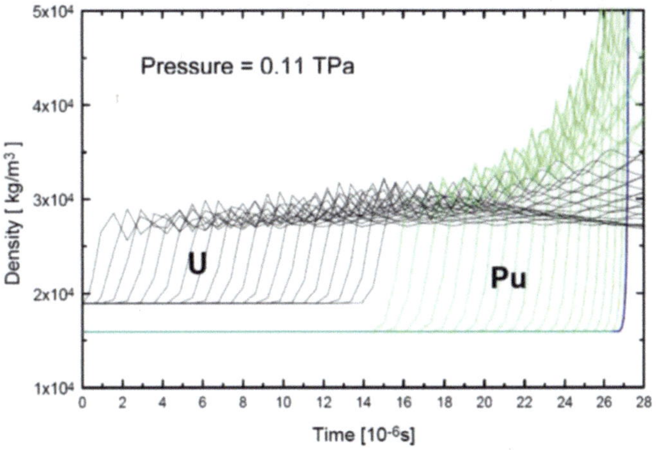

Figure 9.16: Density of each spherical shell as a function of time (10^{-6} s) in a solid plutonium sphere with a natural uranium reflector. The outside pressure at the reflector boundary is a constant 0.11 TPa.

Figure 9.17 shows these particle velocities rising to a maximum of approximately $5x10^3$ m/s as the shock wave progresses to the center of the reactor-grade plutonium sphere.

Finally, Fig. 9.18 indicates materials temperatures (K) in the spherical shells evolving as a result of shock compression. In the plutonium sphere, temperatures increase to more than 3000 K starting from a temperature of about 625 K.

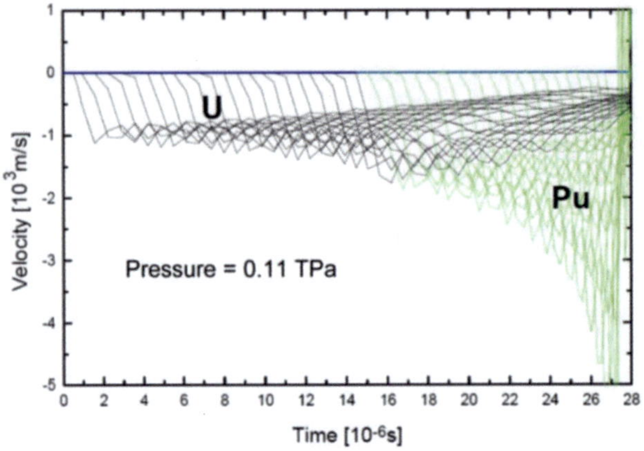

Figure 9.17: Particle velocity, u_P, of the different spherical shells at different times (10^{-6} s) in a solid plutonium sphere with a natural-uranium reflector. The outside pressure at the reflector boundary is a constant 0.11 TPa.

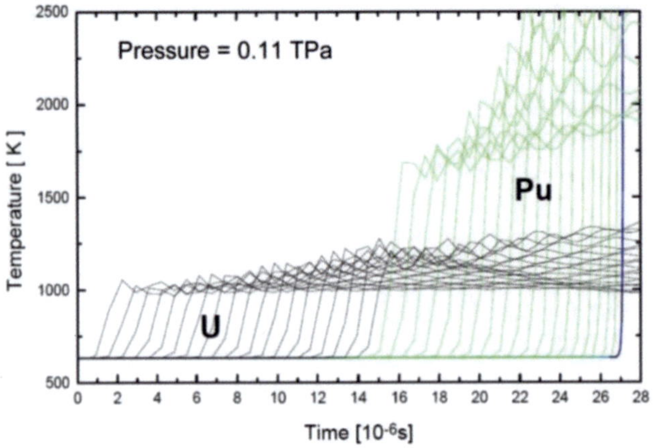

Figure 9.18: Temperature (K) in the different spherical shells as a function of time (10^{-6} s) in a solid plutonium sphere with a natural-uranium reflector. The outside pressure at the reflector boundary is a constant 0.11 TPa.

9.10.5 Shock compression during implosion of a hollow spherical Pu shell with a U_{nat} reflector

Figure 9.19 shows the diagram of implosion of a hollow spherical plutonium shell with a natural uranium reflector, and with the same constant outside pressure of 0.11 TPa exerted by the high-explosive lenses on the outside of the reflector. Shock compression is faster than in the solid Pu sphere. Between $16x10^{-6}$ s and $20x10^{-6}$ s, the innermost shells start moving and flying to the center of the hollow sphere where they are compacted.

More characteristics of HNEDs with hollow spherical fissile parts will be discussed in Section 9.12.7.

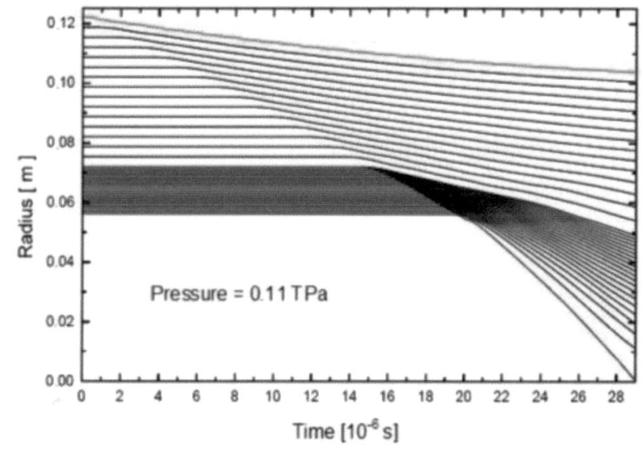

Figure 9.19: The boundaries of each spherical shell as a function of time 10^{-6} s in a hollow plutonium sphere and a natural-uranium reflector. The gradient of the shell boundaries as a function of time represents the particle velocity, u_P. The outside pressure at the reflector boundary is a constant 0.11 TPa.

9.10.6 Effect of Spherical Compression on k_{eff}

The MCNP 4C3 Monte Carlo Code with its incorporated ENDF/B-VI ZAA.600 set of cross sections (Section 4) was used to calculate the reactivity effects of an artificial homogeneous compression of the plutonium sphere plus uranium reflector system. Again the plutonium core and the U_{nat} reflector system were divided each into 20 spherical shells, and the spherical shells were compressed successively (Fig. 9.5) from the outside progressing inward to the center.

Figure 9.20 shows k_{eff} as a function of the compressed thickness or number of spherical shell (only every second result is shown) of the U_{nat} reflector and the Pu-sphere starting from the outside radius of the reflector. Starting from k_{eff} = 0.98, shaped curves are obtained for the three compression ratios, cr = ρ/ρ_o selected:

cr = 1.2 1.5 2.0.

Figure 9.20 applies to plutonium with 8.7% Pu-238, 5.81 cm radius of the Pu-sphere, and a 5 cm U_{nat} reflector (referred to as Pu-(2) in Kessler [12] and Table 9.12). As the reflector consists of natural uranium, the increase in k_{eff} initially is very slight. Compression of the outer layers of the plutonium sphere contribute considerably more to the k_{eff} rise. Finally, as the center is being approached, increasingly smaller volumes and masses are compressed. Despite the effect of higher importance of these central volumes, the increase in k_{eff} becomes smaller and smaller.

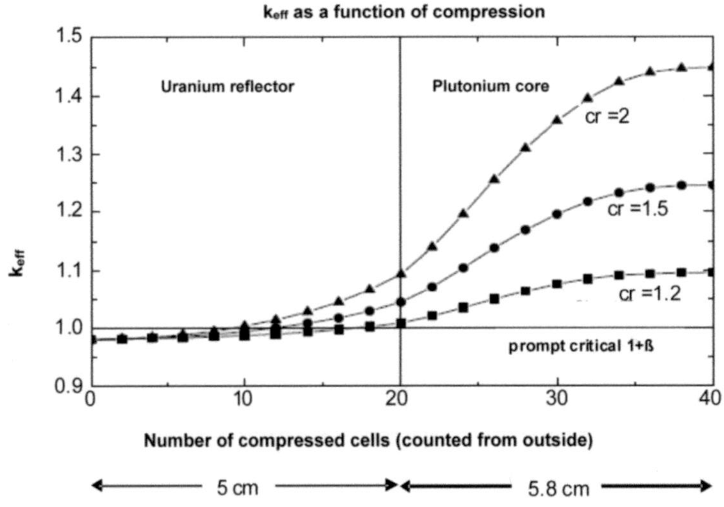

Figure 9.20: k_{eff} as a function of the compression ratio and the number of compressed spherical shells (from outside). The natural-uranium reflector of 5 cm thickness and the plutonium sphere of 5.8 cm radius are subdivided into 20 radial shells each (only every second result shown, cr = compression ratio).

At full compression of the natural uranium reflector and the plutonium sphere, these supercriticalities are obtained (Table 9.11):

Table 9.11: Compression ratios cr and k_{eff} at full compression of the plutonium sphere plus U_{nat} reflector system.

cr	1.2	1.5	2.0
k_{eff}	1.095	1.24	1.45

Criticality and prompt criticality are reached when the compression reaches around 4.3 cm from the outer surface of the uranium reflector for a compression ratio, cr = 1.2. For higher compression ratios, cr = 1.5 or 2.0, the values are approximately 3 cm or 2.25 cm, respectively. Obviously, if the start had been at k_{eff} = 0.90 instead of 0.98, as discussed in Section 9.8.1 above, maximum supercriticality at full homogeneous compression would have been only k_{eff} = 1.015. for cr = 1.2 .This would result in an extremely low nuclear explosive energy yield.

The different plutonium compositions investigated in Kessler [12] are shown in Table 9.12. These plutonium compositions were also used for compression analysis (Table 9.13 and Fig. 9.21).

Table 9.12: Reactor Pu-compositions in weight% used for the parametric analysis [12].

	Pu-238	Pu-239	Pu-240	Pu-241	Pu-242
Pu-(0)	1.6	58.8	20.8	13.8	5
Pu-(0')	2.8	55.8	23.8	9.8	7.8
Pu-(1)	5.5	34.1	31.1	10.6	18.7
Pu-(2)	8.7	30.1	30.6	11.3	19.3
Pu-(3)	12.	26.	30.	12.	20.
Pu-(4)	15.2	27.5	28.6	11.5	17.2
Pu-(5)	20.3	30.	26.3	10.7	12.7
Pu-(6)	24.5	32.	24.5	10.	9.0

Figure 9.21 shows the k_{eff} results as a function of compression for all Pu-compositions from Pu-(0), Pu-(0'), Pu-(1) to Pu-(6) of Table 9.12 for a compression ratio, cr = 2.0 (only every second result is shown). The curves are very similar for all Pu-compositions. This is due to the fact, of course, that Pu-238 and all other Pu-isotopes have similarly high fission cross-sections at neutron energies of approximately 1 MeV [19].

The k_{eff} curves were calculated also for higher compression ratios. The curves are similar, but rise to considerably higher supercriticalities at full compression of the reflector and reactor-grade plutonium sphere (Table 9.13). In Table 9.13, these k_{eff} values for the supercriticality at full homogeneous compression of the reflector and plutonium sphere system are shown for various compression ratios, cr, and Pu compositions. Again, the k_{eff} values are similar for all reactor-grade plutonium compositions.

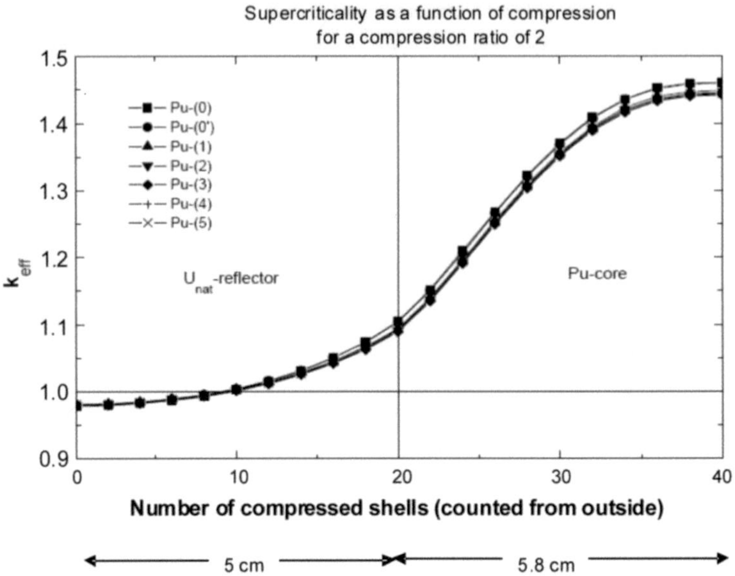

Figure 9.21: k_{eff} for all Pu-mixtures Pu-(0) to Pu-(5), and a compression ratio, cr = 2 (spherical shells successively compressed from the outside; only every second value is shown).

Table 9.13: k_{eff} after compression of the 5 cm U_{nat} reflector and the reactor-grade plutonium sphere for plutonium composition Pu-(0) to Pu-(6) and for the compression ratio, cr.

cr / Composition	1.2	1.5	2	2.5	3	4	5
Pu-(0)	1.0960	1.2508	1.4599	1.6270	1.7602	1.9589	2.0988
Pu-(0')	1.0969	1.2511	1.4599	1.6240	1.7563	1.9526	2.0919
Pu-(1)	1.0955	1.2453	1.4481	1.6063	1.7335	1.9204	2.0531
Pu-(2)	1.0942	1.2429	1.4435	1.6009	1.7267	1.9142	2.0443
Pu-(3)	1.0933	1.2415	1.4417	1.5976	1.7224	1.9086	2.0379
Pu-(4)	1.0927	1.2416	1.4418	1.5976	1.7241	1.9106	2.0436
Pu-(5)	1.0940	1.2432	1.4442	1.6025	1.7296	1.9172	2.0484
Pu-(6)	1.0951	1.2438	1.4461	1.6046	1.7312	1.9195	2.0553

9.10.7 Reactivity increase as a function of compression time

The procedure of calculating the reactivity or the associated value, k_{eff}, and Rossi alpha, $\alpha(t)$, as a function of time during shock compression of the HNED is discussed in this Section.

9.10.7.1 Description of the calculational procedure

The previous section showed the variation of k_{eff} when the spherical system of a Pu-sphere with a U_{nat} reflector is compressed from the outside. The HNED was divided, e.g., into N = 40 spherical shells and compressed, one shell after the other, until the whole Pu-sphere-U_{nat}-reflector system was compressed to a homogeneous compression ratio, cr. Table 9.13 summarizes the k_{eff} values for different reactor-grade plutonium compositions and for the final compression state of the plutonium sphere as well as for compression ratios.

Figures 9.14 and 9.15 Section 9.10.4 demonstrated that particle velocity causes the density (Fig. 9.16) to change differently in each shell as a function of time. These data can be written in the form of a matrix. In this matrix, the densities, ρ_n, of shell n appear in rows of a total length, e.g., N = 40 for each point in time, t, considered. The elements of the matrix are ρ_n^t.

Dividing these matrix elements by their initial density, ρ_n^0, at time zero, ρ_n^t / ρ_n^0, produces a similar matrix for the compression ratios, cr_n^t.

Section 9.10.6 above already described how the k_{eff} values can be determined for the compression of a certain number, n, of shells for different discrete constant compression ratios, e.g., cr = 1.2, 1.5, 2.0, 2.5, 3.0, 4.0, 5.0.

For a specific plutonium composition, e.g., Pu-(0'), Pu-(0) to Pu-(6), k_n is now used for the k_{eff} of the number, n, of shells compressed to different compression ratios starting from the outside of the reflector.These data can also be arranged in a matrix.

As these shock compression calculations yield data mostly lying between our k_n values calculated for cr = 1.2, 1.5, 2.0 etc., we can interpolate each value.

9.10.7.2 Calculation of Δk(t), Neutron Lifetime, and Rossi alpha, α(t)

Interpolation programs then allow the reactivity contribution at each discrete point in time to be calculated by summing up all contributions from the different shells compressed by the shock wave inward traveling. Dividing $\Delta k(t)$ by the neutron lifetime, ℓ_{eff}, provides Rossi alpha, $\alpha_{comp}(t)$, (Section 9.7).

The neutron lifetime, ℓ_{eff}, for the natural-uranium reflector and reactor-grade plutonium sphere system was calculated by the MCNP 4C3 Monte Carlo code and its incorporated ENDF/B-VI ZAA.600 cross-section library. For the uncompressed Pu-sphere-natural-uranium-reflector system, a $\ell_{eff} = 1.4\text{x}10^{-8}$ s was obtained. $\ell_{eff} = 1.3\text{x}10^{-8}$ s was calculated for the compressed reactor-grade plutonium sphere with a natural-uranium reflector and cr = 2. As the influence of the compression ratio on neutron lifetime, ℓ_{eff}, is small (Section 9.8.2), a constant average neutron lifetime, $\ell_{eff} = 1.35 \times 10^{-8}$ s, was assumed for subsequent calculations.

Figure 9.22 shows the development of $\alpha_{comp}(t)$ for a pressure of 0.11 TPa exerted by the chemical high-explosive lenses on the outside surface of the reflector.

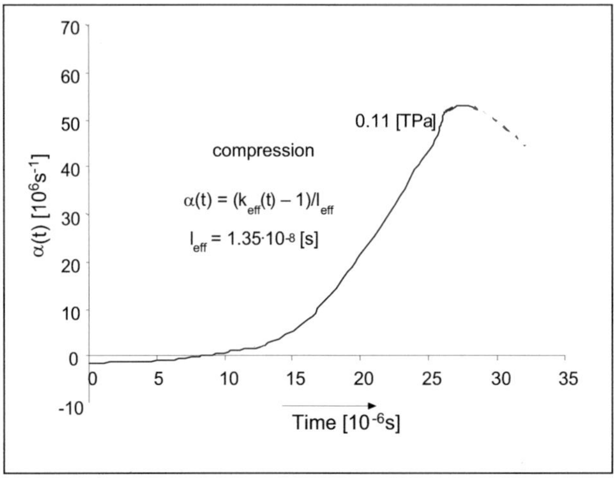

Figure 9.22: Rossi alpha, $\alpha(t)$, as a function of time (10^{-6} s) for reactor-grade plutonium of the composition Pu-(1) and the pressure of 0.11 TPa ed on the outer surface of the reflector.

The curve for $\alpha_{comp}(t)$ starts from t = 0 a negative value according to $k_{eff} = 0.98$, which leads to $\Delta k(0) = k_{eff} - 1 = -0.02$ or $\alpha_{comp}(0) = -1.48 \times 10^6\ s^{-1}$ (assuming $\ell_{eff} = 1.35 \times 10^{-8}$ s). At $t = 8.44 \times 10^{-6}$ s, α_{comp} corresponds to prompt criticality. From then on, α_{comp} raises sharply to values around $40 \times 10^6\ s^{-1}$ at approximately 24×10^{-6} s. The total compression of the HNED would lead to Rossi alpha, $\alpha(t) = 53.03 \times 10^6\ s^{-1}$, at 27.6×10^{-6} s after the onset of compression.

9.10.8 Spontaneous fission neutron source multiplication

The geometric arrangement, with $k_{eff} = 0.98$, of a reactor-grade plutonium sphere surrounded by a natural-uranium reflector and chemical high-explosive lenses is equivalent to a subcritical experiment in a zero-power reactor facility. If this reactor-grade plutonium sphere has an inherent spontaneous fission neutron source, S_{inh}, homogeneously distributed over the radius of the reactor-grade plutonium sphere, this S_{inh} – according to Weinberg et al. [44] – is multiplied by $1/(1 - k_{eff}) = M$.

This leads to the multiplied fission neutron source:

$$S_M = M \cdot S_{inh} = 1/(1-k_{eff}) \cdot S_{inh}.$$

This is well known in reactor physics and follows from the theory of subcritical experiments in zero-power reactor facilities. This feature was first introduced by DeVolpi [8] into the analysis of HNEDs.

This neutron source multiplication factor can be verified also in inhomogeneous neutron transport calculations with, e.g. the ONEDANT code using 30 energy groups [12]. The plutonium composition of Pu-(1) (see Table 9.12) was chosen with a radius of the reactor-grade plutonium sphere of 5.8 cm, and a 5 cm thick reflector of natural uranium at k_{eff} = 0.98, and a total inherent spontaneous fission neutron source of S_{inh} = 9.6x10^6 n/s homogeneously distributed over the volume of the plutonium sphere (Table 9.14). The ONEDANT calculation resulted in a multiplied fission neutron source, S_M = 4.8x10^8 n/s, corresponding, as expected, to a neutron source multiplication, M = 50. This ONEDANT calculation also furnished the initial steady state power generated by the multiplied inherent fission neutron source as

7.73x10^9 fission/s in the HNED or 0.247 W.

This is the initial power L(0) of the HNED for the subsequent calculations in Section 9.12.

9.11 Pre-ignition by spontaneous fission neutrons in HNEDs with higher Pu-238 contents

Spontaneous fission neutrons from the plutonium isotopes, Pu-238, Pu-240, Pu-242, appear with a certain probability function (Hansen [72]). Under certain conditions, they can start the chain reaction (pre-ignition) as early as in the compression phase. Pre-ignition is a stochastic problem requiring treatment by probability theory (Hansen [72] and Seifritz [28,73]).

9.11.1 Pre-ignition as a consequence of strong spontaneous fission neutron sources

The theory of weak spontaneous fission neutron sources by Hansen [72] originally was developed and applied for GODIVA neutron burst experiments at Los Alamos. It can be used as well to explain the pre-ignition probability in nuclear explosive devices with a relatively small spontaneous fission neutron source, e.g., for weapons-grade plutonium. This theory of weak spontaneous fission neutron sources is only valid for spontaneous fission neutron sources very much smaller than 9x10^7 n/s.

So-called strong spontaneous fission neutron sources in the range of >$1.5x10^8$ n/s (see Table 9.14) lead to persistent fission chains, which are possible only after prompt criticality. The delayed neutrons do not play a role, as the delayed neutron precursor atoms have decay times much too long (in the range of milliseconds to seconds), whereas power excursions in HNEDs occur in a time range of microseconds.

The s-shaped or sigmoidal curve for $k_{eff}(t)$ or the Rossi $\alpha(t)$ during shock compression (see Fig. 9.22) was approximated previously by a linear function or ramp starting from prompt criticality (Hansen [72], Seifritz [28], Mark [19]). Such a ramp type increase in the criticality factor, $k_{eff}(t)$ or $\Delta k(t) = k_{eff}(t)-1$, for a compression time, t_0, starting from prompt criticality can be represented in Fig. 9.23.a.

Similarly, Rossi alpha,

$$\alpha(t) = \frac{\Delta k(t)}{\ell_{eff}},$$

can be considered over the time period t_0 (Fig. 9.23.b).

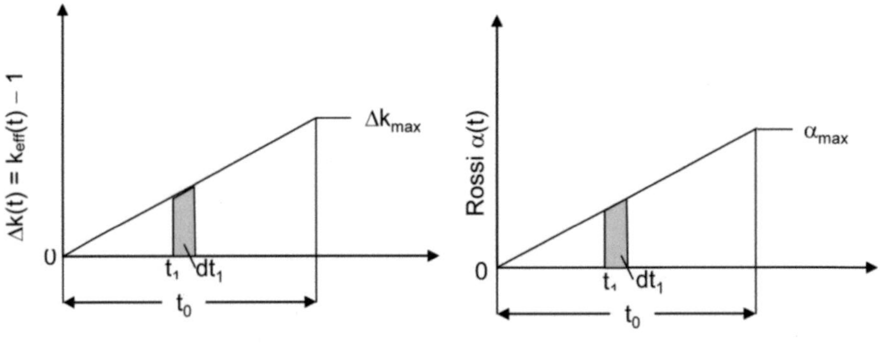

Figure 9.23: (a) $\Delta k(t)$ as a ramp function of time, t_1, starting from prompt criticality, (b) Rossi $\alpha(t)$ as a ramp function of time, t_1, starting from prompt criticality.

Fig 9.23a allows Eqs.(9.13) and (9.14) for the differential probability, $p(t_1)$, and the integral or cumulative probability, $P(t_1)$, of pre-ignition at time, t_1, to be derived (Hansen [72], Seifritz [28]). The differential probability, $p(t_1)$, of a first persistent fission chain being sponsored at time t_1 in the time interval, dt_1, is given by

$$p(t_1) \cdot dt_1 = \frac{2\Delta k_{max}}{\bar{\nu} \cdot \Gamma_2} \cdot \frac{S_M}{t_0} \cdot t_1 \cdot \exp\left[\frac{-\Delta k_{max}}{\bar{\nu} \cdot \Gamma_2} \cdot \frac{S_M}{t_0} \cdot t_1^2\right] dt_1 \; \left[s^{-1}\right] \tag{9.13}$$

and the integral or cumulative probability, $P(t_1)$, that persistent fission chains occurred until the time t_1 (integral of the differential probability between zero and t_1) is given by

$$P(t_1) = 1\text{-}\exp\left[-\frac{\Delta k_{max}}{\bar{\nu}\,\Gamma_2} \bullet \frac{S_M}{t_0} \bullet t_1^2 \right] \tag{9.14}$$

where

Δk_{max} = maximum $k_{eff,max} - 1$ at full compression

$\bar{\nu}$ = average number of prompt neutrons per spontaneous fission

Γ_2 = Diven factor: 0.8 for delta function distribution (weak source) and 1 for Poisson distribution (strong source)

S_M = total multiplied spontaneous fission neutrons per sec

t_1 and t_0 see Fig. 9.23a and b.

As pointed out in Kessler [24], Eq. (9.13) and (9.14) are also consistent with the formalism derived in Mark [19] if $\Delta k_{max}/\nu \cdot \Gamma_2 = ½$. Table 9.14 shows both the inherent and the multiplied total neutron sources for reactor-grade plutonium compositions as considered in Kessler et al. [12]. Below, the focus will be only on Pu-(1) as it has an isotopic concentration of 5.5% Pu-238 and a relatively high critical mass among Pu-(1) to Pu-(6). As the spontaneous fission neutron source for the HNEDs is higher than $9\text{x}10^7$ n/s (see Table 9.14), there will be a Poisson distribution of the spontaneous fission neutrons (Hansen [72], hence $\Gamma_2 = 1$.

Table 9.14: Values of inherent and multiplied spontaneous fission neutron sources for Pu mixtures of Pu-(0) to Pu-(6).

	Pu-238 content [%]	Radius [cm] at $k_{eff} = 0.98$	Subcritical mass $k_{eff} = 0.98$ [kg]	Spontaneous fission neutrons S_{inh} [n/s]	Subcriticality multiplication M = 50 S_M [n/s]
Pu-(0)	1.6	5.19	9.248	$2.92\text{x}10^6$	$1.46\text{x}10^8$
Pu-(0')	2.8	5.30	9.848	$4.15\text{x}10^6$	$2.07\text{x}10^8$
Pu-(1)	5.5	5.80	12.906	$9.6\text{x}10^6$	$4.80\text{x}10^8$
Pu-(2)	8.7	5.81	12.973	$10.8\text{x}10^6$	$5.40\text{x}10^8$
Pu-(3)	12	5.83	13.188	$12.1\text{x}10^6$	$6.00\text{x}10^8$
Pu-(4)	15.2	5.7	13.250	$11.6\text{x}10^6$	$5.80\text{x}10^8$
Pu-(5)	20.3	5.52	11.126	$10.9\text{x}10^6$	$5.45\text{x}10^8$
Pu-(6)	24.5	5.38	10.300	$10.4\text{x}10^6$	$5.20\text{x}10^8$

Eq. (9.14) (cumulative probability of pre-ignition) is now applied to two cases for which the potential explosive yield of HNEDs with reactor-grade plutonium will be calculated later. Fig 9.24 shows Rossi alpha, $\alpha(t)$, for the two cases of 0.06 TPa (Case A) and 0.11 TPa (Case B) representing the pressures exerted in the examples on the outer surface of the reflector (Section 9.10.1.4). Section 9.12 below will show the shock waves to be stopped at 11.47×10^{-6} s for Case A (0.06 TPa), and at 9.307×10^{-6} s for Case B (0.11 TPa). From these points in time on Rossi alpha, $\alpha(t)$, will remain constant at 9.26×10^{6} s^{-1} for Case A (0.06 TPa), and at 12.72×10^{6} for Case B (0.11 TPa). These constant Rossi alpha values will be taken as α_{max} (see Fig. 9.23b and 9.15).

From these constant maximum Rossi alpha values, α_{max}, also Δk_{max} (see Fig. 9.23a) can be calculated by multiplying the Rossi alpha values with the neutron lifetime, $\ell_{eff} = 1.35 \times 10^{-8}$ s furnishing (see Table 9.15) the corresponding $\Delta k_{max} = 0.125$ (Case A with 0.06 TPa and $\Delta k_{max} = 0.1717$ for Case B (0.11 TPa). Now a ramp can be considered between prompt criticality at t = 0 and the values, α_{max} or Δk_{max} (see Fig. 9.24).

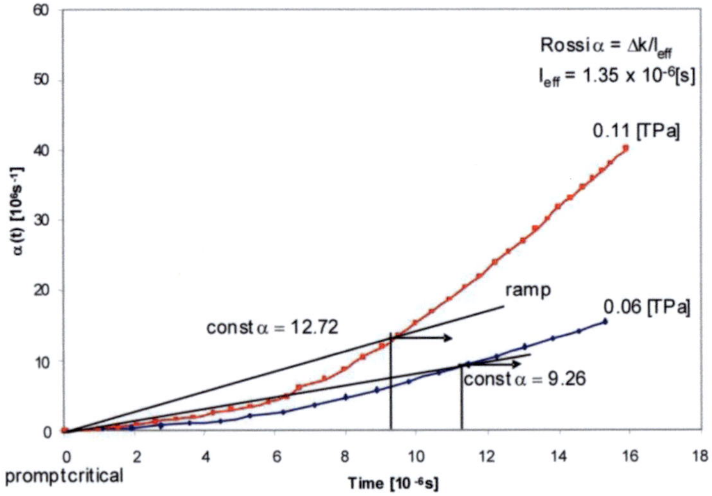

Figure 9.24: Rossi alpha, $\alpha(t)$, for Case A (0.06 TPa) and Case B (0.11 TPa). From prompt criticality to the time when $\alpha(t)$ remains constant, the sigmoidal curve is approximated by a ramp function for pre-ignition analysis.

For both Cases A and B, the minimum and maximum values of the total multiplied spontaneous fission neutron source, S_M, of Table 9.14, are taken, i.e. 1.46×10^{8} n/s and 6×10^{8} n/s. The Δk_{max} or α_{max} and t_0 values define a ramp (see Fig. 9.24 and Table 9.14) (an overestimation compared to the sigmoidal curves in Fig. 9.24).

Table 9.15: Characteristic input values for the cases A and B.

Case A: 0.06 TPa, Fig. 9.24	Case B: 0.11 TPa, Fig. 9.24
$\Delta k_{max} = 0.125$	$\Delta k_{max} = 0.1717$
$\overline{\nu} = 3.1, \Gamma_2 = 1$	$\overline{\nu} = 3.1, \Gamma_2 = 1$
$t_0 = 11.47 \text{x} 10^{-6}$ s	$t_0 = 9.307 \text{x} 10^{-6}$ s
$\alpha_{max} = 9.26 \text{x} 10^6 \text{ s}^{-1}$	$\alpha_{max} = 12.72 \text{x} 10^6 \text{ s}^{-1}$
$\ell_{eff} = 1.35 \text{x} 10^{-8}$ s	$\ell_{eff} = 1.35 \text{x} 10^{-8}$ s

Evaluation of Eq. (9.14) with the input data from Table 9.15 leads to Figs. 9.25 and 9.26 with the cumulative probabilities for pre-ignition for the two cases, A and B. In case A (Fig. 9.25), the cumulative probability can be seen to become 1 between $t_1/t_0 = 0.2$ and 0.3, depending on the spontaneous fission neutron source. In Case B (Fig. 9.26), with the steeper reactivity ramp (faster compression), the cumulative probability of pre-ignition becomes 1 between $t_1/t_0 = 0.15$ and 0.3. The differential probabilities of pre-ignition (Eq. (9.13)) can be seen from Fig. 9.25 and 9.26 as the slope (differentiation) of the curves. It is a maximum at time $t_1 = 0$, dropping to zero when the cumulative probability of pre-ignition approaches 1. The average time $\overline{t_1}$ at which pre-ignition occurs is given by (Hansen [72] and Seifritz [28])

$$\overline{t_1} = \sqrt{\frac{\pi \cdot \overline{\nu} \cdot \Gamma_2 \cdot t_0}{4 \cdot \Delta k_{max} \cdot S}} \tag{9.15}$$

For the two cases, A and B, the average time, $\overline{t_1}$, of pre-ignition after prompt criticality becomes $1.22 \text{x} 10^{-6}$ s or less (see Table 9.16).

Table 9.16: Average time, $\overline{t_1}$, of pre-ignition after prompt criticality.

Case	Spontaneous fission neutron source, S_M (n/s)	Average time of pre-ignition, $\overline{t_1}$ (s)
A	$1.46 \text{x} 10^8$	$1.22 \text{x} 10^{-6}$
	$6 \text{x} 10^8$	$0.61 \text{x} 10^{-6}$
B	$1.46 \text{x} 10^8$	$0.94 \text{x} 10^{-6}$
	$6 \text{x} 10^8$	$0.47 \text{x} 10^{-6}$

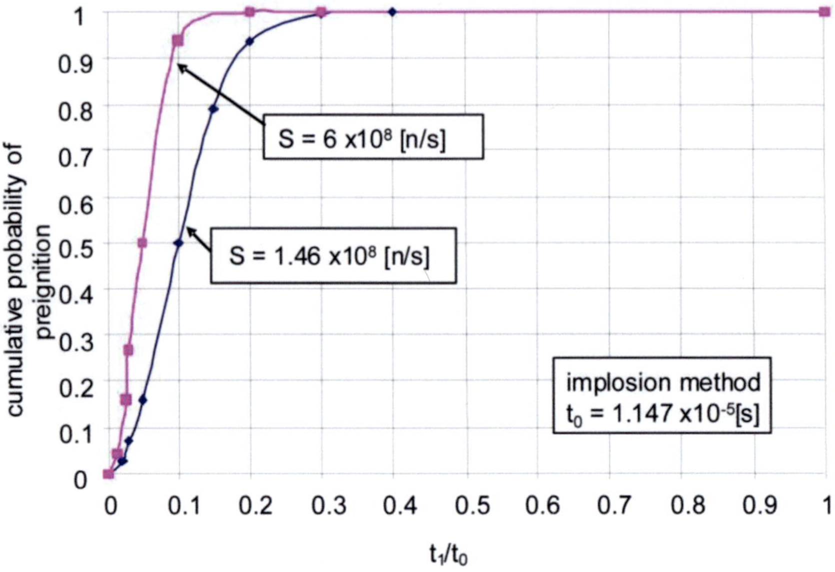

Figure 9.25: Cumulative probability of pre-ignition (Case A for 0.06 TPa) and two different spontaneous fission neutron sources.

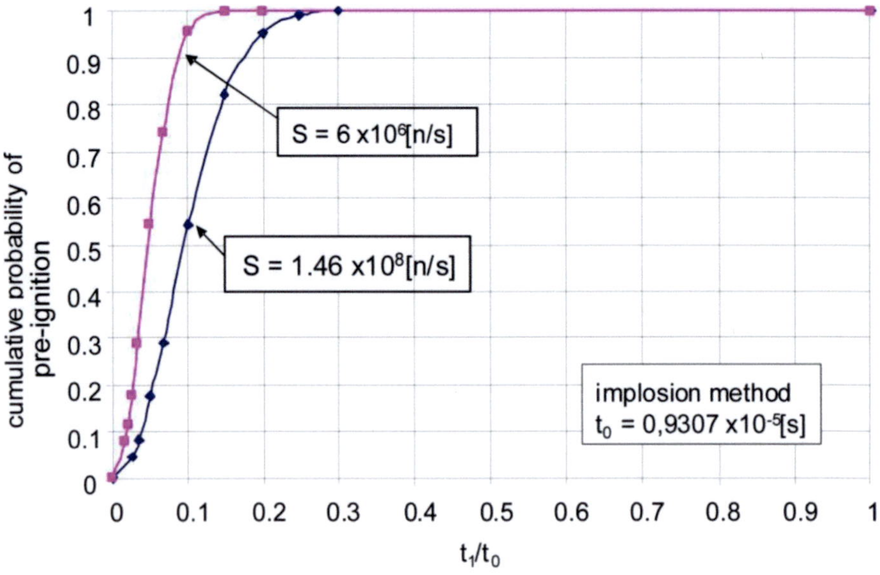

Figure 9.26: Cumulative probability of pre-ignition (Case B for 0.11 TPa) and two different spontaneous fission neutron sources.

9.11.2 Pre-ignition as a consequence of strong spontaneous fission neutron sources and sigmoidal Rossi alpha, $\alpha(t)$

The previous section only covered ramp-type Rossi alpha, $\alpha(t)$, insertions. This is only an approximation of $\alpha(t)$. The case of sigmoidal Rossi alpha, $\alpha(t)$, as shown in Figs. 9.22 or 9.24, was solved comprehensively by Seifritz [73]. The cumulative probability of pre-ignition, for sigmoidal $\alpha(t)$ (see Figs. 9.22 and 9.24) changes then in

$$P(t_1) = 1\text{-}\exp\left[-\frac{2\,l_{eff}}{\bar{\nu}\cdot\Gamma_2}\cdot S_M \cdot \int_0^{t_1} \frac{\Delta k(t')}{l_{eff}} \cdot dt' \right] \tag{9.16}$$

ℓ_{eff} = prompt neutron lifetime (s)

$\bar{\nu}$ = 3.1 average number of fission neutrons for plutonium

Γ_2 = Diven factor = 1 for Poisson distribution (strong source)
= 0.8 for delta function distribution (weak source)

$\Delta k(t')$ = $k_{eff}(t')$-1

S_M = multiplied fission neutron source (subcriticality multiplication of 1/(1-0.98) = 50 taken into account)

t_1 = time after prompt criticality.

The solution of Eq. (9.16) leads to the probability density function for pre-ignition. For reactor-grade plutonium with a multiplied fission neutron source of $S_M = 2\text{x}10^8$ n/s, and $\Delta k_{max} = 0.44$, and for a compression time of 10^{-5} s, an average time of pre-ignition of

$$\bar{t}_1 = 1.02 \text{ x } 10^{-6} \text{ s}$$

and a most probable time, $t_{m,p}$, of pre-ignition,

$$t_{m,p} = 0.88 \text{ x } 10^{-6} \text{ s}$$

are obtained (Seifritz [73]).

These results vary somewhat with Δk_{max} and compression time, but are basically similar to the results in Table 9.16 (Section 9.11.1). Seifritz [73] also shows that the relative time jitter, Δt, around the average time of pre-ignition is only

$$\Delta t = 0.5\text{x}10^{-6} \text{ s}$$

Consequently the spread of possible differential probabilities becomes extremely narrow, and **the expected explosive yields become quasi-deterministic (Seifritz [73]) for the HNEDs with reactor-grade plutonium.**

With the results from Sections 9.11.1 and 9.11.2 the next sections will contain a parametric approach to calculating the explosion yield of reactor-grade plutonium. Pre-ignition is assumed

to occur at 0 or 10^{-6} or $3x10^{-6}$ s after prompt criticality.

9.11.3 Pre-ignition of hybrid HNEDs

Hybrid HNEDs would contain in part reactor-grade plutonium and part highly enriched uranium (HEU) – like hybrid nuclear explosive devices – combining weapon-grade plutonium and HEU (DeVolpi [8]; Podwig [36]). Although this does not address the real proliferation question, for reactor-grade plutonium from civil nuclear energy programs and HEU are only available in nuclear weapon states (NWSs), these cases were analyzed [12].

It is shown there that the total critical mass of reactor-grade plutonium and HEU can be decreased. However, the multiplied fission neutron source becomes as strong as in HNEDs solely employing reactor-grade plutonium (Table 9.14). This again would lead to early pre-ignition immediately upon prompt criticality (as is shown in the sections above) and entail similar consequences for the nuclear explosive yields attainable as will be shown below for HNEDs with reactor-grade pluton

9.12 Calculation of Explosion Yield for HNEDs with Reactor-grade Plutonium

The only explosive yield estimate ever published for reactor-grade plutonium (Mark [19]) indicated a minimum fizzle yield of 0.54 kt TNT (equivalent) for reactor-grade plutonium from spent LWR fuel with a burn-up of 30 GWd/t. Tables 9.12 and 9.14 show the Pu-(0) plutonium composition, which corresponds to the reactor-grade plutonium investigated by Mark [19]. It has a near critical radius of 5.19 cm (with 5 cm U_{nat} reflector) for $k_{eff} = 0.98$ (Table 9.14).

The Pu-(1) reactor-grade plutonium composition was chosen for the nuclear explosion yield analyses by Kessler et al. [12]. This plutonium composition has the highest Pu-238 content of Pu-(0'), Pu-(0), Pu-(1) (Tables 9.12 and 9.14) and the largest near critical radius for $k_{eff} = 0.98$. Consequently, it has also the highest fissile volume and would lead to the highest nuclear explosion yield. This would constitute an upper limit for all plutonium compositions derived from spent LWR UOX fuel up to a burn-up of 60 GWd/t, and for spent LWR MOX fuel up to a burn-up of 50 GWd/t.

9.12.1 Compression Shock Waves and Initial Power

The results for this Pu-(1) reactor-grade plutonium composition and for two levels of pressure exerted by the high-explosive lenses on the outer surface of the reflector of

0.06 TPa (low technology)

and 0.11 TPa (very high technology)

are presented and discussed below. The classification into low and high technology for implosion lenses was defined in Section 9.10.1.4. The Pu-(1) plutonium composition has a near-critical radius of 5.80 cm for $k_{eff} = 0.98$. The reflector of natural-uranium metal is 5 cm thick.

Figures 9.27 and 9.28 show the development of Rossi alpha, $\alpha(t)$, for the two cases of 0.06 TPa and 0.11 TPa. In the case of 0.06 TPa, prompt criticality is attained 16.93×10^{-6} s after the onset of shock compression. In the case of 0.11 TPa, this time span is shorter, prompt criticality being reached 8.44×10^{-6} s after the onset of shock compression.

Only part of the Rossi alpha, $\alpha(t)$, curves is shown in Fig. 9.27 and 9.28 compared to Fig. 9.22 (Section 9.10.7.2) which shows $\alpha(t)$ for the full shock compression time of case B (0.11 TPa). For case A only the characteristic values are reported. Table 9.17 shows these points in time when α_{max} or Δk_{max} and $k_{eff,max}$ would be attained for Case A (0.06 TPa) and Case B (0.11 TPa).

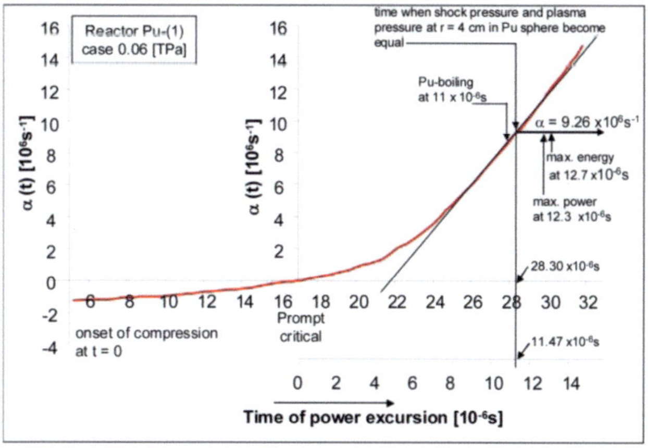

Figure 9.27: Rossi alpha, $\alpha(t)$, for the 0.06 TPa case for two time schedules: onset of compression and onset when prompt criticality is attained.

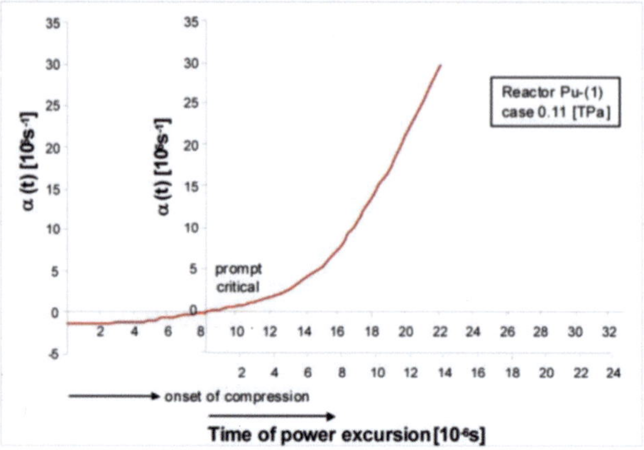

Figure 9.28: Rossi alpha, α(t), for the 0.11 TPa case for two time schedules: onset of compression and onset when prompt criticality is attained.

These maximum values of α_{max} cannot be achieved with reactor-grade plutonium in the HNED, as pre-ignition will start very soon after criticality has been attained and the HNED will explode earlier. This will be shown in Section 9.12.4.

Table 9.17: Maximum α(t) and Δk_{max} or $k_{eff,max}$ reached with different pressures exerted on the outer surface of the reflector.

Pressure exerted on reflector outside (TPa)		Time between compression onset and α_{max} (10^{-6} s)	α_{max} (10^6 s^{-1})	Δk_{max}	$K_{eff,max}$
Case A	0.06	40.2	19.33	0.26	1.26
Case B	0.11	27.6	53.03	0.716	1.716

9.12.2 Initial power at t = 0 for calculation of explosive yield

The initial power was determined in Section 9.10.8 to be

$$L(0) = 0.247 \text{ W}$$

The relative radial distribution of this total power follows from neutron transport calculations. Figure 9.29 shows the relative fission rate, which is proportional to the radial power density distribution. It is equivalent to *w*(r,0) in Eqs. (9.1a) through (9.1c), Section 9.7. The relative fission rate or power density in the reflector is roughly one order of magnitude lower than in the plutonium sphere.

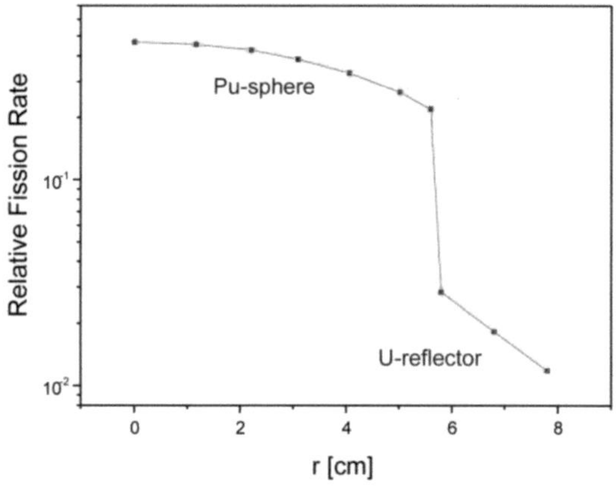

Figure 9.29: Radial distribution of the relative fission rate or power as a function of r at t = 0.

9.12.3 Power Excursion

The Rossi alpha, α(t), of Figs. 9.27 and 9.28 is applied to the two cases of 0.06 TPa and 0.11 TPa, and Eqs. (9.1) through (9.5) are solved by the iterative calculation procedure described in Section 9.7.1 with the materials data and equation of state data taken from Section 9.8.5. Ignition of the chain reaction is assumed to occur at prompt criticality, as in Mark [19]. Section 9.12.6 below will discuss a sensitivity study with time delays for ignition of the chain reaction of $1x10^{-6}$ s and $3x10^{-6}$ s as suggested in Section 9.11.2.

9.12.4 How Far Can the Shock Wave Penetrate into the Pu-sphere?

According to Eqs. (9.1a) through (9.1c), and as a consequence of Rossi alpha, α(t), in Fig. 9.27 and 9.28, power will rise extremely fast. The integral of power over time will be responsible for the temperature increase in the plutonium sphere (Eqs. (9.2) and (9.3)) up to the boiling point. According to the equation of state (Eq. (9.4) and Section 9.8.5), further increases in plutonium temperature will lead to high pressures up to the [TPa] range. This means that the shock wave progressing inward more and more must counteract the rising pressures in the Pu-sphere and will eventually be stopped. From that time on, the compressed volumes cannot be further increased by the shock wave, and Rossi alpha, α(t), remains constant. In the cases under study, this occurs when the shock wave has penetrated the plutonium sphere some 1.3 cm or 1.8 cm from the outside, and the front of the shock wave comes to a halt roughly 4.5 or 4 cm from the center of the plutonium sphere.

This is shown in Fig. 9.30 and Table 9.18. For the 0.06 TPa case, the shock wave progressing inward hits the pressure rising in the plutonium sphere at 11.47×10^{-6} s from prompt criticality and at the radius of approximately 4.5 cm. Plutonium boiling will occur earlier, at 11×10^{-6} s. Rossi alpha stays constant from $\alpha = 9.26 \times 10^{6}$ s^{-1} on.

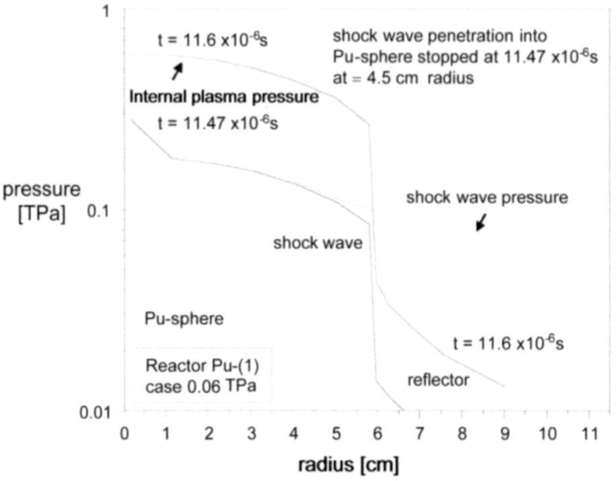

Figure 9.30: Dependence on time and space of shock wave pressure and internal plasma pressure for the 0.06 TPa case.

For the 0.11 TPa case, the time spans are 9.307x10-6 s from prompt criticality when Rossi alpha, $\alpha(t)$, has arrived at $\alpha =$ 12.719x106 s-1. The front of the shock wave is stopped at a radius of approx. 4.0 cm.

Table 9.18: Time spans to prompt criticality and shock wave stopped by higher plasma pressure.

case	Time to prompt criticality $[10^{-6}$ s]	Onset of Pu boiling from criticality $[10^{-6}$ s]	Time from criticality when shock wave stopped $[10^{-6}$ s]	Total time from onset of compression when shock wave stopped $[10^{-6}$ s]	Constant α $[10^{6}$ s]	Radius where shock wave stopped [cm]
0.06 TPa	16.933	11	11.47	28.403	9.26	4.5
0.11 TPa	8.442	8.95	9.307	17.749	12.719	4.0

It should be noted that this criterion, or explanation of the stopping of the shock wave, is not consistent with the criterion applied by Mark [19]. He applied a much cruder criterion: reactivity or α_{comp} stays constant when the power in the plutonium sphere has risen to $L(t) = L(0) \cdot e^{45}$. Then he applies the Serber formula [32].

9.12.5 Detailed results of the calculations of explosion yield for the 0.06 TPa and 0.11 TPa cases

This section contains a presentation of the detailed results of the calculations for both the 0.06 TPa and 0.11 TPa cases. Rossi alpha, $\alpha(t)$, for both cases is shown in Fig. 9.31 and 9.32 from the onset of compression of the reflector up to the point of criticality and, finally, to the shock wave being stopped with α = constant.

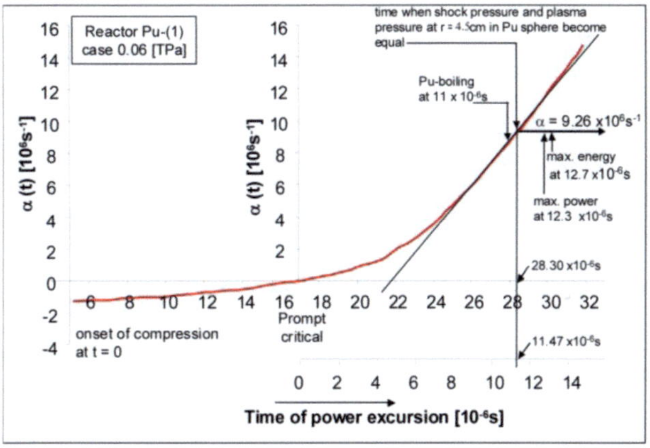

Figure 9.31: Dependence on time of Rossi alpha, $\alpha(t)$, after compression or after transition through prompt criticality (0.06 TPa case).

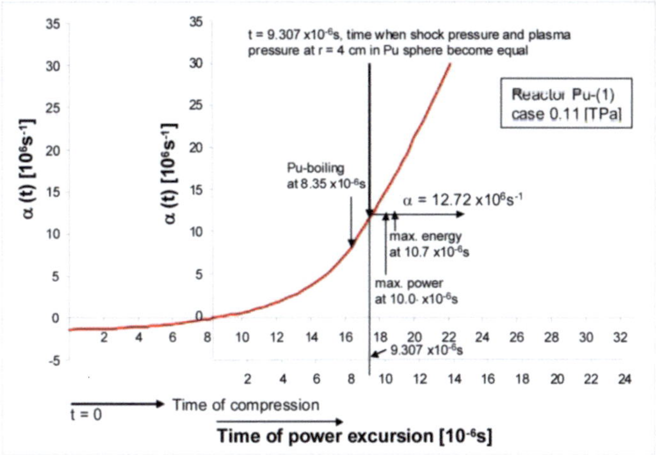

Figure 9.32: Dependence on time of Rossi alpha, $\alpha(t)$, after compression or after transition through prompt criticality (0.11 TPa case).

As a consequence of supercriticality and the relatively high Rossi alpha, α(t), the power rises sharply, as shown in Fig. 9.33, peaking 12.34×10^{-6} s (0.06 TPa case) or 10.0×10^{-6} s (0.11 TPa case) after prompt criticality. In the 0.11 TPa case (Fig. 9.34), the power rises to levels roughly six times higher than in the 0.06 TPa case before the rapid increases in internal energy and internal pressures within the reactor-grade plutonium sphere initiate the explosion. This results in a strong negative feedback. As a consequence, the power drops very sharply.

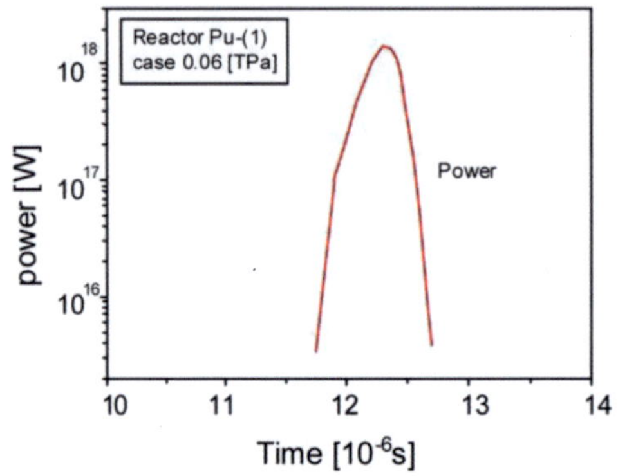

Figure 9.33: Total power as a function of time after prompt criticality (0.06 TPa case).

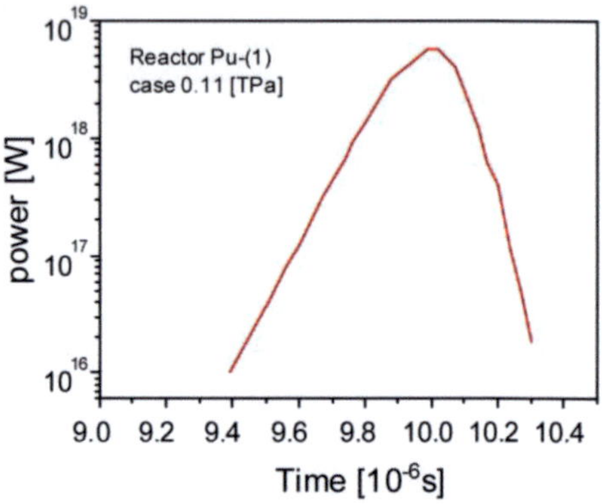

Figure 9.34: Total power as a function of time after prompt criticality (0.11 TPa case).

Energy as an integral over power is shown in Fig. 9.35 and 9.36 for the 0.06 TPa and 0.11 TPa cases. The total energy released, or the explosion yield, results in 0.119 kt of TNT (equivalent) for the 0.06 TPa case. As can be expected, this value is higher for the 0.11 TPa case resulting in 0.34 kt of TNT (equivalent). The values in kt of TNT equivalent are based on the energy equivalence of $4.187 \cdot 10^{12}$ J $\triangleq$ 1 kt TNT [50].

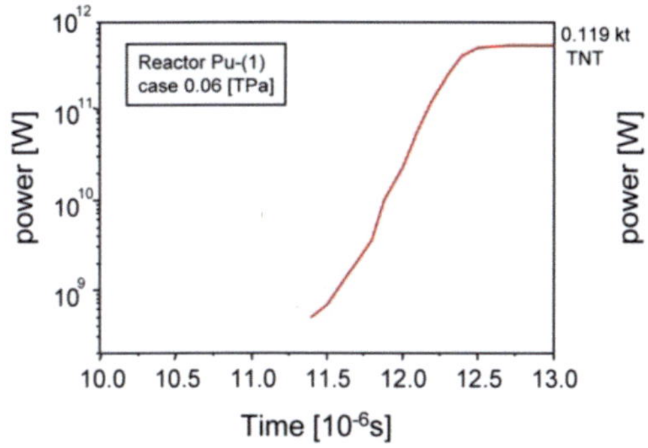

Figure 9.35: Energy released as a function of time after prompt criticality.

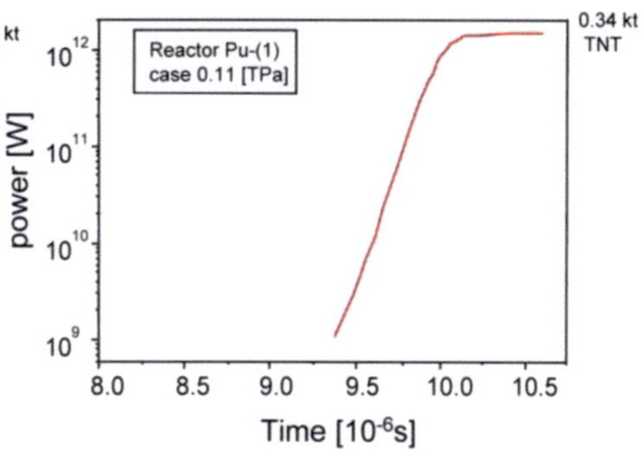

Figure 9.36: Energy released as a function of time after prompt criticality.

Figure 9.37 and 9.38 shows the plasma pressure at the r = 0 and r = 4.04 cm locations of the plutonium sphere as a function of time after prompt criticality. For the 0.06 TPa case, a maximum plasma pressure of approx. 10^2 TPa is attained at 12.34×10^{-6} s. For the 0.16 TPa case, the maximum plasma pressure of approx. 3.4×10^2 TPa is reached roughly 10×10^{-6} s after prompt criticality.

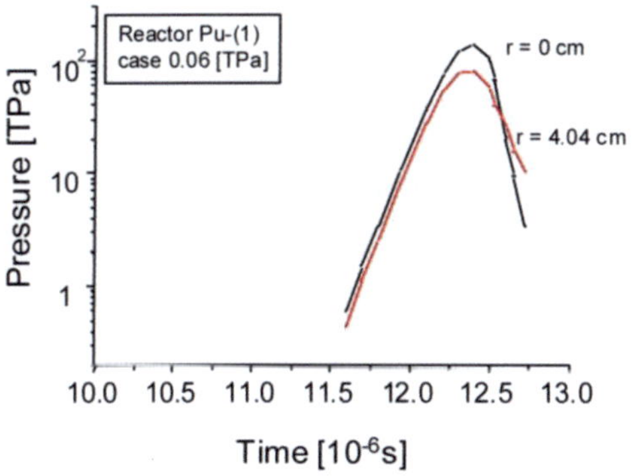

Figure 9.37: Plasma pressure at r = 0 and r = 4.04 cm as a function of time after prompt criticality (0.06 TPa).

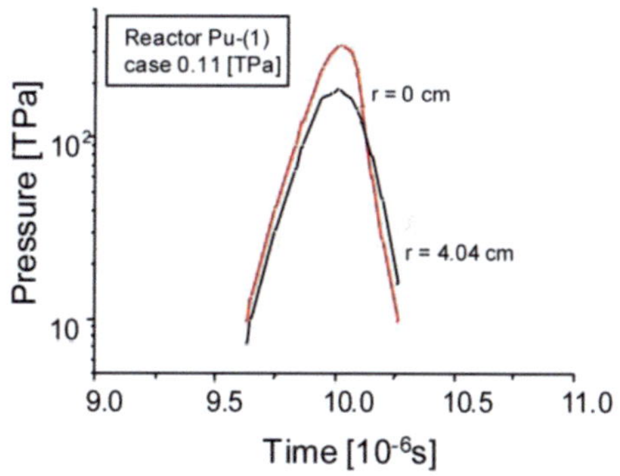

Figure 9.38: Plasma pressure at r = 0 and r = 4.04 cm as a function of time after reaching prompt criticality (0.11 TPa).

To complete the picture, Fig. 9.39 and 9.40 show the plasma pressure as a function of the radius and of the time up to maximum plasma pressure (12.39×10^{-6} s for the 0.06 TPa case and 10.031×10^{-6} s for the 0.11 TPa case). The pressure spikes in the natural uranium reflector are caused by the rapidly expanding plutonium sphere and the inertia of the outer parts of the reflector.

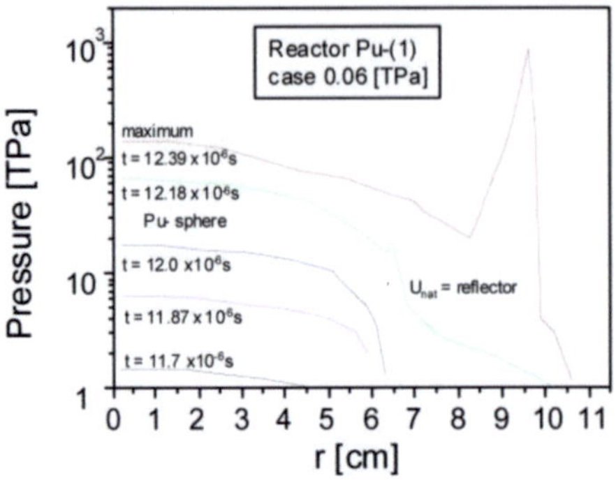

Figure 9.39: Pressure as a function of radius and time after prompt criticality (0.06 TPa case).

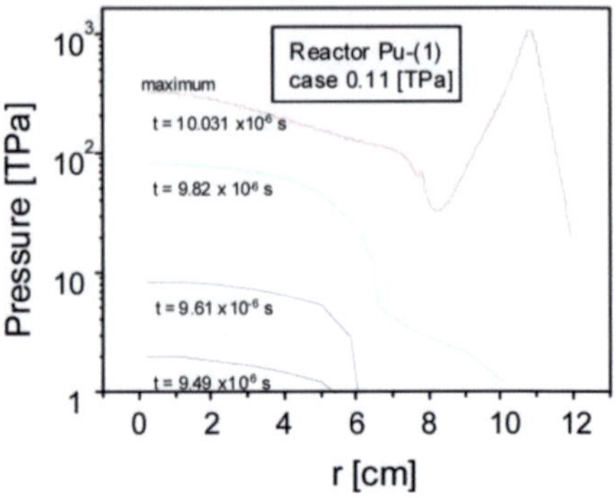

Figure 9.40: Pressure as a function of radius and time after prompt criticality (0.11 TPa case).

Figures 9.41 and 9.42 show the plasma densities in the reactor-grade plutonium sphere and the reflector as a function of space and time for the 0.06 TPa and 0.11 TPa cases. Again, the density spikes at the inner part of the reflector are caused by the expanding plutonium sphere.

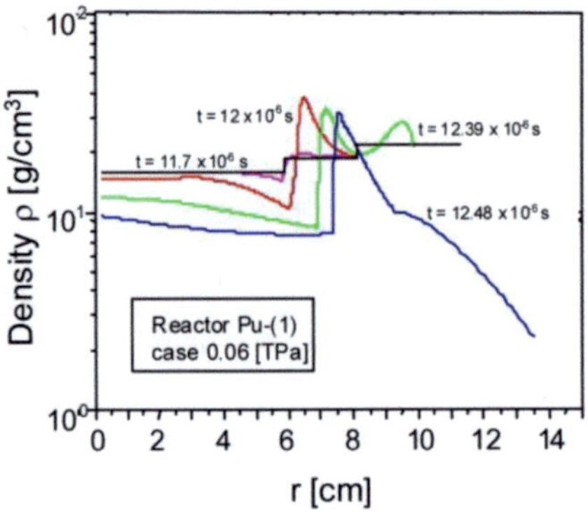

Figure 9.41: Plasma density as a function of space and time after prompt criticality (0.06 TPa case).

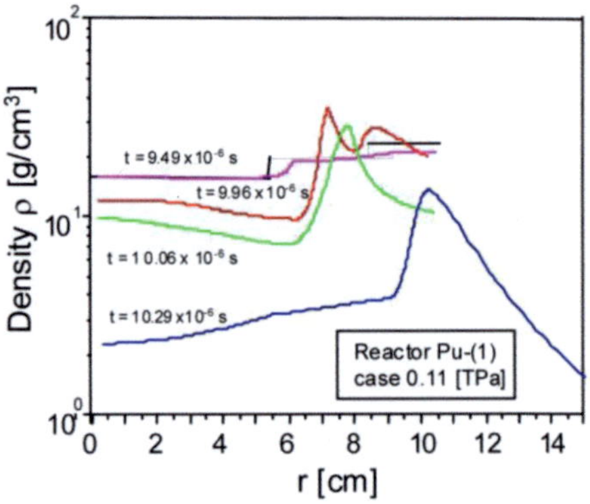

Figure 9.42: Plasma density as a function of space and time after prompt criticality (0.11 TPa case).

Expansion of the plutonium sphere and natural-uranium reflector system as well as the associated density variations give rise to frequent S_N-calculations (see Section 9.7.) to determine $k_{eff}(t)$ or $\alpha_{tot}(t)$. These $\alpha_{tot}(t)$ values determine the power as a function of time (Fig. 9.43). These figures (shown only for 0.06 [TPa]) also indicate the variations in the outer radius of the reactor-grade plutonium sphere as a function of time. They again show the inverse quasi-proportionality of R(t) and α(t) used in the Serber correlation [32].

9.12.6 Sensitivity of calculations of the nuclear explosion yields

Section 9.12.5 contained the results of explosion yields calculated under the assumption of pre-ignition occuring at the earliest possible time when the Rossi alpha, $\alpha(t)$, curve intersects prompt criticality, i.e., pre-ignition occurs at $t = 0$, the onset of the power excursion. Section 9.11, presented the pre-ignition theory, concluding that pre-ignition would occur at an average $\overline{t_1} = 10^{-6}$ s. To get a feeling for the sensitivity of these results, also the explosion yield for pre-ignition at 0, 10^{-6}, or $3x10^{-6}$ s was calculated. The results are shown in Table 9.19.

Table 9.19: Explosion yields for different times of pre-ignition after prompt criticality.

Average time of pre-ignition [10^{-6} s]	0	1	3
[kt] TNT (equivalent)	0.119	0.120	0.130

The results in Table 9.19 show that the impact on nuclear explosion yield is rather small within an average time, $\overline{t_1}$, of pre-ignition between 0 and $3x10^{-6}$ s. The earliest possible time of pre-ignition, $\overline{t_1} = 0$, was assumed also by Mark [19]. The corresponding explosion yield was referred to as the minimum fizzle yield.

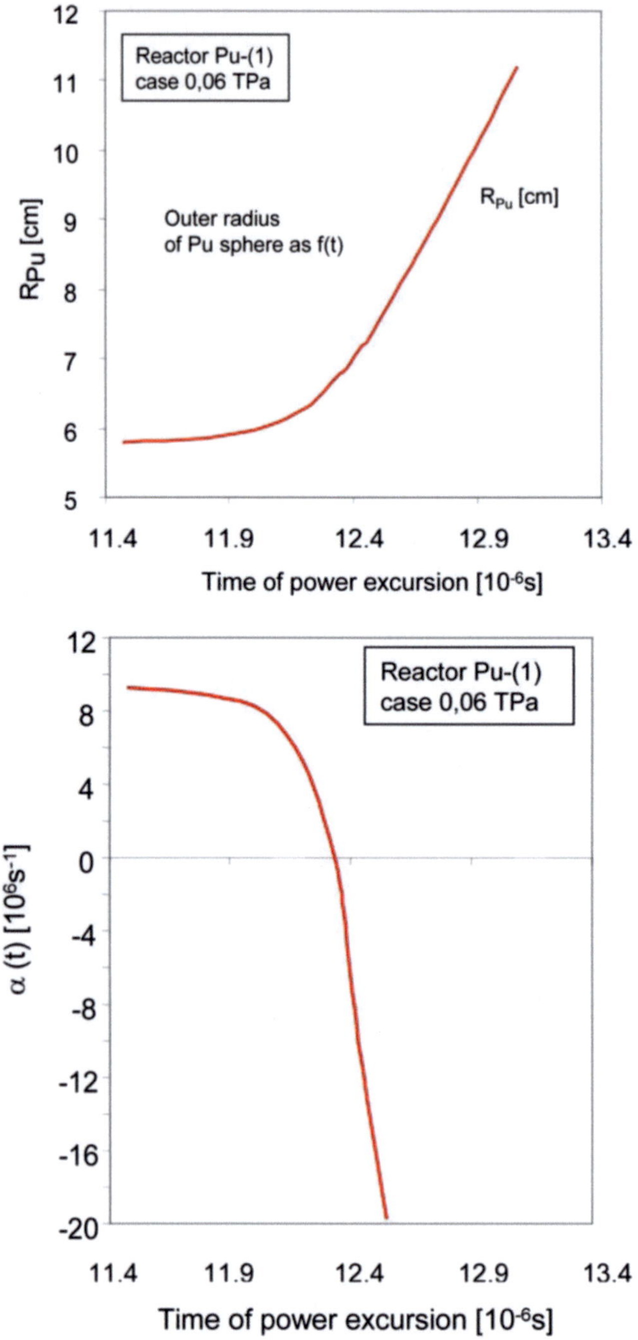

Figure 9.43: Expansion of the outer radius of the Pu-sphere, including internal density changes and the corresponding Rossi alpha change, $\alpha_{tot}(t) = \alpha_{comp} + \alpha_{expa}$, for the 0.06 TPa case.

Another factor studied was the influence of the initial temperature in the reactor-grade plutonium sphere and natural-uranium reflector system. The results in Section 9.12.5 correspond to a constant initial temperature of 300 K over the entire HNED. Other calculations were performed under the assumption of the real temperature in the HNED being caused by alpha heat power in the plutonium sphere (for details, see Kessler [13]) and the temperature increase due to shock wave compression (Section 9.10.4). The results are shown in Table 9.20.

The results are essentially the same because the difference in temperatures caused by alpha heat power and by shock compression is small compared to the temperature increase throughout the power excursion.

Table 9.20: Nuclear explosion yields for the 0.06 TPa and 0.11 TPa cases and for different temperatures in the HNED.

Technology	Case	Initial temperature in the HNED	Yield (kt) of TNT (equivalent)	Const. α ($x10^6$ s^{-1})
Medium technology	0.06 TPa	Real temp. profile Room temp. (300 K)	0.119 0.119	 9.26
Very high technology	0.11 TPa	Real temp. profile Room temp. (300 K)	0.346 0.346	 12.719

9.12.7 Nuclear explosive yield of hollow reactor-grade plutonium HNEDs

Section 9.10.5 showed the results of radial compression of a hollow spherical HNED as a function of time for an assumed pressure of 0.11 TPa acting on the outside of the reflector. The radial shock front hits the outer radius of the spherical reactor-grade plutonium shell at around $15x10^{-6}$ s, arriving at the inner radius of the spherical shell at approx. $20x10^{-6}$ s. From this point in time on, the plutonium particles would fly toward the center where they would be compacted into a highly compressed solid sphere. This would apply to a nuclear explosive device without pre-ignition problems. The neutron chain reaction would be ignited in a nuclear explosive device when the inner solid sphere would be fully compacted.

In a spherical hollow HNED with reactor-grade plutonium, the sequence of steps will be different. The analysis of a hollow reactor-grade plutonium HNED with a Pu-(1) plutonium composition, an inner radius of 5.6 cm, and an outer shell radius of 7.43 cm showed the following results.

When the shock front penetrates into the outer reflector, prompt criticality would be attained when roughly 2.5 cm of the 5 cm natural-uranium reflector are compressed. Pre-ignition will start with a delay of approx. 10^{-6} s. The rising power will increase pressure in the spherical reactor-grade plutonium shell, and the shock front will be stopped after penetrating roughly 1.5 cm into the solid region of the hollow spherical plutonium shell. This is similar to the results for a solid sphere of reactor-grade plutonium as described in detail in Section 9.12. Also the maximum Rossi alpha, α_{max}, will be similar to the values described for the solid reactor-grade plutonium sphere.

As the entire sequence of events in the ensuing nuclear explosion will only take $2x10^{-6}$ s (Section 9.12.5), there will be no time for the plutonium particles to fly toward the center. Internal movement of the plutonium particles or plasma during the nuclear excursion would not increase Rossi alpha, $\alpha(t)$, as the reactivity importance function for the particles is rather flat in the inner hollow spherical part, i.e. the exact position of these particles has only a small influence on the criticality. The nuclear explosive yields would be similar to those reported in Section 9.12.5 for solid reactor-grade plutonium HNEDs. The advantages of hollow spherical implosion in nuclear explosive devices with weapons-grade plutonium cannot be confirmed for hollow spherical reactor-grade plutonium HNEDs.

9.12.8 Discussion of these results compared to those of Mark [19]

It was demonstrated in Section 9.10.8 that the geometrical arrangement of the HNEDs with a $k_{eff} = 0.98$ required a subcriticality multiplication of M = 50 to be taken into account. This leads to a spontaneous fission neutron source of $4.8x10^8$ n/s for the Pu-(1) plutonium composition considered (see Table 9.14). (The other plutonium compositions, Pu-(0) through Pu-(6), would have spontaneous fission neutron sources between $1.5x10^8$ n/s and $6x10^8$ n/s). As a consequence of this high spontaneous fission neutron source, pre-ignition would occur as early as about 10^{-6} s after prompt criticality. Mark [19] assumed that pre-ignition would occur when prompt criticality was attained during shock compression. No pre-ignition theory was applied to that assumption.

As a result of Monte Carlo calculations, Kessler et al. [12] assumed an average neutron life time of $\ell_{eff} = 1.35x10^{-8}$ s (see Section 9.10.7.2). Mark [19] assumed a somewhat shorter neutron lifetime of 10^{-8} s.

Detailed calculations describing shock compression and density variations as a function of space and time as well as reactivity as a function of time show that, during power excursion, the shock wave can penetrate only as far as some 4.5 cm (0.06 TPa case) or 4.0 cm (0.11 TPa case)from the center of the sphere (Table 9.18). The reason is that the rising internal pressure in the plutonium sphere becomes higher during the power excursion than the pressure of the shock wave. This stops the shock wave, and the reactivity, α_{comp}, introduced stays constant until explosion starts. This criterion leads to a nuclear explosion yield of

0.119 kt of TNT (equivalent) for low technology (0.06 TPa),

0.346 kt of TNT (equivalent) for very high technology (0.11 TPa).

Mark [19] applies a cruder criterion: Reactivity, or α_{comp}, stays constant when the power in the plutonium sphere has risen to $L(t) = L(0)\cdot e^{45}$. Application of Serber's formula (Serber [32]) leads to an explosion yield of

0.54 kt of TNT (equivalent).

This value is higher than the results obtained by Kessler et al. [12].

It is pointed out in Mark [19] that a relatively broad probability density function for the differential probability of pre-ignition would allow higher fizzle yields of the nuclear explosion. As pointed out by Seifritz [73] and shown in Section 9.11.2, the probability density function for pre-ignition extends only over a very narrow range of time, and **the results for the explosion yield of Kessler et al. [12] thus become quasi-deterministic**.

Mark [19] also indicates that higher reactivity ramps would be possible as a consequence of higher pressures exerted on the outside surface of the reflector. While this may be possible to a certain extent with weapons-grade plutonium for advanced NWSs, considerable doubts are raised in case of HNEDs based on reactor-grade plutonium by Kessler et al. [23]. Higher shock pressures would require thicker high-explosive lenses of more than 25 cm. Explosive lenses that thick would melt or auto-ignite if reactor-grade plutonium in the HNEDs were used. In addition, it was explained in Section 9.10.1.3 that the shock wave and particle velocity are limited.

It should also be kept in mind that the calculated results discussed above are based on perfect radial symmetry of the shock wave hitting the outer surface of the reflector. In practical reality, however, even quasi-radial symmetry is extremely difficult to achieve.

9.12.9 Conclusions from the analysis of the explosive yields of HNEDs based on reactor-grade plutonium

The physics models developed for early power excursion and disassembly analysis of metal-fueled plutonium criticals are extended to the 10^2 TPa range by additional equation-of-state models for high plasma pressures. These models were applied to the so-called

"Sandmeier case", the only result of a nuclear explosion ever published at least in part [37]. This approach confirmed and validated the theoretical models sufficiently well for further application to determining the nuclear explosive yield potential of HNEDs with reactor-grade plutonium.

Application of the theories of pre-ignition to ramp-type and the real sigmoidal curves of Rossi alpha, $\alpha(t)$, demonstrated that pre-ignition occurs roughly at 10^{-6} s after prompt criticality. This early pre-ignition implies that the shock wave can penetrate not more than 1.3 to 1.8 cm into the sphere of reactor-grade plutonium from the outside. This limits reactivity or the maximum Rossi alpha, α_{max}, to be achieved by shock compression of reactor-grade HNEDs.

As a consequence, nuclear explosive yields were calculated of

0.119 kt of TNT (equivalent) for an outside pressure of 0.06 TPa,

and 0.354 kt of TNT (equivalent) for an outside pressure of 0.11 TPa.

The results which are based on perfect radial symmetry of the shock wave, are quantitatively lower than the 0.54 kt of TNT (equivalent) published by Mark [19]. They are not different qualitatively, as they would still be beyond the explosive yield of any known conventional chemical explosive device.

These are the results of neutronic investigations. However, as shown in Kessler et al. [23], they do not describe the full reality. They must be seen together with a thermal analysis and with the level of technology available or needed to achieve such nuclear explosive yields. When only low technology is used for the geometric dimensions and the type of high explosives, such HNEDs would not be feasible technically with reactor-grade plutonium from reprocessed spent LWR fuel with a burn-up higher than about 35 GWd/t. Only if advanced medium technology can be applied, this limit of technical feasibility must be shifted from 35 GWd/t to a level above 58 GWd/t (see Section 10).

9.13 Categorization of different isotopic compositions of plutonium

The analysis of C. Mark [19] for the potential nuclear explosive yield of HNEDs with reactor-grade plutonium is based mainly on assumptions already described in the Los Alamos Primer [32]. This possibly intentionally simplified model does not follow a detailed analysis as presented by Kessler et al. [12] and summarized in Sections 9.12.5 to 9.12.7. However, it allows to categorize the different plutonium compositions collected in

Tables 9.2-9.6.b of Section 9.3 according to their pre-ignition potential and their resulting potential nuclear yield without doing many detailed calculations.

C. Mark [19] applies the following formalism:

$$L(t) = L(0) \cdot e^{\int_0^{\tau} \alpha(t') \cdot dt'} \quad \text{(corresponding to Eq. 9.9b)}$$

L(t) = power as a function of t

$k_{eff}(t)$ = criticality at a function of t

ℓ_{eff} = effective neutron life time sec (corresponding to Eq. 9.1c)

$\approx 10^{-8}$sec

$$\frac{k_{eff}(t)-1}{\ell_{eff}} = \frac{\Delta k(t)}{\ell_{eff}} = \alpha(t)$$

$\alpha(t)$ called Rossi alpha

L(0) = initial power at t = 0

In addition, he assumes that L(t) increases exponentially. With $L(t) = L(0) \cdot e^{45}$ the plutonium will have vaporized and at a pressure of about 1 Mbar will start the expansion or explosion of the device. At this point in time also the maximum possible criticality or α_{max} will be achieved (Fig. 9.44).

In a further simplified approach it is assumed – as in Los Alamos Primer [32] – that the shock compression leads to a linear increase (instead of a sigmoidal increase (Section 9.11) of $\alpha(t)$

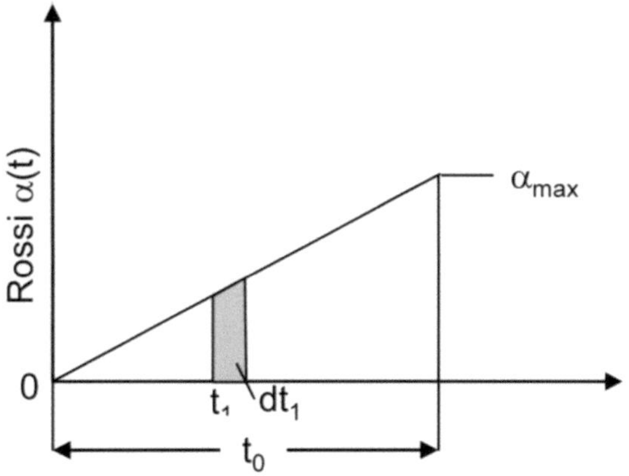

Figure 9.44: Linear increase of Rossi alpha during shock compression.

$$\int_{t_1}^{t_2} \alpha(t')dt' = 45$$

$$t_1 = 0 = \text{initial time}, t_2 = \text{final time}$$

$$t_2 = t_0 = \text{time full compression when C. Mark's criterion achieved}$$

$$\text{with } \alpha(t) = c \cdot t \text{ (linear function)}$$

$$\rightarrow \int_0^{t_0} \alpha(t')dt' = \int_0^{t_0} c \cdot t'dt' = \frac{1}{2} ct_0^2 = 45$$

C. Mark [19] assumes that pre-ignition starts already at $t = 0$ when supercriticality is exceeded. This earliest possible pre-ignition results also in the smallest nuclear explosive yield, called smallest fizzle yield, Y_F, with a Rossi alpha, $\alpha_{max,F}$, at $t_F = t_0$.

Finally the Serber relation (Section 9.9) is applied to two different nuclear explosive devices A and B. In both cases the shock velocity is 5 km/s and the radius of both plutonium spheres is 5 cm.

Case A is an NED based on weapons-grade plutonium with an explosive yield of 20 kt TNT (equivalent) with $l_{eff} = 10^{-8}$ s, $k_{eff,max} = 2$, $\Delta k_{max} = k_{eff,max}-1 = 1$ and $\alpha_{max} = 10^8$ s^{-1}.

Case B is a HNED based on reactor-grade plutonium. It is compressed with the same shock velocity of 5 km/s. Pre-ignition starts at the earliest point in time when supercriticality is exceeded. At t_c the power results in $L(t_F) = L(0) \cdot e^{45}$ and the maximum possible Rossi alpha will be attained and stays constant at $\alpha_{max,F}$. Therefore, Case B can attain only a minimum fizzle yield Y_F. The $\alpha_{max,F}$ is determined by the above assumptions.

The application of the Serber relation (Section 9.9) leads then to C. Mark's [19] estimated minimum explosion fizzle yield for case B of

$$Y_F = 0.45 \text{ kt TNT (equivalent)}$$

9.13.1 Integral pre-ignition probability and nuclear explosive yield

In the Appendix to the paper of C. Mark an analytical relation between the integral probability of pre-ignition and the possible nuclear explosive yield Y is derived by van Hippel and Lyman [19]. They start from Eq. (9.14) Section 9.11.1 for the integral or cumulative probability for pre-ignition.

$$P(t_1) = 1\text{-exp}\left[-\frac{\Delta k_{max}}{\bar{\nu}\,\Gamma_2} \bullet \frac{S_M}{t_0} \bullet t_1^2\right] \qquad (9.17)$$

with Δk_{max} = ($k_{eff,max}$-1) with $k_{eff,max}$-max. supercriticality of the sigmoidal curve
t_0 = compression time
t_1 = time after prompt criticality.

As already mentioned in Section 9.11, C. Mark [19] assumes $\Delta k_{max} = 1$ which corresponds to $\Delta k_{max}/\nu \cdot \Gamma_2 = 1/2$. Using this factor 1/2 and the Serber relation they are able to derive an analytical relationship between the integral or cumulative probability for pre-ignition P and the possible nuclear explosive yield fraction Y/Y_o.

$$P\left(\frac{Y}{Y_0}\right) = 1\text{-exp}\left[-\frac{1}{2}\, S_M t_0 \left(\frac{Y}{Y_0}\right)^{2/3} + 45\, S_M t_0 l_{eff}\right] \qquad (9.18)$$

Y = nuclear explosive yield achieved after pre-ignition at some point in time
Y_o = maximum possible nominal yield without pre-ignition
S = multiplied fission neutron source
t_0 = time for full compaction of the NED, e.g. 10^{-5} s
ℓ_{eff} = effective neutron life time, e.g. 10^{-8} s

9.13.2 Numerical evaluation of the integral probability for pre-ignition for different isotopic compositions of plutonium

The integral probability for pre-ignition can be evaluated, e.g. for two cases with different numerical assumptions.

Case 1 describes roughly the data used in Section 9.12.5

$$\Delta k_{max} = 0.6 \qquad k_{eff,0} = 0.98 \qquad t_0 = 10^{-5}\ \text{s} \qquad \bar{\nu} = 2.5 \qquad \ell_{eff} = 1.35 \cdot 10^{-8}\ \text{s}$$

Case 2 describes roughly the data used by Mark [19]

$$\Delta k_{max} = 1 \qquad k_{eff,0} = 0.90 \qquad t_0 = 10^{-5}\ \text{s} \qquad \bar{\nu} = 2.5 \qquad \ell_{eff} = 10^{-8}\ \text{s}$$

The S_M are taken according to the critical masses and to the spontaneous fission neutron sources of Tables 9.2-9.6.b for case A. For $k_{eff,0} = 0.90$ (Case 2) the critical masses and S_M were corrected accordingly. The plutonium compositions of Tables 9.2 to 9.6.b contain the plutonium compositions of Tables 9.12 and 9.14 in part but also some more realistic compositions of proliferation-proof plutonium for the application of civil nuclear energy.

Figures 9.45a and b show the cumulative or integral probabilities of pre-ignition as a function of t_1/t_0 or of the real time scale assuming $t_0 = 10^{-5}$ s.

The different isotopic compositions of plutonium are given in colors. The denotations A, B, C, D, E correspond to the Tables 9.2 through 9.6.b. It can be seen that reactor-grade plutonium C-1, C-2, C-3, C-4 and proliferation-proof plutonium D-1, E-1, E-2, E-3, E-4 represent a special category for themselves. For these isotopic compositions the cumulative or integral probability for pre-ignition attains 1 for $t_1/t_0 \leq 0.1$ i.e. pre-ignition occurs within or $\leq 10^{-6}$ s in case 1. In case 2 pre-ignition occurs within about $t_1/t_0 \leq 0.2$ or $\leq 2x10^{-6}$ s.

Case 2 corresponds to $\Delta k_{max} = 1$ which would represent a compression ratio of about 4 to 5 of the original plutonium density. A comparison of Case 1 and Case 2 clearly shows that some of the isotopic compositions of the category A and B attain the cumulative or integral probability of 1 rather late between $0.6 \leq t_1/t_0 \leq 0.8$ or between $6x10^{-6}$ s to $8x10^{-6}$ s

These plutonium compositions which correspond to super-grade (A-1), weapons-grade (A-2), low burnup CANDU (A-3) and the CANDU (B-2) and FBR blanket (B-3) confirm the US-statement of Fig. 9.1. Such type of reactor plutonium would pre-ignite over a broad time scale of the spherical shock compaction and therefore attain different unpredictable explosive yields over a broader range. This is not the case for reactor-grade and proliferation-proof plutonium from spent PWR fuel (categories, C, D and E). They show the quasi-deterministic behavior for the corresponding minimum fizzle yields.

An absolute exemption is the Trinity test case which is discussed by C. Mark [19] and shown here only for Case 2. A plutonium composition of 99.2% Pu-239 and 0.8% Pu-240 was chosen for the TRINITY test which corresponds roughly to the cumulative probability data of C. Mark [19]. This is slightly lower than Chebeskov's value of 0.9% Pu-240 [14]. This gives an indication that the Trinity test probably had a plutonium composition with a Pu-240 content better, i.e. smaller than supergrade weapons plutonium.

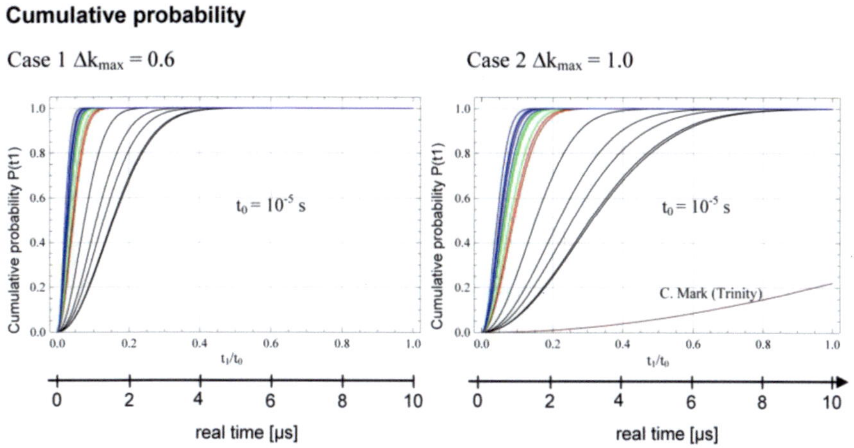

Color	Type of plutonium
black	Weapons Pu: A-1, A-2, A-3, A-4, B-3 (FBR-blanket)
red	Plutonium from low burnup fuel: B-1, B-2
green	Plutonium from LWR spent fuel: C-1, C-2, C-3, C-4
blue	Proliferation-resistant Pu: D-1, E-1, E-2, E-3, E-4
brown	C. Mark, Trinity nuclear explosive test

Figure 9.45: (a) and (b) Cumulative probability for Cases 1 and 2.

Figs. 9.46a and b show the evaluation of Eq. (9.18) for the Cases 1 and 2 and for the different isotopic categories of plutonium of Tables 9.2 through 9.6.b. The left ordinate shows the cumulative or integral probability of pre-ignitions as a function of the possible nuclear explosive yield fraction Y/Y_0. The right ordinate represents 1-P the cumulative or integral probability for **no pre-ignition** which is equivalent to the probability to attain Y/Y_0. Fig. 9.46a shows the results for Case 1 whereas Fig. 9.46b represents Case 2, the highest state of implosion technology. Both figures show that reactor-grade plutonium from LWRs and especially proliferation-proof plutonium lead to **quasi deterministic minimum fizzle yields**. Only the isotopic plutonium composition from low burnup CANDU or AGR spent fuel or FBR blanket plutonium offer the possibility to attain small ratios of Y/Y_0 for small cumulative or integral probabilities of success. Only these plutonium compositions (super-grade (A-1), weapons-grade (A-2), very low burnup CANDU (A-3), US test (A4, Pellaud [9]) and the AGR (B-1), CANDU (B-2) and FBR blanket (B-3) plutonium composition confirm the US-statement of Fig. 9.1. Different explosion yields can be attained for different cumulative probabilities of pre-ignition. Again the Trinity test device represents the absolute exemption attaining $Y/Y_0 = 1$ with an about 80% cumulative probability for success. The data for the TRINITY test device (brown curve) are thoroughly discussed by C. Mark [19].

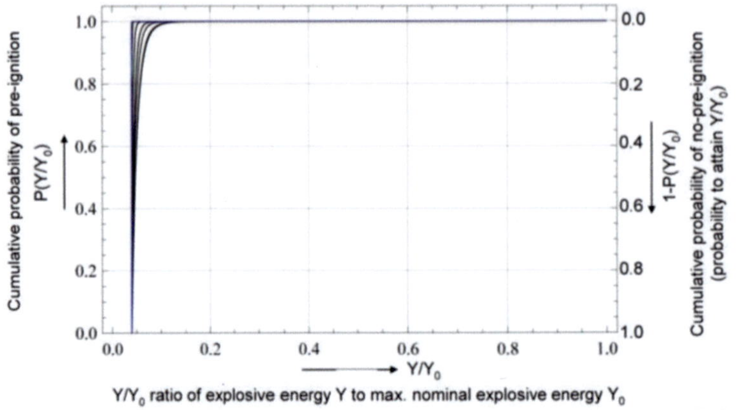

Color	Type of plutonium
black	Weapons Pu: A-1, A-2, A-3, A-4, B-3 (FBR-blanket)
red	Plutonium from low burnup fuel: B-1, B-2
green	Plutonium from LWR spent fuel: C-1, C-2, C-3, C-4
blue	Proliferation-resistant Pu: D-1, E-1, E-2, E-3, E-4

Figure 9.46a: Cumulative probability $P(Y/Y_0)$ for pre-ignition and cumulative probability $1\text{-}P(Y/Y_0)$ to attain Y/Y_0 for Case 1.

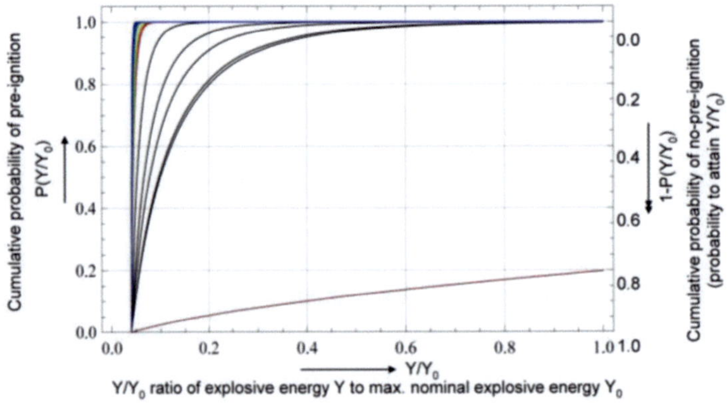

Color	Type of plutonium
black	Weapons Pu: A-1, A-2, A-3, A-4, B-3 (FBR-blanket)
red	Plutonium from low burnup fuel: B-1, B-2
green	Plutonium from LWR spent fuel: C-1, C-2, C-3, C-4
blue	Proliferation-resistant Pu: D-1, E-1, E-2, E-3, E-4
brown	C. Mark, Trinity nuclear explosive test

Figure 9.46b: Cumulative probability $P(Y/Y_0)$ for pre-ignition and cumulative probability $1\text{-}P(Y/Y_0)$ to attain Y/Y_0 for Case 2.

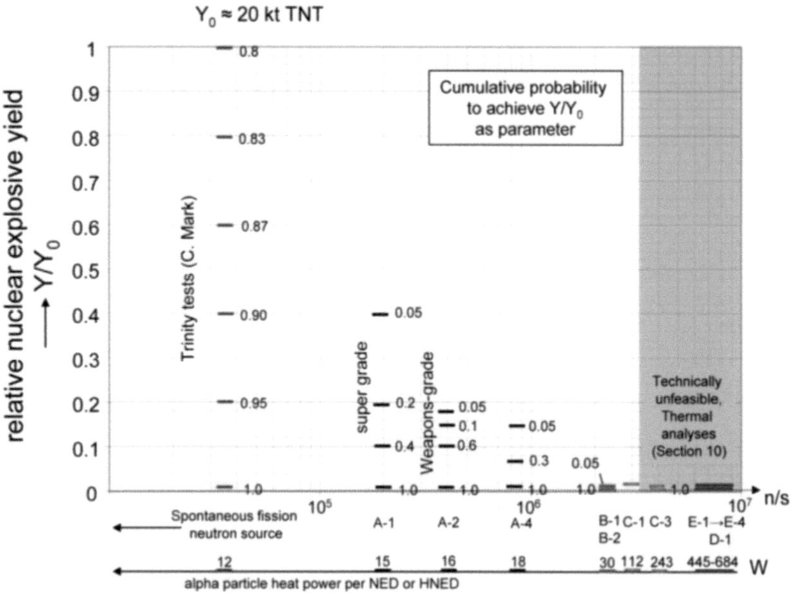

Pu-238	0	0	0	0	0	0.016	0.038	0.06
Pu-239	99.2	0.97	0.94	0.88	0.68	0.57	0.52	0.41
Pu-240	0.08	0.03	0.06	0.12	0.26	0.23	0.23	0.26
Pu-241	0	0	0	0	0.05	0.13	0.14	0.18
Pu-242	0	0	0	0	0.01	0.05	0.07	0.09

Figure 9.47: Relative nuclear explosive Y/Y_0 with cumulative probability for achievement for the different plutonium compositions (weapon-grade and reactor-grade).

It is again understood that the Trinity test device is the absolute exemption. Trinity is followed by supergrade, weapons-grade and the test of US DOE with the 12% Pu-240 plutonium (A-4) as suggested by Pellaud [9]. The latter plutonium composition could attain $Y/Y_0 = 0.15$ with a cumulative probability of 5%.

The results for the low burnup CANDU plutonium (A-3 in Table 4.2) and the results for the FBR blanket plutonium (B-3 in Table 9.3) would be found between the results for A-2 and A-4 (Table 9.2)

All reactor-grade plutonium from LWR spent fuel would be found right of the cases B1 and B2 (Table 9.3) in the range of $Y/Y_0 = 0.027$ (C. Mark [19]) case 1) or 0.006 to 0.017 (Kessler et al. [12]) cases C2 through C4 (Table 9.4), case D1 (Table 9.5) and E1 through E4 (Table 9.6b).

Also shown on a second abscissa in Fig. 9.47 are the alpha-particle heat powers for each nuclear explosive device starting from left with Trinity (12 W), US test of 1962 (Pellaud

(18 W), AGR and CANDU plutonium (30 W). The right side of this figure shows the reactor grade plutonium composition C-1 through C-3 with 112 to 243 W per HNED and proliferation-proof plutonium D-1 as well as E-1 through E-4 with 445 to 684 W (Table 9.2 through 9.6b). It will be shown in the next Section 10 that HNEDs with more than 120 W alpha-particle heat power for low technology HNEDs and with more than 240 W for medium technology HNEDs are technically unfeasible.

These data underline convincingly that different categories of plutonium must be considered – as suggested in Section 9.3 - for IAEA safeguards and for discussions of the proliferation issue.

9.13.3 The US test of 1962 with reactor-grade plutonium

The US-DOE [75] states that the reactor-grade plutonium used for the US-underground test in 1962 was plutonium from gas cooled reactors of the UK. The four 50 MW(e) gas cooled reactors at Calder Hall – the first British nuclear reactors – were connected to the electrical grid in 1956. The next four MAGNOX reactors at Chapelcross started operation in 1959. Therefore, the plutonium for the US-underground test of 1962 must have been from these British nuclear reactors. Pellaud [9] estimated that this plutonium from the gas cooled graphite moderated reactors of the UK had a Pu-240 content of 12%. According to Chesson [76] – who calculated the isotopic content of plutonium from MAGNOX fuel of the Calder Hall reactors – a fuel burnup of 3000 MWd/t leads to a Pu-240 content of 12% in the plutonium as estimated by Pellaud [9].

Even if the Pu-240 content of this plutonium of the US-test in reality was slightly different from the estimate of Pellaud [9] with 12% it can be understood from Fig. 9.47 for a Pu-240 content of less than 12% the related nuclear explosive yield would be found left of case A4 between A2 and A4. In case the real Pu-240 was higher than 12% the relative nuclear explosive yield would be found right of A4 between A4 and B1. The maximum burnup of MAGNOX fuel is 5000 to 6000 MWd/t which corresponds to case B1 (Fig. 9.47 and Table 9.2).

All these possible results between case A2 and B1 confirm the statement of US-DOE [75]

- the plutonium of this weapon test was reactor-grade because it had a higher Pu-240 content than weapon-grade plutonium with 5.8% Pu-240 (see Tab. 9.7)
- the nuclear explosive test was successful and the nuclear explosive yield was less than 20 kt TNT (see Fig. 9.47).
- However, this reactor-grade plutonium used in the US underground test of 1962 does not at all correspond to reactor-grade plutonium from spent fuel of LWRs with a burnup of 40-60 GWd/t.

References Section 9

[1] INFCE – Summary volume and reports of INFCE working groups, IAEA, Vienna (1980) (ISBN 92-0-159880-1 through 9).

[2] National Academy of Sciences, Management and disposition of excess weapons plutonium, reactor related options, National Academic Press, Washington D.C. (1995).

[3] Garwin, R.L., http://www.fas.org/rlg/90-96.htm (1998).

[4] American Nuclear Society, Special report on protection and management of plutonium, LaGrange Park (1995).

[5] GNEP: Roll out means big jump for fuel cycle, Nuclear News, American Nuclear Society, Advanced fuel cycle initiative, www.ne.anl.gov/research (2006).

[6] Laidler, J.J., AFCI separations overview, AFCI semiannual review, Santa Fe, NM, USA (2003).

[7] Additional Information Concerning Underground Nuclear Weapon Test of Reactor-Grade Plutonium, DOE Facts (1994), 186-7.

[8] DeVolpi, A., Proliferation, Plutonium and policy/institutional and technological impediments to nuclear weapons propagation, Pergamon Press, New York (1979).

[9] Pellaud, B., Proliferation aspects of plutonium recycling, J. of Nuclear Material Management, 21, No. 1, 30 (2002).

[10] DeVolpi, A. et al., Nuclear shadow boxing, Vol. 1 and 2, Fidler Doubleday, Michigan (2004).

[11] Dautray, R., L'énergie nucléaire civile dans le cadre temporel des changements climatiques. Rapport à l'Académie des Sciences, Institut de France, Editing TECDOC, Paris (2001) (ISBN-10 2-7430-0534-3).

[12] Kessler, G., et al., Potential nuclear explosive yield of reactor-grade plutonium using the disassembly theory of early reactor safety analysis, Nucl. Eng. and Design, 238, 3475-3499 (2008).

[13] Kessler, G., Plutonium denaturing by Pu-238, Nucl. Sci. Eng., 155, 53-73 (2007).

[14] Broeders, C.H.M. et al., Fuel cycle options for the production and utilization of denatured plutonium, Nucl. Sci. Eng., 156, 1-23 (2007).

[15] Kessler, G., Requirements for nuclear energy in the 21st Century. Progress in Nuclear Energy, 40, No. 3-4, 309-325 (2002).

[16] Herbig, R., et al., Vibrocompacted fuel for the liquid metal reactor BOR-60, Journal of Nuclear Materials, 204, 93 (1993).

[17] Laidler, J.J., et al., Development of pyro-processing technology, Progress in Nuclear Energy, 31 (1/2), 131-140 (1997).

[18] Conocar, O., et al., Promising pyro-chemical actinides/lanthanides separation process using aluminum, Nucl. Sci. Eng., 153, 253-261 (2006).

[19] Mark, J.C., Explosive properties of reactor-grade plutonium, Science and Global Security, 4, 111-128 (1993).

[20] Carlson, J. et al., Plutonium isotopics – Non-proliferation and safeguards issues, IAEA-SM-351/64 (1997)

[21] Chebeskov, A., A quantiative approach to evaluate the attractiveness of enriched uranium and civilian plutonium, Personal communication at Int. Workshop on Non-Proliferation of Nuclear Materials, Obninsk, Sept 29 – Oct 3, 2008.

[22] Broeders, C.H.M., Investigations related to the build-up of transurania in pressurized water reactors, FZKA 5784, Forschungszentrum Karlsruhe (1996).

[23] Kessler, G. et al., A new scientific solution for preventing the misuse of reactor-grade plutonium as nuclear explosive, Nucl. Eng. Design, 238, 3429-3444 (2008).

[24] Kessler, G., Proliferation resistance of americium originating from spent irradiated reactor fuel of pressurized water reactors, fast reactors and accelerator driven systems with different fuel cycle options, Nucl. Sci. Eng., 159, 56-82 (2008).

[25] Rineiski, A., Kessler, G., Proliferation-resistant fuel options for thermal and fast reactors avoiding neptunium production, Nucl. Eng. Design, 240, 500-510 (2010).

[26] IAEA, Multilateral approaches to the Nuclear Fuel Cycle, INFCIRC/640 (2005).

[27] DeVolpi, A., Fissile materials and nuclear weapons proliferation, Ann. Rev. Nucl. Part. Sci., 36, 83-114 (1986).

[28] Seifritz, W., Nukleare Sprengkörper – Bedrohung oder Energieversorgung für die Menschheit, Karl Thiemig AG, München (1984).

[29] Bethe, H.A. et al., An estimate of the order of magnitude of the explosion when the core of a fast reactor collapses, RHM(56)/113, UKAEA Research Establishment Harwell (1956).

[30] Jackson, J.F., et al., VENUS-II: An LMFBR disassembly Program, ANL-7952, Argonne National Laboratory (1972).

[31] Smith, L.L., et al., SIMMER-I, An LMFBR disrupted core analysis code. In: Proceedings of the International Meeting on Fast Reactor Safety and Related Physics, Chicago, IL, CONF-871001, p. 1195 (1976).

[32] Serber, R., Los Alamos Primer, edited with an introduction by R. Rhodes, University of California Press (1992).

[33] Rhodes, R., The Making of the Atomic Bomb. Simon & Schuster, New York (1986).

[34] Leuthäuser, K.D., Kernwaffen und Kernwaffenwirkungen: Analysen, Daten, Trends. Teil 1 und Teil 2. Report INT 72, Fraunhofer Gesellschaft, Institut für Naturwissenschaftlich-Technische Trendanalysen, Kiel (1975).

[35] Rhodes, R., Dark Sun, The making of the hydrogen bomb, Touchstone Book, Simon & Schuster, New York (1995).

[36] Podwig, P., Russian Strategic Nuclear Forces. MIT Press, Cambridge, Mass. (2004).

[37] Sandmeier, H.A., et al., Electromagnetic pulse and time-dependent escape of neutrons and gamma rays from nuclear explosions. Nucl. Sci. Eng., 48, 343 (1972).

[38] Paxton, H.C., GODIVA, TOPSY, JEZEBEL, critical assemblies at Los Alamos, Nucleonics, Vol. 13, No. 10 (1955).

[39] Wimett, T.F., et al., Applications of GODIVA II neutron pulses. In: Proceedings of the Second UN International Conference on the Peaceful Use of Atomic Energy. Vol. 10, Geneva, p. 449 (1958).

[40] McCarthy, W.J., et al., Studies of nuclear accidents in fast power reactors. In: Proceedings of the Second UN International Conference on the Peaceful Use of Atomic Energy. Vol. 12, Geneva, p. 207 (1958).

[41] Nicholson, R.B., Methods for determining the energy release in hypothetical fast reactor meltdown accidents. Nucl. Sci. Eng., 18, 207-219 (1958).

[42] Stacey, W., Nuclear reactor physics, John Wiley & Sons Inc., New York (2007).

[43] Waltar, A.E., et al., Fast Breeder Reactors. Pergamon press, New York (1981).

[44] Weinberg, A.M., et al., The physical theory of neutron chain reactors. The University of Chicago Press (1958).

[45] Kiefhaber, E., Cross sections library (6 energy groups) for fast reactor analysis, private communication, FZK (2004).

[46] Briesmeister, J.F. (Ed.) MCNP – A general Monte Carlo N-particle transport code, Version 4B, LA-12625-M, (1997).

[47] Hummel, H.H., et al., Reactivity coefficients in large fast power reactors. American Nuclear Society (1960).

[48] Harlow, F.H., et al., Fluid dynamics, LA-4700, UC34 (1971).

[49] von Neumann, J., et al., A method for the numerical calculation of hydrodynamic shocks. J. Appl. Phys., 21, 232 (1950).

[50] Benedict, U. et al., Transuranium materials under extreme pressure. Journal of Nuclear Materials, 166, 48 (1980).

[51] Seifritz, W., A simple excursion model for a nuclear explosive device. Nucl. Eng. Design, 239, 80-86 (2009).

[52] Tipton, C.R. (Ed.), Reactor Handbook, Materials, Vol. 1, Interscience Publishers Inc., New York (1960).

[53] Skidmore, I. et al., Experimental equation of state data for uranium and its interpretation in the critical region. Thermodynamics of Nuclear Materials, IAEA, Vienna (1962).

[54] Fischer, E.A., A new evaluation of the urania equation of state based on recent vapor pressure measurements. Nucl. Sci. Eng. 101, 97-116 (1989).

[55] Fischer, E.A., Fuel equation of state data for use in fast reactor accidents analysis codes, KFK 4889 (1992).

[56] Duderstadt, J.J., et al., Inertial Confinement Fusion. John Wiley & Sons, New York (1982).

[57] Locke, G., et al., Die Wirkung von Kernwaffen, Fraunhofer Gesellschaft, Institut für Naturwissenschaftlich-Technische Trendanalysen. Report INT 71, Kiel (1974).

[58] Stratton, W.R., et al., Analysis of prompt excursions in simple systems and idealized fast reactors. In: Proceedings of the Second UN International Conference on the Peaceful Uses of Atomic Energy. Vol. 12, Geneva, p. 196 (1958).

[59] Zel'dovich, Ya., et al., Theory of detonation. Academic Press, New York, London (1960).

[60] Johansson, C.H., Persson, P.A., Detonics of High Explosives, Academic Press, London, New York (1970).

[61] Dobratz, B.M., Properties of chemical explosives and explosive simulants, UCRL-51319/Rev. 1 (1972).

[62] Gibbs, T.R., Popolato, A., LASL Explosive Property Data, University of California Press, Berkeley (1981).

[63] Mader, C.I. et al., Los Alamos Explosion Performance Data. University of California Press (1982).

[64] Bushmann, A.V. et al., Intense dynamic loading of condensed matter. Taylor & Francis, Washington, D.C. (1983).

[65] Kinslow, R., High-velocity impact phenomena. Academic Press, New York (1970).

[66] Chabildas, L.C. et al., Enhanced hypervelocity launcher-capabilities to 16 km/s. Int. J. Impact Eng., 17, 183-194 (1995).

[67] Al'tshuler, L.V., et al., Equation of State for aluminum, copper and lead in the high pressure region, Soviet Physics, JETP 11, 573 (1960).

[68] Benedict, U., et al., Transuranium materials under extreme pressure. J. Nucl. Mater., 166, 48-55 (1989).

[69] Baumung, K., Institute for High Power and Microwave Technology, FZK, personal communication (2003).

[70] Kirzhnits, D.A., On the limits of applicability of the semiclassical Equation of State. Sov. Phys. JETP 8, No. 6, 1081 (1959).

[71] Hoebel, W., et al., Numerische Simulation von Teilchen- und Strahlungstransportprozessen, KfK-Nachrichten, 20, 169 (1989).

[72] Hansen, G.E., Assembly of fissionable material in the presence of a weak neutron source. Nucl. Sci. Eng., 8, 709-719 (1960).

[73] Seifritz, W., The pre-ignition problem in nuclear explosive devices (NEDs) for a sigmoidal Rossi alpha and high neutron background, Kerntechnik, 72, 309-313 (2007).

[74] Chebeskov, A.N., Are there insuperable obstacles towards nuclear weapon creation if a nation has in hands nuclear fuel cycle technologies? Personal communication at Int. Workshop on Non-Proliferation of Nuclear Materials, Obninsk, Sept 29 – Oct 3, 2008.

[75] Grizzle, S., DOE FACTS, Additional information concerning underground nuclear weapon test of reactor-grade plutonium. http://www.ccnr.org/plute_bomb.html

[76] Chesson, K.E., A one group parametric sensitivity analysis for the graphite isotope ratio method and other related techniques using ORIGEN 2.2.

[77] http://repository.tamu.edu/bitstream/handle/1969.1/ETD-TAMU-1944/CHESSON-THESIS.pdf?sequence=1

10 Thermal analysis of HNEDs at different levels of technology

In this Section the favorable results of neutronic analyses will be shown to not necessarily imply that such HNEDs are technically feasible. This point will be proved by thermal analysis. The heat produced by HNEDs will be used as the leading parameter. This alpha-particle heat power is associated with a specific composition of reactor-grade plutonium, in which the Pu-238 isotope dominates the heat output. Moreover, various levels of technology required to design HNEDs will be defined and discussed in the light of the neutronic and thermal analyses.

10.1 Definition of different levels of technology

The geometric dimensions of the reactor-grade plutonium sphere and its reflector (Figure 10.1) are determined by their nuclear characteristics. The radius of the plutonium sphere as the central part of HNEDs varies between 5.3 cm and 5.8-cm. The subcritical mass (k_{eff} = 0.98) ranges between 9.2 to 13.2 kg for a solid reactor-grade plutonium sphere when reflected by a 5-cm thick natural-uranium reflector (see Tables 9.4 and 9.5 but also 9.6b as well as Tables 9.12 and 9.14). Hollow reactor-grade plutonium spheres would have an outer diameter of about 7.2 cm.

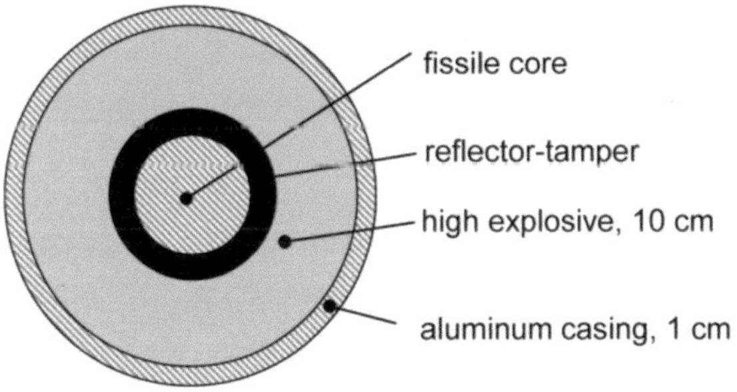

Figure 10.1: Schematic geometric arrangement of HNEDs and dimensions for thermal analyses.

Kessler et al. [1,7,16,32] show the levels of technology of HNEDs to be based not only on geometric dimensions but also on the thermodynamic data of the materials used, e.g. the high explosive lenses. These thermodynamic data are much more important in reactor-grade plutonium HNEDs than in weapon-grade plutonium nuclear explosive devices.

The heat produced by Pu-238 is negligible (less than 20 W) in a weapon-grade plutonium based nuclear explosive device (NED), thus requiring no thermal analysis. For reactor-grade plutonium HNEDs with higher Pu-238 content, however, these thermodynamic data are decisive, as will be seen below. The most critical parts are the high explosive lenses.

As the geometric shape of the high explosive lenses is classified, a so-called one-dimensional conservative approach must be chosen here for the thermal analysis. With this approach the real temperature profile in the explosive lenses tends to be underestimated which is conservative (Kessler [1]). This will be explained in Section 10.3.1.

The history of nuclear weapons development (Rhodes [3], Cochran et al. [4], and Podwig [5]) shows that a Non-Nuclear Weapon State (NNWS) about to design and build HNEDs from reactor-grade plutonium would have to proceed from low to medium technology levels. Only current Nuclear Weapon States (NWSs) would be able to master the very high level of technology because of their know-how in research and experiments accumulated over long periods of time.

10.2 Geometric dimensions for different levels of technology

Accounts of the development of nuclear weapons in Rhodes [3], Cochran et al. [4] and Podwig [5] indicate the outside radius of nuclear weapons to have decreased from 75 cm (earliest nuclear weapon) to some 22 cm (Fetter et al. [6]). At the same time, the thickness of high explosive lenses decreased from 43 cm (Rhodes [3]) to approx. 25 cm and then to some 10 cm (Fetter et al. [6]).

For the thermal analysis three classes of technological development (low, medium and very high technology) are selected (Fig. 10.2).

Kessler [1] selected an outer radius of 65 cm for the <u>low-technology</u> case. This is somewhat smaller than the 75 cm for the earliest nuclear explosive device. Similarly for the size of <u>medium-technology</u> (HNEDs) the outer radius was estimated to be 42 cm from pictures in Fetter et al. [6], Cochran et al. [4], and Blechman et al. [21]. The <u>high-technology</u> case is chosen to have an outer radius of 21 cm from pictures in Podwig et al. [5], Fetter et al. [6], and Cochran et al. [4]. These are reasonable guesses only for the following thermal analysis of HNEDs with reactor-grade plutonium (Fig. 10.2).

Whereas the Pu-sphere, the U_{nat} reflector and the Al-layer are of metallic structure, the high explosive hollow sphere consists of a complicated geometric lense structure of at least 2 different high explosives with different detonation velocities (Rhodes [3,22]), but also different thermal conductivities.

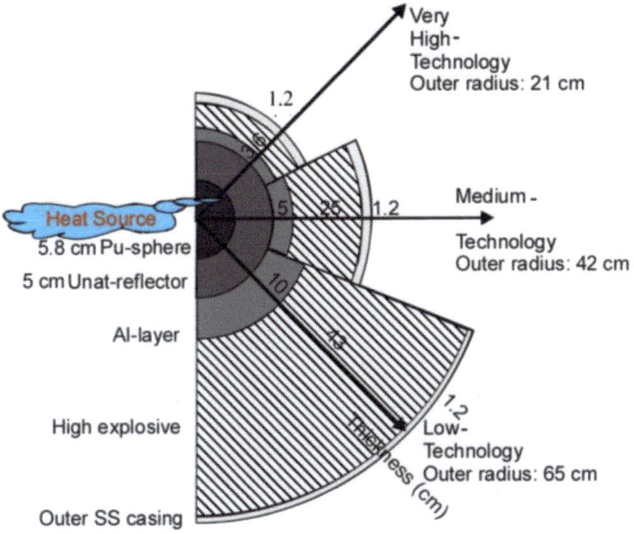

Figure 10.2: Assumed geometric dimensions for three classes of technology.

10.3 High explosives for different classes of technology

10.3.1 Low technology high explosives

According to Section 9 the first nuclear explosive device had Baratol and Composition B as high explosives. Table 10.1 shows the thermal characteristic data of these low technology high explosives, together with those of TNT. Their melting point is 79-81 °C representing the limiting characteristic of these high explosives.

These data are collected from Gibbs et al. [18], Mader et al. [19] and Dobratz [24].

Table 10.1: Characteristic materials data for low technology high explosives.

High explosive	Density [g/cm^3]	Thermal conductivity [W/m °C]	Detonation velocity [km/s]	Melting point [°C]	Onset of pyrolysis [°C] / Temperature for onset of chem. self-explosion T_e [°C]
Baratol	2,61	0,494	4,9	79-80	180
					300
Composition B	1,74	0,219	7,8-8,0	79	200
					214
TNT	1,45	0,259	6,9	81	260
					288

10.3.2 Medium technology high explosives

Medium technology high explosives have higher thermal conductivities, higher melting points and higher temperatures T_e for onset of pyrolysis and onset of chemical self-explosion. Table 10.2 shows these thermal characteristic data for DATB, HMX, PBX 9011, PBX 9404, PBX 9407 and PBX 9501. These data are collected from Gibbs et al. [18], Mader et al. [19] and Dobratz [24]. Kessler [1,16] assumed these same temperatures T_e for the onset of self-explosion for all thermal analyses of the low, medium, and high technology cases because it is a temperature determined by experiments for each high explosive (Gibbs et al. [18], Mader et al. [19] and Dobratz [24]). Shmelev et al. [51] used a relation for the pyrolysis of 2% of the high explosive.

Table 10.2: Characteristic materials data for medium technology high explosives.

High explosive	Density [g/cm^3]	Thermal conductivity [W/m °C]	Detonation velocity [km/s]	Melting point [°C]	Onset of pyrolysis [°C]	Temperature for onset of chem. self-explosion T_e [°C]
DATB	1.83	0.259	7.6	286	300	322
HMX	1.84	0.406	9.11	256-286	285	259
PBX9011	1.77	0.381	8.5	190	260	
PBX 9404	1.84	0.385	8.8	190	290	236
PBX 9407	1.65	0.335	8.4	204	240	
PBX 9501	1.86	0.452	8.7	190	275	235

10.3.3 Very high technology high explosives

Very high technology high explosives have the highest melting points and the highest temperatures T_e for onset of pyrolysis or onset for chemical self-explosion (Table 10.3). These data are collected from Gibbs et al. [18], Mader et al. [19] and Dobratz [24]. They are only available to advanced NWSs, e.g. the USA. As both very high technology high explosives have similar detonation velocities and thermal characteristics, they will have to be combined with other high explosives or materials in the implosion lenses. For the following calculations of the temperature profiles in the very high technology HNEDs, only the temperature curve for the two very high technology explosives TATB or PBX-9502 will be shown.

Table 10.3: Characteristic materials data for very high technology high explosives.

High explosive	Density [g/cm³]	Thermal conductivity [W/m °C]	Detonation velocity [km/s]	Melting point [°C]	Onset of pyrolysis [°C]	Temperature for onset of chem. self-explosion T_e [°C]
PBX 9502	1.89	0.561	7.6	448	395	331
TATB	1,89	0.544	7.6	448	395	347

10.4 The one-dimensional conservative approach for the thermal analyses

The spherical lenses in the first NED consisted of two different fast and slow reacting high explosives (Baratol and Composition B) and had a complicated explosive lense structure [3,22]. As the complicated high explosive lense structure is not known (classified), the following one-dimensional conservative approach is applied.

Imagine radial sections through the high explosive lense structure or small elements which consist e.g. of 2 different high explosives 1 (Composition B) and 2 (Baratol). Along such sections (Fig. 10.3) interchanging parts of the two different high explosives Composition B and Baratol may be found which – apart from different detonation velocity – have different thermal conductivity λ_1 and λ_2. Not knowing in which radial position the different parts of the two high explosives are located and what dimensions they have, it can nevertheless be concluded that the unknown radial temperature profile in the unknown structure can only vary between two limiting radial temperature profiles of Composition B and Baratol. These are determined by two simple cases. Each of these two simple cases would consist of only one high explosive 1 (Composition B) or 2 (Baratol) with its thermal conductivity λ_1 or λ_2 (Fig. 10.3). It is assumed that high enough gap heat transfer coefficients exist at the interfaces between the two high explosives such that the temperature change across the interface can be neglected.

In the azimuthal direction there will also be interfaces between the two high explosives and there will be quasi-oscillations of the azimuthal temperature profile at constant radius. In this azimuthal direction the conservative approach of the two limiting temperature profiles will also be valid.

The real temperature profiles will always be located within the limiting temperature profiles for Composition B and Baratol.

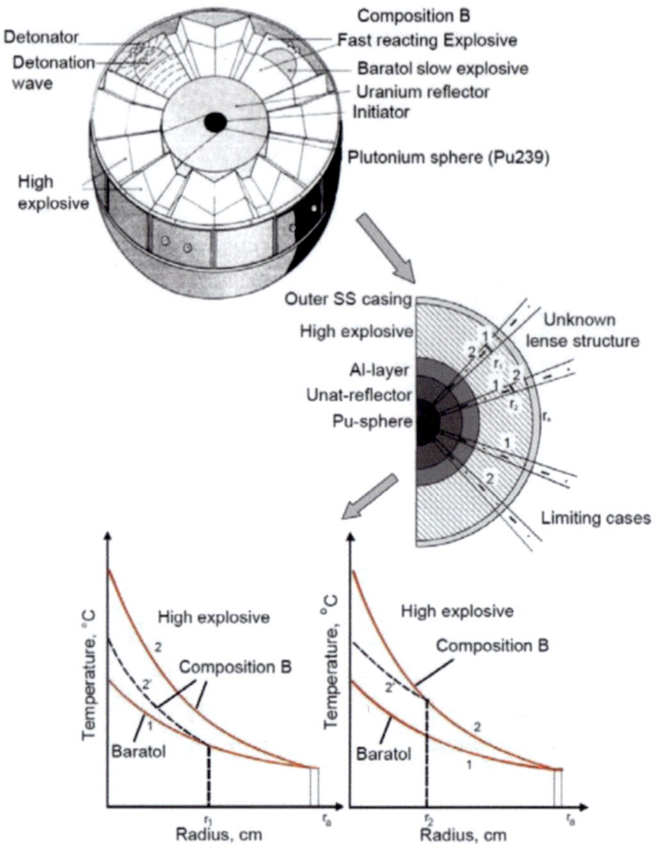

Figure 10.3: Explanation of the conservative one-dimensional approach for an unknown 3-dimensional high explosive lense structure.

In cases where limiting temperatures of the high explosives are exceeded, e.g. melting point or temperature T_e for initiation of self-explosion, always the lower limiting temperature profile is considered which is determined by the high explosive with the higher thermal conductivity. This is always conservative, because the explosive with the lower thermal conductivity has the higher temperatures in the interior regions. In addition, the real limiting temperatures (melting point, temperature T_e for initiation of self-explosion) are accounted for (see Tables 10.1 through 10.3).

For very high technology cases the two high explosives are assumed to be the same and have the highest thermal conductivities and limiting temperatures which can be found in the open scientific literature [Gibbs et al. [18], Mader et al. [19] and Dobratz [24]] (again to be conservative).

If three or more different high explosives would be used in the high explosive lense structure the approach would be the same, but the limiting temperature curves would be represented by the high explosives with the highest and the lowest thermal conductivity. This is e.g. shown in cases for medium technology.

It is assumed that an HNED would not work properly if its high explosive lenses would be either partially molten or if the temperature for start of pyrolysis and/or the temperature T_e for initiation of self explosion would be exceeded.

For low technology it is assumed that the HNED would not work if the melting temperature of Baratol or Composition B of 79-80 °C would be exceeded (Table 10.1).

For medium technology it is assumed that both the fast- and the slow-reacting high explosives will melt at about 190 to 286 °C, pyrolysis will start at about 260 °C to 300 °C and have a temperature T_e for initiation of self-explosion around 235 °C to 322 °C. This would correspond to high explosives like DATB, HMX, PBX 9011, PBX 9407, PBX 9501. Again it is assumed that an HNED would not work properly if its high explosive lenses would be partially molten or if the temperatures for start of pyrolysis and chemical self-explosion would be exceeded (Table 10.2).

For very high technology PBX 9502 or TATB are assumed. They have a melting point of 448 °C. The start of pyrolysis will be at 395 °C, but the limiting temperatures will be the initiation of self-explosion at $T_e = 331$ °C (PBX 9502) or $T_e = 347$ °C (TATB) (see Table 10.3). As is stated in Zinn et al. [23]: a high explosive will start self-explosion, if the temperature at its surface exceeds T_e.

10.5 Temperature profile within an HNED

The alpha-particle heat power generated in the plutonium core must be transferred through the different material layers to the outer steel casing by heat conduction and from there to the atmosphere by natural convection and radiation. This can be described by:

$$\dot{Q} = \dot{Q}_N + \dot{Q}_R = h \cdot A \cdot (T_a - T) + C_{1,2} \cdot A \cdot \left[T_a^4 - T_w^4\right] \qquad (10.1)$$

where

$\dot{Q}$ = alpha-particle heat power in the Pu-core

$\dot{Q}_N$ = heat power transferred by natural convection

$\dot{Q}_R$ = heat power transferred by radiation

h = heat transfer coefficient for natural convection to be determined by the relation of [25], given below

A = outer steel casing surface area of the HNED

T_a = outer steel casing temperature of the HNED

T = ambient gas temperature around the HNED

T_w = wall temperature of a room or structures surrounding the HNED

$C_{1,2}$ = is defined by Eq. 10.1c

According to Yuge [25] the heat transfer coefficient at the surface of a heated sphere can be determined from

$$\mathrm{Nu} = 2 + 0.43 \cdot \mathrm{Ra}^{1/4} \tag{10.1.a}$$

for Pr = 1 and $1 \leq \mathrm{Ra} \leq 10^5$

where:

Pr = Prandtl number, Nu = Nusselt number, Ra = Raleigh number and

$$\mathrm{Ra} = \frac{g \cdot ß\,(T_a - T)\,D^3}{\nu \cdot \alpha} \tag{10.1.b}$$

with

α = thermal diffusivity

ν = kinematic viscosity

ß = thermal expansion coefficient

D = diameter of HNED

g = gravity constant

The constant $C_{1,2}$ for concentric radiating surfaces is:

$$C_{1,2} = \frac{\sigma}{\frac{1}{\varepsilon_a} + \frac{A}{A_w}\left(\frac{1}{\varepsilon_w} - 1\right)} \tag{10.1.c}$$

where

σ = Stefan-Boltzmann constant = 5.67×10^{-8} W/m²K⁴

A = surface of HNED outer casing, A_w = surface of room or structures surrounding the HNED

$$\frac{A}{A_w} << 1$$

ε_a = emissivity of the outer steel casing of the HNED

ε_w = emissivity of room walls or structures surrounding the HNED

It can further be assumed $\frac{A}{A_w}\left(\frac{1}{\varepsilon_w} - 1\right) \rightarrow 0$;

then

$$\dot{Q} = \varepsilon_a \cdot \sigma \cdot A \left(T_a^4 - T_w^4\right) \tag{10.1.d}$$

The emissivities ε_a of different materials, collected by Kuchling [9], are given in Table 10.4.

Table 10.4: Some emissivities of different materials [9].

Black body	1	Silver polished	0.035
Gold polished at 500 °C	0.03	Steel polished 20 °C	0.286
Aluminum polished	0.04	Aluminum rolled	0.07

As can be seen from the above equation for radiation transfer, the highest radiation heat transfer occurs at $\varepsilon_a = 1$ (black body radiation). This also leads to the lowest surface temperature of the outer steel casing T_a.

To be on the conservative side, black body radiation is assumed for the further considerations. With an emissivity of e.g. 0.286 for polished steel the surface temperature difference (T_a-T_w) of the outer casing would be by $0.286^{-1/4}$ or about a factor of ~1.3 higher.

Equation (10.1) for the total heat transfer from the outer steel casing of the HNED to the atmosphere is solved inversely by assuming temperatures for T, T_a and T_W and calculating $\dot{Q}$ which allows the outer wall temperature T_a to be determined by interpolation tables.

10.6 Outer temperature at the casing of the HNED

For the different Pu-compositions of Section 9 the temperatures in degrees Celsius for the three assumed cases of technology are given in Table 10.5. It clearly shows the trends of higher outer casing temperatures for higher alpha-particle heat powers (higher Pu-238 content) and the increase of outer-casing temperatures as a function of lower outer radius going from low → medium → high technology.

Table 10.5: Outer Steel Casing Temperature T_a of HNEDs for different alpha-particle heat power and different technology.

HNED	Technology		
α-heat power (kW)	low	medium	high
	Outlet radius of HNED 65 (cm)	Outlet radius of HNED 42 (cm)	Outlet radius of HNED 21 (cm)
	Ta (°C)	Ta (°C)	Ta (°C)
0.144	32	36	58
0.240	34	42	76
0.614	43	62	129
0.858	49	71	154
1.121	55	85	183
1.244	58	89	187
1.416	61	96	199
1.530	64	100	208

The calculations showed also that the radiation heat transfer is dominant. In many cases the radiation heat transfer amounts to 90% and more, whereas the natural convection heat transfer is only several percent up to 10%.

After determination of the temperature T_a of the outer steel casing, the temperature profile in the spherical shells without internal heat sources (high explosives, aluminum, U_{nat}-reflector etc.) can be adressed.

10.7 Radial temperature distribution within the HNED for constant thermal conductivity

The radial temperature distribution within the HNED is obtained by the solution of the heat conduction equation with internal and external boundary conditions and constant thermal conductivities.

The temperature difference ϑ_1 - ϑ_{n+1} over n shells with constant thermal conductivities λ_1 to λ_n and inner and outer radii r_n, r_{n+1} for n=1....N is given by Eq. (10.2):

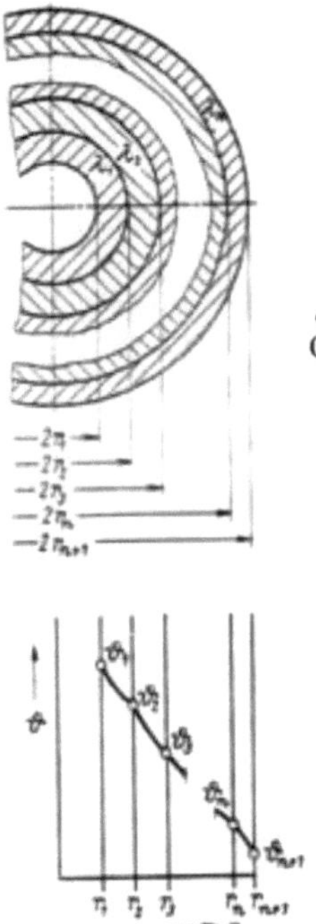

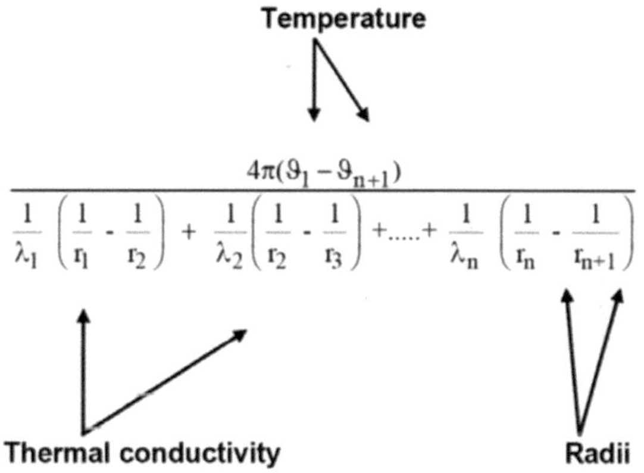

$$\dot{Q} = \frac{4\pi(\vartheta_1 - \vartheta_{n+1})}{\frac{1}{\lambda_1}\left(\frac{1}{r_1} - \frac{1}{r_2}\right) + \frac{1}{\lambda_2}\left(\frac{1}{r_2} - \frac{1}{r_3}\right) + \ldots + \frac{1}{\lambda_n}\left(\frac{1}{r_n} - \frac{1}{r_{n+1}}\right)} \quad (10.2)$$

As will be seen below from Table 10.7 as well as already shown in Tables 10.1 through 10.3 the high explosives have the lowest thermal conductivities of all spherical shells without internal heat sources. It is therefore important to have reliable experimental data for the high explosives.

The heat transfer through a gap, e.g., between the heat producing Pu-sphere and the U_{nat} reflector can be described by

$$\dot{Q} = \alpha_{gap} \cdot A \cdot \Delta T \qquad (10.3)$$

α_{gap} = gap heat transfer coefficient (Kämpf [27,28]

A = surface of Pu-sphere or Pu-spherical shell

ΔT = Temperature difference between outer Pu-sphere and inner U_{nat} reflector surface

The temperature difference due to alpha-particle heat power in a solid Pu-sphere is described by Eq. (10.4):

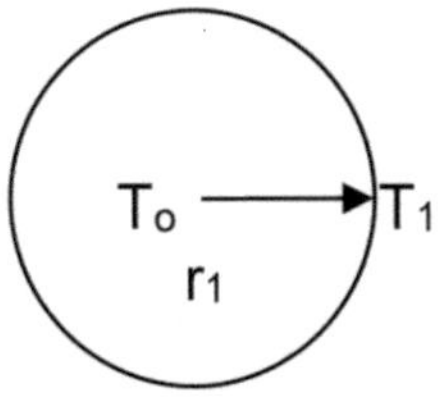

r_1 - solid sphere radius

T_0 - center temperature

T_1 - surface temperature

$$T_0 - T_1 = \frac{L_0 \bullet r_1^2}{6 \bullet \lambda_{Pu}} \qquad (10.4)$$

where

L_0 heat power rate in the sphere (cal/cm^3 · s)

λ_{Pu} thermal conductivity of δ-phase plutonium which is 0.0458 (cal/cm·s·°C) (Blank et al. [34]).

For cylindrical geometry of the reactor-grade plutonium, Eq. (10.4a) holds

$$T_0 - T_1 = \frac{L_0 \bullet r_1^2}{4 \bullet \lambda_{Pu}} \qquad (10.4a)$$

The temperature difference (T_0-T_1) would be by a factor 1.5 higher.

If the reactor-grade Pu would be arranged in a hollow spherical shell, then Eq. (10.5) holds:

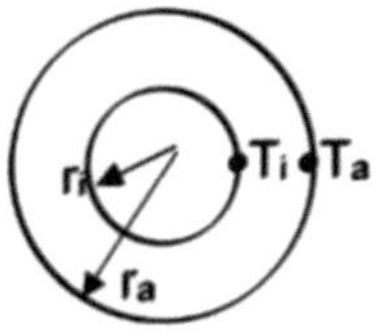

$$T_i - T_a = \frac{L_o}{3\lambda_{Pu}}\left[\frac{r_a^2}{2} - \frac{3}{2}r_i^2 + \frac{r_i^3}{r_a}\right] \tag{10.5}$$

where: r_i - inner radius of the Pu-spherical shell

r_a - outer radius of the Pu-spherical shell

T_i - inner temperature of the Pu spherical shell

T_a - outer temperature of the Pu spherical shell

10.8 Radial temperature distribution in a bare solid Pu-sphere

As discussed earlier the bare solid sphere or two hemispheres of it must be machined, handled etc.. This must be done in glove boxes or hot cells with a gas temperature of about room temperature.

Eq. (10.1) together with Eqs. (10.1a) through (10.1d) describing natural convection and radiation must be solved by assuming temperatures for T, Ta and TW and inverse calculation of the alpha-particle heat power $\dot{Q}$ with interpolation tables. The results are given in Table 10.6 for the plutonium compositions shown already in Table 9.14. Natural convection only contributes by a few percent whereas radiation is predominant. The outer surface temperatures increase steadily for increasing Pu-238 contents and alpha-particle heat power.

Table 10.6: Outer temperature, T_a, and central temperature, T_0, of bare plutonium spheres as a function of alpha-particle heat power.

radius of Pu-sphere (cm)	α-heat power (kW)	outer Temperature T_a (°C)	central Temperature T_0 (°C)
5.2	0.144	251	257
5.3	0.240	213	322
5.8	0.614	434	456
5.8	0.858	496	527
5.8	1.121	550	590
5.7	1.244	571	616
5.5	1.417	618	669
5.4	1.530	molten	molten

The central temperature of the bare plutonium spheres is calculated from Eq. 10.4. The thermal conductivity of reactor grade plutonium is taken from Blank et al. [34] to be 19 W/mK. Fig. 10.4 shows the outer and central temperatures of a bare metallic plutonium sphere as a function of the total alpha-particle heat power or of the Pu-238 content [%] of the reactor grade plutonium.

For an alpha heat power of 1417 [W] the surface temperature would be 618 °C. Its central temperature would be 669 °C (above the melting point for Pu-metal of 640 °C). The inner 3.85 cm of the Pu-sphere would be molten, the outer spherical shell of 1.65 cm would be at a temperature between 618 °C and 640 °C (melting point). The sphere would collapse (lose the geometrical form) under its own gravity.

Subcritical plutonium spheres (k_{eff} = 0.98) with a Pu-238 content of >20% would melt and collapse at about 23%. For a hollow spherical shell design the conclusions would be very similar, although the outer radius would be somewhat larger (e.g. 7.2 cm instead of 5.8 cm).

The outside temperatures of Table 10.6 and Fig. 10.4 are conservative as they are valid for black body radiation $\varepsilon_a = 1$. The real emissivity of polished plutonium metal cannot be found in the literature. However, if the plutonium sphere would be plated with gold at the outside (see Section 9.5) the emissivity would be 0.03 (Table 10.4). This would increase the temperature difference (T_a-T_0) by $0.04^{-1/4}$ or by a factor of 2.24.

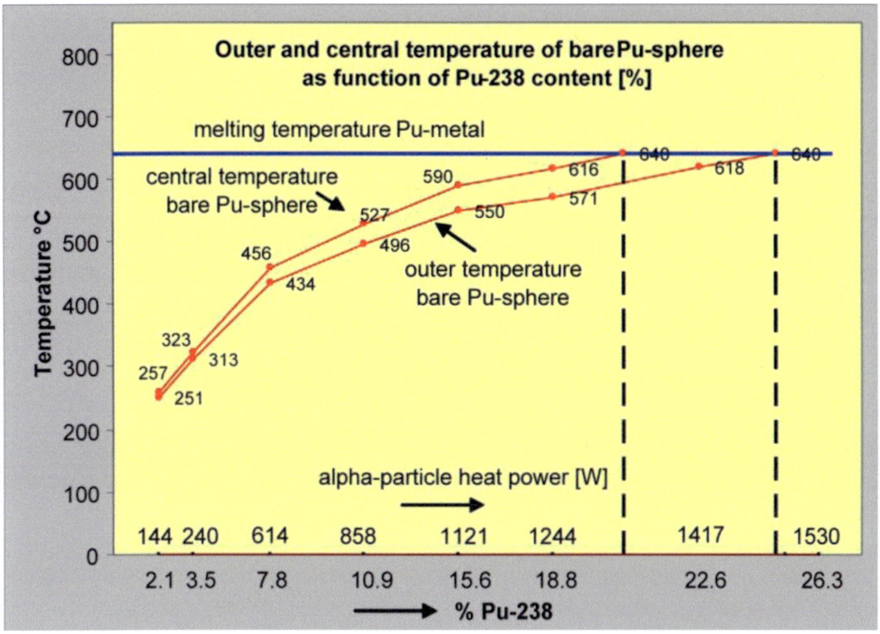

Figure 10.4: Outer and central steady state temperature of a bare metallic Pu-sphere as function of Pu-238 content [%] of reactor-grade plutonium and its alpha heat power.

Figure 10.5: A 100 W heat source of about 3 cm diameter containing 250 g of 238-PuO_2. The oxide glowed at red heat. It was manufactured by Los Alamos Laboratory, USA for the thermoelectric generators of the Voyager space mission [30].

10.8.1 Comparison with IAEA definitions

With the above results the IAEA definitions (INFCIRC 153 Section 8) cannot be understood. IAEA requires that only plutonium with more than 80% Pu-238 can be exempt of IAEA safeguards or is not to be considered weapon-grade plutonium. Subritical (k_{eff} = 0.98) plutonium metal spheres operated with 80% Pu-238 content cannot exist (molten already for 23% or even lower Pu-238 content).

As an illustration of these facts Fig. 10.5 taken from [30] shows a sphere of plutonium dioxide with a diameter of about 3 cm. The heat source which was part of the Voyager space missions thermoelectric generators contains 250 g of 238-PuO_2.

A subcritical (k_{eff}=0.91) reactor grade plutonium metal sphere with 80% Pu-238 enrichment would have an outer diameter of about 11 cm, weigh about 11 kg (Table 9.14) and contain about 8.2 kg Pu-238. It would produce about 5 kW of alpha-particle heat power. Such plutonium quantities can only exist above the melting point of plutonium.

In the same context also publications as Bathge et al. [31] must be seen. They are based on rules of US-DOE defining also the 80% Pu-238 criterion similar to IAEA (INFCIRC 153).

10.9 Temperature profile in an assembled HNED

After determination of the outer casing temperatures T_a (see Table 10.5) for the different cases of HNEDs, equations (10.2) through (10.5) are applied for calculating the associated inner radial temperature profiles. The following constant thermal conductivity data (Table 10.7) are used.

Table 10.7: Thermal conductivities for the different materials within an HNED.

Pu-metal 0.0458 [cal/cm s·K]	19 [W/m K] [34]
reflector (U-metal)	34 [W/m K] [29]
Aluminum	146.5 [W/m K] [29]
Beryllium	117.2 [W/m K] [29]
Tungsten	117.2 [W/m K] [29]
High explosive	see Tables 10.1 through 10.3
Steel casing	13.6 W/m K [29]
Gap coefficient Pu-sphere-reflector	700 W/m^2 K [27,28]

There will be a thin gap between the Pu-sphere or hollow spherical shell and the U_{nat} reflector. For the gap transfer coefficient a value is assumed as it was determined for fresh mixed oxide plutonium/uranium fuel elements (Kämpf [27,28]). For the other spherical shells (aluminum, high explosives), gap coefficients were not considered in order to remain conservative. Aluminum, Beryllium or Tungsten can also surround the U_{nat} reflector. They have similar high thermal conductivity, but different melting points. For each class of technology the thermal conductivity values for the corresponding high explosive are applied (Table 10.1 through 10.3). For Pu-metal the measured thermal conductivity (Blank et al. [34]) which is valid for 1% Gallium-stabilized plutonium-metal up to temperatures of 500 °C is used. As will be seen later, many of the investigated cases will lead to even higher Pu temperatures and melting of the reactor Pu.

10.10 Results of thermal analyses [1,32]

In this section the inner temperature profile for low, medium, and very high technology HNEDs will be discussed for increasing alpha-particle heat power.

10.10.1 Radial temperature profiles in an HNED with reactor-grade plutonium with an alpha-particle heat power of 0.144 kW

10.10.1.1 Low technology

As has been explained above, the class "low technology" has the largest overall dimensions (Fig. 10.3) and contains the high explosives Baratol and Composition B. In Fig. 10.6 it is shown that the radial temperature profile rises from 32 °C at the outside casing to 107 °C in Baratol and 201.8 °C in Composition B. According to the one-dimensional conservative approach (Section 10.4.) the real radial temperature profile would be varying between the two limiting radial temperature profiles which are determined by the thermal conductivities of Baratol and Composition B.

In the low technology case with an alpha-particle heat power of 0.144 kW about one-sixth (radius) of the Baratol and one-half (radius) of Composition B would be molten and the explosive lenses could not work properly. Any higher alpha-particle heat power than 0.144 kW would lead to higher temperatures and increase of melt zones.

It can be shown by calculation that for 0.120 kW the high explosion Baratol would just be molten at the inner surface. The high explosive lenses could, therefore, not function any more.

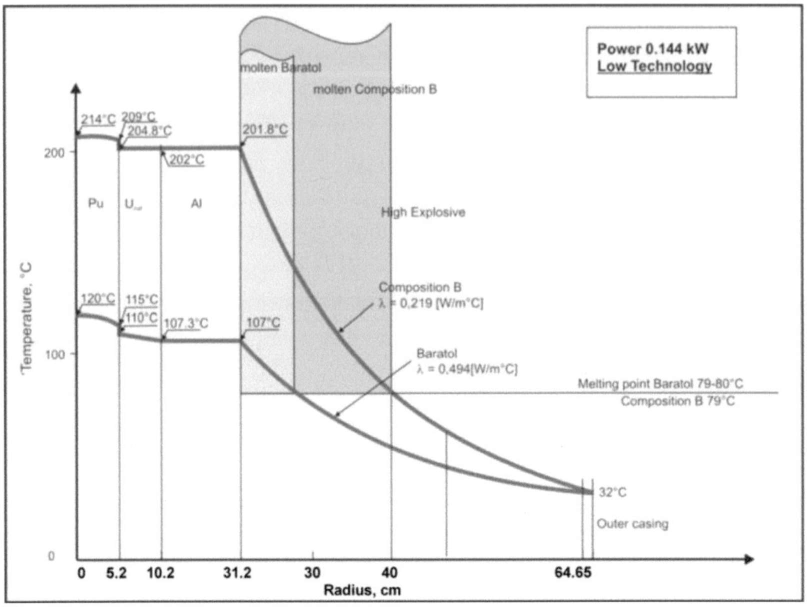

Figure 10.6: Temperature profiles in a low technology HNED with an alpha-particle heat power of 0.144 kW.

10.10.1.2 Medium Technology

For medium technology smaller overall dimensions (42 cm outer-casing radius) are assumed. For this case: DATB with λ = 0.259 W/cm°C, PBX 9011 and PBX 9404 with λ = 0.38 W/cm°C and PBX 9501 with λ = 0.452 W/cm°C, different high explosives with better thermal conductivities and higher melting points (DATB, HMX, PBX 9011, PBX 9407, PBX 9501) are assumed. Again the principle of limiting temperature profiles is applied. Obviously the high explosives with a higher melting point have also higher thermal conductivities so that at no spatial point in the high-explosive lenses would the melting point be attained (Fig. 10.7). The explosive lenses could function properly.

For comparison Fig. 10.7 shows also the temperature profiles which would evolve with low technology high explosives (Baratol and Composition B) and medium-technology dimension, see dashed lines. Again large volumes of Baratol and Composition B would be molten.

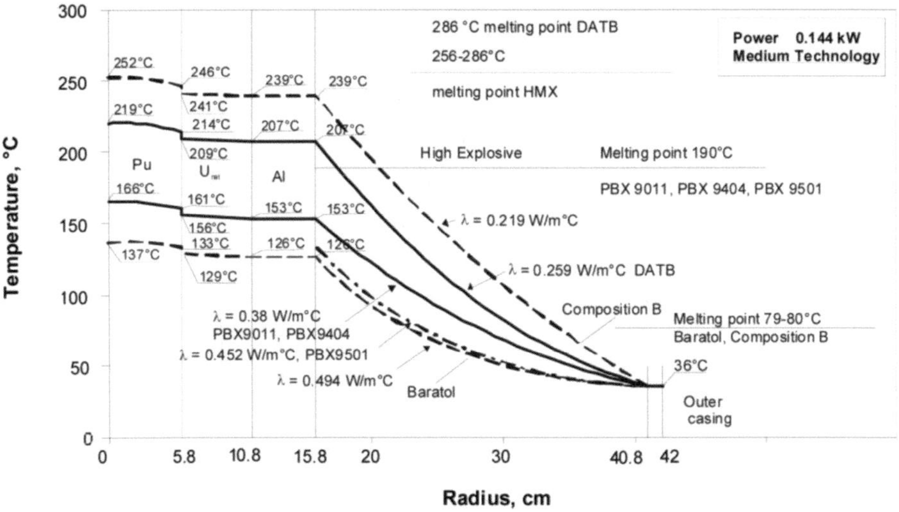

Figure 10.7: Temperature profiles in a medium technology HNED with reactor Pu (0.144 kW alpha-particle heat power).

10.10.1.3 Very High Technology

With a relatively small radius of 21 cm and the outer-casing temperature of 58 °C, the high explosives PBX 9502 or TATB with the highest thermal conductivity of 0.561 W/cm °C and the highest melting point 448 °C are assumed (Table 10.3) for that option. The maximum temperature within the Pu sphere would be 113 °C. The high technology HNED could function properly.

10.10.2 Radial temperature profiles for reactor plutonium from spent fuel with an alpha-particle heat power of 0.240 kW

10.10.2.1 Low Technology

With a higher alpha-particle heat power, namely 240 W, the temperatures in such an HNED would be higher than shown in Fig. 10.6. The explosive lenses would not work.

10.10.2.2 Medium Technology

This case is shown in Fig. 10.8. The outer-casing temperature would be 42 °C at an outer radius of 42 cm. It can be realized that the melting point would be attained at about 2 to 4 cm from the inner edge of the chemical high explosives DATB, PBX 9011, PBX 9404 and HMX. Also for the high explosive PBX 9501 the melting plant would be reached in a thin zone.

Therefore, the medium technology high explosive lenses would not function properly for this limiting alpha-particle heat power of 0.240 kW.

For comparison Fig. 10.8 shows also the temperature profiles which would evolve with low technology explosives and medium technology dimensions. Again a large volume of the high explosive lenses would be molten in such case.

10.10.2.3 Very High Technology

Under the assumptions explained above and PBX 9502 or TATB as high explosives with the highest thermal conductivity, there will be an outer-casing temperature of 76 °C and a maximum temperature in the very high explosive of 151 °C and of 170 °C in the plutonium. The very high technology HNED could still function properly.

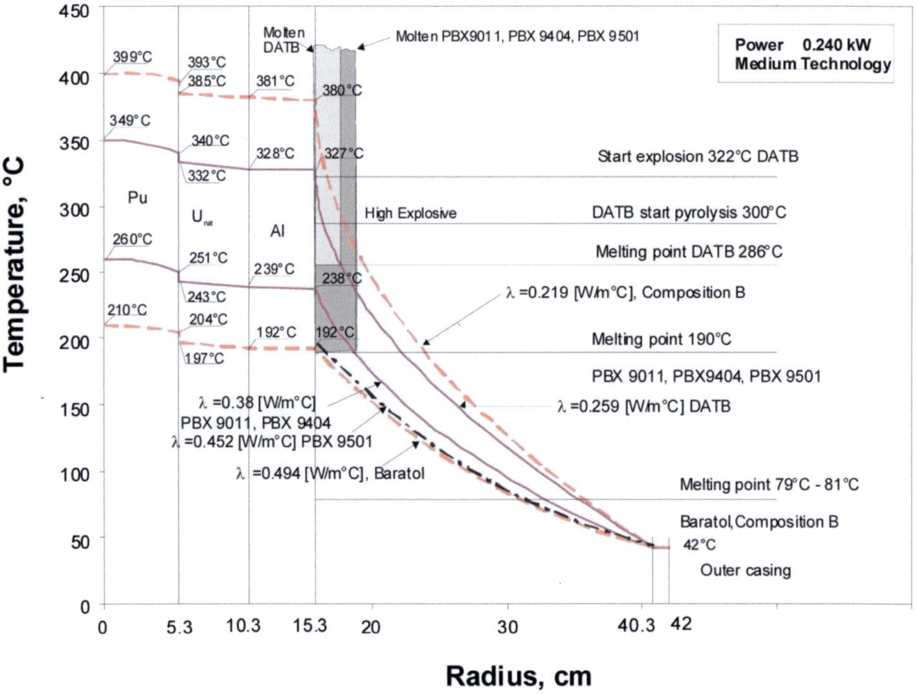

Figure 10.8: Temperature profiles in a medium-technology HNED with an alpha-particle heat power of 0.240 kW.

10.10.3 Radial temperature profiles for reactor plutonium with an alpha-particle heat power between 0.375 and 0.562 kW

10.10.3.1 Low Technology

With a higher alpha-particle heat power between 0.375 and 0.562 kW, the temperatures would be such high that most of the high explosive lenses would be molten. Such an HNED could not work at all.

10.10.3.2 Medium Technology

Also in the medium technology case with an alpha-particle heat power between 0.375 and 0.562 kW the temperatures would be so high that such an HNED could not function at all.

10.10.3.3 Very High Technology [16]

The high-technology explosives PBX 9502 and TATB would attain 347 °C at the inner radius of the high explosive lenses for an alpha-particle heat power of 0.375 kW (see also Section 10.13.9.1). This is above the temperature for initiation of explosion T_e = 331 °C (PBX 9502) or equal to T_e = 347 °C (TATB). Under the assumption that the real temperatures in the high explosive lenses would be higher than those determined by the one-dimensional conservative approach such a very high-technology HNED would not function properly. At an alpha-particle heat power of 0.562 kW both PBX 9502 and TATB would reach the melting point of 448 °C.

10.10.4 HNEDs with other implosion geometries [1,32]

10.10.4.1 Very High technology HNED with hollow spherical reactor Pu shell

The hollow spherical shell has an inner radius r_i = 5.6 cm, an outer radius of 7.2 cm. The outer dimensions of the HNED were the same as in the solid-sphere. Only the thickness of the aluminum spherical shell was reduced from 3 to 1.6 cm.

The temperatures in the high explosive are the same as in the solid Pu-sphere case. Only the temperature difference in the spherical plutonium shell would be somewhat lower.

The limits for alpha-particle heat power would be the same as in the case for a solid plutonium sphere (Section 10.10.3.7).

10.10.4.2 Very High technology Pu-sphere combined with spherical shell design [1,32]

In this case an inner solid reactor-grade plutonium sphere of 4 cm radius and a Pu spherical shell of 5.8 cm inner and 6.8 cm outer radius were assumed. The resulting temperature profiles in the explosive lenses remain essentially equal to those of the solid sphere with reactor-grade plutonium. Only the temperatures in the small Pu sphere become considerably higher.

Again as in the cases of solid or hollow plutonium spheres such an HNED would have the same limits of alpha-particle heat power when the HNED would not function any more.

10.10.4.3 HNEDs with smaller outer diameter of 15.5 cm

For the sake of completeness also HNEDs with reactor-grade plutonium and an outer-casing diameter down to 15.5 cm were analyzed. Such cases belong only to the very high technology class. Applying Eqs. (10.1) through (10.1d) for natural convection and radiation, the outer-casing temperatures were determined.

Table 10.8 shows that the outer-casing temperature for a 15.5 cm HNED design, e.g. for Pu with 0.614 kW alpha-particle heat power would be 338 °C. This means that most part of the high explosives inside would exceed the temperature T_e = 331 °C or 347 °C for initiation of self-explosion of PBX-9502 or TATB, respectively. Such designs would not be possible with reactor-grade plutonium at least from an alpha-particle heat power of 0.55 kW upwards.

Table 10.8: Outer-Casing Temperature T_a of HNEDs with 15.5 cm outer diameter.

α-heat (kW)	155 mm diameter casing outer temperature (°C)
0.144	168
0.240	183
0.614	338
0.858	392
1.121	451
1.244	456
1.416	479
1.530	494

10.11 Conclusions for the results of the thermal analyses

Following the conservative approach described in Section 10.4, always the lower limiting temperature profile is taken to define a limit for denatured proliferation-proof plutonium. The limiting temperatures are

1. the melting point of the high explosives
2. the temperature T_e for initiation of self-explosion of the high explosive
3. the melting temperature of the Pu metal

The reactor-grade plutonium can be considered to be denatured or proliferation-proof if at least one of these limits is exceeded, because the high explosive lenses and, therefore, the whole HNED would not work.

10.11.1 Low Technology

As was shown already in Section 10.10.1.1, and can be seen from Fig. 10.9 the melting point of both Baratol and Composition B are exceeded already for 0.120 kW. This corresponds to reactor-grade Pu from spent LWR fuel with about 35 GWd/t burnup (Table 9.4). The temperature T_e for initiation of self-explosion for Composition B would be exceeded for a slightly higher alpha-particle heat power. Thus, reactor-grade plutonium from LWR spent fuel with 1.8% Pu-238 or about 35 GWd/t burn-up (Table 9.4) if used in low technology HNEDs is proliferation-proof. [1]

10.11.2 Medium Technology

As was shown already in Section 10.10.2.2 and can be seen from Fig. 10.10 the melting points are exceeded for an alpha-particle heat power of 0.240 kW for the high explosives PBX 9011, PBX 9404, PBX 9501, as well as for HMX and DATB.

Reactor-grade plutonium from LWR spent fuel with about 3.6% Pu-238 or a burnup of about 58 GWd/t (Table 9.4) if used in HNEDs of medium technology can be considered as proliferation-proof.

[1] The 1.8% Pu-238 in the reactor grade plutonium and the 0.120 kW are not fully consistent with the 1.6% Pu-238 and 0.144 kW of Pu(0) in [1] as the alpha-particle heat for Pu-242 had to the corrected in later publications [16,32].

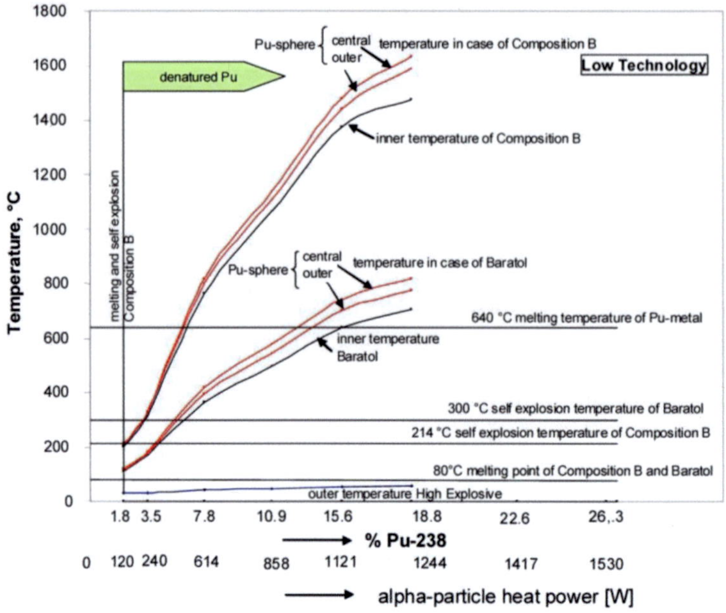

Figure 10.9: Temperature profiles in an HNED with low technology and different values of alpha-particle heat power.

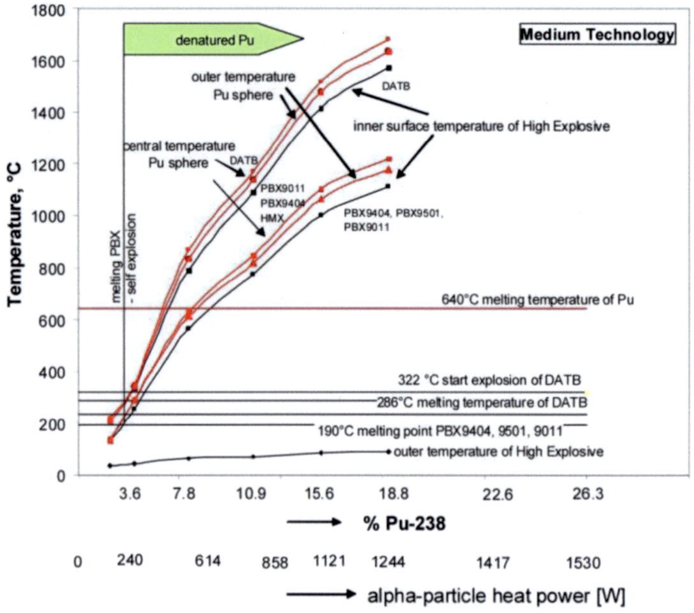

Figure 10.10: Temperature profiles in HNEDs for reactor-grade plutonium for the medium technology case.

10.11.3 Common assessment of the neutronic and the thermal analysis

In Section 9.10.1.4 and 9.12.1, two levels of technology and the associated shock pressures acting on the outside of the natural uranium reflector had been defined:

low technology	0.06 TPa
very high technology	0.11 TPa.

These levels of technology incorporate the art of designing and building the high explosive lenses and are directly related to the **thickness of the explosive lenses**. The increase in pressure applied to the outside of the reflector is a function of convergent spherical flow and has been shown (Kessler et al. [2]) to be the higher the thicker the high explosive lenses are. The higher the thickness of the explosive lenses the higher the magnification factor for the shock pressure applied to the outer surface of the reflector, the higher are the particle velocities caused by the shock wave, the maximum Rossi alpha attained and, finally, also the nuclear explosive yields.

According to the shock compression results of [2,26], an increase of the thickness of the implosion lenses up to 40 cm would allow to increase the concentric shock pressure from about 0.11 (TPa) to about 0.18 (TPa) acting on the outer surface of the reflector. This would theoretically result in even higher nuclear explosive yields than those reported in Section 9.12.5. However, such HNEDs would not be technically feasible above a certain alpha-particle heat power of reactor-grade plutonium. This is born out by the results of the thermal analysis of Section 10.10.

The results of the thermal analysis shown in Figs. 10.6 and 10.7 show that for both 43 cm thick (low technology) and 25 cm thick implosion lenses (low technology of high explosives and medium technology for thickness of high explosive lenses) (Table 10.9) the implosion lenses would be partially molten for an alpha-particle heat of 0.144 (kW). One can calculate that the corresponding limit would be 0.120 kW where the high explosives begin to melt. This corresponds to plutonium in LWR spent fuel with about 1.8% Pu-238 or a burnup of about 35 GWd/t. Similarly, this limit would be 0.240 kW or a burnup of 58 GWd/t for medium technology with 25 cm thick lenses as shown in Fig. 10.8.

Table 10.9: Geometric dimensions and thermodynamic data of implosion lenses for different levels of technology.

Level of technology	Thickness of implosion lenses (cm)	Thermal conductivity (W/m °C)	Melting temperature (°C)	Temperature start of self-explosion (°C)
Low technology	43	0.219-0.494	79-80	180-288
Medium technology	25	0.259-0.452	190-286	260-322
Very high technology	10	0.561	448	331-347

10.11.4 Limits of alpha-particle heat power for proliferation-proof plutonium

All reactor-grade plutonium generating an alpha-particle heat power of more than 0.120 kW or reactor-grade plutonium from LWR-UOX or LWR-MOX fuel with a burnup of more than 35 GWd/t or more than 1.8% Pu-238 in the reactor plutonium (see Table 9.6) will render a **low technology HNED technically unfeasible** (Fig. 10.9). This means that the explosive yields of 0.12 or 0.35 kt TNT (equivalent) of Section 9.12.5 would not be feasible either.

For medium technology HNEDs this limit for the alpha-particle heat power is 0.240 kW. Temperatures in this case would be so high that medium technology high explosives (see Table 10.2) would melt or even reach the temperature limit above which a chemical self-explosion would be initiated. The thermal limit of 0.240 kW is determined by the high explosive PBX-9501 which has the highest thermal conductivity with 0.452 W/cm °C out of the medium technology high explosives (Table 10.2). From Fig. 10.8 it can be seen that it would just be molten at the inner boundary. This thermal limit of 0.240 kW corresponds to reactor-grade plutonium from reprocessed UOX or MOX spent LWR fuel with a burnup above 58 GWd/t (roughly 3.6% Pu-238 in the plutonium – see Table 9.6.b). It would render a medium technology HNED technically unfeasible. This means that the nuclear explosive yield of 0.12 or 0.35 kt TNT (equivalent) of Section 9.12.5 would not be feasible either.

These results do not confirm the statement in Fig. 9.1 which reads:

.........plutonium of the quality produced in current nuclear reactors................can be used to make nuclear explosives, using technology comparable to the earliest plutonium weapons.

This statement seems to be based on the neutronic analysis only but disregards any results from a thermal analysis (see also Fig. 9.47).

To bring the above results in direct relationship with practical reality the current burnup of LWR spent fuel is close to 60 GWd/t.

The first nuclear explosive device (Fig. 9.3) had somewhat larger geometric dimensions as used in Fig. 10.6, and contained the same chemical explosives as shown in Fig. 10.6 or Table 10.1. However, it incorporated a weapons-grade plutonium sphere with an alpha-particle heat power of less about 12 W (see Fig. 9.47). It did work with weapon-grade plutonium, but it would not work with reactor-grade plutonium producing more than 0.120 kW in its fissile part. This corresponds to reactor grade plutonium from spent PWR fuel with more than 35 GWd/t_{HM} or more than 1.8% Pu-238 (Table 9.4).

The above results could confirm the statement in Fig. 9.1 only for plutonium from LWR spent fuel with less than about 35 GWd/t burnup (1.8% Pu-238) and low technology (earliest plutonium weapons). In reality, the above limits of 0.120 kW and 0.240 kW are even time-dependent. The Pu-241 of the reactor-grade plutonium decays to Am-241 with a halflife of 14.4 years. Whereas the Pu-241 produces 3.3 W/kg alpha-particle heat power, the Am-241 produces 110 W/kg. A typical HNED with about 10 kg of aged reactor-grade plutonium (Tables 9.4 through 9.6b) and a typical Pu-241 isotopic composition of about 14% would produce the following alpha-particle heat power from Am-241 in addition to that of the reactor-grade plutonium:

time after reprocessing (years)	2	4	6	8	10	12	14
alpha-particle heat power (W) of Am-241	14	26	38	48	57	66	75

This decreases somewhat the above thermal limits of 0.120 kW and 0.240 kW, which are given for the alpha-particle heat power of the reactor-grade plutonium.

A technically still feasible HNED would have to be designed for a lower thermal limit than the above 0.120 kW (low technology) and 0.240 kW (medium technology) or the Am-241 would have to be chemically separated prior to applications.

In addition impurities, e.g. carbon, boron etc. in the metallic reactor-grade plutonium undergo (n,α) reactions which also contribute somewhat to the alpha-particle heat power [48]. This contribution is not accounted for in the above considerations.

10.11.5 Additional remarks on the low and medium-technology cases

The above thermal analysis is based on conservative assumptions:

- Black-body radiation at the outer surface of the HNED casing (Kessler [1]).
- If the radiation emissivities measured, e.g., for steel or aluminum at the outer casing were taken into account, the difference between the temperature of the outer casing and that of the ambient air (30 °C) would increase by $\varepsilon^{-1/4}$ (ε being the emissivity of the outer casing). For steel as the outer casing, $\varepsilon = 0.22$ (Paloposki et al. [8] and Kuchling [9]). This leads to a temperature difference higher by a factor of 1.46. For aluminum, $\varepsilon = 0.07$ is reported, and a temperature difference higher by a factor of 1.94 is obtained.

The low technology and medium technology cases represent the highest levels of technology to be achieved in a NNWS. Subnational groups would not even be able to manage the low technology case (see also Younger [10]). NWSs never used reactor-grade plutonium for nuclear weapons [11,12].

Grizzle [50] gives the following reasons "Reactor-grade plutonium is significantly more radioactive which complicates the design, manufacture and stockpiling of weapons. Use of reactor-grade plutonium would require large expenditures for remote manufacturing facilities to minimize radiation explosure to workers. Reactor-grade plutonium use in weapons would cause concern over radiation exposure to military service personnel."

10.12 Outside cooling of the HNEDs

10.12.1 Coolability of HNEDs

Statements in the literature asserted that the problem of temperature limits, e.g. melting temperatures or temperatures initiating explosion of high explosives, could be overcome by means of cooling (strips of conducting materials or external cooling) (Garwin [11], Mark [13]).

10.12.2 Metal strips of high thermal conductivity

Garwin [11] refers to the use of so called "in-flight insertion devices". Nikitin [14] showed in a preliminary analysis that a relatively large number of metal strips arranged symmetrically within the explosive lenses would have to be used for cooling. In addition, an insulating layer would have to separate the metal strips adjacent to the high explosives. Three-dimensional analysis would have to show the absence of azimuthal temperature increases between strips. Finally, the metal strips would have to be replaced by strips of high explosives assuring precise three-dimensional internal structures of the high explosive lenses before an HNED could be fired (otherwise detonation physics would not work). All of this would have to be done in a short span of time because interruption of cooling would raise temperatures in the HNED (Kessler [16] and Section 10.14).

10.12.3 Coolability of very high technology HNEDs

Another idea for overcoming the temperature problem in the high explosive lenses was presented recently by (Shmelev et al. [17]). A very high-technology HNED could be surrounded by a hollow spherical shell of aluminum 43 cm thick. The good thermal conductivity of aluminum and the larger outer surface would lower the outer temperature. This would also lower the whole temperature profile. Thus decreasing the temperatures at the high explosive lenses. Subsequent cooling by liquid nitrogen to -200 °C would again lower the whole temperature profile for the inner temperature of the high explosive lenses. This would be below the temperature at which self-explosion (331 or 347 °C) of the TATB or PBX-9502 high explosives sets in (Table 10.3). Such HNED could still function. However, it would have to be assembled remotely under liquid nitrogen. High technology high explosives are only available in advanced NNWs (Gibbs et al. [18] and Mader et al. [19]). A hollow sphere with an outer radius of 7.2 cm at k_{eff} = 0.98 would have an outside temperature around 400 °C (at 30 °C outside air temperature). This gives an idea of some of the technical difficulties to be overcome (see Kessler [16] and Section 10.13.8). Discussions of such hypothetical examples must stick to reality [10,15].

10.12.4 Effects of cooling low-technology and medium-technology HNEDs

Cooling a low-technology and medium-technology HNED requires submerging them in liquid nitrogen or helium. Submerging the HNED in liquid nitrogen or liquid helium has the consequence that the lower temperatures of all HNED materials give rise to lower thermal heat conductivities raising the above temperatures at the inner surface of the high explosives. This is shown by Kessler [16]. Such more detailed analyses will be described in Section 10.13. The thermal analysis for the description of cooling HNEDs down to cryogenic temperatures requires the use of temperature dependent thermal heat conductivities and specific heats. Three different cooling possibilities were treated consistently with numerical methods for comparison by Kessler [12]:

- cooling the outside casing of the HNEDs by thermal radiation and natural convection of air,
- cooling the outside casing of the HNEDs down to cryogenic temperatures, e.g. to -200 °C by liquid nitrogen or to -270 °C by liquid helium. This was first proposed by Shmelev [17] as a cooling possibility to overcome the problem of high temperatures in HNEDs based on reactor-grade plutonium,

- cooling by internal metal rods which would transfer the alpha-particle heat power from the inner aluminum spherical shell through the high explosive lenses to the outer casing of the HNED. This was discussed, e.g. by Mark [13] to overcome the high temperatures in HNEDs based on reactor-grade plutonium.

Cooling of the HNED by gas or liquids through internal cooling channels or gaps does not need to be not considered. This would destroy the high precision spherical symmetry of implosion shock physics and cause severe hydrodynamic instabilities during shock compression.

10.13 Solution of the steady state and transient heat conduction problem with temperature dependent thermal conductivities and specific heats

Cooling the outside of the HNEDs by cryogenic liquids and calculation of the resulting cryogenic temperatures requires a thermal analysis with the solution of the heat conduction equations with temperature dependent thermal conductivities, specific heats and densities. The assessment of cooling the HNEDs by internal metal rods (thermal bridges) and replacement of these devices prior to the application of the HNED requires a thermal analysis with the solution of the transient heat conduction equation. This can only be achieved by using numerical methods in combination with the one-dimensional conservative approach described in Section 10.4.

10.13.1 Formulation of the heat conduction problem [16]

The thermal conduction problem within the HNED can be described by the following 1-D diffusion equation in the polar coordinate system,

$$\rho\, c_p \frac{\partial}{\partial t} T = \frac{1}{r^2}\frac{\partial}{\partial r}\left(r^2 k \frac{\partial T}{\partial r} \right) + \dot{q} \quad \text{within each material layer (Fig. 10.1)} \tag{10.6}$$

T is the temperature, $\dot{q}$ the power density, which is equal to the power divided by the active volume, ρ the material density, c_P the thermal heat capacity and k the thermal conductivity. The material thermal physical properties, ρ, c_P and k are functions of temperature. The boundary conditions are set as follows:

The symmetry condition is valid at r = 0, i.e.,

$$\frac{\partial}{\partial r} T = 0, \text{ at } r = 0 \tag{10.6a}$$

Inner boundary condition between material layers i and i+1 is the heat flux continuity condition, i.e.,

$$k_i r_i^2 \frac{\partial}{\partial r} T_i = k_{i+1} r_{i+1}^2 \frac{\partial}{\partial r} T_{i+1} \tag{10.6b}$$

Between the plutonium sphere and the natural uranium spherical reflector layer there is a gap with a gap conductance α_{gap}. The temperature jump through this gap can be described by

$$\alpha_{gap} \Delta T_{gap} = \frac{P}{4\pi R_{gap}^2} \tag{10.6c}$$

P is the power, ΔT_{gap} the temperature difference over the gap and R_{gap} the gap radius (changes of the gap width are neglected). At the outer casing surface of the HNED a temperature boundary condition has to be adopted. If the outer surface temperature T_s is known, e.g. the HNED is in a liquid medium, it is simply

$$T = T_{out} \text{ at } r = R \tag{10.6d}$$

If the outer surface temperature of the casing is determined by natural convection of air and by thermal radiation (Kessler [1]), the following equations have to be taken into account. For the natural convection the correlation (10.1a) is used as coefficient with 0.43. The coefficient 0.43 can also be 0.5 as given by Kakac et al. [33].

Nu is the Nusselt number, *Gr* the Grashof, and *Pr* the Prandtl number. They are defined as

$$Nu = \frac{\alpha_{conv}\, D}{k}, \quad Gr = \frac{g\beta D^3\, \Delta T}{\nu^2\, T}, \quad \Pr = \frac{\nu\rho c_p}{k} \tag{10.6e}$$

a_{conv} is the heat transfer coefficient due to the natural convection, $D = 2\ R$ is the sphere diameter, β the volumetric thermal expansion coefficient, g the acceleration due to gravity, ν the kinematic viscosity and ΔT is the temperature difference between the surface temperature and the air.

The thermal radiation is described by (see Eq. 10.1d)

$$\varepsilon\sigma(T_s^4 - T_{env}^4) = \frac{P_{rad}}{4\pi R^2} \tag{10.7}$$

T_s is the temperature at the surface of the HNED, T_{env} is the temperature of the environment, Prad is the power transferred by radiation, ε the emissivity (Table 10.5) and ε = 5.67 x 10-8 W/(m2K4) the Stefan-Boltzmann radiation constant. An equivalent heat transfer coefficient αrad can be derived for the thermal radiation process as

$$\alpha_{rad}(T_s, T_{env}) = \varepsilon\sigma(T_s + T_{env})(T_s^2 + T_{env}^2)$$

so that (10.7) becomes

$$\alpha_{rad}(T_s, T_{env})\Delta T = \frac{P_{rad}}{4\pi R^2} \text{ with } \Delta T = T_s - T_{env}$$

Thus, with Eqs. (10.1) as well as (10.1a) and (10.1b) the outer casing surface temperature is obtained by

$$\alpha_{eff}(\Delta T, T_s)\Delta T = \frac{P}{4\pi R^2} \tag{10.8}$$

$$\alpha_{eff}(\Delta T, T_s) = \alpha_{conv} + \alpha_{rad} \tag{10.9}$$

10.13.2 Numerical solution for the transient temperature distribution

The code system Mathematica is applied to solve the steady state and time dependent problem. Mathematica has a good solver for ordinary differential equation (ODE) systems. If a finite difference approximation in space is applied to Eq. (10.6), an ODE system in time is obtained.

Thus, within layer n, where Δr_n is uniform, Eq. (10.6) can be discretized at $r = r_i$ as

$$(\rho c_p)_i \frac{dT_i}{dt} = \frac{1}{4r_i^2 \Delta r_n^2}\left(\left(r_{i+1}^2 + r_i^2\right)\left(k_{i+1} + k_i\right)\left(T_{i+1} - T_i\right) - \left(r_i^2 + r_{i-1}^2\right)\left(k_i + k_{i-1}\right)\left(T_i - T_{i-1}\right)\right) + \dot{q}_n \tag{10.10}$$

with $\Delta r_n = r_n - r_{n-1}$.

At the inner boundary between layers n and n+1, the diffusion equation is discretized as

$$(\rho c_p)_i \frac{dT_i}{dt} = \frac{1}{2r_i^2(\Delta r_{n+1} + \Delta r_n)} \times$$

$$\left((r_{i+1}^2 + r_i^2)(k_{i+1} + k_i)\frac{(T_{i+1} - T_i)}{\Delta r_{n+1}} - (r_i^2 + r_{i-1}^2)(k_i + k_{i-1})\frac{(T_i - T_{i-1})}{\Delta r_n} \right) + \frac{\Delta r_n}{(\Delta r_{n+1} + \Delta r_n)} \dot{q}_n$$

(10.11)

At the gap ($r = r_g$ or $i = i_g$), the diffusion equation is discretized as

$$(\rho c_p)_i \frac{dT_i}{dt} = \frac{1}{r_i^2 \Delta r} \left((r_{i+1}^2 + r_i^2)\frac{(k_{i+1} + k_i)}{2}\frac{(T_{i+1} - T_i)}{\Delta r} - (r_i^2 + r_{i-1}^2)\alpha_{gap}(T_i - T_{i-1}) \right)$$

(10.12)

where Δr is the gap width.

On the outer casing surface ($r = R$ or $i = i_R$), the temperature boundary condition is assumed as,

$$T_i = T_s = T_{out} \quad \text{or} \quad T_{i+1} = T_{env} \tag{10.13}$$

This condition means that the outer casing surface temperature is either given as the cooling temperature as Eq. (10.6d) or calculated by Eq. (10.8) where natural convection and thermal radiation are considered.

10.13.2.1 Numerical solution for the steady state temperature distribution

In case of the steady state problem the time derivative dT_i/dt on the left side becomes zero. The Eqs. (10.10) through (10.13) become a system of linear equations which is solved by Mathematica.

10.13.3 Thermal conductivity and specific heat at cryogenic temperatures

According to cryogenic temperature physics both the thermal conductivity and the specific heat decrease for metals and polymers to smaller and smaller values, when the temperature approaches low Kelvin values.

10.13.3.1 Thermal conductivity for plutonium metal

Plutonium metal exists in 5 allotropic phases and has a melting temperature of 640 °C. Its thermal conductivity was measured down to very low temperatures. The thermal conductivity data for plutonium are taken from the Reactor Handbook [29] and supplemented by measured data of Blank et al. [34] (Fig. 10.11).

10.13.3.2 Thermal conductivity of uranium metal

Similar data can be found for uranium metal. It has 3 allotropic phases and a melting point of 1132 °C. Its thermal conductivity was also measured down to very low temperatures. For the calculation the thermal conductivity data are taken from Gebhardt et al. [35].

10.13.3.3 Thermal conductivities for aluminum, beryllium and stainless steel

Thermal conductivity data for aluminum, beryllium and stainless steel down to cryogenic temperatures can be found in Marquardt et al. [36]. They are displayed in Fig. 10.11.

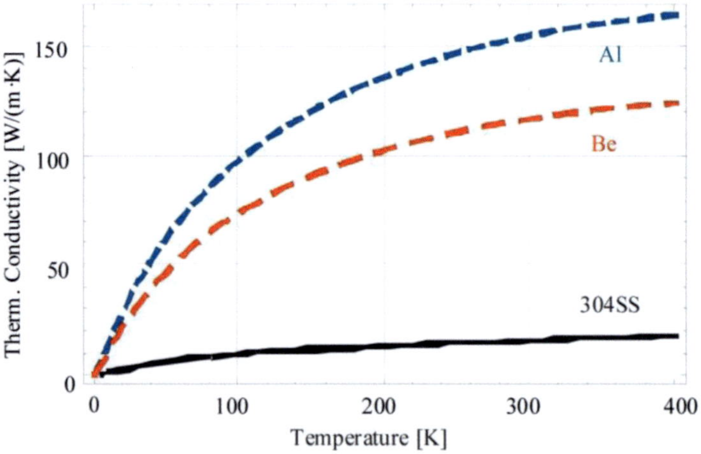

Figure 10.11: Thermal conductivity of aluminum, beryllium and stainless steel SS 304 down to cryogenic temperatures.

10.13.3.4 Thermal conductivity for chemical high explosives

For chemical high explosives no measured data can be found in the open literature for the temperature range down to cryogenic temperatures. Only thermal conductivity data for 20 °C are given by Dobratz [18], Gibbs et al. [19] and Mader et al. [24].

For the temperature range down to cryogenic temperatures, therefore, a theoretical analogy was applied which was reported by Shchetinin [37]. This theoretical analogy was combined with data for temperature dependent thermal conductivities of Hartwig [38], Barron et al. [39] and Marquardt et al. [36] for different polymers. In addition, they were adapted to the measured data at 20 °C of Gibbs et al. [18] and Mader et al. [19]. This is shown for chemical high explosives in Fig. 10.12.

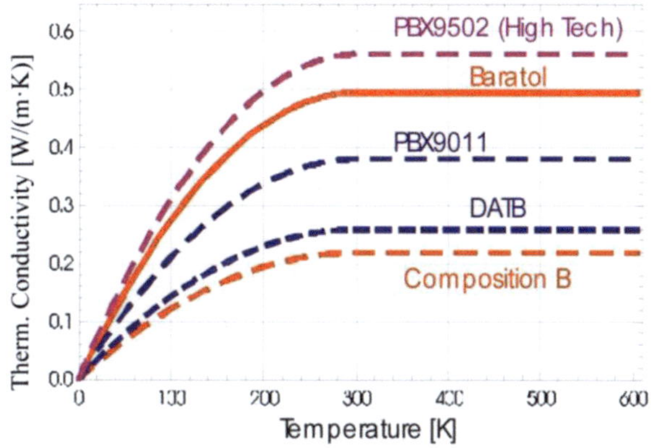

Figure 10.12: Thermal conductivity of chemical high explosives down to cryogenic temperatures.

10.13.4 Specific heat data at cryogenic temperatures

10.13.4.1 Specific heat for the metals applied

The data for the specific heat of plutonium metal, uranium metal, beryllium, aluminum and steel are shown by Fig. 10.13 for a temperature range down to cryogenic temperatures. They are based on Shmelev [40], Lashley et al. [41] and Marquardt et al. [36].

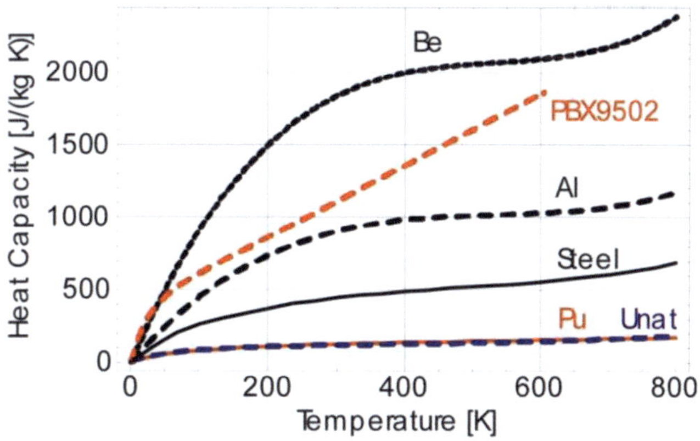

Figure 10.13: Heat capacity of metals down to cryogenic temperatures.

10.13.4.2 Specific heat for high explosives

The specific heat data for high explosive materials are taken from Hartwig [38], Barron et al. [39], Marquard et al. [36], Gibbs et al. [18] and Mader et al. [19]. Again the theoretical analogy by Shchetinin [37] and Shmelev [40] is applied. The data are displayed in Fig. 10.14.

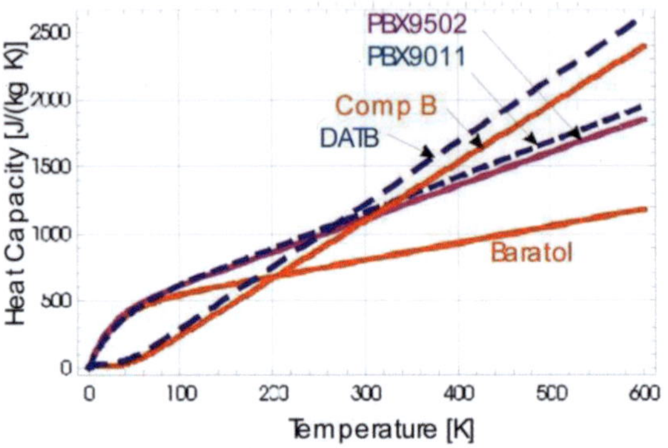

Figure 10.14: Heat capacity of chemical high explosives down to cryogenic temperatures.

10.13.5 Cooling of low technology HNEDs by liquid nitrogen or liquid helium

Fig. 10.15 shows the temperature profile for the high explosive Baratol and an alpha-particle heat power of 0.24 kW (for a constant thermal conductivity k). If the temperature-dependent thermal conductivities k(T) of Figs. 10.11 and 10.12 are applied, the temperature profile is raised as shown by the upper curve of Fig. 10.15. This accounts for the smaller thermal conductivities at lower temperatures.

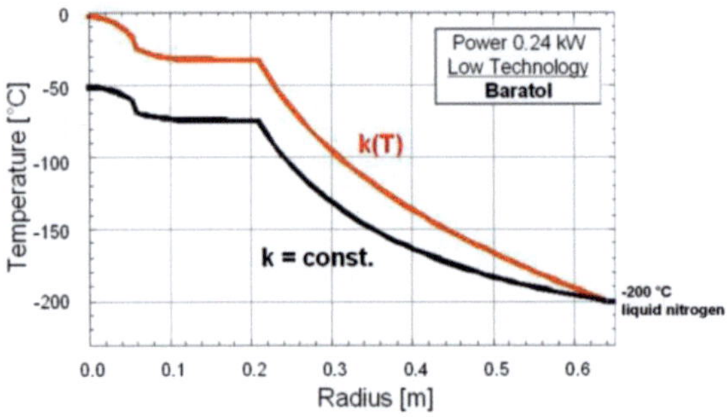

Figure 10.15: Temperature distribution in a low technology HNED with reactor-grade plutonium of 0.240 kW alpha-particle heat power (cooling by liquid nitrogen).

More results for low technology HNEDs with higher alpha-particle heat power are given in Table 10.10. As can be understood from Table 10.10 the maximum temperatures at the inner border of the high explosive lense systems and Baratol (temperature-dependent thermal conductivity) would be -32 °C for an alpha-particle heat power of 0.24 kW and cooling the outer casing by liquid nitrogen. It would be -50 °C for cooling by liquid helium (Table 10.10).

For Composition B due to the significantly smaller heat conductivity compared to Baratol these temperatures would be 128 and 111 °C, respectively. The melting point of 79 °C for Composition B (Table 10.2) would be already exceeded for this alpha-particle heat power of 0.24 kW.

Table 10.10: Temperature at the inner border of the implosion lenses as a function of the alpha-particle heat power and cooling of the outside by liquid nitrogen or liquid helium.

Alpha-particle heat power (kW)	With Baratol in -200 °C nitrogen (°C)	With Baratol in -270 °C helium (°C)	Composition B in -200 °C nitrogen (°C)	Composition B in -270 °C helium (°C)
0.240 kW	-32	-50	**128**	**111**
0.350 kW	26	10	**258**	**242**
0.460 kW	**97**	**79**		

Due to the high thermal conductivity of Baratol the melting point of 79 °C is reached only for an alpha-particle heat power of about 0.460 kW. Already for an alpha-particle heat power of 0.35 MW a considerable part of the volume of the high explosive Composition B would be molten and have exceeded the limiting temperature for start of chemical self-explosion at 214 °C (Table 10.10).

Therefore, a range between 0.24 kW (melting of Composition B) and 0.35 kW alpha-particle heat power was determined for which a low technology HNED will be technically questionable or very probably technically unfeasible. The alpha-particle heat of 0.46 kW (melting of Baratol) would certainly be a conservative upper limit for low technology HNEDs [16]. There are only relatively small differences in temperature for cooling by liquid nitrogen or by liquid helium. This is due to the temperature-dependence of the thermal conductivities of the high explosives at cryogenic temperatures.

10.13.6 Numerical results for medium-technology HNEDs [16]

10.13.6.1 Cooling of medium-technology HNEDs by liquid nitrogen or liquid helium

The limiting temperatures for the medium technology high explosives DATB and PBX-9011 (representative for PBX-9011, PBX-9404, 9501 and HMX) were shown in Table 10.2.

Table 10.11: Temperature at the inner border of the implosion lenses as a function of alpha-particle heat power and cooling of the outer casing by liquid nitrogen or liquid helium.

Alpha-particle heat power [kW]	cooling by liquid nitrogen		cooling by liquid helium	
	DATB [°C]	PBX 9011 [°C]	DATB [°C]	PBX 9011 [°C]
0.4	**321**	169	**305**	153
0.415	**333**	175	**323**	165
0.44	**369**	**202**	**353**	185
0.50	**441**	**250**	**425**	**234**

The alpha-particle power limits at which the temperature limits, e.g. melting points or temperatures for start of self-explosion at the inner radial boundary of the high explosive lenses, will just be attained, can be found by varying the alpha-particle heat power in the range between 0.4-0.5 kW. Table 10.11 gives the temperatures at the inner border of the implosion lenses as a function of the alpha-particle heat power for both high explosives DATB and PBX9011 and for both cases of cooling the outer cooling of the HNED either by liquid nitrogen or liquid helium.

According to Zinn et al. [23] and Mader et al. [19] chemical self-explosion of the high explosives starts when a critical temperature T_e is exceeded. For DATB this critical T_e temperature is 322 °C. The melting point of DATB would be exceeded already at 286 °C. For PBX9011 the melting point is 190 °C and the critical temperature for start of self explosion is somewhat above 260 °C.

Therefore, a medium technology HNED would become questionable and very probably technically unfeasible for an alpha-particle heat power range of 0.39 to 0.46 kW for cooling by liquid helium.

10.13.7 Conclusions for low and medium technology HNEDs

The results of Sections 10.11 and 10.12 show that

- any low technology HNED based on reactor-grade plutonium producing more than 0.12 kW of alpha-particle heat power (if cooled by natural convection of air and thermal radiation) and more than 0.46 kW, respectively, (if cooled at the outside by liquid nitrogen or helium) and
- any medium technology HNED with reactor-grade plutonium producing more than 0.24 kW of alpha-particle heat power (if cooled by natural convection of air and thermal radiation) and more than 0.46 kW alpha-particle heat power, respectively, (if cooled by liquid nitrogen or helium

will be technically unfeasible (Figs. 10.16 and 10.17). The high explosives would either melt in considerable parts of the implosion lenses or exceed the temperatures for start of chemical self-explosion. Outside casing temperatures lower than -270 °C (liquid helium) are technically impossible. Any increase of the outside casing temperature would increase these temperatures in the high explosives again. Figs. 10.16 and 10.17 also show the results for high technology HNEDs. This analysis will be presented below.

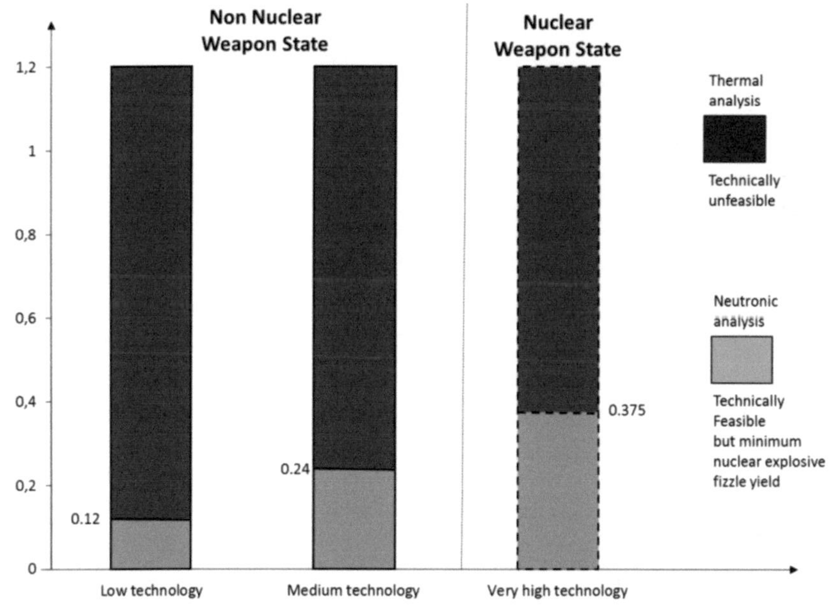

Figure 10.16: Limits for alpha-particle heat power above which HNEDs with reactor-grade plutonium become technically unfeasible for different levels of technology when cooled by air and radiation (Kessler et al. [16]).

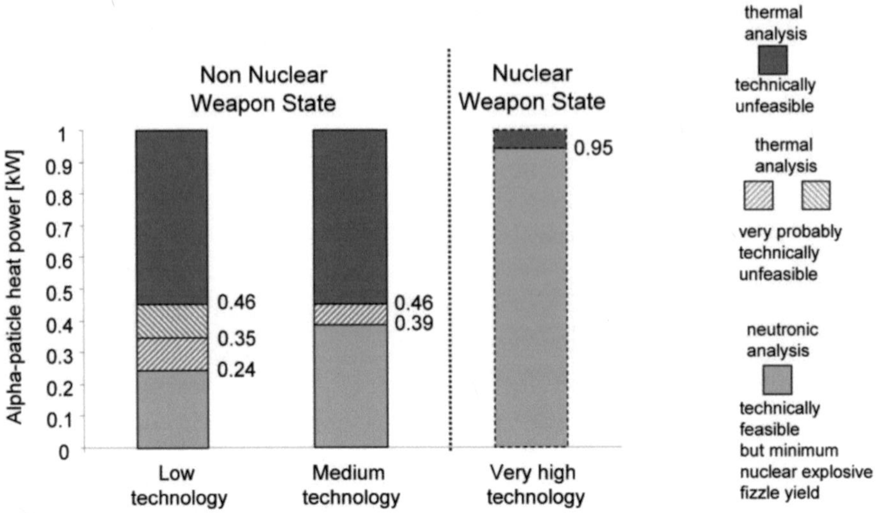

Figure 10.17: Limits for alpha-particle heat power above which HNEDs with reactor grade plutonium become technically unfeasible for different levels of technology and cooling at the outside by liquid helium (Kessler [16]).

As explained already in Section 10.11.4 the above limits are time-dependent and will decrease with ageing because of the buildup and α-decay of Am-241 which is a consequence of the beta-decay of Pu-241.

10.13.8 Technical difficulties for cooling by liquid nitrogen or liquid helium

Cooling of the HNEDs by submerging them into liquid nitrogen or helium would require heavy transport vehicles or ships for transportation of the cryogenic cooling devices. These transport possibilities are discussed by Hecker [42]. The following technical difficulties would have to be overcome for cooling the HNEDs by liquid nitrogen or helium.

A bare sphere of reactor-grade plutonium with an alpha-particle heat power of e.g. 0.35 kW would have an outside temperature in air of about 350 °C (black body radiation assumed). For an alpha-particle heat power of e.g. 0.42 kW the bare sphere would have an outside temperature of about 380 °C. According to Chebeskov [43], or Campbell et al. [44] such bare reactor-grade plutonium spheres would cause a radiation exposure (γ-radiation and neutrons) at 1 m distance of >5 Sv/hr. The whole HNED would have to be assembled by remote technology under liquid nitrogen or liquid helium. High precision machining of the solid or hollow reactor-grade plutonium metal sphere at temperatures of about 350-380 °C and radiation levels of several Sv/h would be a very high technological challenge. Experimental programs for the determination of the exact thermal conductivity down to cryogenic temperatures and of the thermal expansion coefficients of the high explosives would be needed. Measurements of the temperature field in the implosion lenses would have to assure that the melting points or the critical temperature for chemical self-explosion would not be exceeded.

Plutonium metal has 5 allotropic phases in solid state with different densities over certain temperature ranges. Long term cooling to -200 °C or -270 °C of the outer casing of the HNED would bring the metallic plutonium through temperature ranges from about 350-370 °C down to the range of about 0°C. Metallic plutonium applied in nuclear explosive devices is usually stabilized by about 1% gallium to extend the δ-phase (Mark [13]). One reason is – among others – the almost zero thermal expansion coefficient of the δ-phase plutonium. Moreover, the δ-phase leads to favorable ductile properties of the metallic plutonium (Morss et al. [45]). According to Timofeeva [46] this δ-phase can change into the α-phase below about 100 °C. This α-phase metallic plutonium has a higher density. Long term changes of the density, e.g. from 15.8 [g/cm^3] (δ-plutonium) to 18.9 [g/cm^3] (α-plutonium), might, therefore, have to be considered. This would lead to an increase of about 0.1 in k_{eff} (criticality factor).

The difficulties would become even higher as a consequence of impurities and buildup of americium in the metallic plutonium. They would change the allotropic phase diagrams (Cleveland [47]). In addition, impurities like boron or beryllium lead to (n,α)-reactions and higher alpha-particle heat powers (Shmelev et al. [48]).

At least two different high explosives must be used in the explosive lenses (Rhodes [3]). They have different thermal expansion coefficients (Gibbs et al. [38]). Together with the temperature difference of about 300 °C across the radius of the explosive lenses this would create tremendous difficulties for the very high precision needed in assembling the high explosive lense system.

These are only some of the expected technical difficulties which would have to be overcome. It is hardly conceivable that a Non-Nuclear Weapon State would make such extreme high-technology efforts. Even if a Non-Nuclear Weapon State would be able to overcome all these technical difficulties the results could only be a fizzle explosive yield under the most favorable conditions (Kessler et al. [32] and Section 9). This holds only for HNEDs with an alpha-particle heat power of less than 0.24 kW for low technology and for less than 0.39 kW for medium technology. For the alpha-particle heat powers above these alpha-particle power levels for HNEDs based on reactor-grade plutonium either the melting points or the critical temperatures for self-explosion of the high explosives would be exceeded (Figs. 10.16 and 10.17).

10.13.9 Numerical results for high technology HNEDs

10.13.9.1 Cooling of high technology HNEDs by air and radiation

Despite the above results and arguments there is some theoretical interest in analyzing also so-called very high technology cases. However, one should remind that such high technology cases could – if at all – only be mastered by present advanced Nuclear-Weapon-States having performed many years of research, experiments and development (Kessler et al. [32]).

Garwin [11], deVolpi [12] and Grizzle [50] stated that no Nulcear Weapon State has ever used reactor-grade plutonium for its weapons arsenal.

Nevertheless, this section contains a thermal analysis of a high-technology case which was presented in Kessler [1] and Kessler et al. [32]. The same geometric dimensions and materials were used by a US-Russian group (Fetter et al. [6]). This HNED has an outer diameter of 22 cm and a thickness of the spherical high explosive lenses of 10 cm. The reactor-grade plutonium (density 15.8 g/cm^3) is arranged as a hollow spherical shell of 7.2 cm outer diameter and 5.6 cm inner diameter. Its mass is around 13 kg (k_{eff} = 0.98). The HNED is cooled by natural convection air and thermal radiation (black body radiation assumed). The high technology high explosives TATB or PBX 9502 (see Table 10.3) have the highest thermal conductivities, the highest melting point, and the highest temperature for initiating chemical self-explosion to be found in the open scientific literature (Dobratz [24], Gibbs et al. [18], and Mader et al. [19]).

The thermal analysis of Kessler [16] shows that the melting point of the high explosives PBX 9502 or TATB of 448 °C would be exceeded for 0.562 kW. Similarly the limiting temperature for start of self-explosion of the chemical explosives (347 °C) would be already exceeded for 0.375 kW. Again it must be emphasized that these calculated temperatures are conservative as the one-dimensional conservative approach (Kessler [1]) and black-body radiation at the outer casing were assumed. In addition, these thermal limits are time-dependent and decrease because of the buildup of americium (Section 10.11.4).

10.13.9.2 Cooling of high technology HNEDs by liquid nitrogen or liquid helium

Analysis [16] shows that the alpha-particle heat power of reactor-grade plutonium must be increased to 0.95 kW to obtain a temperature of 347 °C at the inner boundary of the high explosives PBX9502 or TATB. In this case chemical self-explosion would be initiated. The high technology HNED would then become technically unfeasible even for the case of cooling its outside boundary by liquid helium.

10.13.9.3 Improved coolability of high technology HNEDs

An interesting idea for improving the coolability of high technology HNEDs has been presented by Shmelev [17]. The high-technology HNED could be surrounded by a hollow spherical shell of aluminum of 23 cm thickness. The good thermal conductivity of aluminum and the larger outer surface would decrease the outer temperature from 144 °C to some 46 °C (cooling by free convection of air and thermal radiation). This would also lower the whole temperature profile. Subsequent cooling by liquid helium to -270 °C would lower the whole temperature profile further. The alpha-particle heat power would have to be raised to 0.96 kW (cooling the outside by liquid helium). In this case the temperature would be 347 °C at the inner border of the high explosives PBX9502 or TATB and chemical self-explosion would start. The melting point of 448 °C at the inner border of the high explosive lenses would be attained for an alpha-particle heat power of 1.146 kW.

This high technology HNED would have to be assembled remotely – as discussed in the previous section – under liquid nitrogen or liquid helium. A hollow sphere with an outer radius of 7.2 cm (k_{eff} = 0.98) would have an outside temperature around 440 °C. This illustrates some important technical difficulties to be overcome.

The results of these sections on high technology HNEDs are shown in Figs. 10.18 and 10.17. Again it should be emphasized that only advanced Nuclear Weapon States could perhaps master these technological difficulties if at all.

10.14 Steady state and transient temperature distributions for cooling of the HNED by internal rods of high thermal conductivity

There are assertions (Mark [13]) or speculations that the problem of limiting temperatures, e.g. melting temperatures or the temperatures for the start of self-explosion of the high explosives could also be overcome by other cooling possibilities (strips of conducting materials as thermal bridges).

In a preliminary analysis Nikitin [14] showed that a relatively high number of symmetrically arranged 1 cm thick aluminum rods would have to be used for internal cooling (Fig. 10.18). However, an insulation layer would have to separate the aluminum rods, wherever they border the chemical high explosives. A three-dimensional thermal analysis would have to show which azimuthal temperature oscillations would occur between the rods within the high explosive lenses. Finally, prior to the use of the HNED as a nuclear explosive the aluminum rods would have to be removed and replaced by rods of high explosive material. This explosive material consisting of rods would have to fit exactly into the three-dimensional structure of the high explosive lenses. Only the exact internal structure can assure the symmetrical shock implosion needed for technical feasibility (Rhodes [3]). Otherwise the detonation physics would not work. This replacement would have to be done in a short time period, since a stop of the cooling (by removing the aluminum rods with high thermal conductivity) would again increase the temperatures fairly rapidly within the high explosive lenses of the HNED.

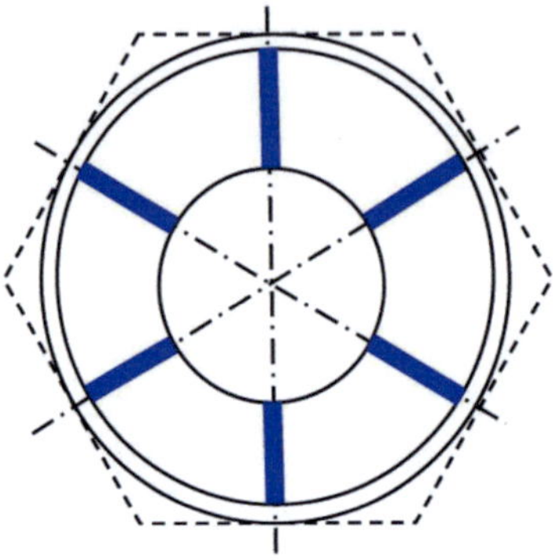

Figure 10.18: Schematic cross-section through a spherical HNED with thermal bridges.

In Nikitin [14] it was shown that for a low technology HNED with an alpha-particle heat power of 0.144 kW 20 aluminum rods of 1 cm diameter would be needed to keep the temperature at the inner boundary of the high explosive lenses 11°C below the melting temperature of 79°C.

An approximate method for the determination of the steady state temperature distribution in an HNED with aluminum rods was described by Kessler [16].

10.14.1 Outline of the approximate method for determining the steady state temperature distribution in an HNED with cooling by aluminum rods

Two heat conducting materials are put together for the approximate method. They form a parallel heat conducting channel. If k_1 and k_2 are the thermal conductivities for the two materials and A_1 and A_2 are their conducting areas, the heat conducting process can be described by

$$Q = k_1 A_1 \frac{dT_1}{dx} + k_2 A_2 \frac{dT_2}{dx} = \bar{k}\, A \frac{d\bar{T}}{dx}, \text{ with } A = A_1 + A_2 \tag{10.14}$$

If it holds approximately

$$\frac{dT_1}{dx} \approx \frac{dT_2}{dx} \approx \frac{d\bar{T}}{dx},$$

an effective thermal conductivity $\bar{k}$ can be expressed as

$$\bar{k} = k_1 A_1 / A + k_2 A_2 / A = k_1 a_1 + k_2 a_2 \text{ with } a_1 + a_2 = 1 \tag{10.15}$$

This linear volumetric weighting is known as Vegard's law in the literature.

In the case of aluminum cylindrical rods inserted in the spherical high explosive (HE) material the effective thermal conductivity can be written as

$$\bar{k}(T,r) = k_{Al}(T)\alpha_{Al}(r) + k_{HE}(T)\left[1 - \alpha_{Al}(r)\right] \tag{10.16}$$

where

$$a_{Al}(r) = N_{rod} r_{rod}^2 / \left(4r^2\right) \tag{10.17}$$

$\boldsymbol{N_{rod}}$ the number of aluminum rods and $\boldsymbol{r_{rod}}$ the radius of the aluminum rod.

This approximate model of Vegard underestimates the effective thermal conductivity and its temperature reduction effect. It overestimates the temperature distribution in the high explosives [20]. This weakness can be partially made up again by calibrating the results of the approximate model to the accurate results of Nikitin [14].

10.14.2 Calculated results for low technology HNEDs (steady state temperature profile)

Fig. 10.19 shows temperature profiles of a low technology HNED (with Baratol and Composition B as high explosives) with an outer casing radius of 0.65 m and an alpha-particle heat power of 0.144 kW (Kessler [1] without aluminum rods). Also presented are the calculated results for 20 aluminum rods inserted in the high explosive lenses (Nikitin [14]).

As mentioned before, the approximate homogeneous model would overestimate the temperature in the high explosive lenses significantly. However, the same temperature as in Nikitin [14] is roughly obtained if the aluminum rod diameter is increased from 1 cm (Nikitin) to 1.73 cm (calibration). The maximum temperature at the inner border of the chemical high explosive (Baratol) is about 11°C below its melting point of 79°C. Only Baratol is considered as high explosive, because it has a higher thermal conductivity than Composition B (low-technology case). The temperature profile of Baratol is therefore the lowest possible one. For Composition B the number of aluminum rods would have to be increased if the same low temperature distribution as for Baratol should be achieved.

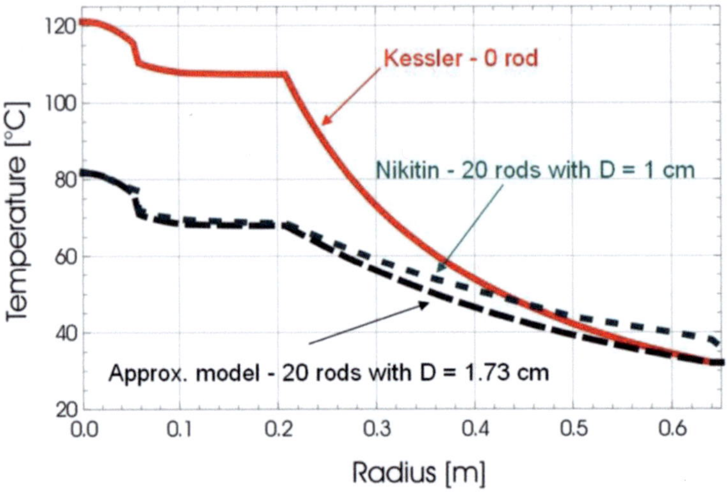

Figure 10.19: The radial temperature distribution of the cases of 0 and 20 aluminum rods of 1 cm and 1.73 cm diameter inserted. Alpha-particle heat power P = 0.144 kW (low-technology case).

10.14.3 Transient temperature distribution if the aluminum rods will be replaced by high explosive material

It is hypothesized that the aluminium rods have to be removed from the HNED and the remaining cylindrical holes will be refilled with the high explosive material before the HNED could become active. The question is how much time remains for this action before the high explosive material would be melting at its inner border. The HNED would then become technically unfeasible. The transient temperature calculation is performed by solving the equations of Section 10.13 with the temperature distribution of Fig. 10.21 as initial condition.

The result is that the transient temperature development for Baratol at the inner border of the high explosive lenses needs about 5.8 hours and for Composition B about 6.5 hours to reach the melting point of 79°C.

These calculations were repeated for an alpha-particle heat power 0.480 kW. Of course more aluminum rods are needed for the heat removal in the steady state case. A similar temperature distribution (11 °C below its melting point) is obtained for 170 aluminum rods with a diameter of D = 1.73 cm. A transient temperature calculation is also performed starting from these steady state conditions. The time period for the temperature at the inner border of the high explosive lenses to reach the melting point of 79°C would be 0.8 hours for Baratol and 0.85 hours for Composition B. Exceeding the melting point of Baratol and Composition B will render this HNED technically unfeasible.

It can be concluded that 20 aluminum rods can probably be replaced within 5.8 h. However, replacing 170 aluminum rods within a time period of about 1 h under the existing nuclear radiation and temperatures with the high precision needed within the three-dimensional internal structures of the implosion lenses could be an almost impossible undertaking. Therefore, the limiting case for the alpha-particle heat power can be estimated to be below 0.4 kW for a number of aluminum rods between 20 and about 170.

For this alpha-particle heat power a low-technology HNED would become technically unfeasible even if cooled internally by aluminum rods. Again it must be emphasized that also this thermal limit is time-dependent and decreases because of the buildup of americium (Section 10.11.4).

10.14.4 Technical difficulties

A bare sphere of metallic reactor-grade plutonium with an alpha-particle heat power of 0.4 kW would have a surface temperature of about 370°C and cause a radiation exposure of more than 5 Sv/hr (Campbell et al. [44] and Chebeskov [43]). High precision machining of the plutonium sphere could only be done by remote technology. Assembling of the HNED together with the explosive lenses and all aluminum-rods would probably have to occur in radial sections. The radial plutonium metal sphere sections would have to be placed at last. All radial sections would then have to be clamped together with high precision.

When the heat conducting aluminum rods would be replaced by high explosive materials, the latter cylindrical parts would have to be machined with very high precision. They must exactly fit into the 3-dimensional structure of the explosive lenses (in radial and azimuthal direction). Any inaccuracies at interfaces of different high explosive materials in the high explosive lenses would lead to hydrodynamic instabilities during the implosion process and disturbances of the symmetry of the shock waves and deteriorate to the nuclear explosive yield (Rhodes [3]). Such inaccuracies would, e.g. be caused by the radial temperature differences and by the different linear expansion coefficients (Gibbs et al. [18]) of the different high explosive materials in the high explosive lenses.

These are only some of the technical difficulties which would have to be overcome. It is hardly conceivable that a Non Nuclear Weapon State would make such high technology efforts. As a result it could at best only produce HNEDs based on reactor-grade plutonium, which would lead to a quasi-deterministic fizzle explosive yield (Kessler et al. [2,32]).

10.14.5 Installing the reactor-grade-plutonium sphere prior to detonation

Kang et al. [49] emphasized that the reactor-grade plutonium sphere could be installed into the high explosive lense system in a short time before detonation. However, the statement is based on the assembling of the first NED (Fat Man) which lasted more than one day (Rhodes [3]).

Calculations show that the installation of a reactor-grade plutonium sphere with an alpha-particle heat of 0.48 kW and a surface temperature of 400 °C would bring the temperature of the high explosive lenses at the inner border above the melting point within 10 min.

This reactor-grade plutonium sphere could also be cooled in liquid nitrogen to -200 °C. If this reactor grade plutonium sphere would be suddenly removed from the liquid nitrogen it would take 27 min to heat up to an outer surface temperature of 68 °C. If it could be installed with this temperature under the existing neutron and gamma radiation of 5 Sv/h into the HNED with its high explosive lense system, than it would take another 0.8 h for the high explosive to attain the melting temperature of 79 °C (see previous Section 10.13.4). If the inner border of the high explosive lenses would be thermally shielded by a layer of very low thermal conductivity (0.035 W/cmK) then the reactor-grade plutonium sphere would start to melt about 7 h after removal from the liquid nitrogen (Kessler [16]).

10.15 Conclusions

Although details of the arrangements of high explosive materials in the high explosive lenses are classified, the applied conservative approach (Kessler [1]) leads to scientifically reliable data with respect to the tolerable alpha-particle heat power and the associated temperature distributions in HNEDs.

Three different technologies: low, medium and very high technology for the multishell spherical systems (HNEDs) were defined. They differ in geometrical dimensions and different chemical high explosives. For the case of cooling the outside casing by natural convection of air and thermal radiation, the analysis shows that above an alpha-particle heat power of 0.12 kW for low technology, 0.24 kW for medium technology and 0.37 kW for very high technology the HNEDs would become technically unfeasible. These thermal limits related to alpha-partiele heat power are time-dependent and decrease as a consequence of the buildup of Am-241.

For the case of cooling the outside casing by submerging the HNED into liquid nitrogen or liquid helium the analysis becomes more sophisticated. The thermal conductivities of all materials become temperature dependent going to zero at zero Kelvin. In case of cooling the outside casing by liquid helium the limits above which the HNED become technically unfeasible raise to the limiting alpha-particle heat power of 0.46 kW for low-technology, the range of 0.46 kW for medium-technology and to 0.95 kW for very high technology. If the very high-technology HNEDs would be surrounded by a spherical aluminum shell of 23 cm thickness, the alpha-particle heat power would be raised to 0.96 kW. Cooling by liquid helium leads only to small differences if compared to cooling by liquid nitrogen. All these thermal limits are time-dependent and decrease because of the buildup of Am-241.

For the case of cooling the HNED by thermal conduction through cylindrical aluminum rods reaching from the inside aluminum shell to the outside casing, different analyses are required. Depending on the number of aluminum rods this leads to a steady state temperature distribution that is below the melting point of the high explosives. However, the thermally conducting aluminum rods must be replaced by rods of chemical high explosives fitting exactly into the three-dimensional structure of the high explosive lenses. When cooling by the thermally conducting rods is stopped, a transient thermal analysis must show within which time period the melting temperature at the inner border of the high explosive lenses would be attained or exceeded.

For low-technology HNEDs the results allow the conclusion that the limit, above which the HNED could become technically unfeasible, is below an alpha-particle heat power of about 0.4 kW, i.e. below the range which was determined for the case of cooling the outside casing by liquid helium to -270°C.

In both cases of cooling either the outside casing by liquid helium or by cooling the high explosive lenses by aluminum rods, tremendous technical difficulties would have to be overcome. Doubts are raised whether a Non Nuclear Weapon State would ever undertake such technical efforts. Any attempts to improve the cooling of the reactor grade-plutonium based HNEDs are colliding with the extreme requirements for high geometrical precision of all parts within the HNEDs. If all these technical difficulties could be overcome the results would still be a low quasi-deterministic fizzle explosive yield (Kessler et al. [2,32]).

High technology cases could only be mastered by advanced Nuclear Weapon States. But these have never used reactor-grade plutonium for their weapons arsenal (Garwin [11], deVolpi [12], Grizzle [50]).

References Section 10

[1] Kessler, G., Plutonium denaturing by Pu-238, Nucl. Sci. Eng., 155, 53-73 (2007).

[2] Kessler, G., et al., Potential nuclear explosive yield of reactor-grade plutonium using the disassembly theory of early reactor safety analysis, Nucl. Eng. and Design, 238, 3475-3499 (2008).

[3] Rhodes, R., The Making of the Atomic Bomb. Simon & Schuster, New York (1986).

[4] Cochran, Th.B. et al., Nuclear Weapons Databook, Vol. 1/2, Ballinger Publishing Company, Cambridge, MA, USA (1987).

[5] Podwig, P., Russian Strategic Nuclear Forces. MIT Press, Cambridge, Mass. (2004).

[6] Fetter, S. et al., Detecting Nuclear Warheads, Vol. 1, Science and Global Security, p. 225 (1990).

[7] Kessler, G., Analysis for a future proliferation-resistant plutonium fuel cycle. Atomwirtschaft 51, 337-340 (2005).

[8] Paloposki, T., et al., Steel Emissivity at High Temperatures, VTT TIEDOTTEITA Research Notes 2299, Espo, Sweden (2005).

[9] Kuchling, H., Taschenbuch der Physik, Verlag Deutsch Harri, Frankfurt (1982).

[10] Younger, S., The bomb, a new history. Harper Collings Publishers, New York (2009).

[11] Garwin, R., http://www.fas.org/rlg/90-96.htm (1998).

[12] deVolpi, A. et al., Nuclear Shadow Boxing, Vol. 1/2, Fidler Doubleday, Michigan (2004).

[13] Mark, J., Explosive properties of reactor-grade plutonium. Sci. Glob. Security, 4, 111-128 (1993).

[14] Nikitin, K., On plutonium proliferation resistance criteria: effect of thermal bridges on nuclear explosive device cooling. In: Proceedings of the 10th Int. Conf. on "Nuclear Safety and Nuclear Education", October 1-4, 2007, Obninsk, Russia (2007).

[15] Biello, O., A need for new warheads. Sci. Am. 297(5), 54-59 (2007).

[16] Kessler, G., Steady state and transient temperature profiles in a multishell spherical system heated internally by reactor-grade plutonium, Nucl. Eng. Design, 239, 2430-2443 (2009).

[17] Shmelev, A., et al., An approach to quantitative evaluation of proliferation for fissionable materials in nuclear fuel cycle. In: Proceedings of the 10th Int. Conf. on "Nuclear Power Safety and Nuclear Education", October 1-7, 2007, Obninsk, Russia (2007).

[18] Gibbs, T., et al., Explosive property data, University of California Press, Berkeley (1981).

[19] Mader, C.L., et al., Los Alamos Explosion Performance Data, University of California Press (1982).

[20] Chen, X-N., Forschungszentrum Karlsruhe, personal information (2008).

[21] Blechman, B.M. et al., A Nuclear-Weapon-Free Zone in Europe, Sci. Am., 248, 4 (Apr. 1983).

[22] Rhodes, R., Dark Sun, The Making of the Hydrogen Bomb, Simon & Schuster, New York (1995).

[23] Zinn, J., et al., Thermal Initiation of Explosives, J. Appl. Phys., 31, 2 (1960).

[24] Dobratz, B.M., Properties of chemical Explosives and Explosive Simulations, UCRL-51319/Rev. 1, Lawrence Livermore National Laboratory (Dec. 1972).

[25] Yuge, T., Experiments on heat transfer from spheres including combined natural convection and forced convection, J. Heat Transfer, 82, 214 (1960).

[26] Zelldovich, Ya. et al., Theory of Detonation, Academic Press, New York (1960).

[27] Kämpf, H., Allgemeine Spaltgleichung für den Wärmedurchgang Brennstoff-Hülle in Kernbrennelementen mit Tablettenbrennstoff, KfK 604, Kernforschungszentrum Karlsruhe (Juni 1967).

[28] Kämpf, H. et al., Effects of Different Types of Void Volumes on the Radial Temperature Distribution of Fuel Pins, Nucl. Applications Techn. 9 (Sept. 1970).

[29] Tipton, C.R. (Editor), Reactor Handbook: Vol. 1, Materials, Interscience Publishers, New York (1960).

[30] Clark, D. L. et al., Plutonium (chapter 7) in: The Chemistry of the Actinides and Transactinide Elements, Springer Dordrecht (2006).

[31] Bathge, C. et al., The attractiveness of materials in advanced fuel cycles for various proliferation and theft scenarios, Proc. of Global 2009, Paris (2009).

[32] Kessler, G., et al., A new scientific solution for preventing the misuse of reactor-grade plutonium as nuclear explosive, Nucl. Eng. and Design, 238, 3429-3444 (2008).

[33] Kakac, S., et al., Handbook of Single-Phase Convective Heat Transfer. John Wiley & Sons (1987).

[34] Blank, H., et al., Heat transport properties of delta-stabilized plutonium, 5th Int. Conference on Plutonium and Other Actinides, Baden-Baden, Germany (1975).

[35] Gebhardt, F., et al., Reaktorwerkstoffe, Teil 1, Metallische Werkstoffe, Teubner, Stuttgart (1964).

[36] Marquardt, E.D., et al., Cryogenic material properties database, 11th Int. Cryocooler Conference, Keystone, Co (2000).

[37] Shchetinin, V.G., Calculation of specific heats of organic material behind shock and detonation waves, Chem. Phys. Reports, Vol. 18/5, Gordon and Breach Publishing House (1999).

[38] Hartwig, G., Polymer properties at room and cryogenic temperatures, Plenum Press, New York (1994).

[39] Barron, T.H.K., et al., Heat capacity and thermal expansion at low temperatures, Kluwer Academic/Plenum Publishers, New York (1999).

[40] Shmelev, A., MEPHI, Moscow, Personal communication (2007).

[41] Lashley, J.C. et al., Low temperature specific heat and critical magnetic field of α-uranium single crystals, Physical Review B, Vol. 63, 224510 (2001) [7 pages].

[42] Hecker, S.S., Why we need a comprehensive safeguards system to keep fissile materials out of the hand of terrorists. In: Doyle, J.E. (Ed.), Nuclear Safeguards, Security and Nonproliferation. Elsevier (2008).

[43] Chebeskov, A., A quantitative approach to evaluate attractiveness of enriched uranium and civilian plutonium, International Workshop on Non-Proliferation of Nuclear Materials, Obninsk, Russia (2008).

[44] Campbell, D.O., et al., Proliferation-resistant nuclear fuel cycles, ORNL/TM-6392, Oak Ridge National Laboratory (1978).

[45] Morss, L.R., et al., The Chemistry of the Actinide and Transactinide Elements, Springer, Dordrecht (2006).

[46] Timofeeva, L.F., Phase diagrams, in Ageing Studies and Lifetime Extension of Materials (Ed. L.G. Mallinson), Kluwer Academic Plenum Publishers, New York (2001).

[47] Cleveland, J.M., The chemistry of plutonium, American Nuclear Society, Chicago (1991).

[48] Shmelev, A., Some physical aspects of plutonium proliferation resistance. Personal communication at 1st Int. Science and Technology Forum on Protected Plutonium Utilization for Peace and Sustainable Prosperity. Tokyo Institute of Technology, Japan (2004).

[49] Kang, J., et al., Limited proliferation-resistance benefits from recycling unseparated transuranides and lanthanides from light water reactor spent fuel, Sci. Global Security (2005).

[50] Grizzle, S., DOE FACTS, Additional information concerning underground nuclear weapon test of reactor-grade plutonium. http://www.ccnr.org/plute_bomb.html

11 Proliferation Resistance of Americium Originating from Spent Irradiated Reactor Fuel

11.1 Introduction

The three most important americium isotopes Am-241, Am-242m and Am-243 originate from neutron irradiation during nuclear fuel burn-up in nuclear reactors, e.g. LWR fuel, FR fuel or fuel of accelerator driven systems (ADSs). The isotope Am-241 alone can originate from the beta decay of the isotope Pu-241 of stored spent fuel elements or of chemically separated plutonium. Separated americium together with separated neptunium has been of concern in proliferation and safeguards discussions, see Albright et al. [1]. The total amounts of americium in spent fuel elements and high level waste was estimated by IAEA [50] to be 160 tonnes in 2010 (Section 1).

11.2 Some nuclear physics data of the three americium isotopes Am-241, Am-242m and Am-243

The americium isotopes Am-241, Am-242m, and Am-243 arise from nuclear reactions [2,3,4] as shown in Fig. 11.1. Only the main production paths for the isotopes are shown. Their abundance in irradiated fuel depends on the cross sections and the neutron energy spectra in the different reactors, e.g. LWRs, FRs or ADSs. Due to its short half life (16 h), Am-242 is not relevant here for the further considerations.

Some nuclear properties, e.g. alpha decay or gamma decay, half lives, related energies and branching ratios are given in Table 11.1 (Cesana et al. [2]). The alpha decay is responsible for the considerable alpha particle heat rate, which is also listed in Table 11.1 for all three americium isotopes. Because of the branching ratio of only 0.459% for alpha-decay, Am-242m causes only a relatively low alpha particle heat rate (Holden et al. [3] and Kocharov et al. [4]). The high gamma activity of all three americium isotopes is often cited as a proliferation barrier (Ronen et al. [5,6]. Also, Am-241, Am-242m, and Am-243 are spontaneous fission neutron emitters [7-12]. Their half lives for spontaneous neutron emission and their spontaneous fission neutron emission rates [n/(g·s)] [3,4] are indicated in Table 11.2.

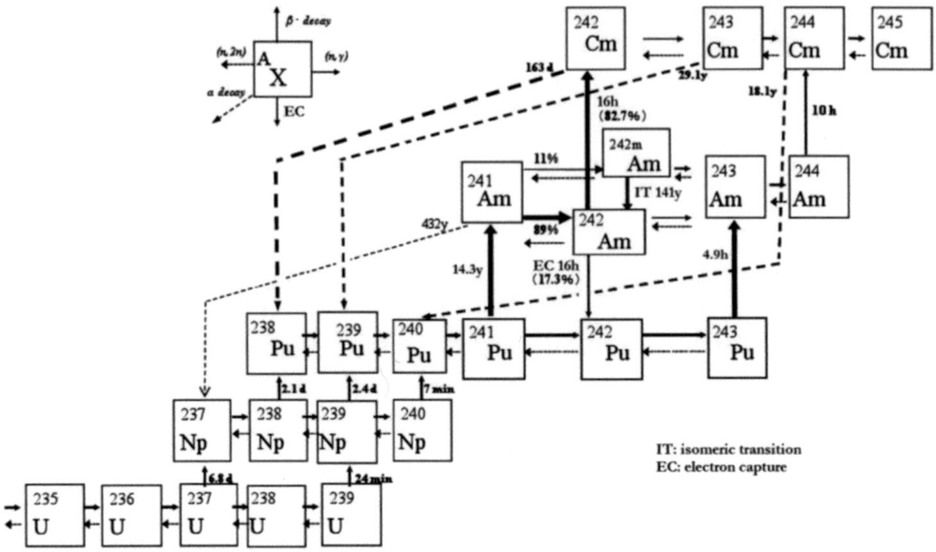

Figure 11.1: Production paths for the three americium isotopes Am-241, Am-242m, andAm-243.

Table 11.1: Alpha and gamma-decay for Am-241, Am-242m, and Am-243 [2].

Nuclide	Half-life (y)	Decay Products	Energy (keV)	Branching (%)	alpha particle heat power (W/kg)
Am-241	432	Alpha	5486	85.2	110
			5443	13	
			5388	1.4	
		Gamma	59.5	35.9	
Am-242m	141	Alpha	5207	0.459	1.5
		Gamma	984	0.128	
			1028.5	0.093	
Am-243	7370	Alpha	5275	87.4	6.4
			5233	11	
			5181	1.1	
		Gamma	74.7	68.2	

Table 11.2: Spontaneous fission neutron rates for Am-241, Am-242, Am-242m and Am-243 [7-12].

Isotope	Half-life spontaneous fission (y)	ν spontaneous fission	Spontaneous fission neutron emission (n/(g·s))
Am-241	$1.2 \cdot 10^{14}$	2.85	1.3
Am-242m	$3 \cdot 10^{12}$	2.45	44.6
Am-243	$2 \cdot 10^{14}$	2.57	0.7

The critical mass of Am-241 metal reflected by 20 cm of steel was reported to be between 33.6 and 43.6 kg depending on the codes and nuclear cross-section sets applied. With the same codes and nuclear cross-section sets a critical mass of 3.7 to 5.2 kg was calculated for metallic Am-242m. For Am-243 metal a critical mass between 111 to 193 kg was determined. In all cases 20 cm of steel was used as a reflector (Diaz et al. [13]).

11.3 Isotopic ratio of americium isotopes generated in spent fuel of different fuel cycle options of Pressurized Water Reactors and Fast Reactors

Fig. 11.2 shows the isotopic ratios of Am-241, Am-242m, and Am-243 for a variety of different fuel cycle options in the spent-fuel elements of a modern PWR for a cooling time of 10 y after discharge [14-20]. Options A and B represent low-enriched-uranium (LEU) fuel after a burnup between 33 and 50 GWd/t. The ratios of 39.7% Am-241, 0.3% Am-242m and 60% Am-243 (option A) or 23.2% Am-241, 0.3% Am-242m and 76.5% Am-243 (option B) would exist at discharge of the spent fuel after a burnup of 33 or 50 GWd/t. During a subsequent cooling time of 10 y, Pu-241 (half life for beta decay of 14.4 y) would decay into Am-241 and change the isotopic ratio to 85.6% Am-241, 0.08% Am-242m and 14.32% Am-243 (option A) or 81% Am-241, 0.1% Am-242m and 18.9% Am-243 (option B). These isotopic ratios of Am-241, Am-242m, and Am-243 after 10 y cooling time would be representative for americium which could be misused for a nuclear explosive device (NED).

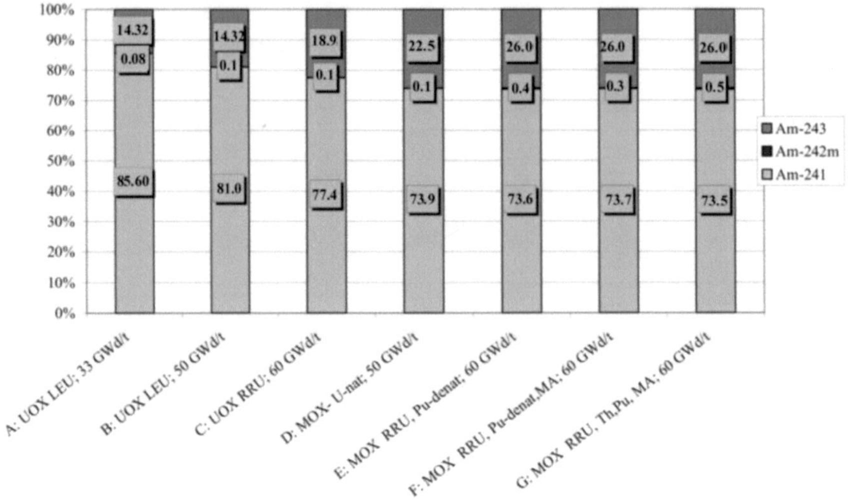

A Low enriched uranium oxide fuel (LEU-UOX) with 33 GWd/t burn-up and 10 years cooling time
B Low enriched uranium oxide fuel (LEU-UOX) with 50 GWd/t burn-up and 10 years cooling time
C Reenriched reprocessed uranium (RRU) oxide fuel with 60 GWd/t and 10 years cooling time
D Natural uranium plutonium oxide fuel (MOX) with 50 GWd/t and 10 years cooling time
E Reenriched reprocessed uranium mixed with denatured plutonium RRU-MOX with 60 GWd/t and 10 years cooling time
F Reenriched reprocessed uranium mixed with plutonium and minor actinides (RRU-Pu-MA-MOX) with 60 GWd/t and 10 years cooling time
G Thorium and reenriched reprocessed uranium (RRU) mixed with plutonium and minor actinides (RRU-Th-Pu-MA-MOX) with 60 GWd/t and 10 years cooling time

Figure 11.2: Isotopic compositions of americium separated from spent fuel (10 years after discharge) for different fuel cycle options of PWRs.

In the options C, D, E either reenriched reprocessed uranium (RRU) (option C) or natural uranium or RRU both together with plutonium (option D and E) are used. The isotopic ratios of Am-241, Am-242m, and Am-243 vary from 21.2% Am-241, 0.3% Am-242m and 78.5% Am-243 up to 40.7% Am-241, 0.3% Am-242m and 59% Am-243 at discharge after a burnup of 60 GWd/t. After a cooling time of 10 y, these ratios are modified to range from 73.6% Am-241, 0.4% Am-242m and 26% Am243 to 77.4% Am-241, 0.1% Am-242m and 22.5% Am-243 (only option E has 0.4% Am-242m).

In options F and G, RRU or thorium are mixed with plutonium and MAs in order to incinerate both plutonium and MAs in PWRs. In these cases the isotopic ratio of Am-241, Am-242m, and Am-243 changes from 46% Am-241, 0.7% Am-242m and 53.3% Am-243 up to 51% Am-241, 1.1% Am-242m and 47.9% Am-243 at discharge after a burnup of 60 GWd/t.

After a cooling time of 10 y this ratio is modified to range from 73.7% Am-241, 0.3% Am-242m and 26% Am-243 up to 73.5% Am-241, 0.5% Am-242m and 26% Am-243.

In sodium cooled FRs (Fig. 11.3) the ratio of the americium isotopes Am-241, Am-242m, and Am-243 is 51.5% Am-241, 3.6% Am-242m and 44.9% Am-243 with metallic fuel after a burnup of 140 GWd/t (option H) and a cooling time of 2 y. In the spent fuel of mixed oxide (MOX) fuel FRs (option I), americium can be found with 55% Am-241, 1.5% Am-242m and 43.5% Am-243 for a cooling time of 7 y after discharge of the fuel having a burnup of 185 GWd/t. Plutonium and MA incinerating ADSs with metallic fuel (option J) contain americium with 45% Am-241, 3.5% Am-242m and 51.5% Am-243 in the spent fuel after a burnup of 250 GWd/t and 2 y cooling time. ADSs which incinerate both plutonium and MAs in nitride fuel (case H) will have americium with 49% Am-241, 4.5% Am-242m and 46.5% Am-243 after a cooling time of 2 y after discharge of the spent fuel having a burn-up of 150 GWd/t (Hill [21], Messaoudi [22]).

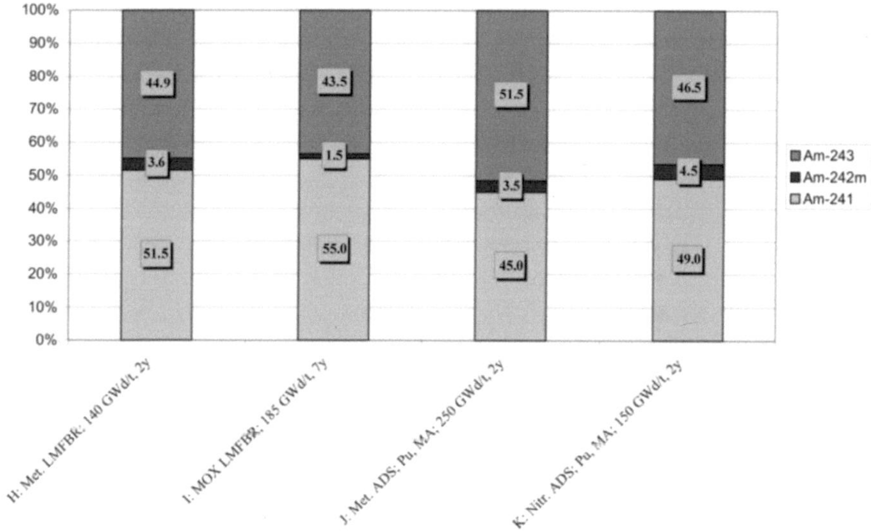

H Liquid Metal Cooled Fast Breeders Reactor (LMFBR) with metallic fuel after 140 GWd/t burn-up and 2 years cooling time

I Liquid Metal cooled Fast Reactor (FR) with MOX fuel after 185 GWd/t burnup and 7 years cooling time

J ADS (accelerator driven system) with metallic fuel with Pu and MAs having a burn-up of 250 GWd/t and 2 years cooling time

K ADS (accelerator driven system) with nitride fuel containing Pu and MAs having a burn-up of 150 GWd/t and 2 years cooling time.

Figure 11.3: Isotopic compositions of americium separated from spent fuel (2 years or 7 years after discharge) for different fuel cycle options of FRs.

11.3.1 Am-241 from the decay of Pu-241

Chemically separated reactor plutonium, which has been intermediately stored for several years, will contain Am-241 from the beta decay of Pu-241 (Fig. 11.1). This pure Am-241 isotope (in the later analyses called option L) can be chemically separated from the plutonium. This is for example presently being done before MOX refabrication in order to avoid the high gamma radiation of 59.5 keV (Table 11.1) during hands-on refabrication of the presently applied glove-box technique. As shown in Tables 11.1. and 11.2., this pure Am-241 has a relatively low spontaneous fission rate but a relatively high alpha-particle heat production.

11.3.2 Am-242m production

As reported above, the Am-242m generation is <1% for PWR fuel cycles but several percent in FR and ADS fuel cycles. It has a small spherical critical mass of 3.7 to 5.2 kg metal (Section 11.2) if reflected by steel and, therefore, has become of importance for special purpose applications, e.g. space applications [2,5,6,23,24]. In the latter case, it would have to be especially produced by applying special neutron filters in high flux thermal reactors (Cesana et al. [2]) or in FR blankets (Ronen [6]). Further isotopic enrichment from the accompanying isotopes Am-241 and Am-243 would become necessary to achieve Am-242m enrichments up to 70% but would be particularly difficult and costly (Cesana et al. [2], Ronen et al. [6]).

For the following analysis a case with 92.37% Am-241, 7.13% Am-242m and 0.5% Am-243 from (Ronen [6]) is added (option M) for the further analysis. This isotopic composition corresponds to 18 y of Am-241 irradiation in outer core and blanket fuel elements of Fast spectrum Reactors (FRs).

11.4 Considerations on pre-ignition, alpha-particle heat power and critical mass of americium

The motivation and incentive of the following analysis is to find out whether or not metallic reactor americium with the composition of isotopes as they originate in spent reactor fuel e.g. from PWRs, FRs, or ADSs would be suitable for a nuclear explosive device (NED). Reactor americium with the composition of isotopes as they appear in spent reactor fuel of, e.g. PWRs, FRs or ADSs would have a high spontaneous fission neutron source rate such that early pre-ignition could occur during the compaction of the fissile material. It will be shown in Section 11.7 that the use of the gun system method [25-29] would lead to extremely low explosion yields (fizzle yields) that are of no interest for HNEDs. The pre-ignition results would be much better if the implosion method

[25-29] would be used. But, because of the compact geometrical (spherical) arrangement that must be used for the implosion method, the subcriticality multiplication of the spontaneous fission neutron source must be accounted for. This will also lead to relatively early pre-ignition during the compaction process.

In addition, it will be shown, that the high alpha-particle heat power of Am-241 (Table 11.1), together with the relatively high critical mass of metallic reactor americium originating from reactor spent fuel of PWRs, FRs, or ADSs, will lead to very high temperatures in the high explosive lenses necessarily surrounding the fissile americium metal in an implosion-type device. As a consequence, the melting point or temperature for initiation of self-explosion of the high explosive material will be exceeded.

11.5 Critical mass of reactor americium metal based Hypothetical Nuclear Explosive Devices

The gun-type system (Fig. 11.4) and the spherical implosion type system (Figs. 9.3 and 11.5) are considered for the subsequent calculations of the critical mass for an HNED based on reactor americium. From the critical mass data, the spontaneous fission neutron source and the alpha-particle heat power can be determined.

For the gun-type system and for the implosion type system, the following assumptions are taken for the further analysis:

11.5.1 Gun-type HNED with metallic americium

For the gun-type system, a cylindrical arrangement of reactor-americium metal is assumed. A cylinder with a diameter of 16 cm is selected surrounded by a cylindrical and axial reflector of metallic U_{nat} with 5-cm thickness leading to an inner barrel diameter of 26 cm. This is in accordance with data given in [25,26,27] where gun barrel diameters of ~16 cm up to 28 cm are reported for U-235 highly enriched uranium. The critical masses or critical lengths of this reflected fissile material arrangement will be determined for $k_{eff} = 1$ in Section 11.6.1. However, this critical length L_{crit} for $k_{eff} = 1$ determined in Section 11.6.1 does not represent the length of the critical assembly that will be responsible for determining the time period t_0 between reaching prompt criticality and the final maximum $k_{eff,max}$ or for the determination of the total spontaneous fission neutron source (Section 11.7). The critical mass or critical length for $k_{eff} = 1$ underestimates the mass needed for gun-type device by a factor of ~2. The gun-type systems described in [25-27] have typically about two or somewhat more critical masses or a $k_{eff,max}$ »1.

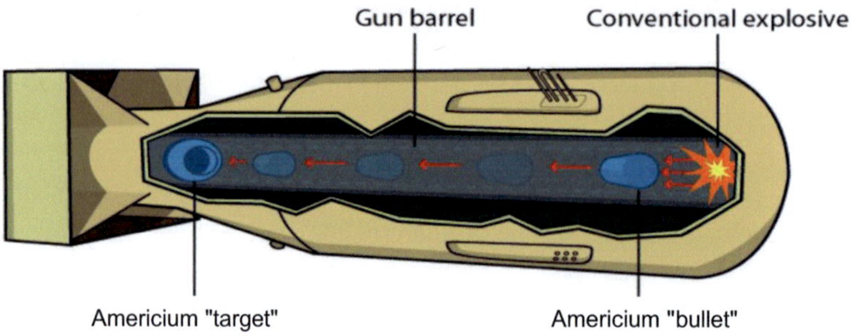

Figure 11.4: Gun-type HNED based on reactor-americium [47,48] (adapted).

Cooling of the cylindrical fissile parts with their high alpha-particle heat production rate of reactor-americium is not further discussed because the gun-type system will lead to extremely low explosion yields. This will be of no further interest for HNEDs (Section 11.7)

11.5.2 Spherical implosion type HNED with metallic americium

For the spherical implosion type system, the geometric arrangement is shown in Figs. 9.3 and 11.5. A reactor-americium metal sphere with a U_{nat} reflector/tamper is surrounded by high explosive lenses and an outer casing. Such an HNED would have to be subcritical prior to the start of the implosion process. Similarly as in Kessler [31], a $k_{eff} \approx 0.98$ is selected for the neutronic calculations.

Internal cooling of the solid reactor-americium sphere with its considerable alpha-particle heat power (Section 11.8.1) is not considered. Either cooling channels or heat-conducting metallic sheets with proper heat insulation would be required (see Section 10.14). The concerns are that such cooling measures would lead to reflections and perturbations of the shock waves on their way through the high explosives and would cause deterioration of symmetry and increase of hydrodynamic instabilities. The shock waves within the HE lenses would be deformed [22,23,29] and the resulting nuclear explosive yield would be strongly reduced. In addition, it will be shown in Sections 11.10 and 11.11 that the temperatures in americium based implosion type HNEDs would be so high that cooling by heat conducting sheets would become impossible.

The high explosive lenses must have a specific geometry with at least two explosives with different detonation velocities. They must be very accurately machined to produce a precisely spherical shock wave [25,26,32,33]. The so-called one-dimensional conservative approach described in Section 10.4 will be applied later in Sections 11.8 and 11.8.3 for the thermal analysis of the HNEDs.

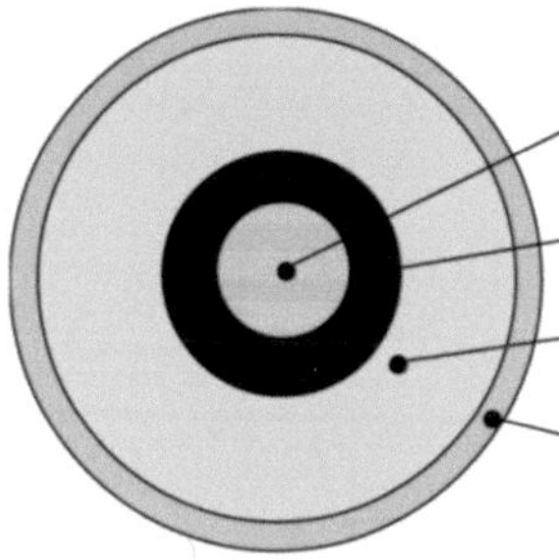

Figure 11.5: Geometric arrangement and dimensions for scoping studies of critical masses for reactor americium.

11.6 Critical masses for gun-type HNEDs and spherical implosion type HNEDS

11.6.1 Critical masses for gun-type systems

The TWODANT neutron transport code with S_{16} order and P_3 scattering matrices, 100 neutron energy groups, and ENDF/B-VII cross sections are applied to calculate the critical length of the cylindrical metallic americium arrangement reflected by 5-cm thick metallic U_{nat}, which was described above. Table 11.3 shows the results for all different reactor-americium options A through K for PWRs and FRs or ADSs fuel cycle options which were described in Section 11.3. In addition, option L for 100% Am-241 and one case for dedicated Am-242m production (option M) (Section 11.3.2) are listed.

11.6.2 Critical masses for spherical implosion type systems

The ONEDANT neutron transport code with S_{16} order and P_3 scattering matrices, 100 neutron energy groups, and ENDF/B-VII cross sections was applied to calculate the critical radius of the solid metallic americium sphere reflected by 5-cm thick metallic U_{nat}. This calculation route had been validated before by benchmark calculations for one-dimensional models of GODIVA and JEZEBEL where agreement to reference values was observed within ~0.1% Δk. Although even better agreement might be achieved by higher S_N- or P_N-order, this scheme was considered to be sufficiently accurate for the current purpose. Of course, one has to be aware that the uncertainty that has to be attributed to the nuclear data of the americium isotopes is considerably larger than that of the nuclear data of the fissile isotopes U-235 and Pu-239, relevant for GODIVA and JEZEBEL.

The data listed in Table 11.4 are valid for $k_{eff} \approx 0.98$ as described above. Table 11.4 shows the results for all different reactor-americium options A through K from PWR and FR or ADS spent fuel described in Section 11.3. In addition the option L for 100% Am-241 and option M for dedicated Am-242m production, described in Section 11.3.2 are listed.

Table 11.3: Gun-type system: critical length L_{crit} (cm) for $k_{eff} = 1$ of cylinder of 16-cm diameter of reactor-americium from different fuel cycle options of PWRs, FRs and ADSs.

PWR fuel cycle options	Isotopic composition			critical length (cm) L_{crit} for d = 16 (cm) radial refl.: 5 (cm) U-nat $k_{eff} = 1.0$	critical mass of cylinder of reactor-americium (kg)	mass of U_{nat} reflector (kg)
	Am-241 %	Am-242m %	Am-243 %			
A: UOX LEU; 33 GWd/t	85,6	0,08	14,32	21,59	59,3	234,9
B: UOX LEU; 50 GWd/t	81,0	0,1	18,9	22,48	61,7	240,5
C: UOX RRU; 60 GWd/t	77,4	0,1	22,5	23,27	63,9	245,4
D: MOX-U-nat; 50 GWd/t	73,9	0,1	23,0	24,09	66,2	250,6
E: MOX RRU, Pu-denat; 60 GWd/t	73,6	0,4	26,0	23,46	64,4	246,6
F:MOX RRU, Pu-denat,MA; 60 GWd/t	73,7	0,3	26,0	23,67	65,0	247,9
G: MOX RRU, Th,Pu,MA; 60 GWd/t	73,5	0,5	26,0	23,26	63,9	145,3
H: Met.LMFBR; 140 GWd/t, 2y	51,5	3,6	44,9	21,41	58,8	133,8
I: MOX LMFBR; 185 GWd/t, 7y	55,0	1,5	43,5	25,33	69,6	257,6
J: Met.ADS;Pu,MA; 250 GWd/t, 2y	45,0	3,5	51,5	22,84	62,7	242,8
K: Nitr. ADS; Pu,MA: 150 GWd/t, 2y	49,0	4,5	46,5	20,28	55,7	226,8
L: 100% Am-241	100	–	–	19,27	52,9	220,5
M: MOX LMFBR; Am-242m breeding	92,37	7,13	0,5	13,40	36,8	183,9

Table 11.4: Spherical implosion method: critical radius for $k_{eff} \approx 0.98$ and critical mass for sphere of reactor-americium (reflected by 5 cm U_{nat}) and different fuel cycle options or PWRs, FRs and ADSs.

PWR fuel cycle options	Isotopic composition			critical radius (cm) $k_{eff} \approx 0.98$	critical mass of cylinder of reactor-americium (kg)	mass of U_{nat} reflector (kg)
	Am-241 %	Am-242m %	Am-243 %			
A: UOX LEU; 33 GWd/t	85,6	0,08	14,32	9,45	48,4	172,2
B: UOX LEU; 50 GWd/t	81,0	0,1	18,9	9,54	49,8	174,8
C: UOX RRU; 60 GWd/t	77,4	0,1	22,5	9,62	51,0	177,0
D: MOX-U-nat; 50 GWd/t	73,9	0,1	23,0	9,69	52,2	179,2
E: MOX RRU, Pu-denat; 60 GWd/t	73,6	0,4	26,0	9,64	52,0	177,6
F:MOX RRU, Pu-denat,MA; 60 GWd/t	73,7	0,3	26,0	9,66	51,6	179,3
G: MOX RRU, Th,Pu,MA; 60 GWd/t	73,5	0,5	26,0	9,62	51,0	177,0
H: Met.LMFBR; 140 GWd/t, 2y	51,5	3,6	44,9	9,46	48,5	171,4
I: MOX LMFBR; 185 GWd/t, 7y	55,0	1,5	43,5	9,81	54,2	187,8
J: Met.ADS;Pu,MA; 250 GWd/t, 2y	45,0	3,5	51,5	9,61	58,3	176,6
K: Nitr. ADS; Pu,MA: 150 GWd/t, 2y	49,0	4,5	46,5	9,34	46,6	168,9
L: 100% Am-241	100	–	–	9,18	44,4	164,6
M: MOX LMFBR; Am-242m breeding	92,37	7,13	0,5	8,26	32,3	140,0

11.7 Pre-ignition for reactor-americium based gun-type and spherical implosion type HNEDs

A discussion of the proliferation resistance of reactor-americium requires analyzing the importance of pre-ignition – caused by the high number of spontaneous fission neutrons – on the potential nuclear explosive yield of the HNED. In addition, also a thermal analysis is needed because of the high alpha-particle heat power of the reactor-americium. In this section the pre-ignition problem of metallic reactor-americium will be investigated for both the gun and the spherical implosion system.

The theory of pre-ignition based on ramp-type reactivity increases during compaction of the fissile material arrangement [34-37] was applied by Kessler [30]. The main calculational parameters needed for the pre-ignition problem are the total spontaneous fission neutron source S (n/s) of the HNED, the neutron life time l_{eff} (s), the ramp rate and the associated time period t_0 (s) from reaching prompt criticality up to the end of compaction of the fissile material.

11.7.1 Pre-ignition of metallic americium based gun-type systems

Table 11.5 shows that the total spontaneous fission neutron source for gun systems and for $k_{eff} = 1$ ranges from $0.72 \cdot 10^5$ n/s for 100% Am-241 (option L) to ~$1.68 \cdot 10^5$ n/s for case K (Table 11.4). For spherical implosion systems and $k_{eff} \approx 0.98$, they range from ~$0.6 \cdot 10^5$ n/s for 100% Am-241 (option L) to $1.48 \cdot 10^5$ n/s for option J.

For the case with Am-242m breeding in a liquid-metal fast breeder reactor (Ronen [6]) (LMFBR) (option M), the spontaneous fission neutron source is $1.64 \cdot 10^5$ n/s for gun-type systems and for $k_{eff} = 1$ and $1.43 \cdot 10^5$ n/s for spherical implosion systems.

Gun-type systems – as reported in [25-27] – had a fissile material mass of about two or more critical masses. Therefore, values of a factor of 2 higher for the spontaneous neutron source should be accounted for gun-type HNEDs in the discussion following. Table 11.5 shows values for the spontaneous fission source S that represent values based on the critical mass ($k_{eff} = 1$) multiplied by the factors 1.35 (conservative) and 2.

In Kessler [30] a parametric approach with $S = 10^5$ n/s and $S = 1.5 \cdot 10^5$ n/s was chosen. For the neutron life time, $l_{eff} = 10^{-8}$ s, was selected as it was used also by Mark [37] for reactor plutonium metal. For the time period, t_0, two parametric values are investigated: $t = 5 \cdot 10^{-4}$ s and $t_0 = 10^{-3}$ s. They are based on Mark [37], on data from [25-29], and critical length data of Table 11.3.

Table 11.5: Spontaneous fission neutron source (n/s) (fissile core and reflector) for different reactor americium mixtures as well as for gun-type and spherical implosion type HNEDs.

gun-type system $k_{eff} = 1$ (mass)	gun-type HNED* factor 1.35 to 2.0	Isotopic composition of reactor americium (option)	spherical implosion type HNEDs $k_{eff} \approx 0.98$
$0.765 \cdot 10^5$	10^5 to $1.53 \cdot 10^5$	A	$0.622 \cdot 10^5$
$0.788 \cdot 10^5$	$1.06 \cdot 10^5$ to $1.57 \cdot 10^5$	B	$0.624 \cdot 10^5$
$0.801 \cdot 10^5$	$1.08 \cdot 10^5$ to $1.60 \cdot 10^5$	C	$0.637 \cdot 10^5$
$0.816 \cdot 10^5$	$1.10 \cdot 10^5$ to $1.63 \cdot 10^5$	D	$0.641 \cdot 10^5$
$0.876 \cdot 10^5$	$1.18 \cdot 10^5$ to $1.75 \cdot 10^5$	E	$0.706 \cdot 10^5$
$0.858 \cdot 10^5$	$1.16 \cdot 10^5$ to $1.72 \cdot 10^5$	F	$0.678 \cdot 10^5$
$0.898 \cdot 10^5$	$1.21 \cdot 10^5$ to $1.8 \cdot 10^5$	G	$0.715 \cdot 10^5$
$1.522 \cdot 10^5$	$2.02 \cdot 10^5$ to $3 \cdot 10^5$	H	$1.276 \cdot 10^5$
$1.20 \cdot 10^5$	$1.62 \cdot 10^5$ to $2.4 \cdot 10^5$	I	$0.937 \cdot 10^5$
$1.60 \cdot 10^5$	$2.16 \cdot 10^5$ to $3.2 \cdot 10^5$	J	$1.482 \cdot 10^5$
$1.68 \cdot 10^5$	$2.27 \cdot 10^5$ to $3.36 \cdot 10^5$	K	$1.404 \cdot 10^5$
$0.720 \cdot 10^5$	$0.98 \cdot 10^5$ to $1.43 \cdot 10^5$	L	$0.60 \cdot 10^5$
$1.64 \cdot 10^5$	$2.21 \cdot 10^5$ to $3.28 \cdot 10^5$	M	$1.43 \cdot 10^5$

*gun-type HNED with a total of 1.35 to 2 times the mass calculated for $k_{eff} = 1$.

11.7.2 Results of pre-ignition analysis for gun-type systems

For these chosen parametric values of the gun HNEDs the cumulative probabilities for pre-ignition were calculated by Kessler [30]. Pre-ignition occurs stochastically already early during the compaction phase between $t_I/t_0 = 0$ and t_I/t_0 at ~0.3 for a spontaneous neutron source of 10^5 n/s or $1.5 \cdot 10^5$ n/s and a total compaction time of $t_0 = 10^{-3}$ s (t_I = time when pre-ignition occurs). For a compaction time of $t_0 = 5 \cdot 10^{-4}$ s the values are only slightly larger.

The cumulative probabilities for pre-ignition as a function of the ratio x of the attainable explosive yield, Y, relative to the maximum nominal yield, Y_0, were also determined by Kessler [30].

The minimum relative explosive fizzle yield would be

$x_{F,min} = (Y/Y_0)_{F,min} = 2.7 \cdot 10^{-5}$ for the time period of compaction of $t_0 = 10^{-3}$ (s)

and

$x_{F,min} = (Y/Y_0)_{F,min} = 7.6 \cdot 10^{-5}$ for the time period of compaction of $t_0 = 5 \cdot 10^{-4}$ (s)

(for Y_0 typically a value of 20 kt TNT can be considered [36,37]).

It is also understood that each value $x = Y/Y_0$ belongs to a certain cumulative probability for pre-ignition (integral of the differential probability) [30].

It can be concluded that for the gun-type systems the attainable nuclear explosive yield, Y, would be extremely low. On the basis of $Y_0 = 20$ kt TNT, the minimum fizzle yields would be 0.54 t TNT for $t_0 = 10^{-3}$ s and 1.5 t TNT for $t_0 = 5 \cdot 10^{-4}$ s **and therefore be of no interest.**

This conclusion is also valid for an HNED with 100% Am-241 (option L in Table 11.3), having a spontaneous fission neutron source of $\sim 10^5$ to $1.4 \cdot 10^5$ n/s.

The same conclusion also holds for the isotopic composition of option M (Am-242m generated by breeding in outer core and blanket assemblies of an LMFBR [6]) as the spontaneous fission neutron source would be between $2.2 \cdot 10^5$ n/s and $3.2 \cdot 10^5$ n/s.

11.7.3 Results of pre-ignition analysis for spherical implosion HNEDs

For the subcritical spherical fissile assembly (Fig. 11.5) with $k_{eff} \approx 0.98$ the subcriticality multiplication (Keepin [10], deVolpi [26], Weinberg [38]

$$M \approx \frac{1}{1-k_{eff}}$$

must be accounted for which is $M \approx 50$ for $k_{eff} \approx 0.98$. This was also confirmed, e.g., for option A by inhomogeneous ONEDANT calculations with the corresponding internal spontaneous fission source [30].

This means that all values for the spontaneous fission neutron source S of Table 11.5 for the spherical implosion system will have to be multiplied by 50. This leads to spontaneous fission neutron sources that can be represented by the parameters $3 \cdot 10^6$ n/s and $7 \cdot 10^6$ n/s.

For the spherical implosion system much shorter time periods of $t_0 = 2 \cdot 10^{-5}$ s down to $t_0 = 10^{-5}$ s can be achieved [25,26,37,46].

The relative minimum fizzle yields, $x_{F,min}$ are now

$$x_{F,min} = (Y/Y_0)_{F,min} = 9.5 \cdot 10^{-3} \text{ for the time period for compaction of } t_0 = 2 \cdot 10^{-5} \text{ s}$$

and

$$x_{F,min} = \left(\frac{Y}{Y_0}\right)_{F,min} = 0.027 \text{ for the time period of compaction of } t_0 = 10^{-5} \text{ s}$$

These are by a factor of $\sim 10^2$ and 10^3 higher than for the gun-type HNED and would lead to 0.54 kt TNT for $t_0 = 10^{-5}$ s or 0.19 kt TNT for $t_0 = 2 \cdot 10^{-5}$ s. These results for the attainable minimum fizzle yield are the same for $t_0 = 10^{-5}$ s as given for reactor-grade plutonium by Mark [37]. There is, however, also the relatively high alpha-heat rate of the isotope Am-241 (Table 10.1) to be considered. This needs, in addition, a thermal analysis of the implosion type HNED.

11.8 Geometric dimensions, alpha particle heat power and material characteristics for the thermal analysis of spherical americium based implosion type HNEDs

11.8.1 Geometric dimensions of a reactor-americium based spherical implosion type HNED for the thermal analyses

Based on the analysis of Kessler [31] for reactor-grade plutonium the following geometric dimensions, material for the reflector/tamper [47], and material characteristics of the high explosives are assumed. The geometric dimensions of the metallic americium sphere with 5-cm thick metallic U_{nat} reflector/tamper are given in Table 10.6 and Fig. 11.4 for fuel cycle options A through M. The thickness of the high explosives spherical lense system is assumed to be 10 cm. The outer casing shall be of 1 cm thick aluminum (Fig. 11.4). The material characteristics of the high explosives will be defined in Section 11.8.2. All assumptions for HNEDs are merely intended to represent a range for which the thermal analysis can be performed [45].

Table 11.6: Geometric dimensions and alpha-particle heat power of HNEDs for reactor-americium from fuel cycle options A through M.

Case	Radius of americium sphere (cm)	Outer radius of reflector/tamper (cm)	outer radius high explosives (cm)	Outer radius aluminum casing (cm)	Alpha particle heat power (W)
A	9,45	14.45	24.45	25.45	4575
B	9,54	14.54	24.54	25.54	4497
C	9.62	14.62	24.62	25.62	4415
D	9.69	14.69	24.69	25.69	4330
E	9.64	14.64	24.64	25.64	4296
F	9.66	14.66	24.66	25.66	4269
G	9.62	14.62	24.62	25.62	4208
H	9.46	14.46	24.46	25.46	2889
I	9.81	14.81	24.81	25.81	3431
J	9.61	14.61	24.61	25.61	3027
K	9.34	14.34	24.34	25.34	2443
L	9.18	14.18	24.18	25.18	4884
M	8.26	13.26	23.26	24.26	3286

11.8.2 Material properties for high explosives

The high explosives represent the limiting thermal material characteristics for HNEDs with reactor-grade americium in their central region. High explosives have the lowest thermal conductivities of all material layers in the considered HNED designs. In addition, they melt at relatively low temperatures or undergo transition to pyrolysis, to self-ignition or self-explosion if certain limiting temperatures are exceeded. The real design of high explosives lenses is classified. They have a sophisticated geometry and consist of at least two different high explosives with different detonation velocities (Rhodes [25,26]). In Table 11.7 a selection of materials data for high explosives is shown. The very high-technology high explosives have the highest thermal conductivity values, the highest melting points, and the highest temperatures for start of pyrolysis or start of self-ignition and self-explosion that can be found in the literature [39-42]. Just for demonstration, data for two very high technology and one medium technology high explosive material are given as defined in Kessler [30].

As the complicated three-dimensional geometric structure for the high explosive lenses is unknown, the so-called one-dimensional conservative approach – as explained in detail in Section 10.4 – is applied. This leads to limiting temperature curves. To generate final results of the thermal analysis, always the lower of these limiting temperature curves with the highest thermal conductivity for the high explosives will be considered. This is always on the conservative side.

Table 11.7: Material Data of High Technology High Explosives [39-42].

Level of technology	High Explosive	Density (g/cm^3)	Thermal conductivity (W/cm K)	Melting point (°C)	Start pyrolysis	self explosion T_e
medium technology	PBX 9501	1.86	$0.452 \cdot 10^{-2}$	190	275 °C	235 °C
very high technology	PBX 9502	1.89	$0.561 \cdot 10^{-2}$	448	395 °C	331 °C
	TATB	1,89	$0.544 \cdot 10^{-2}$	448	395 °C	347 °C

11.9 Outside temperature of the reactor americium based HNED

The alpha-particle heat produced in the solid americium metallic sphere must be transferred through the different spherical material layers by thermal conduction to the outside casing, where it is transferred to the ambient atmosphere by natural convection and radiation.

The highest radiation heat transfer and therefore the lowest surface temperature can be found for black body radiation. Black body radiation is assumed although, e.g., for rolled aluminum with an emissivity of 0.07 [43], the temperature difference between the outer casing and the environment would be about a factor of ~1.94 higher. For polished aluminum with this factor would be 2.24.

11.9.1 Temperatures of a metallic reactor americium bare sphere and gamma radiation problems

The temperatures in a bare metallic americium sphere are of interest for the manufacturing process. For different outside radii of the bare metallic reactor americium sphere (options A through M) and black body radiation the following outside surface temperatures are calculated as shown in Table 11.8. Also the central temperatures are given on the basis of a thermal conductivity of 0.1 (W/cmK) for metallic reactor americium [44].

These temperatures are very high and would make manufacturing and assembling of a reactor americium based bare sphere extremely difficult.

In addition to the high temperatures, there is the high gamma radiation to be considered. Table 11.1 lists the gamma-ray energies of the three americium isotopes. Together with the relatively high critical masses of such HNEDs (see Table 11.4) of ~32 to 58 kg reactor americium, the associated gamma activities with

59.5 keV	for Am-241
984 keV 1028.5 keV	for Am-242m
74.7 keV	for Am-243

would represent extreme impediments for manufacturing and handling of such devices. This high gamma activity is therefore often cited as a proliferation barrier [28,30].

Table 11.8: Outside temperature and central temperature of a bare sphere of reactor americium for fuel cycle options A through M.

Case	Radius of americium bare sphere (cm)	Outside temperature of americium bare sphere, T_a (°C)	Central temperature of americium bare sphere, T_0 (°C)
A	9.45	645	838
B	9.54	636	824
C	9.62	628	811
D	9.69	628	806
E	9.64	621	798
F	9.66	619	795
G	9.22	619	793
H	9.46	544	666
I	9.81	565	704
J	9.61	548	673
K	9.34	515	619
L	9.18	677	889
M	8.26	629	787

11.9.2 Outside casing temperature of americium based HNEDs

For an HNED consisting of the metallic reactor-americium sphere surrounded by a 5 cm thick reflector/tamper of metallic U_{nat} a 10 cm thick HE and a 1 cm thick aluminum casing (Fig. 11.4, Table 11.6) the outside surface temperature of the casing would be lower than for the bare sphere (Table 11.9) because the heat power is rejected from a larger surface. These outside casing temperatures are again given for black body radiation.

11.9.3 Inside temperature profile in the americium based HNEDs

Having determined the outer temperature T_a of the aluminum casing (Table 11.9), one can now evaluate the temperature profile in the hollow spherical shells without internal heat sources (aluminum casing, high explosives, U_{nat} reflector/tamper).

Table 11.9: Outside casing temperature of reactor americium based HNEDs (fuel cycle options A through M).

Case	Outer radius of aluminum casing (cm)	Outside casing temperature (°C)
A	25.45	288
B	25.54	285
C	25.62	281
D	25.69	279
E	25.64	278
F	25.66	277
G	25.62	274
H	25.46	232
I	25.81	249
J	25.61	236
K	25.34	214
L	25.18	299
M	24.26	258

11.9.4 Radial temperature profile for a reactor-americium sphere HNED (option G, PWR)

Fig. 11.6 shows the radial temperature profile within the HNED described as option G. Option G has the lowest alpha-heat power of PWR fuel cycle options A through G (Table 11.6). As the casing outer radius does not vary much (Table 11.6), it would therefore have the lowest temperature profile. The temperature of the outer aluminum casing at 25.62 cm would be 274 °C. (For the real emissivity of rolled aluminum, it would be considerably higher).

Because of the relatively low thermal conductivity of $\lambda = 0.561 \cdot 10^{-2}$ W/cm °C for TATB or PBX 9502, the radial temperature would rise in these high explosives from 274.6 °C at their outer radial boundary to 1933 °C at their inner radial boundary at 14.62 cm. For comparison also the radial temperature profile for another high explosive (PBX 9501 with a lower thermal conductivity of $\lambda = 0.452 \cdot 10^{-2}$ W/cm °C) is shown (see Table 11.7) as dotted line in Fig. 11.6, for which the temperature in the high explosives would rise even up to 2332 °C. The melting temperature of 448 °C and the temperature for initiation of pyrolysis (395 °C) or self-explosion (331 °C or 347 °C) of the high explosives would be exceeded in >90% of the volume of the high explosives, which would melt or self-explode. The HNED described as option G could not function technically. The inside temperature of the 5-cm thick spherical shell of U_{nat} metal would be 1968 °C or 2367 °C. For all other PWR fuel cycle options A through F, the temperatures would be somewhat higher (Table 11.10 in Section 11.10). The U_{nat} metal would also be molten since its melting temperature of 1132 °C is exceeded. Finally, the temperatures in the metallic reactor-americium sphere would also be so high that it would be fully molten (melting temperature 1176 °C). The HNED defined as option G would not be feasible technically.

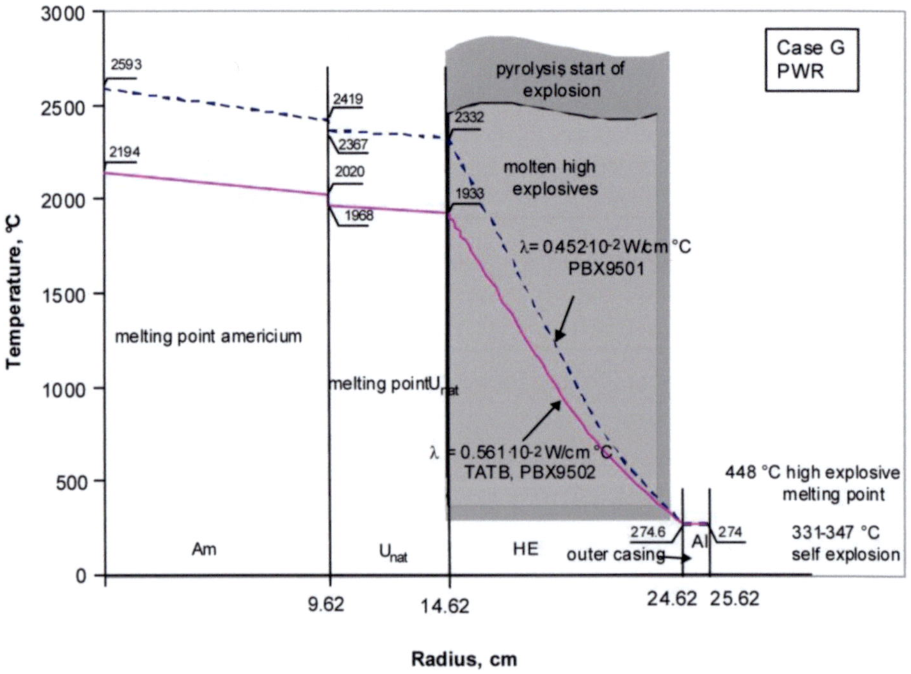

Figure 11.6: Radial temperature profile for a reactor-americium based HNED (option G PWR).

11.9.5 Radial temperature profile for a reactor-americium HNED (Option H LMFBR)

Fig.11.7 shows the radial temperature profile within the HNED described as option H. Option H represents the metallic LMFBR fuel cycle (Fig. 11.3) and has the lowest alpha-particle heat rate of the two considered FR fuel cycle option options H and I. As the outer-casing radius of these two options is almost equal (Table 11.7), it would have the lowest temperatures among the HNEDs (see Table 11.12). Assuming black body radiation, the temperature of the outer aluminum casing at 25.46 cm would be 232 °C. (For an emissivity of 0.07 for rolled aluminum it would be considerably higher). Because of the relatively low thermal conductivity of λ = 0.561·10-2 W/cm °C for the high explosive PBX 9502, the radial temperature would rise in these high explosives from 232.5 °C at their outer radial boundary to 1391 °C at their inner radial boundary at 14.46 cm. For comparison also the radial temperature profile for another high explosive (PBX 9501 with a lower thermal conductivity of λ = 0.452·10-2 W/cm °C) is shown as dotted line, for which the temperature in the high explosives would rise to 1671 °C. The melting temperature of 448 °C and the temperature for initiation of pyrolysis (395 °C) or self-explosion (331 °C or 347 °C) of the high explosives would be exceeded in >80% of the volume of the high explosives, which would melt or self-explode. The HNED described as option H could not function technically. The inside temperature of the 5-cm-thick spherical shell of Unat metal would be 1416 °C or 1695 °C. For LMFBR fuel cycle option I, the temperatures would be somewhat higher. The Unat metal would also be molten, and the temperatures in the metallic reactor-americium sphere would also be so high (Fig. 11.7) that it would be fully molten. The HNED defined as option H would not be feasible technically.

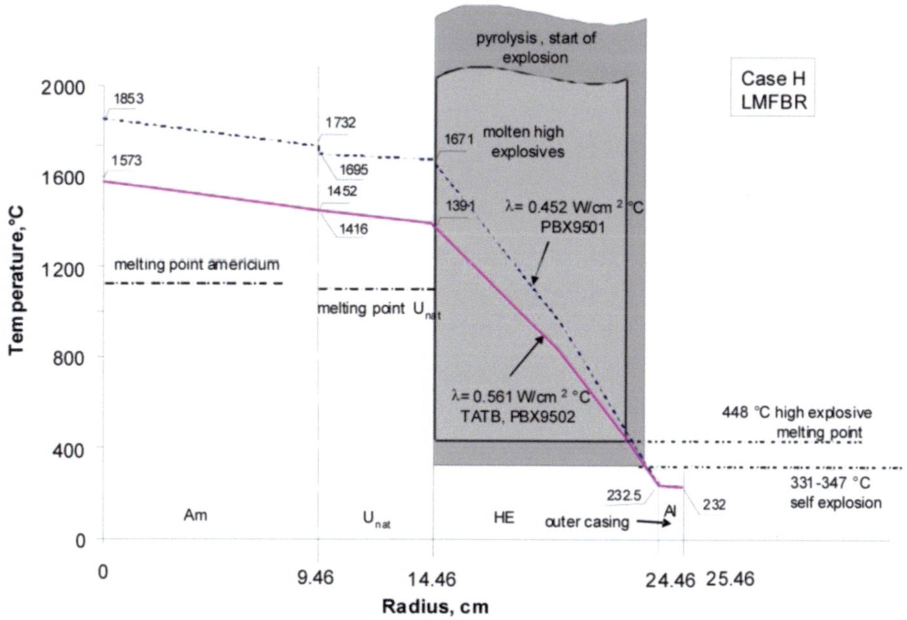

Figure 11.7: Radial temperature profile for a reactor americium based HNED (option H LMFBR).

11.9.6 Radial temperature profile for a reactor-americium HNED (option K, ADS)

Fig. 11.8 shows the radial temperature profile within the HNED described as option K. Option K represents the ADS-nitride fuel cycle (Fig. 11.3) and has the lowest alpha-particle heat rate of the two considered ADS fuel cycle options J and K. As the outer-casing radius of these two options is almost equal (see Table 11.7), it would therefore have the lowest temperatures among the HNEDs (see Table 11.11). Assuming black body radiation, the temperature of the outer aluminum casing at 25.34 cm would be 214 °C. (For an emissivity of 0.07 for rolled aluminum, it would be considerably higher). Because of the relatively low thermal conductivity of $\lambda = 0.561 \cdot 10^{-2}$ W/cm °C for the high explosive PBX 9502, the radial temperature would rise in these high explosives from 214.2 °C at their outer radial boundary up to 1207 °C at their inner radial boundary at 14.34 cm.

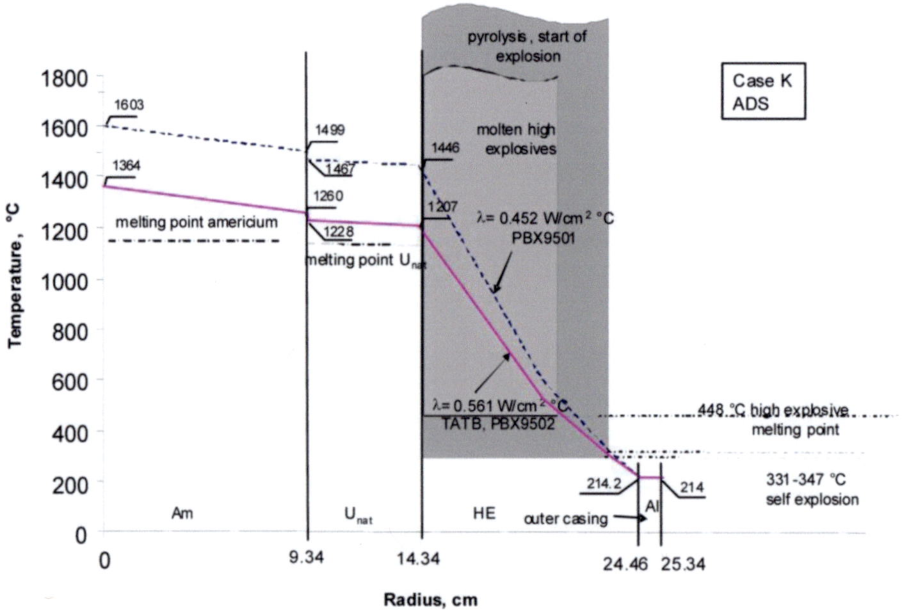

Figure 11.8: Radial temperature profile for a reactor-americium HNED (option K ADS).

For comparison the radial temperature profile for another high explosives (PBX 9501 with a lower thermal conductivity of $\lambda = 0.452 \cdot 10^{-2}$ W/cm °C) is shown (see Table 11.8) as dotted line, for which the temperature in the HE would rise even up to 1446 °C. The melting temperature of 448 °C and the temperature for initiation of pyrolysis (395 °C) or self-explosion (331 °C or 347 °C) of the high explosives would be exceeded in >60% of the volume of the high explosives, which would melt or self-explode. The HNED described as option K could not function technically. The inside temperature of the 5-cm-thick spherical shell of U_{nat} metal would be 1228 °C or 1467 °C. For the other ADS fuel cycle option J, the temperatures would be somewhat higher. The U_{nat} metal would also be molten, and the temperatures in the metallic reactor-americium sphere would also be so high (Fig. 11.8) that it would be fully molten. The HNED defined as option K would not be feasible technically.

11.9.7 Radial temperature profile for a reactor-americium HNED (option L, 100% Am-241)

Fig. 11.9 shows the radial temperature profile within the HNED described as option L. This option represents 100% Am-241 from the decay of Pu-241 (section 11.3.1).

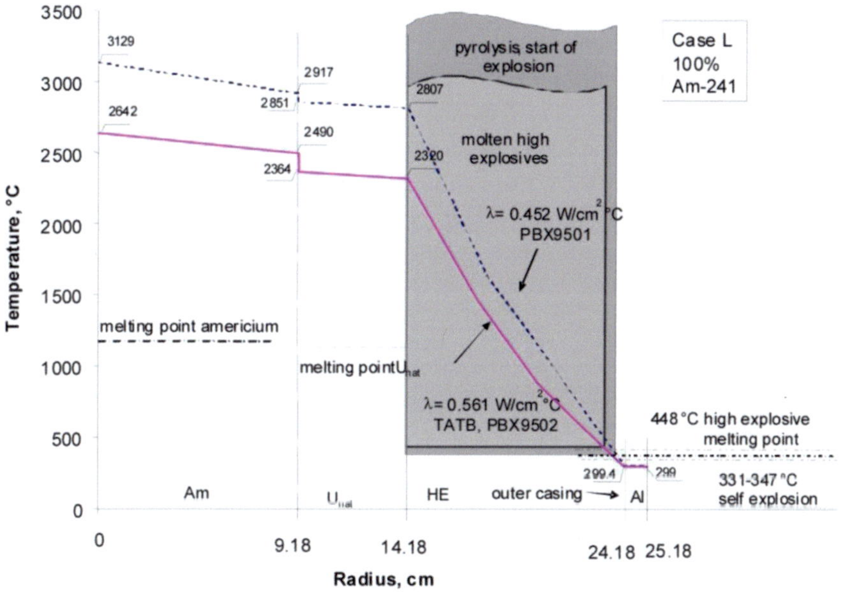

Figure 11.9: Radial temperature profile for a reactor-americium based HNED (option L 100% Am-241).

Assuming black body radiation, the temperature of the outer aluminum casing at 25.18 cm would be 299 °C. (For an emissivity of 0.07 for rolled aluminum it would be considerably higher). Because the thermal conductivity of $\lambda = 0.561 \cdot 10^{-2}$ W/cm °C for the high explosive PBX 9502, the radial temperature would rise in these high explosives from 299.4 °C at their outer radial boundary to 2320 °C at their inner radial boundary at 14.68 cm. For comparison also the radial temperature profile for another high explosive

(PBX 9501 with a lower thermal conductivity of $\lambda = 0.452 \cdot 10^{-2}$ W/cm °C) is shown (see Table 11.7) as dotted line, for which the temperature in the high explosives would rise even up to 2807 °C. The melting temperature of 448 °C and the temperature for initiation of pyrolysis (395 °C) or self-explosion (331 °C or 347 °C) of the high explosives would be exceeded in >90% of the volume of the high explosives, which would melt or self-explode. The HNED described as option L could not function technically. The inside temperature of the 5-cm-thick spherical shell of U_{nat} metal would be 2364 °C or 2851 °C. Option L would have the highest temperatures in the HNED. The U_{nat} metal would also be molten, and the temperatures in the metallic reactor americium sphere would also be so high (Fig. 11.9) that it would be fully molten. The HNED defined as option L would not be feasible technically.

11.9.8 Radial temperature profile for a reactor-americium HNED (option M Am-242m breeding)

The radial temperature profile within the HNED described as option M represents the fuel cycle option Am-242m breeding in an FR (Section 11.3.2). Assuming black body radiation, the temperature of the outer aluminum casing at 25.26 cm would be 258 °C. The melting temperature of 448 °C and the temperature for initiation of pyrolysis (395 °C) or self-explosion (331 °C or 347 °C) of the high explosive PBX 9502 would be exceeded in >85% of the volume of the HE, which would melt or self-explode. The HNED described as option M could not function technically (Kessler [30]).

11.10 Discussion of the results of the thermal analyses and uncertainties

Table 11.10 collects the data for the temperature profile in the HNEDs for options A through M. It can be seen that the main temperature rise is always occuring within the high explosives lenses, the material with the lowest thermal conductivity. Most parts of the high explosives lenses would be molten and would exceed the critical temperatures for start of pyrolysis and initiation of explosion. The americium metal and the metallic U_{nat} would also be molten. Options A through G show very similar temperature profiles among the HNEDs. As can be expected, among them option A with the highest alpha-heat power – due to the highest percentage of Am-241 (Fig. 11.2) – has the highest temperatures, whereas option G, shown in Fig. 11.6, with an Am-241 content of 73.5% (Table 11.4) has the lowest temperatures. LMFBR options H and I show somewhat lower temperatures due to the lower percentages of Am-241, but still the critical temperatures in the high explosives lenses would by far be exceeded, and the americium metal and the metallic U_{nat} would be molten (Fig. 11.6 for option H). Similar results are valid for the two ADS options J and K (Fig. 11.7 for option H).

Table 11.10: Temperatures within HNEDs for fuel cycle options A through M.

option	Central/outer temperature americium °C	Inner temperature U_{nat} °C	Inner/outer temperature high explosives °C	Outer temperature Al-casing °C
A	2415/2223	2164	2125/288,4	288
B	2355/2167	2111	2073/285,4	285
C	2295/2112	2058	2021/281,4	281
D	2239/2060	2008	1973/279,4	279
E	2233/2056	2003	1968/278,4	278
F	2215/2038	1988	1952/277,4	277
G	2193/2019	1968	1933/274,4	274
H	1574/1452	1416	1391/232,4	232
I	1781/1642	1601	1574/249,3	249
J	1618/1493	1456	1430/236,3	236
K	1364/1260	1228	1207/214,2	214
L	2641/2430	2364	2320/299,4	299
M	2018/1859	1805	1770/258,3	258

Option L, which represents the case of 100% Am-241 originating from the decay of Pu-241 has, as expected, the highest temperatures of all cases investigated in an HNED (Fig. 11.9).

Option M, representing americium with a high content of Am-242m, has similar high temperatures as the PWR, LMFBR, and ADS fuel cycle options A through K.

11.11 Characteristics of material data

If the reflector U_{nat} were replaced by beryllium and tungsten, the temperature difference within the reflector tamper would be decreased by a factor of only ~3.5 according to the different thermal conductivities of U_{nat} on the one side and beryllium and tungsten on the other side (see Table 10.7).

In addition, one should clearly keep in mind that hat a conservative approach was applied with the assumption of the black body radiation.

If a hollow americium sphere for the application of the implosion process would have to be investigated instead of the solid americium sphere, the above results would not be very different since the main temperature rise again would be caused by the HE lenses with their low thermal conductivity.

11.12 Coolability of the reactor americium HNED

The possibility of external cooling of the HNED, e.g., by forced convection or by liquid nitrogen or liquid helium, was analyzed with the following results. Cooling the outside casing down to the lowest temperature of -270 °C lowers the whole temperature profile (Fig. 11.6 through 11.9 and Table 11.11) accordingly. If, as an example, the temperature profile of option K with an outside-casing temperature of 214 °C, which is the lowest of options A through M would be lowered to -270 °C, then the temperature at the inner radius of the high explosives would be lowered from 1207 °C to 723 °C. This is still by 275 °C higher than the melting temperature, 328 °C higher than the starting temperature for pyrolysis, and 376 °C higher than the limiting temperature for start of explosion of, e.g. the high explosive TATB. All other options listed in Table 11.11 would have even higher temperatures.

11.13 Conclusions

Reactor-americium originates during irradiation of the fuel in nuclear reactors to presently applied burnup values of 50 to 60 GWd/t as a mixture of the isotopes Am-241, Am-242m, and Am-243. In PWR spent fuel – depending on the fuel cycle strategies considered in this paper – this isotopic mixture varies between 73.5% Am-241, 0.5% Am-242m and 26% Am-243 up to 81% Am-241, 0.1% Am-242m and 18.9% Am-243 after a burnup of 50 to 60 GWd/t and a cooling time of 10 y (Fig. 11.2). In the spent fuel of FRs after 2 or 7 y cooling time, the composition varies between ~51.5% Am-241, 3.6% Am-242m and 44.9% Am-243 up to 55% Am-241, 1.5% Am-242m and 43.5% Am-243 for the considered options (Fig. 11.3). For longer cooling times of the spent fuel, the content of Am-241 increases because of the decay of Pu-241. Pure Am-241 can be chemically separated from plutonium being stored over longer times.

The spontaneous fission neutron rates of Am-241, Am-242m, and Am-243 require a detailed pre-ignition analysis for the gun-type system and the implosion method. After determination of the critical masses for all fuel cycle options (options A through G for PWRs, options H and I for FRs, and options J and K for ADSs as well as for pure Am-241 and a dedicated Am-242m breeding case) and calculation of the spontaneous fission sources, a detailed pre-ignition analysis was performed (Kessler [30]). It was shown that for all cases considered, the gun-type system would lead to very early pre-ignition during the compaction process and to extremely low explosion yields, which are of no further interest. A thermal analysis was therefore not performed.

The implosion method, however, would lead – despite of early pre-ignition – to minimum fizzle yields which are in the same range as they were determined in Section 9 for reactor-grade plutonium. A subsequent thermal analysis for implosion-type HNEDs shows, however, that the high alpha-particle heat production of Am-241 and the relatively high near-critical mass between 46.6 and 58.3 kg of reactor americium lead to an alpha-particle heat power between 2.4 and 4.5 kW (for pure Am-241 and dedicated Am-242m breeding, it would be 4.9 kW (option L) and 3.3 kW (option M)).

For the detailed thermal analyses, the same calculational procedures and the same one-dimensional conservative approach, were applied as described in Section 10. The results of the thermal analyses are that in all cases the limiting temperatures for melting and the start of self-explosion of the high explosives are exceeded in 60% or 90% of the volume of the high explosive lenses. Also the melting temperature of the metallic reactor americium would be exceeded.

If external cooling even down to cryogenic temperatures of –270 °C would be applied, the limiting temperatures for the high explosives would still be exceeded for all options A through M considered in this paper.

It is therefore concluded that HNEDs based on the gun system or on the implosion system using reactor americium as fissile material would be technically unfeasible.

Reactor americium can be considered proliferation-proof.

References Section 11

[1] Albright, D., et al., Troubles tomorrow? Separated neptunium-237 and americium, chapter V in "The challenges of fissile material control". The Institute for Science and International Security, Washington (1999).

[2] Cesana, A., et al., Some considerations on Am-242m production in thermal reactors, Nucl. Techn., 148, 97-101 (2004).

[3] Holden, N.E., et al., Spontaneous fission half-lives for ground state nuclides, Pure Appl. Chem., Vol. 72, No. 8, p. 1525 (2000).

[4] Kocherov, N., et al., Handbook of nuclear data for safeguards, IAEA Nuclear Data Section, IAEA, Dec. 1998, Vienna

[5] Ronen, Y., et al., A novel method for energy production using Am-242m as a nuclear fuel, Nucl. Techn., 129, 407-417 (2000).

[6] Ronen, Y., et al., Breeding of Am-242m in a fast reactor, Nucl. Techn., 153, 224-233 (2006).

[7] Aleksandrov, B.M., et al., Determination of the probabilities of spontaneous fission in U-233, U-235 and Am-243, Atomic Energy, Vol. 20, No. 4 (April 1966).

[8] Zelenkov, A.G., et al., Measurement of spontaneous fission half-lives of Cm-242 and Am-242m, Atomic Energy, Vol. 60, No. 6 (June 1986).

[9] Caldwell, J.T., et al., Spontaneous fission half-life of Am-242m, Physical Review, Vol. 155, No. 4 (March 1967).

[10] Keepin, G.R., Physics of Nuclear Kinetics, Addison-Wesley Publ. Inc., Reading, Mass. (1965).

[11] Gold, R., Spontaneous-fission decay constant of ^{241}Am, Physics Rev. C1, p. 738 (1971).

[12] Artisyuk, V., Review of the data on spontaneous fission, Obninsk, Russia, personal information (2007).

[13] Diaz, M., et al., Critical mass calculations for Am-241, Am-242m and Am-243, 7th International Conference on Nuclear Criticality Safety (ICNC 2003), Oct. 24-24, 2003, Tokai-mura, Japan.

[14] Broeders, C.H.M., Investigations related to the build-up of transurania in pressurized water reactors, FZKA 5784 (Dec. 1996).

[15] Herring, J.S., et al., Thorium-based transmuter fuels for light water reactors, Nucl. Techn., 147, 84-101 (2004).

[16] Shwageraus, E., et al., Use of thorium for transmutation of plutonium and minor actinides in PWRs, Nucl. Techn., 147, 53-68 (2004).

[17] Tommasi, J., et al., Long-lived waste transmutation in reactors, Nucl. Techn., 111, 133-148 (1995).

[18] Tommasi, J., et al., A coherent strategy for plutonium and actinide recycling, Transactions Vol. 2, p. 358, – International Nuclear Congress – Atoms for Energy, ENC 94, Oct 2-6. 1994, Lyon, France.

[19] Advanced nuclear fuel cycles and radioactive waste management, NEA No. 5990, OECD (2006).

[20] Broeders, C.H.M., Kessler, G., Fuel cycle options for the production and utilization of denatured plutonium, Nucl. Sci. Eng., 156, 1-23 (2007).

[21] Hill, R.N. et al., Physics studies of weapons plutonium disposition in the Integral Fast Reactor Closed Fuel Cycle, Nucl. Sci. Eng. 121, 17 (1995).

[22] Messaoudi, et al., Fast Breeder Reactor devoted to minor actinide incineration, Nucl. Techn., 137, 84-96 (2002).

[23] Ronen, Y., et al., Ultra-thin Am-242m fuel elements in nuclear reactors, Nucl. Instrum. Methods A, 455, 442 (2000).

[24] Benetti, P., et al., Am-242m and its potential use in space applications, Journal of Physics, Conference Series 41, 161-168 (2006).

[25] Rhodes, R., The making of the atomic bomb. A touchstone book, published by Simon & Schuster, New York (1988).

[26] Rhodes, R., Dark Sun, The making of the hydrogen bomb. A touchstone book, published by Simon & Schuster, New York (1995).

[27] Cochran, Th.B., et al., Nuclear Weapons Databook, Volume 1 and 2. Ballinger Publishing Company, Cambridge, Mass., USA (1987).

[28] de Volpi, A., et al., Nuclear shadow boxing. Contempory threats from Cold War Weaponing, Fidler Doubleday, Kalamazoo, Mich. (2005).

[29] deVolpi, A., Proliferation, Plutonium and Policy, Pergamon Press (1979).

[30] Kessler, G., Proliferation resistance of americium originating from spent irradiated reactor fuel of pressurized water reactors, fast reactors and accelerator driven systems with different fuel cycles. Nucl. Sci. Eng., 159, 56-82 (2008).

[31] Kessler, G., Plutonium denaturing by Pu-238, Nucl. Sci. Eng., 155, 53-73 (2007).

[32] Neal, T.R., AGEX I, the explosives regime of weapons physics, Los Alamos Science, No. 21, p. 54 (1993).

[33] Collins, G.P., Kim's Big Fizzle, the physics behind a nuclear dud, Scientific American, p. 8-9 (Jan. 2007).

[34] Hansen, G.E., Assembly of fissionable material in the presence of a weak neutron source, Nucl. Sci. Eng., 8, p.700 (1960).

[35] van Hippel, F., et al., Appendix: Probabilities of different yields, Mark, C., Explosive properties of reactor grade plutonium, Science Global Security, Vol. 4, p. 111, Gordon and Breach Science Publishers (1993).

[36] Seifritz, W., Nukleare Sprengkörper – Bedrohung oder Energieversorgung für die Menschheit, Karl Thiemig, Munich (1984).

[37] Mark, J. C., Explosive properties of reactor-grade plutonium, Science & Global Security, Vol. 4, p. 111, Gordon and Breach Science Publishers (1993).

[38] Weinberg, A.M., Wigner, E.P., The physical theory of neutron chain reactors, University of Chicago Press, Chicago (1958).

[39] Zinn, J., Mader, C.L., Thermal initiation of explosives, J. of Appl. Physics, 31, No. 2 (1960).

[40] Dobratz, B.M., Properties of chemical explosives and explosive simulants, UCRL-51319/Rev. 1 (Dec. 1972).

[41] Mader, C.L., et al., Los Alamos explosion performance data, University of California Press (1982).

[42] Gibbs, T.R., Popolato, A., LASL Explosive Property Data, University of California Press, Berkeley (1981).

[43] Kuchling, H., Taschenbuch der Physik, Verlag Harri Deutsch, Frankfurt (1982).

[44] Periodic Table of Elements: Americium, available on http://environmentalchemistry.com

[45] Fetter, St., et al., Detecting Nuclear Warheads. Science and Global Security, 1, 225 (1990).

[46] Serber, R., The Los Alamos Primer, edited with an introduction by R. Rhodes, University of California Press, Berkeley and Los Alamos (1992).

[47] Tipton (Editor), Reactor Handbook, Vol. 1, Materials, Interscience Publishers, New York (1960).

[48] http://en.wikipedia.org/wiki/Gun-type_fission_weapon

[49] Chebeskov, A.N., Are there insuperable obstacles towards nuclear weapon creation if a nation has in hands nuclear fuel cycle technologies? Personal communication at Int. Workshop on Non-proliferation of Nuclear Materials, Obninsk (September 29 – October 3, 2008).

[50] Fukuda, K. et al., Prospects of Inventories of Uranium, Plutonium and Minor Actinides and Mass Balance, presented at 2nd Consultancy Meeting. Protected Plutonium Production Project, Vienna, June 15-16, 2006, International Atomic Energy Agency.

12 Fuel cycle options for the production of denatured, proliferation-proof plutonium

12.1 Introduction

In Section 10 it is demonstrated that reactor plutonium with an increased isotopic content of Pu-238, e.g. more than 1.8% Pu-238 corresponding to more than 0.12 kW in a low technology HNED or more than 3.5 Pu-238 leading to more than 0.24 kW in an HNED of medium technology can be regarded as proliferation-proof. The utilization of such so-called denatured proliferation-proof plutonium would be unsuitable for a nuclear explosive device (NED), because the chemical high explosive lenses surrounding the plutonium would partially melt, or their elevated temperature would lead to self-ignition and chemical explosion.

The utilization of such proliferation-proof plutonium will require the generation of Pu-238 in sufficient percentages in the reactor-grade, proliferation-proof plutonium isotopic composition in an adapted fuel cycle. The incineration of this proliferation-proof plutonium requires modern reprocessing and, e.g., mixed-oxide (MOX) refabrication technologies which are already under development [1 through 6]. The evolution of proliferation-proof, reactor-grade plutonium during irradiation will be discussed in Sections 13 and 14.

12.2 Review of earlier research

During and just after the International Nuclear Fuel Cycle Evaluation Program [7] research results were published [8-11] indicating that reactor plutonium with isotopic contents of up to 8-10% Pu-238 or more can be generated. They were aiming at a reactor-grade plutonium composition with 5% Pu-238 and more as proposed by Heising-Goodman [8]. However, this lower limit had not been derived from detailed criticality calculations for the hypothetical nuclear explosive device (HNED) and also differed considerably in thermal conductivity data for the high explosive lenses as applied by Kessler [1].

More recent results for Pu-238 production were obtained, e.g., in the context of studies on plutonium and actinide transmutation and incineration analyses using plutonium and minor actinides, U_{nat} and thorium in pressurized water reactors. This resulted in a Pu-238 isotopic content of 8 to 10% (Shwageraus et al. [12]). Extensive studies with neptunium- and americium-doped enriched uranium fuel in PWRs yielded Pu-238 contents in the reactor-grade plutonium of 30% and more [13 through 15]. Doping of the blanket fuel elements of the fast reactor (FR) JOYO with neptunium lead to Pu-238 isotopic contents of higher than 10% [16].

However, as will be shown in Section 13, neptunium must be avoided in a future proliferation-proof civil nuclear fuel cycle.

In all these studies the Pu-238 production was enhanced, e.g., by

- recycling of the U-235/U-236 from reprocessed spent fuel; also re-enriched U-235/U-236 from reprocessed uranium can be utilized.
- recycling of recovered MAs: neptunium, americium, and curium from reprocessed spent fuel
- reduction of the U-238 content in the fuel by substitution of thorium.

Pu-238 is produced in nuclear reactors principally through several routes: from U-235/ U-236 neutron capture, to a smaller extent from U-238 (n,2n) reactions, more directly from Np-237 (if separated chemically from the MAs), and from the decay of Cm-242, which will be of importance if plutonium recycling is utilized. Fig. 12.1 displays different possible routes for an increased Pu-238 production.

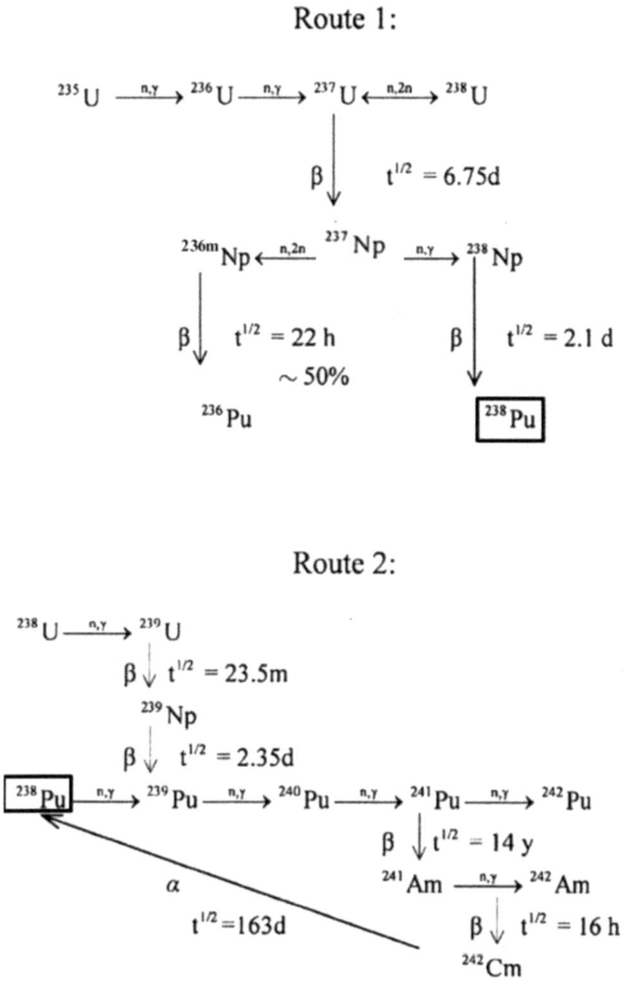

Figure 12.1: Different production routes for Pu-238.

12.3 Analysis of fuel cycle options for the production of proliferation-proof plutonium

Broeders and Kessler [17] demonstrated that proliferation-proof, denatured reactor plutonium can be directly generated by loading re-enriched reprocessed uranium into PWR cores. In addition they presented three different options how the presently already existing plutonium in spent light water reactor (LWR) fuel, after reprocessing, can be converted into proliferation-proof, denatured reactor plutonium. This can be done by mixing the presently existing reactor plutonium with reenriched reprocessed uranium, refabricating it to, e.g., MOX fuel, and subsequently irradiating this fuel in a PWR over a burnup cycle of 60 GWd/t. Similarly existing reactor plutonium can be mixed with reenriched reprocessed uranium and TRU. Partial replacement of uranium by thorium leads to similar results for proliferation-proof, denatured reactor plutonium, but additional U-233 denatured in U-238 is generated.

12.4 Fuel cycle options for the production of denatured, proliferation-proof plutonium

Based on the results of earlier investigations [8,9,10,18,23] Broeders and Kessler [17] selected different fuel types for the production of plutonium with a Pu-238 fraction of 6% or somewhat more, as shown in Table 12.1.

Table 12.1: Fresh fuel compositions for fuel cycle calculations leading to denatured, proliferation-proof plutonium.

Fuel type	P/D Ratio	M/F Ratio	Fuel composition	Th (wt%)	U (wt%)		Plutonium (wt %)		MA (wt%)
					Total	Fissile Fraction	Total	Fissile Fraction	
A	1.4427	2.2039	Re-enriched recycled U	0	100	5.52	0	0	0
B	1.3389	1.7132	Re-enriched recycled U + Pu (Table 12.2)	0	93.9	5.00	6.1	64.5	0
C	1.3389	1.7132	Re-enriched recycled U + Pu (Table 12.3)	0	94.9	5.00	5.1	54.3	0
D	1.4068	2.0302	Re-enriched recycled U + Pu/MA (Table 12.4)	0	92.5	5.00	6.5	64.5	1.0
E	1.5926	2.9780	Enriched U + Th + Pu/MA (Table 12.4)	52.6	35.1	6.00	10.7	64.5	1.6

Fuel type A is re-enriched uranium from reprocessed UO_2 fuel (see Sec. 12.5.1). The re-enriched recycled uranium will generate plutonium with a sufficiently high Pu-238 content (denatured, proliferation-proof plutonium) after a burnup of ~60 GWd/t. Fuel type A is close to the presently used standard LEU fuel with a pitch/diameter (P/D) ratio of 1.4427, which corresponds to a moderator-to-fuel (M/F) ratio of 2.2039 [19,20].

For the subsequently chosen MOX fuel types B through E, fully MOX fuel-loaded cores are assumed, as they lead to simpler fuel assembly design [21,22,23]. In addition, the M/F ratio is varied from 1.7132 to 2.9780 in order to obtain adequate safety characteristics. The technical and economic implications of such higher M/F ratios are discussed in Section 12.8.

In the case of fuel type B, recycling of presently existing plutonium (Table 12.2) together with re-enriched reprocessed uranium in a PWR is considered. Fuel type C contains already recycled denatured, proliferation-proof plutonium (Table 12.3) together with re-enriched reprocessed uranium. In both cases, MAs would have to be separated from the spent nuclear fuel (SNF) and stored for later incineration [24].

Table 12.2: Plutonium composition from SNF with 50 GWd/t, 10 years after unloading. (This plutonium composition differs somewhat from the plutonium of Table 7.5 due to different cross sections used).

Isotope	wt %
Pu-238	3.2
Pu-239	56.4
Pu-240	26.6
Pu-241	8.0
Pu-242	5.8

Table 12.3: Plutonium composition of recycled, proliferation-proof PWR MOX fuel with 50 GWd/t, 10 years after unloading.

Isotope	wt %
Pu 238	7.7
Pu-239	44.0
Pu-240	31.0
Pu-241	10.3
Pu-242	7.0

Fuel types D and E initially contain plutonium and MAs of isotopic composition in the proportion in which they arise from present spent fuel with a burnup of 50 GWd/t after 8 y cooling time in intermediate storage and 2 y reprocessing time (Tables 12.2 and 12.4). In the fuel type E the use of thorium, together with plutonium and enriched uranium, is introduced, similarly to [12].

Table 12.4: Plutonium and MA composition from SNF with 50 GWd/t, 10 years after unloading.

Isotope	wt %
Np-237	6.65
Pu-238	2.75
Pu-239	48.73
Pu-240	23.02
Pu-241	6.94
Pu-242	5.04
Am-241	4.64
Am-242m	0.19
Am-243	1.48
Cm-243	0.00
Cm-244	0.50
Cm-245	0.06
Cm-246	0.00

All fuel types are assumed to be dioxides with 96% of their theoretical density. The assumed theoretical density of thorium is 9.6 g/cm^3, of plutonium and MAs is 11.0 g/cm^3, and of uranium is 10.5 g/cm^3. Table 12.2 shows the isotopic composition of the plutonium as it arises from present PWR spent nuclear fuel after 50 GWd/t burnup and a cooling and reprocessing time of 10 y. Table 12.3 shows the isotopic composition of denatured plutonium as it arises after several times recycling of the plutonium. In Table 12.4 the isotopic composition of the plutonium and the MAs are given as they arise from present PWR SNF after 50 GWd/t burnup and a cooling and reprocessing time of 10 y.

12.5 Results of physics calculations for the selected fuel types

The following results were obtained with the KAPROS modular program system [25] for pin/ cell calculations and the related cross-section sets based on ENDF/B6.5. Pin/cell calculations are adequate for such investigation as long as the recommendations of Driscoll [26] are followed.

12.5.1 Results for fuel type A; UO_2 from reenriched recycled uranium

In the case of fuel type A, denatured plutonium is produced with UO_2 fuel from re-enriched recycled uranium coming from PWR spent fuel (50 GWd/t burnup). Reprocessed uranium contains ~0.8-0.9% U-235 and 0.6 to 0.7% U-236 depending on the irradiation history of the spent reprocessed fuel. This reprocessed uranium can be reenriched, e.g., by centrifuge enrichment technology. According to the different atomic masses of U-235 and U-236, the ratio of enrichments of both uranium isotopes would be about 4:3 [27,28]. The re-enriched reprocessed uranium considered in these investigations is based on the results of Broeders [20]. Using the factor 4/3 mentioned above leads to a uranium vector of 5.52% U-235, 3.0% U-236, 91.48% U-238 (U-234 is neglected) [17].

A satisfactory burnup behaviour could be obtained with a PWR lattice with P/D=1.4427.

Figs. 12.2 and 12.3 show the buildup of the isotopes Pu-238 and Pu-242 for fuel type A. (For a simple presentation only these two plutonium isotopes are shown.) It can be seen that after a burnup of 60 GWd/t already 12% Pu-238 and 7% Pu-242 are attained. This means that a lower U-236 content in the fresh fuel would already lead to about 6% Pu-238 isotopic content in the spent fuel after 60 GWd/t burnup.

12.5.2 Results for fuel type B

In the case of fuel type B, a lattice with a P/D = 1.3389 could be applied for the production of denatured proliferation-proof plutonium with MOX fuel from the re-enriched recycled uranium of fuel type A (UO_2 from PWR spent fuel with 50 GWd/t burnup), mixed with plutonium from spent LWR UO_2 fuel with a burnup of 50 GWd/t (Table 12.2).

The buildup of the isotopes Pu-238 and Pu-242 is shown in Figs. 12.2 and 12.3. It can be seen that 6% Pu-238 and 9% Pu-242 in the denatured, proliferation-proof plutonium are attained after a burnup of 60 GWd/t.

12.5.3 Results for fuel type C

In the case of fuel type C, a lattice with a P/D = 1.3389 can be applied for the production of denatured plutonium with MOX fuel from the re-enriched recycled uranium of fuel type A (UO_2 from PWR spent fuel with 50 GWd/t burnup), mixed with dedicated denatured plutonium from spent LWR fuel with a burnup of 50 GWd/t (Table 12.3). The main results for the burnup-dependent isotopic compositions for Pu-238 with 9% and Pu-242 with 10.3% after a burnup of 60 GWd/t can be seen in Figs. 12.2 and 12.3.

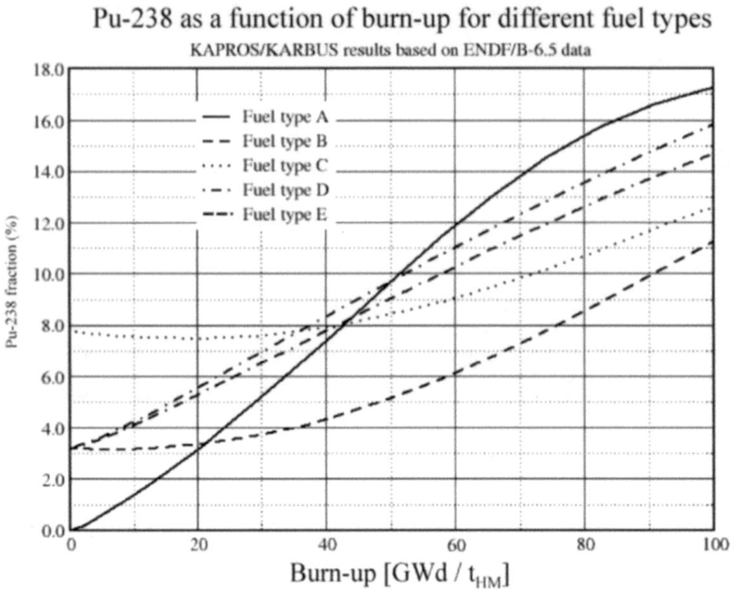

Figure 12.2: Changes of Pu-238, Pu-242, U-233 and U-235 fractions of optimized MOX fuel from enriched uranium, plutonium, MAs, and thorium (fuel type E).

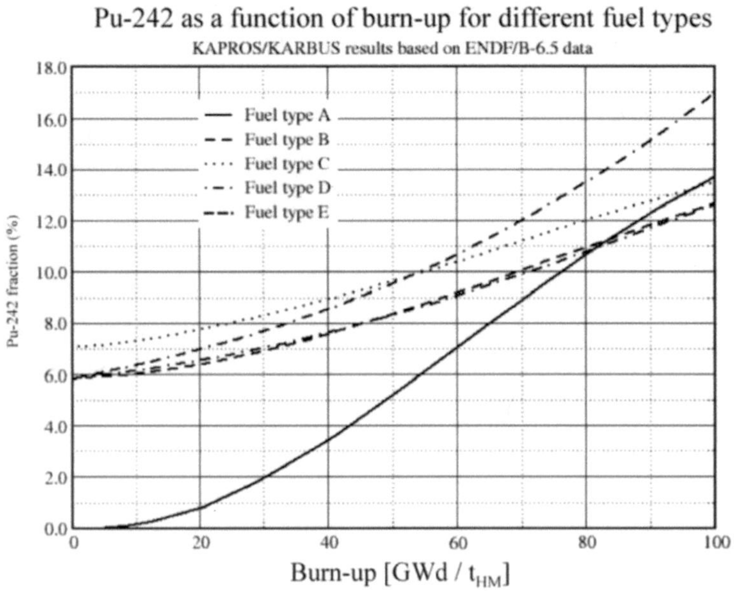

Figure 12.3: Buildup of Pu-242 in wt% for different fuel types.

12.5.4 Results for fuel type D

In the case of fuel composition D, a lattice with a P/D = 1.4068 can be applied for the production of denatured, proliferation-proof plutonium with MOX fuel from the re-enriched recycled uranium of fuel type A (UO_2 from PWR spent fuel with 50 GWd/t burnup), mixed with TRU from spent LWR UO_2 fuel with a burnup of 50 GWd/t (Table 12.4). The MAs are in the proportion in which they appear, together with the plutonium, after a burnup of 50 GWd/t in UO_2 PWR fuel (Table 12.4), as determined with the KAPROS system [25]. The main results for the burnup-dependent fraction of the plutonium isotopes Pu-238 and Pu-242 are shown in Figs. 12.2 and 12.3. After a burnup of 60 GWd/t 11% Pu-238 and 9% Pu-242 are attained for the isotopic composition vector of denatured, proliferation-proof plutonium.

12.5.5 Results for fuel type E; MOX fuel with thorium, uranium, plutonium and minor actinides

In the case of fuel type E, denatured plutonium is produced in a PWR lattice with P/D = 1.5926 with MOX fuel from enriched natural uranium, plutonium, and MAs in the proportion in which they appear after a burnup of 50 GWd/t in UO_2 PWR fuel, and additional thorium. The main results for the burnup-dependent isotopic compositions for Pu-238 and Pu-242 are again given in Figs. 12.2 and 12.3. Pu-238 attains 11% and Pu-242 about 10.8% in the isotopic composition vector of denatured, proliferation-proof plutonium after a burnup of 60 GWd/t.

Fig. 12.4 shows the buildup of the mainly interesting heavy metal isotopes (U-233, U-235, Pu-238, Pu-242) for fuel type E with thorium. Because of the presence of thorium, fissile U-233 is built up. The consequences are a concentration of ~3% U-233 together with ~3.5% remaining U-235 in U-238 (remaining below the limits set by IAEA (Section 8)) after a burnup of 60 GWd/t. At this burnup of 60 GWd/t, the Pu-238 and Pu-242 isotopic contents would rise to somewhat more than 10% each.

The above results show that reactor plutonium from spent fuel of 50 GWd/t, as described by Table 12.2, can be modified to denatured plutonium in one single burnup cycle of ~60 GWd/t for fuel types A, B, D, and E. This denatured plutonium would be, according to Section 10 and Kessler [1,38] unsuitable as fissile material for an HNED. The results of fuel type C show that reactor-grade plutonium which is already proliferation-proof at the beginning of the burnup cycle remains proliferatioin-proof over the full burnup cycle of 60 GWd/t. It is obvious that a proper Pu-238 content around 6% can be tailored by adequate selection of the U-236 content or of the content of neptunium and americium in the fresh fuel of options A, B, D and E.

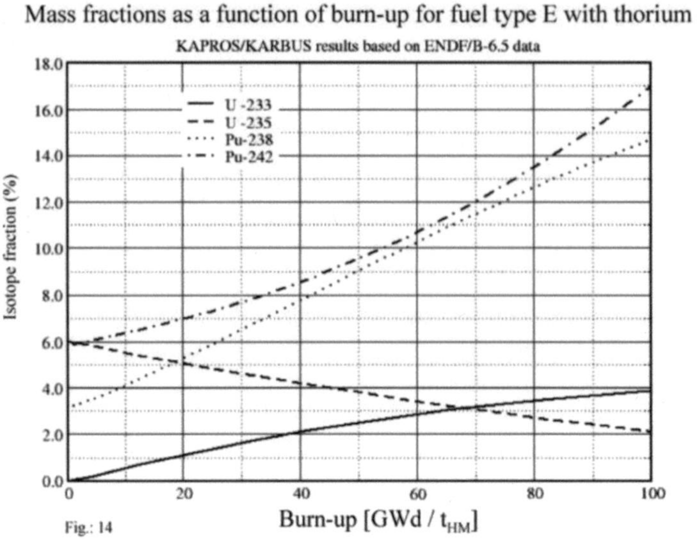

Figure 12.4: Changes of Pu-238, Pu-242, U-233 and U-235 fractions of optimized MOX fuel from enriched uranium, plutonium, MAs, and thorium (fuel type E).

12.6 Moderator density and Doppler reactivity coefficients for the fuel type A, B, C, D, E

Broeders and Kessler [17] also presented an analysis of the safety-related reactivity coefficients MDC (moderator density coefficient), MTC (moderator temperature coefficient) and Doppler Coefficient (DC). In Table 12.5 a summary of the results is presented. The MDC is calculated from a 10% density decrease at the nominal coolant density. The MTC is based on the water coolant properties at pressure 15.6 MPa and temperature 583 K, leading to a value of -3.22×10^{-3} for the MDC. For the DC investigations the lattice reactivity was calculated at three temperatures: 300, 900 and 2100 K. In all cases a good fit of these values could be obtained, utilizing the following formula for the DC:

$$dk / dT = A_D / T^X,$$

where A_D = Doppler constant, and T = mean fuel temperature in kelvin. It was found that the value, X = 1.0, leads to a good fitting to the calculated temperature-dependent data, although X = 1/2 should be expected for theoretical reasons for PWR cores (with fairly thermal neutron spectra (Section 4.10.2.1).

The data of Table 12.5 show satisfactory results for the voiding effects and coolant temperature changes. The required boron concentrations for natural boron are in most cases too high for practical application. Therefore, more refined solutions are required, e.g. B^{10} enrichment and the use of burnable poisons like gadolinium or erbium.

Table 12.5: Summary of reactivity coefficients for LWR lattices with dedicated fuels.

Reactivity coefficients		Fuel Type				
		A	B	C	D	E
Moderator Density Coefficient (MDC)	BOL	0.1257	0.1542	0.1692	0.1616	0.1744
	EOL	0.1336	0.1706	0.1840	0.1721	0.1677
Moderator Temperature Coefficient (MTC)	BOL	-4.048x10^{-4}	-5.449x10^{-4}	-4.965x10^{-4}	-5.204x10^{-4}	-5.617x10^{-4}
	EOL	-4.302x10^{-4}	-5.923x10^{-4}	-5.494x10^{-4}	-5.541x10^{-4}	-5.400x10^{-4}
Doppler Constant (AD)	BOL	-2.113x0^{-2}	-2.212x10^{-2}	-2.203x10^{-2}	-1.911x10^{-2}	-1.330x10^{-2}
	EOL	-1.172x10^{-2}	-2.012x10^{-2}	-2.001x10^{-2}	-1.796x10^{-2}	-1.269x10^{-2}
BOC boron in ppm		4600	6100	4800	4000	2100
Boron efficiency (pcm/ppm)	BOL	-5.73	-1.80	-1.64	-1.85	-2.60
	EOL	-5.20	-1.81	-1.64	-1.92	-2.67

MDC: Moderator Density Coefficient calculated from 10% reduction at nominal density
MTC: Moderator Temperature Coefficient, being –3.22x10^{-3} x MDC at nominal coolant conditions
DC: Doppler Coefficient calculated from fit of $dk/dT = A_D / T$ for T=300, 900 and 2100K
Boron at BOL: The boron concentration to obtain $k_\infty \approx 1.03$ at BOL
BOL: Begin of cycle; EOL: End of cycle

The absolute values for the MTC and Doppler constant in Table 12.5 are only slightly different from those for present PWRs (Tommasi et al. [22] and Kloosterman [19]).

12.7 Long term behavior of denatured, proliferation-proof fuel in PWRs and FRs

Once denatured reactor plutonium would be introduced into the denatured plutonium fuel cycle, it could be incinerated by further recycling in PWRs or in either integral fast reactors [29,30] (IFRs) or FRs of type Consommation Améliorée du Plutonium dans les Réacteurs Avancés (CAPRA) [31-34]. The time periods for storage of denatured, proliferation-proof reactor plutonium should be smaller than several decades, because Pu-238 has an α-decay half-life of 87.7 y. The question then arises: how the Pu-isotopic composition would change during irradiation up to a certain burnup in these reactors with different neutron energy spectra (PWRs or FRs).

This question and the following analyses or considerations – will be also discussed in Sections 13 and 14. They shall be valid here for NWSs, because neptunium – posing a serious proliferation problem – is produced together with plutonium and other actinides. Therefore it is proposed in Section 14 to do further recycling (two or more recycles) of proliferation-proof plutonium during a transition period only in NWSs or future multilateral fuel cycle centers.

12.7.1 Long term behavior of denatured plutonium in LWRs

If MOX fuel with denatured proliferation-proof reactor plutonium and reenriched reprocessed uranium is loaded in a PWR core with the pin-cell parameters as reported above (Table 12.1), the results are as follows. Fig. 12.5 shows the time evolution of the Pu-isotopic composition over ~100 GWd/t. For fuel type C (already denatured proliferation-proof reactor plutonium) the Pu-238 percentage is only slightly increasing in a PWR core from 7.7% to ~9% after burnup of 60 GWd/t (see also Fig. 12.2, Section 12.5. For comparison also the Pu-238 time evolution for fuel type B (see Fig. 12.2) is shown.

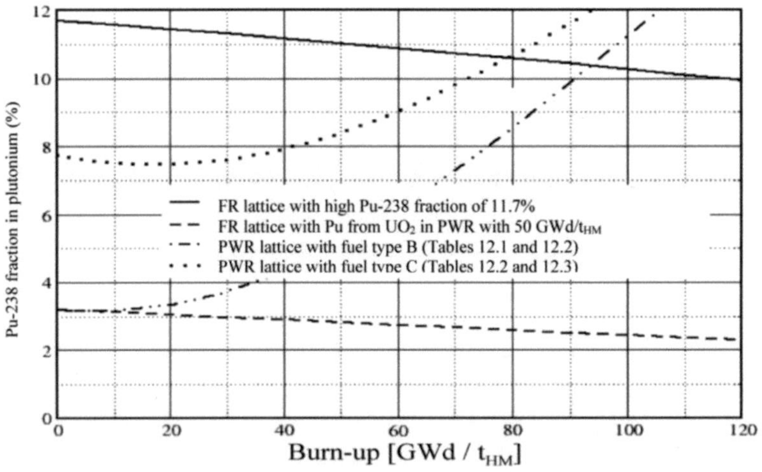

Figure 12.5: Plutonium-238 fraction in fuel in thermal and fast neutron spectrum as a functiion of burnup.

12.7.2 Long-term irradiation behavior of denatured plutonium in fast reactors

If MOX fuel with denatured, proliferation-proof reactor plutonium mixed with natural uranium would be loaded into a typical FR core, (e.g., German SNR-300 typical lattice, see also Table 12.6), the Pu-238 isotope concentration, because of the higher fission/absorption ratio in the fast neutron spectrum [29,30], decreases as a function of burnup. However, over a typical possible burnup of 150 GWd/t in a typical FR core this is only a ~2% net decrease (Fig. 12.5). This means that, if the Pu-238 content is raised to

~11% at the beginning of the burnup cycle, it decreases to ~9% at the end of the burnup cycle; i.e., the reactor plutonium remains proliferation-proof during the full burnup cycle until unloading of the fuel. The decrease of the Pu-238 content is a well-known characteristic of the FR neutron spectrum [29,30]. Fast reactors can incinerate plutonium more efficiently than LWRs.

Fig. 12.5 also shows for comparison the Pu-238 concentration decrease in the same FR neutron spectrum for plutonium coming from a PWR after 50 GWd/t burnup as it would be the case for Pu-recycle scenarios discussed in Section 7.7 (Table 12.2).

Table 12.6: Data for a fast reactor SNR-300 fuel pin-cell geometry.

fuel pin diameter [mm]	5.24
outer fuel pin diameter [mm]	6.08
cladding thickness [mm]	0.38
pin diameter/pitch (P/D)	1.32
fuel average temperature [K]	1183
average power rating [W/cm]	300

However, FRs, due to their more efficient destruction of Pu-238, will permanently need a certain feed of this plutonium isotope from LWRs or – as described in Section 13 – by admixing of certain percentage of americium to the fresh fuel. LWRs and FRs can operate in symbiosis.

12.7.3 Destruction of denatured fuel type C in a PWR

Figure 12.6 shows the denatured reactor plutonium inventory changes during burnup in kg/t for the considered fuel types, loaded as MOX fuel in a PWR core. At a burnup of 60 GWd/t, the net inventory decreases (destruction) for the fuel types B, C, and D by ~9.5 to 10 kg plutonium/t. Higher destruction rates above 30 kg plutonium/t may be obtained with thorium-based fuel type E. The production rate for denatured, proliferation-proof plutonium with fuel type A amounts to ~13 kg plutonium/t.

Fig. 12.7 displays the concentration (in %) for the 5 plutonium isotopes of fuel type C as a function of burnup up to 100 GWd/t. It can be seen that the Pu-238 concentration, which is responsible for the denaturing of reactor plutonium, is even slightly increasing over the burnup. Only the Pu-239 content is decreasing from 44% to 35.6%. The denatured reactor plutonium remains denatured during burnup.

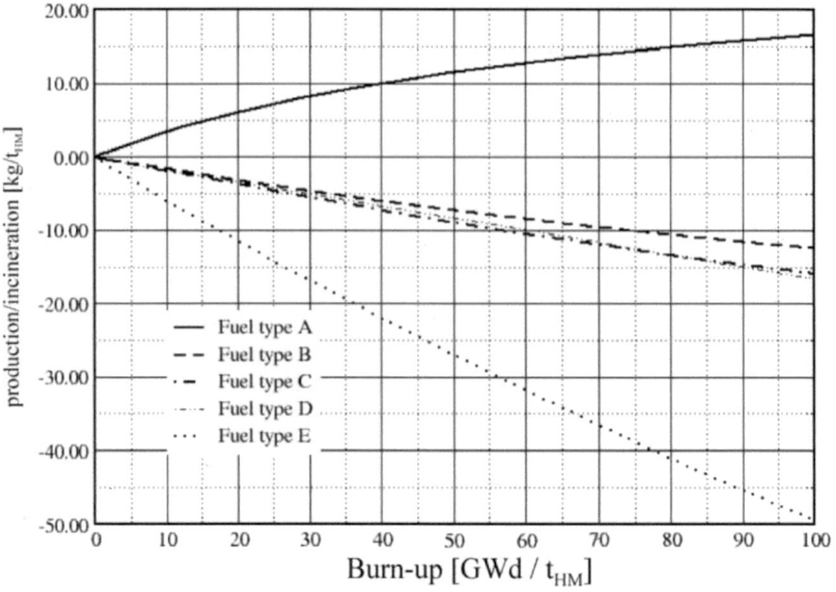

Figure 12.6: Denatured reactor plutonium inventory changes in 1 tonne fuel when loaded into PWR core as a function of burnup of fuel types considered.

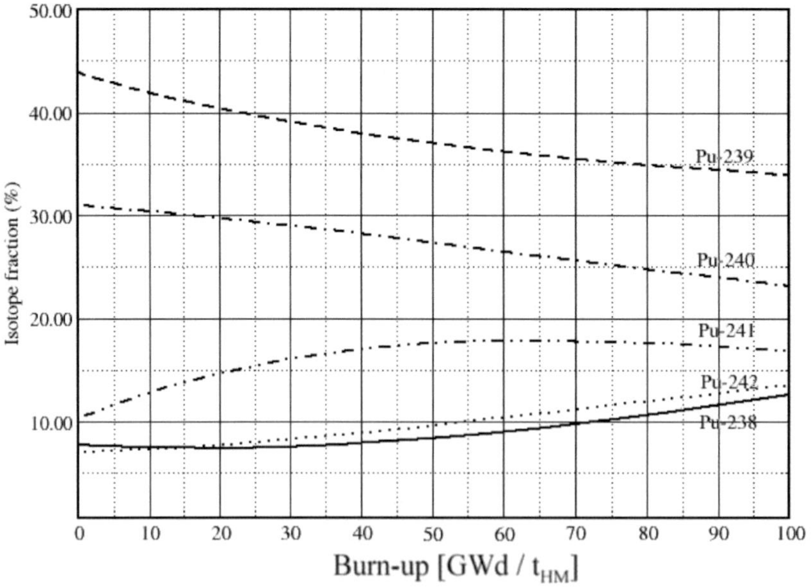

Figure 12.7: Fraction development of plutonium isotopes of denatured plutonium when loaded into PWR core with fuel type C.

12.8 Peculiarities of the fuel cycle and of the PWR design for production and recycling of denatured proliferation-proof plutonium

The separated plutonium and minor actinides (neptunium, americium, curium) together with uranium and thorium would have to be fabricated as fuel type B through E in a future refabrication plant and loaded into a full MOX PWR core. As americium and curium-doped fuel becomes too difficult for standard refabrication, both americium and curium have to be separated and the curium stored [22].

The PWR core design would need small changes and adaptations in the control rod systems, e.g. more control rods, poison rods with high-enriched boron or fuel doped with gadolinium or erbium [20,21]. The moderator to fuel ratio can be increased from 2 to 2.5 or 3 [19,23] by either

- reducing the fuel pin diameter
- increasing the P/D ratio
- replacing a certain number of fuel rods by water rods.

The latter design possibility was suggested for full MOX cores by Barbrault [23]. If a larger P/D ratio or water rods are used, either a smaller electrical power output is obtained or the diameter of the pressure vessel must be increased.

After a burnup of 60 GWd/t the plutonium will be converted to proliferation-proof denatured reactor plutonium, as shown in Figs. 12.2 and 12.3 as well as 12.4 for fuel type A through E.

Fuel types D and E containing MAs still need further development in fuel refabrication technologies. While there is some fabrication and irradiation experience available for neptunium-doped fuel [24,29,37], there is still research and development necessary for fuel containing americium and curium [22,24,36,37] (if curium is not separated and stored for decay) [22,36]. The heat conductivity and irradiation behavior of such fuel up to high burnups will have to be investigated. The safe design and operation of fuel assemblies accounting for thermal hydraulic and hot-spot effects must be assured [11].

Because of the higher Pu-238 content in the converted denatured, proliferation-proof reactor plutonium with its resulting high heat production and spontaneous neutron radiation and γ-radiation, the present aqueous reprocessing technology would have to be slightly modified and the present MOX refabrication technology would not be feasible any more. Present glove-box type MOX refabrication technology is limited to a Pu-238 isotopic content of ~4% and present aqueous reprocessing technology to ~5% [11].

Advanced aqueous and pyrochemical reprocessing [4-6,24,36,39] and related refabrication technology for metallic fuel [4], applying remote handling, would become necessary. These advanced technologies are currently being developed in the United States, Russia, Japan, and Europe in the context of actinide transmutation research and would have to be applied [4-6, 36].

There will certainly be some penalties in fuel cycle costs compared to present MOX fuel according to the degree of severity given by the difficult reprocessing and refabrication technology going from fuel types A, B and C to fuel types D, and E.

Also, if a larger P/D ratio of the fuel rods becomes necessary for reasons of assuring an adequate negative reactivity coefficient, this will increase the electricity generating costs, since either a smaller power output from the same core volume or the higher capital cost for an increased diameter of the pressure vessel must be accounted for (if the same presently chosen fuel rod diameter should be kept). More burnable poison rods and high enriched boron for control rods and boron acid in the coolant will also increase the electricity generating cost, somewhat.

12.9 Conclusions

Denatured, proliferation-proof reactor plutonium can be generated in a number of different fuel cycle options. First, denatured reactor plutonium can be obtained if instead of low-enriched U-235 PWR fuel, re-enriched U-235/U-236 from reprocessed uranium is used (fuel type A). Also the envisaged existing 2500 tonnes of reactor plutonium (being generated worldwide up to the year 2010 (Section 1)), mostly stored in intermediate fuel storage facilities at present, could be converted during a transition phase into denatured proliferation-proof reactor plutonium by the options fuel type B and D. Denatured, proliferation-proof reactor plutonium could have the same safeguards standard as present low-enriched (<20% U-235) LWR fuel. It could be incinerated by recycling once or twice in PWRs and subsequently by multirecycling in FRs, e.g. of CAPRA-type or IFRs. Once denatured or proliferation-proof, the reactor plutonium would remain denatured and proliferation-proof during multiple recycling. In a PWR, e.g., denatured reactor plutonium could be destroyed at a rate of ~250 kg/GW(e)·y. While the denatured, proliferation-proof reactor plutonium could be recycled and incinerated, the generated neptunium would still have to be monitored by the IAEA for all cases in which considerable amounts of neptunium are produced.

Therefore it is proposed in Section 14 to perform the conversion into proliferation-proof plutonium preferably first in NWS where most of the PWRs and reprocessing/ refabrication facilities operate. Also the neptunium could be incinerated there after co-separation of plutonium/neptunium and utilizing the neptunium for the production/conversion to proliferation-proof plutonium. This proliferation-proof plutonium can then be used and incinerated in NNWSs.

References Section 12

[1] Kessler, G., Plutonium denaturing by Pu-238, Nucl. Sci. Eng., 155, 53-73 (2007).

[2] Kessler, G. et al., Moderne Strategien zur Beseitigung von Plutonium. Atomwirtschaft, 46, 132 (2001).

[3] Kessler, G., Requirements for nuclear energy in the 21st Century. Progress in Nuclear Energy, 40, 3-4, 309-325 (2002).

[4] Laidler, J.J. et al., Development of pyroprocessing technology, Progress in Nuclear Energy, 31, 1-2, 131-140 (1997).

[5] Conocar, O., et al., Promising pyrochemical actinide/lanthanide separation process using aluminum, Nucl. Sci. Eng., 153, 253-261 (2006).

[6] Herbig, R. et al., Vibrocompacted fuel for the liquid metal reactor BOR 60, J. of Nuclear Materials, 204, 93-101 (1993).

[7] INFCE – Summary volume and reports of INFCE working groups, IAEA, Vienna (1980).

[8] Heising-Goodman, C.D., An evaluation of the plutonium denaturing concept in an effective safeguards method, Nucl. Techn., 50, 242-251 (1980).

[9] Campbell, D. et al., Proliferation resistant nuclear fuel cycles, ORNL/TM-6392 (1978).

[10] Massey, J.V. et al., The role of Plutonium-238 in nuclear fuel cycles, Nucl. Techn., 56, 55 (1982).

[11] Ronen, Y. et al., A "nonproliferating" nuclear fuel for Light Water Reactors, Nucl. Techn., 96, 133-138 (1991).

[12] Shwageraus, E. et al., Use of thorium for transmutation of plutonium and minor actinides in PWRs, Nucl. Techn., 147, 53-68 (2004).

[13] Saito, M. et al., Innovative nuclear energy systems for inherently protected plutonium production, Int. Conf. on Innovative Technologies for Nuclear Fuel Cycles and Nuclear Power, Vienna, June 23-26, 2003.

[14] Saito, M., Advanced core concepts with enhanced proliferation resistance by transmutation of minor actinides, paper No. 172, p. 533, Proceedings of Global 2005, Oct. 9-13, 2005, Tsukuba, Japan.

[15] Saito, M., The technical barriers for proliferation resistance in nuclear material management, IAEA technical meeting on fissile material management strategies for sustainable nuclear energy, Sept. 12-15, 2005, Vienna, Austria.

[16] Osaka, M. et al., Aspects of Pu-238 production in the experimental fast reactor JOYO, Annals of Nuclear Energy, 32, 1023-1031 (2005).

[17] Broeders, C.H.M., and Kessler, G., Fuel cycle options for the production and utilization of denatured plutonium, Nucl. Sci. Eng., 156, 1-23 (2007).

[18] Sagara, H. et al., Denaturing of plutonium by transmutation of minor actinides for enhancement of proliferation resistance, J. Nucl. Sci. Technol., 42, 2, 161-168 (2005).

[19] Kloosterman, J.L., Plutonium recycling in pressurized water reactors: Influence of moderator-to-fuel ratio, Nucl. Techn., 130, 227-241, (2000).

[20] Broeders, C.H.M., Investigations related to the build-up of transurania in pressurized water reactors, FZKA 5784, Forschungszentrum Karlsruhe (1996).

[21] Aniel-Buchheit, S. et al., Plutonium recycling in a Full-MOX 900-MW (electric) PWR: Physical analysis of accident behaviors, Nucl. Techn., 128, 245-256 (1999).

[22] Tommasi, J. et al., Long-lived waste transmutation in reactors, Nucl. Techn., 111, 133-148 (1995).

[23] Barbrault, P., A plutonium-fueled high-moderated pressurized water reactor for the next century, Nucl. Sci. Eng., 122, 240-246 (1996).

[24] Salvatores, M., Scenarios using P/T and major challenges, Nuclear Fuel Cycle Workshop, Forschungszentrum Karlsruhe, Sept. 1-3, 2005.

[25] Broeders et al, C.H.M. KAPROS-E:, A Modular program system for nuclear reactor analysis, status and results of selected applications, Jahrestagung Kerntechnik, Düsseldorf, 2004.

[26] Driscoll, M.J. et al., The linear reactivity model for nuclear fuel management, American Nuclear Society, LaGrange, Illinois, USA, Febr. 1992.

[27] Müller, F.J., Beitrag zur Theorie der Mehrkomponenten-Isotopentrennung und deren Anwendung insbesondere auf die Untersuchung des U-236-Aufbaus im Brennstoff-kreislauf, Dissertation RWTH Aachen, Brosch. 1540 (1975).

[28] Hoppe, H.O., Urenco personal communication (Oct. 2005).

[29] Wade, D., et al., The design rationale for the IFR, Progress in Nuclear Energy, 31, 1-2, 13-42 (1997).

[30] Hill, R.N. et al., Physics studies of weapons plutonium disposition in the integral fast reactor closed fuel cycle, Nucl. Sci. Eng., 121, 17-31 (1995).

[31] Languille, A. et al., CAPRA core studies, the oxide reference option, GLOBAL 1995, Versailles, France, 874, Sept. 1995.

[32] Messaoudi, N. et al., Fast burner reactor devoted to minor actinide incineration, Nucl. Techn., 137, 84-96 (2002).

[33] Tommasi, J. et al., A coherent strategy for plutonium and actinide recycling, Transactions of the International Nuclear Congress – Atoms for Energy –, ENC 94, Vol. 2, p358, Oct. 2-6, 1994, Lyon, France.

[34] Wehmann, U. et al., : Studies on plutonium burning in the prototype fast breeder reactor MONJU, Nucl. Sci. Eng., 140, 205-222 (2002).

[35] Status of liquid metal cooled fast breeder reactors, Technical report series No. 246, IAEA, Vienna (1985).

[36] Bouchard, J., The closed fuel cycle and non proliferation issues, Proc. Int. Conf. Sustainable Nuclear Energy Systems for Future Generation (Global 2005), Tsukuba, Japan, October 9-13, 2005, Atomic Energy Society of Japan (2005).

[37] Blake, E.M., GNEP rollout means big jump for fuel cycle, Nucl. News, 49, 3, 64-69 (2006); see also "Advanced fuel Cycle Initiative", available on the Internet at www.ne.anl.gov/research.

[38] Rineiski, A., Kessler, G., Proliferation-resistant fuel options for thermal and fast reactors avoiding neptunium production, Nucl. Eng. Design, 240, 500-510 (2010).

[39] Benedict, R.W., Status of reprocessing development, Nuclear Fuel Cycle Workshop, Karlsruhe, September 1-3, 2005, Forschungszentrum Karlsruhe (2005).

13 Neptunium as a proliferation problem and fuel cycle options for avoiding neptunium production

13.1 Neptunium as a proliferation problem

Neptunium is considered usable in nuclear explosive devices (Loaiza et al. [1,2] and Albright et al. [3]). It has a bare critical mass of 57±4 kg (Sanchez et al. [4]). A reflector, e.g., beryllium can reduce this bare critical mass to approx. 45 kg. It produces virtually no alpha-particle heat and has a very low spontaneous fission neutron rate of 0.11 n/kg·s, which is lower than for U-235 (0.29 n/kg s) (Holden et al. [5]). The IAEA has begun to adopt measures to monitor neptunium (Albright et al. [3], Ottmar et al. [6], Morgenstern et al. [7]). The amount of neptunium available in civil nuclear energy programs is estimated by IAEA to be around 90 tonnes (Fukuda et al. [8]). Neptunium, therefore, should be incinerated as early as possible by the NWSs and its production should be avoided as far as possible in future denatured, proliferation-proof plutonium fuel cycles.

13.2 Neptunium-free nuclear fuel cycle

Broeders and Kessler [9] showed that reactor-grade plutonium can be converted so as to become denatured and proliferation-proof. Such reactor-grade plutonium can also be incinerated (almost completely, except the unavoidable losses of about 1% during recycling) [9,10]. This also holds for the incineration potential of neptunium and americium. Minor actinide incineration is being discussed also with the aim of minimizing the radioactive inventory of nuclear waste disposal sites (Kessler [10], Wigeland et al. [11]).

Neptunium, however, poses a problem because of its usability in nuclear explosive devices. Therefore, neptunium should be avoided in a future civil denatured, proliferation-proof nuclear fuel cycle, in which denatured, proliferation-proof plutonium and reactor-grade americium are incinerated.

13.2.1 Model of a neptunium-free nuclear fuel cycle

Neptunium cannot be denatured with other neptunium isotopes. Therefore, it should be avoided in a future denatured, proliferation-proof civil nuclear fuel cycle.

Galperin et al. [12] and Sagara [13] offered some indications that this is possible in PWR cores by combining denatured, proliferation-proof plutonium with depleted U-238 (only 0.2% U-235) and thorium. In this way, neptunium can be avoided as only protactinium isotopes and uranium isotopes up to U-234 (Fig. 13.1) or plutonium isotopes as well as americium and curium (Fig. 13.2), but no significant amounts of U-236 or neptunium are produced. The U-238 is necessary to keep the originating U-233 denatured ≤12% U-233 in U-238 (Section 8.1). Neptunium can only be produced in tiny amounts via neutron capture in U-236 or α-decay of Am-241 (Section 13.4.1).

Americium can be used together with U-238 and reactor-grade plutonium in FR cores. This produces some Pu-238. This occurs by alpha decay of Cm-242 (Fig. 11.1 in Section 11.2). Irradiation experiments in a fast reactor core have shown that Pu-238 can be produced by converting americium into Cm-242 (Walker et al. [14] and Sagara et al. [15]). The use of thorium and U-238 allows producing new U-233 denatured in U-238.

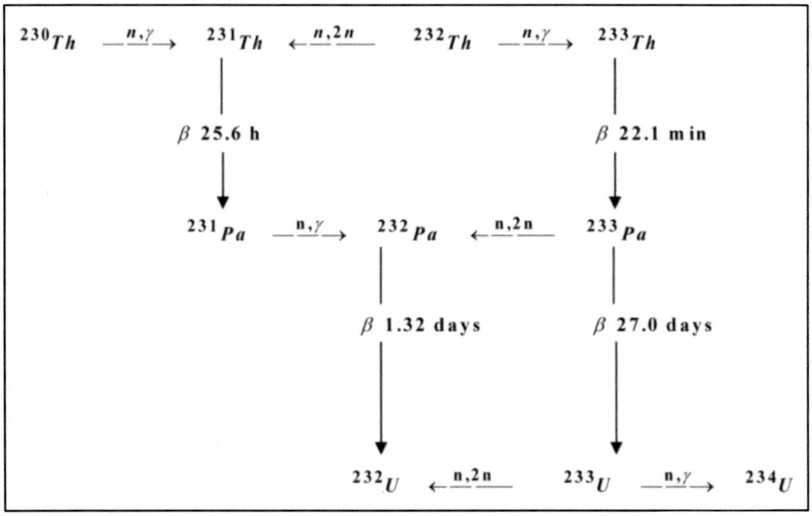

Figure 13.1: Buildup of isotopes in the thorium fuel cycle.

13.2.2 Future proliferation-proof, neptunium-free fuel cycles

Starting from the earlier results of Galperin et al. [12] and Sagara [13] a scientific concept for a future proliferation-proof fuel cycle was proposed by Kessler [16] and Rineiski et al. [17]. The initial amount of proliferation-resistant plutonium would have to be produced by using existing PWR reactors, as well as reprocessing and refabrication facilities of the NWSs or multilateral reprocessing and refabrication centers (MLRCs) as suggested by IAEA [18].

After the initial amount of denatured, proliferation-proof plutonium is available, it can be utilized together with small amounts of proliferation-proof americium in nuclear reactors outside of NWSs, provided that the neptunium production is drastically minimized by proper fuel cycle options. This can be accomplished, e.g. by minimizing the U-235 content and avoiding U-236 in the fresh fuel through the use of depleted uranium (hereafter it is assumed that depleted uranium contains 99.8% of U-238 and 0.2% of U-235). This uranium can be (1) mixed with U-233, thorium, denatured, proliferation-proof plutonium and americium and be used in PWRs or (2) mixed with denatured, proliferation-proof plutonium and americium and used in fast reactors (FRs) or accelerator driven systems (ADSs). The ADSs are not considered in detail here, but would also be characterized by a fast neutron spectrum.

13.3 Initial fuel composition for proliferation-proof plutonium and neptunium-free fuel cycles

Several constraints must be taken into account if proliferation-proof plutonium shall be recycled and incinerated in PWRs without generation of neptunium. Proliferation-proof plutonium would lead to positive coolant temperature coefficients in PWR cores. This was demonstrated by Broeders [19] who showed that a plutonium composition above 5.5% Pu-238 (corresponding to D1 in Table 9.5) leads to positive coolant temperature coefficients in PWR cores.

Rineiski and Kessler [17] also found strong positive coolant temperature coefficients for a plutonium composition with 7.7% Pu-238 shown in Table 13.1. Such PWR cores can not be licensed by regulatory authorities.

Table 13.1: Plutonium composition in recycled denatured PWR MOX fuel with burnup of 50 GWd/t, after 10 years cooling.

Pu isotope	at.%
Pu-238	7.7
Pu-239	44.0
Pu-240	31.0
Pu-241	10.3
Pu-242	7.0

U-235 cannot be admixed to the plutonium – it would lead to neptunium production via U-236. The only solution is the admixture of several percent of U-233 in U-238 and thorium to the proliferation-proof MOX fuel. An increase of the moderator to fuel ratio by wider spacing of the fuel rods in the fuel element will also ameliorate the coolant temperature coefficient.

13.4 Selection of fuel composition for neptunium-free proliferation-proof fuel cycles

The following four cases of fuel compositions and moderator to fuel ratios M/F were analyzed by Rineiski and Kessler [17] (Table 13.2). The uranium (about one third of the fuel content) consists of depleted uranium mixed with U-233 to keep the uranium denatured. Depending upon the case, a different amount of U-233 was mixed with depleted U in order to obtain proper criticality values. The assumed isotopic composition of the proliferation-proof plutonium is given in Table 13.1. The isotopic composition of americium is Am-241 to Am-243 in the ratio 3:1.

The four cases with different percentages of americium: 0% (case 1), 0.5% (cases 2 and 3) and 1.5% (case 4) of americium in the fresh fuel shall show the effect of americium on the build-up of Pu-238 (via Cm-242 decay, see Fig. 11.1, Section 11).

Table 13.2: Fuel compositions and M/F ratios for four investigated cases.

case	P/D ratio	M/F ratio	Thorium wt%	Uranium, including U-233 wt%	U-233 wt%	Denatured Plutonium wt/%	Americium wt%
1	1.40	2.0	54.67	34.52	1.69	10.81	0
2	1.40	2.0	54.67	34.02	2.03	10.81	0.50
3	1.686	3.50	54.66	34.03	1.25	10.81	0.50
4	1.686	3.50	54.66	33.02	3.17	10.81	1.51

For the analysis it was assumed that fresh fuel contains no curium. It is a strong spontaneous neutron emitter (mainly due to Cm-244 with half life of about 18.1 years, Cm-242 decays relatively fast). Its presence would make fuel fabrication very difficult. To avoid handling curium during PWR fuel fabrication, the minor actinides, americium and curium, must be separated from each other during spent fuel reprocessing [20]. Curium could be stored in special storage facilities (where Cm-244 would decay to Pu-240) (Section 7.9.7).

The four cases of Table 13.2 assure a sufficiently strong negative coolant temperature coefficient, and allow a burnup of 60 GWd/t (Section 13.5).

13.4.1 Isotopic compositions of the fuel during burnup [17]

In Fig. 13.2 the variations of the plutonium isotopic compositions during burnup for cases 1 and 2 (0% and 0.5% americium, M/F ratio of 2.0) are shown. The corresponding results for cases 3 and 4 (0.5% and 1.5% americium, M/F ratio of 3.5) are presented in Fig. 13.3. In both cases, the lines with markers show the cases with the higher americium content. The percentage of Pu-239 is strongly decreasing, whereas the percentages of Pu-238, Pu-240, Pu-241 and Pu-242 are increasing.

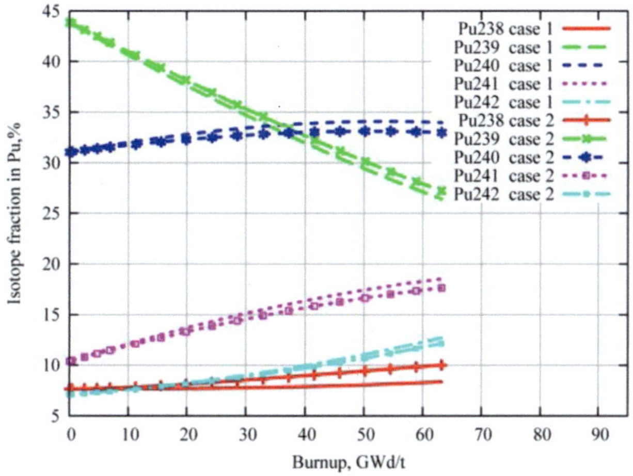

Figure 13.2: Plutonium isotopic fraction variations during burnup, cases 1 and 2 [17].

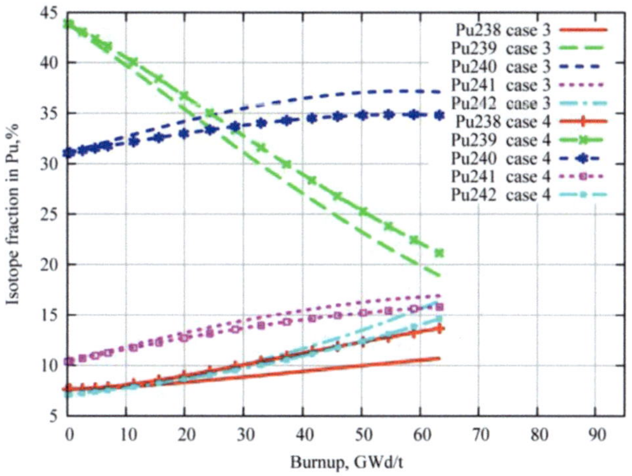

Figure 13.3: Plutonium isotopic fraction variations during burnup, cases 3 and 4 [17].

One may conclude that the initial americium content affects appreciably the Pu-238 build-up, whereas the relative variations vs. time for the content of the other plutonium isotopes can be considered as less dependent on the americium content. Table 13.3 shows the isotopic compositions of the denatured plutonium for the begin of the burnup cycle (BOC) and the end of the burnup cycle (EOC) at about 60 GWd/t for all 4 cases of different americium content and different M/F ratios. It can be seen that the admixture of americium increases the Pu-238 percentage during burnup. However, the Pu-238 percentage increases also slightly during burnup even if no americium is added to the fresh fuel in accordance with pronounced decrease of the Pu-239 content.

Table 13.3: Plutonium isotopic fraction at BOC and EOC (at about 60 GWd/t) for different americium content in the fresh fuel.

Pu isotope wt% at BOL		wt% at EOL			
		Case 1	Case 2	Case 3	Case 4
Pu-238	7.7	8.2	9.8	10.5	13.2
Pu-239	44.0	27.4	28.2	20.3	22.4
Pu-240	31.0	34.0	33.1	37.2	33.9
Pu-241	10.3	18.2	17.3	16.8	15.6
Pu-242	7.0	12.1	11.6	15.3	13.8

If no americium is put into the fresh fuel, an appreciable amount of it is produced up to EOC (Fig. 13.4). A net production can be avoided if about 0.5% of americium is added to the fuel at BOC. For the higher initial americium content (1.5%), the incineration of americium is about 6 kg/t of fuel after a burnup of 60 GWd/t (see Fig. 13.4).

The higher is the americium content, the higher is the curium production and the lower is the plutonium incineration, see Figs 13.5 and 13.6. At EOC, the fraction of Am-241 in americium is between 33% and 45%, but after cooling for a few years it approaches or exceeds 70% due to decay of Pu-241

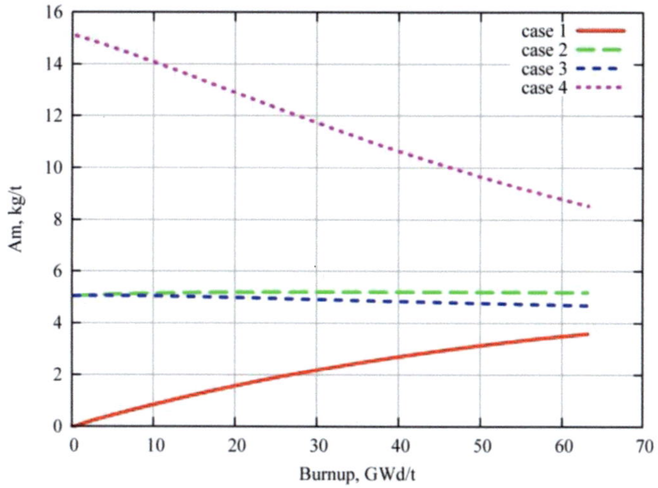

Figure 13.4: Americium content in the fuel during burnup [17].

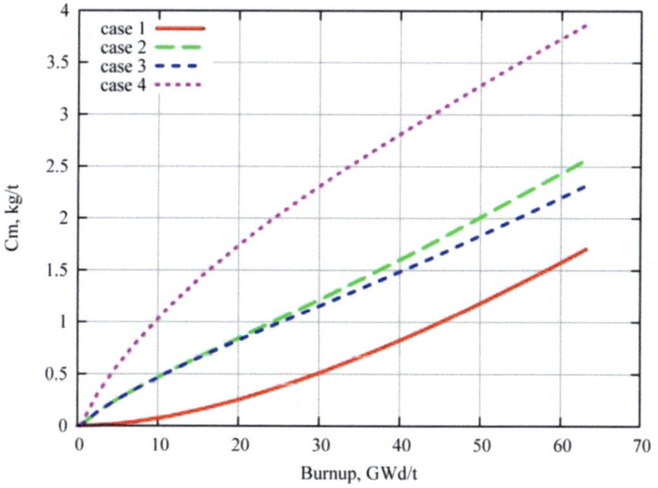

Figure 13.5: Curium content in the fuel during burnup [17].

Fig. 13.6 shows the plutonium incineration for all four cases. For case 3 (0.5% of americium, M/F=3.5) the plutonium incineration rate is the highest (about 40 kg/t of spent fuel) over a burnup period of 60 GWd/t. The pronounced difference between the curves of case 3 and case 4 in Fig. 13.6 is due to the fact that a higher content of U-233 is present in the fresh fuel of case 4 which is fissioned during burnup, thus reducing the contribution of plutonium isotopes to energy production.

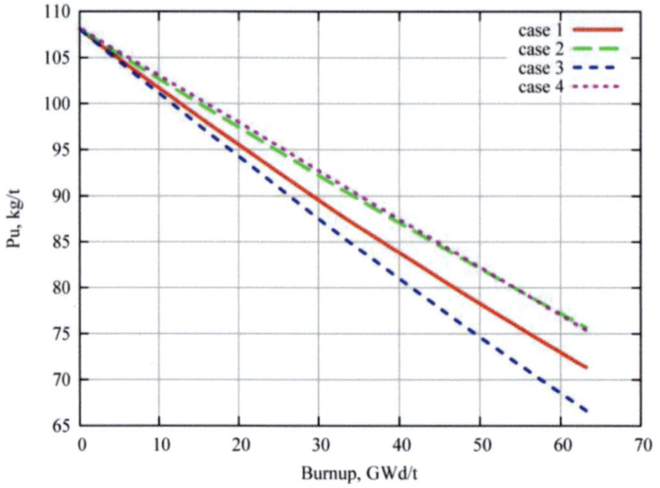

Figure 13.6: Plutonium content in the fuel during burnup [17].

The U-233 content in the fresh fuel at BOC depends upon the amount of americium and M/F ratio (see Table 13.2). The fraction of U-233 in uranium is higher for higher americium content (provided that the M/F ratio is the same). This is shown in Fig. 13.7. It is necessary to compensate the strong negative influence of americium on criticality in the PWR. The fraction is lower for a larger M/F ratio (provided that the Am content is similar); this is possible due to higher contribution of U-233 to the neutron balance in a better moderated environment.

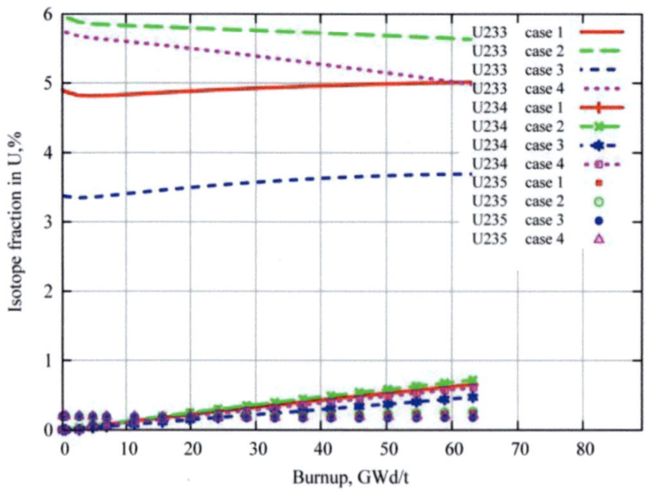

Figure 13.7: Uranium isotopic fraction variations during burnup [17].

During burnup, the U-233 fraction (denatured in U-238) does not vary appreciably (due to the large thorium content in the fuel) except for case 4 (see Fig. 13.7). The uranium content in the fuel decreases only slightly, by ca. 1-2% (e.g. from ca. 34% to ca. 32%) for the considered burnup of ca. 60 GWd/t. The U-233 inventory (including Pa-233) increases by ca. 3% in case 3 and by ca. 0.5% in case 1, but decreases by ca. 11% in case 2 and by ca. 21% in case 4. Fig. 13.7 also shows the fractions of the isotopes U-234 in uranium for cases 1 to 4. The U-235 inventory slightly increases, but remains quite small.

The results of calculations [17] show a negligible amount of neptunium in the spent fuel after a burnup of 63 GWd/t and 10 years cooling time (Table 13.4). This is a consequence of the use of depleted uranium that almost excludes the production of U-236, that is converted into Np-237 in a nuclear reactor.

The neptunium appears mainly due to the alpha-decay of Am-241, the latter being a decay product of Pu-241 (see Fig. 11.1 in Section 11).

Table 13.4: Neptunium content in spent fuel after a burnup of 63 GWd/t and 10 years cooling time.

case	1	2	3	4
Np content %	0.014	0.016	0.013	0.018

13.5 Reactivity coefficients relevant to PWR safety

The reactivity coefficients required for the safety analysis of a PWR are shown in Table 13.5 for cases 1-4. The moderator density and temperature coefficients (MDC and MTC) are obtained by pin-cell calculations. The MDC is the reactivity effect due to 10% density decrease of the nominal coolant density (including boron). The MTC is obtained by multiplying MDC by -3.22×10^{-3} that is based on the water coolant properties at pressure of 15.6 MPa and temperatue of 583 K.

For computing the Doppler contant, the lattice reactivity values were calculated at three fuel temperatures: 300, 773, and 2100 K. A good fit of the results could be obtained, by

$$\frac{d\rho}{dT} = \frac{A_D}{\sqrt{T}}.$$

with ρ Doppler reactivity
T fuel temperature
A_D Doppler constant.

The Doppler constant is given in Table 13.5 separately for the temperature ranges 300-773 and 773-2100 K. The absolute values of the MTC and Doppler constant are only slightly lower than those for present PWRs (Tommasi et al. [21]; Kloosterman et al. [22]).

The required boron concentrations at BOL are in all cases too high for application of conventional soluble boron acid. More refined solutions applying solid burnable poisons, e.g. gadolinium or erbium are required.

From the above results it can be concluded that the objective of a neptunium-free PWR fuel cycle with proliferation-proof reactor-grade plutonium can be fulfilled, in particular by case 1. If, in addition, americium shall be incinerated, cases 2 to 4 could be applied.

Table 13.5: Reactivity coefficients relative to safety of PWR [17].

		Case 1, 0% americium	Case 2, 0.5% americium	Case 3, 0.5% americium	Case 4, 1.5% americium
Moderator density coefficient [MDC]	BOL EOL	0.100 0.154	0.004 0.142	0.091 0.121	0.089 0.116
Moderator Temperature Coefficient [MTC] [pcm/K]	BOL EOL	-32.2 -49.7	-32.6 -45.7	-29.2 -38.5	-28.8 -37.2
Doppler constant (A_D) between 300-773 [K]	BOL EOL	$-2.33x10^{-3}$ $-2.84x10^{-3}$	$-2.71x10^{-3}$ $-2.81x10^{-3}$	$-1.65x10^{-3}$ $-2.11x10^{-3}$	$-2.71x10^{-3}$ $-2.10x10^{-3}$
Doppler constant A_D between 773–2100 [K]	BOL EOL	$-2.18x10^{-3}$ $-2.67x10^{-3}$	$-2.16x10^{-3}$ $-2.62x10^{-3}$	$-1.57x10^{-3}$ $-2.00x10^{-3}$	$-1.61x10^{-3}$ $-1.97x10^{-3}$
Boron efficiency [pcm/ppm]	BOL EOL	-1.26 -1.74	-1.13 -1.62	-2.49 -4.29	-2.13 -3.13
Boron concentration to bring k_∞ to 1.03 [ppm]	BOL	10930	13360	8050	10670

MDC: Moderator Density Coefficient calculated from 10% reduction at nominal density
MTC: Moderator Temperature Coefficient, being $-3.22.10^{-3}$ x MDC at nominal coolant conditions
DC: Doppler Coefficient calculated from fit of results for T=300, 773 and 2100K
Boron at BOL: The boron concentration to obtain $k_\infty \approx 1.03$ at BOL

13.6 Peculiarities and technical modifications required for the PWR design

In comparison to present PWR core designs with low enriched uranium fuel the above PWRs – incinerating proliferation-proof plutonium and americium – would need several design modifications. They may need adaptations of the control and shim rod system, as well as a different number of solid burnable poison rods at begin of the burnup cycle. The relatively high (as compared to conventional values near 2.0) M/F ratio of 3.5 can be attained in the fuel assembly design by reducing the number of fuel rods per assembly, e.g. through replacement of a part of them by water rods (Barbrault [23]). If a larger M/F (or P/D) ratio is chosen, either a smaller electrical power output is obtained or the diameter of the pressure vessel must be increased (if the active core height remains the same). If the fresh fuel is doped with americium, more research will be needed for fuel fabrication and more experimental experience regarding fuel behavior is required.

Because of the higher Pu-238 content in the proliferation-resistant reactor plutonium with its resulting high heat production and spontaneous neutron radiation and γ-radiation, the present aqueous reprocessing technology would have to be modified and the present MOX refabrication technology would not be feasible any more. Present glove-box-type MOX refabrication technology is limited to a Pu-238 isotopic content of ~4% and present aqueous reprocessing technology to ~5% (Broeders and Kessler [9]). Advanced aqueous or pyrochemical reprocessing for plutonium/thorium/uranium fuel and related fuel refabrication technology applying remote handling will become necessary. These advanced technologies are currently being developed in the Europe, Japan, Russia, and the United States in the context of actinide transmutation research and would have to be applied.

As was shown in Fig. 13.3 the Pu-239 isotopic content in the plutonium/thorium/uranium/americium fuel will decrease from 44% to about 20% over a burnup of 60 GWd/t. For further recycling steps this fuel can be mixed always with fresh proliferation-resistant plutonium with about 40% Pu-239 coming from the NWSs. After several steps this proliferation-proof plutonium may have to be loaded into fast reactors (due to variations in the isotopic composition that will make fuel fabrication difficult and may worsen PWR safety coefficients). There will certainly be small penalties in fuel cycle costs compared to present MOX fuel according to the degree of complexity given by the fuel reprocessing and refabrication technology.

13.7 Conclusion for incineration of proliferation-proof plutonium in PWR cores [17]

Due to the isotopic content of >5-6% Pu-238 in proliferation-proof plutonium the mixed oxide plutonium-uranium fuel would lead to positive coolant temperature coefficients. Such PWR cores cannot be operated for safety and licensing reasons. However, a mixture of proliferation-proof plutonium with depleted uranium, thorium and low enriched with U-233 is feasible. This leads to acceptable safety coefficients for the PWR cores and minimizes to production of neptunium to less than 0.02% in the spent fuel after a burnup of about 60 GWd/t. Incineration rates of 30-40 kg proliferation-proof plutonium per tonne of fuel are possible.

13.8 Fast reactor fuel cycle for utilizing americium as well as denatured proliferation-proof plutonium but avoiding neptunium production [17]

Fast spectrum reactors (FRs) or fast breeder reactors (FBRs) can incinerate some of the plutonium isotopes and of the minor actinides much better than LWRs with their thermal neutron spectrum. In a PWR-MOX core most of the neutron fission reactions occur in the 0.1 eV (average) range of neutron kinetic energy. In a FR core fission occurs in the 0.2 MeV range of neutron kinetic energy.

The ratio of fission to absorption in a PWR core and an SFR core is given by Fig. 13.8. It can be understood that in an SFR core the isotopes Np-237, Pu-238, Pu-240, Pu-242, Am-241, Am-243 and Cm-244 are fissioned at a much higher rate than in thermal spectrum reactors, e.g. PWRs. The net result is that more excess neutrons are available and less higher actinides are generated.

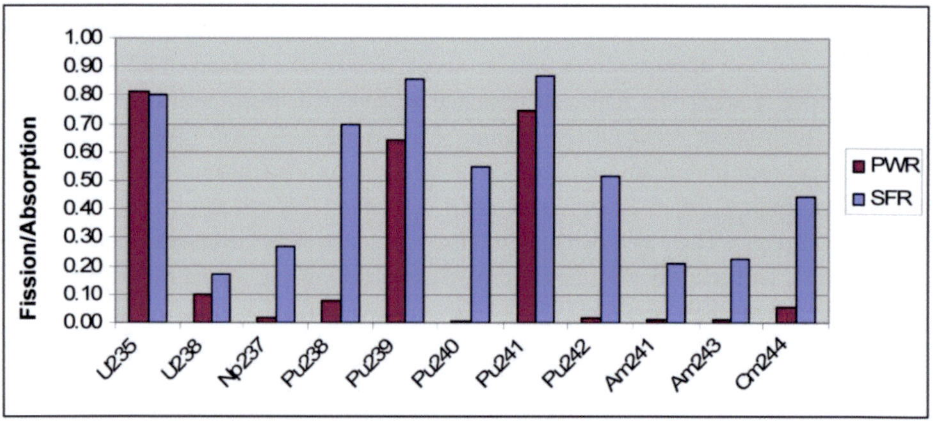

Figure 13.8: Impact of energy spectrum on incineration and transmutation performance [26].

Rineiski and Kessler [17] used the BN-600-core as an example for their analyses of a plutonium proliferation-proof and neptunium-free fuel cycle for fast spectrum reactors.

Fig. 13.9 shows a cross section of a BN-600 type MOX-fuelled core with lower fertile blanket, an internal fertile blanket (a 5 cm layer at axial mid-plane in the inner core), a radial steel reflector and a sodium plenum (a region with a large sodium volume fraction) above the core [17,24]. The core fuel contains depleted uranium mixed with denatured proliferation-proof plutonium (Table 13.1) and americium (4% wt. of total heavy metal content). The isotopic composition of americium in Am-241 to Am-243 shall be in the ratio 3:2.

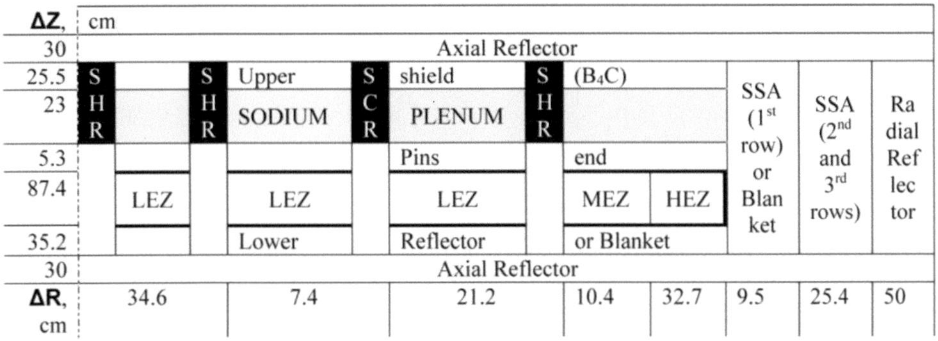

Figure 13.9: R-Z model for the BN-600 type reactor [17,24].

The core is subdivided into a low enrichment inner zone (LEZ), a middle enrichment zone (MEZ), and a high enrichment outer zone (HEZ). The control and shim rods (SCR/SHR) are radially interspersed in the LEZ region.

The core zones contain 24.6 wt% fuel and americium, 52.7 wt% steel and 22.7wt% sodium. They have the following plutonium enrichments and fractions (wt%) of uranium, plutonium and americium given in Table 13.6. The isotopic composition of the plutonium is given by Table 13.1.

Table 13.6: Composition (fractions) of the different core zones LEZ, MEZ, HEZ.

Fraction	LEZ	MEZ	HEZ
uranium wt%	76.8	74.0	71.0
plutonium wt.%	19.2	22.0	25.0
americium wt%	4.0	4.0	4.0

Two options are considered: (1) axial lower and radial steel reflectors (SSA) and (2) a core surrounded by axial lower and radial fertile blankets containing depleted uranium mixed with 4.2% of americium (isotopic ratio Am-241:Am-243 equal to 3:1) and 1.08% proliferation-proof plutonium (Table 13.1) (at BOL).

Case 1 is chosen for plutonium incineration. Case 2 represents an option for breeding proliferation-proof, denatured plutonium. The americium is added to the core or core and blankets (case 2) to generate Pu-238. This assures that the plutonium fuel in the core remains denatured or proliferation-proof. Plutonium produced in the blankets becomes also denatured or proliferation-proof (although Pu-239 is generated in the core and the blankets through neutron capture in U-238). Only proliferation-proof plutonium is present in the reactor core during operation. In the breeding blankets proliferation-proof plutonium develops via decay of Cm-242 (see Section 13.8.3). Neptunium production is minimized by the use of depleted uranium with only 0.2% U-235.

Similar to the PWR cases investigated in Section 13.4 americium consists of Am-241 and Am-243 in the ratio of about 3 to 1. No curium is assumed in the fresh fuel. For both options (burner and breeder) core operation at a thermal power of 1470 MW_{th} and 1450 full power days at end of burnup (EOL end of life) are assumed. This leads to an average burnup of about 185 or 113 MWd/t, for the burner or breeder case. The total actinide mass in the burner is about 115 t, while in the breeder case it is about 189 t (the difference amounting to the mass in the blankets). The control rod positions are kept constant during burnup. This modeling is certainly an approximation allowing to see the trends in core reactivity, void effect and isotopic composition during core operation.

13.8.1 Results of the FR core calculations

The k_{eff} values as well as core void effects at the BOL and EOL and the conversion (for burner or incinerator) and breeding (for breeder) ratios are shown in Table 13.7. The effective delayed neutron fraction is about 340 pcm, the Doppler constant is -485 pcm (in the range from 1500 to 2100 K) in case of burner at BOL.

Table 13.7: Criticality, void effect and breeding ratio for the BN-600 type reactor operating with denatured proliferation-proof plutonium [17].

	k_{eff}, BOL	k_{eff}, EOL	Void effect BOL, pcm	Void effect EOL, pcm	Conversion or breeding ratio
Burner	1.0190	0.8689	1777	2691	0.78
Breeder	1.0068	0.9044	1716	2731	1.07

The conversion or breeding ratios (CR and BR) are defined in Section 4.7.3. The void effect given in Table 13.7 is only due to the core voiding. If both the core and the region (including sodium plenum) above the core were voided, the void effect would be lower by 530 pcm, e.g. for the burner case.

13.8.1.1 Isotopic composition of plutonium in case of fast reactor burner (incinerator)

Variations of the plutonium isotopic composition (averaged in space over the core in case of burner and over the core and blanket in case of breeder) during burnup (followed by 5 years of cooling after unloading) are shown in Figs. 13.10 and 13.11. The sudden variations in concentrations of Pu isotopes at the end of burnup are due to decay of unstable isotopes after cooling, that increases, in particular, the fraction of Pu-238, thus decreasing (relatively) other fractions.

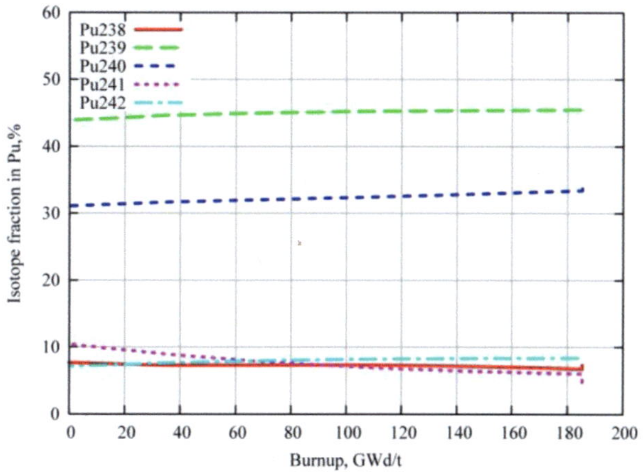

Figure 13.10: Plutonium isotopic composition in the core of the FR operating as burner [17].

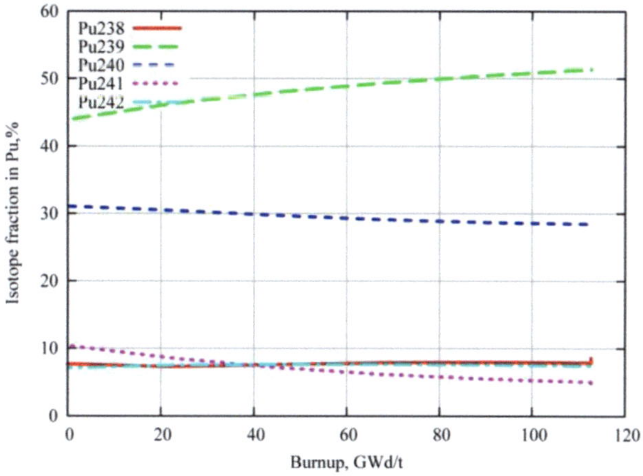

Figure 13.11: Average plutonium isotopic composition in the FR breeder (core and breeder fuel mixed after unloading) [17].

It can be seen that the plutonium remains proliferation-proof in average in both cases (burner or breeder) as long as the fuel of the core and the blankets are mixed after unloading in case of the FR breeder. However, the blanket plutonium isotopic composition differs considerably from the one in the breeder core. This needs a detailed discussion and design modifications in view of future proliferation-proof blanket fuel.

13.8.1.2 Proliferation-proof blanket fuel for FR-breeders

The fresh fuel elements of axial and radial blankets of FR-breeder prototype reactors contain either natural uranium or depleted uranium (~0.2% U-235) as UO_2 fuel. It is commonly assumed that both core and blanket fuel after unloading and cooling are mixed together in the head end (fuel element chopping and dissolver tank) of the reprocessing plant. This can be controlled and verified by IAEA inspectors (Section 8.4).

However, if the blanket elements, especially those of the radial blanket, would be unloaded separately and chemically reprocessed (violation of the NPT and IAEA safeguards) the resulting plutonium is weapon-grade. This is demonstrated by Fig. 13.12 which shows the Pu-239 content of the plutonium produced in the axial and radial blankets of the Indian Prototype Fast Breeder Reactors (PFBR) [27,28]. For this PFBR the axial blankets remain in the core for about 540 full power days, whereas the radial blanket elements remain 1440 full power days, before they are unloaded. All plutonium which is produced in the axial blankets and in the radial blankets remains weapon-grade over the entire operation time of 540 or 1440 full power days which is demonstrated by comparison with the classification of US-DOE and US-NRC [29] (Tab. 9.2).

Table 13.8: Isotopic composition of supergrade weapon plutonium and weapon-grade plutonium [29].

Pu-Isotope	Pu-238	Pu-239	Pu-240	Pu-241	Pu-242
Supergrade weapon-Pu	0% Pu-238	97% Pu-239	3% Pu-240	0% Pu-241	0%-242
Weapon-grade plutonium	0.01%	93.8%	5.8%	0.35%	0.022%

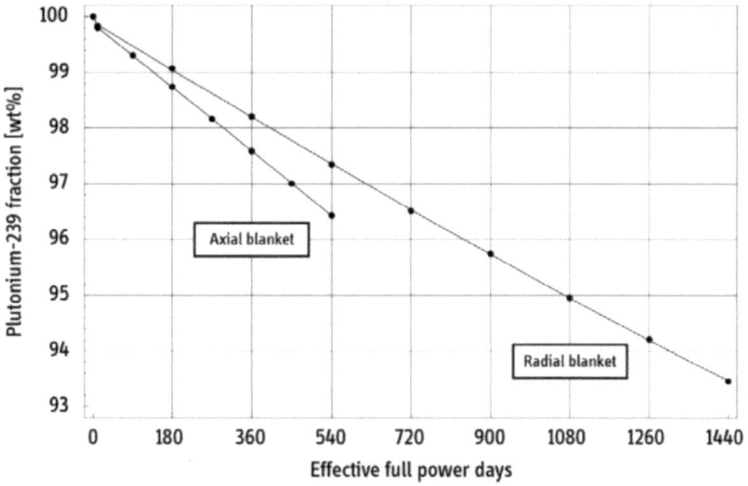

Figure 13.12: Plutonium-239 fraction of the plutonium produced in the Indian Prototype Fast Breeder Reactor (PFBR) as a function of irradiation time in the axial or radial blanket fuel elements [27].

The quality of the plutonium in the axial blankets even remains supergrade. This can be modified by admixing to the UO_2 blanket fuel about 4.2% reactor-grade americium and about 1.08% proliferation-proof plutonium. The results of such calculations [30] are shown in Fig. 13.13 which shows the development of the fractions of americium, plutonium and Cm-242 as a function of core full power days in the radial blanket fuel elements.

Fig. 13.14 shows the isotopic fractions of the different plutonium isotopes Pu-238, Pu-239, Pu-240, Pu-241 and Pu-242 in the radial blanket fuel elements for the case of proliferation-proof plutonium with initially 11.4% Pu-238.

During the initial time period of about 200 to 300 days the initial proliferation-proof plutonium will be diluted by the fresh plutonium with more than 98% Pu-239 (the Pu-238 as decay product of Cm-242 appears later). Therefore, the initial 1.08% proliferation-proof plutonium must have a Pu-238 content of about 11.4% (see the plutonium composition E1 in Table 9.6b). In this case the blanket plutonium will always remain above 7.8% Pu-238 (Fig. 13.14), i.e. proliferation-proof (non-proliferation level II being defined in Section 14).

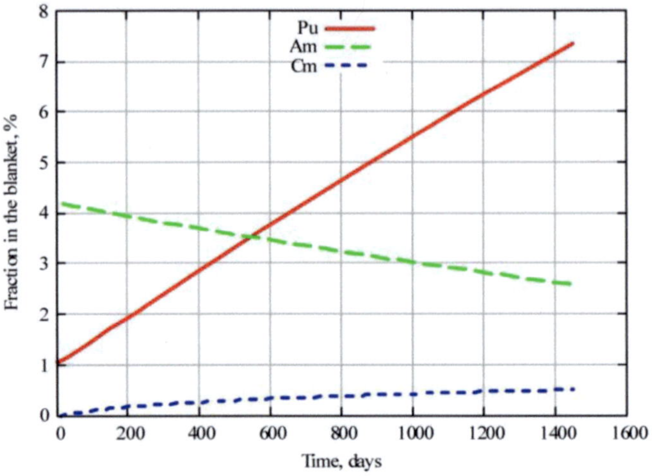

Figure 13.13: Fraction (%) of americium, curium and plutonium in radial blanket fuel elements of BN-600 type FBR (Initial fractions 4.2% reactor americium, 1.08% proliferation-proof plutonium with 11.4% Pu-238) [30].

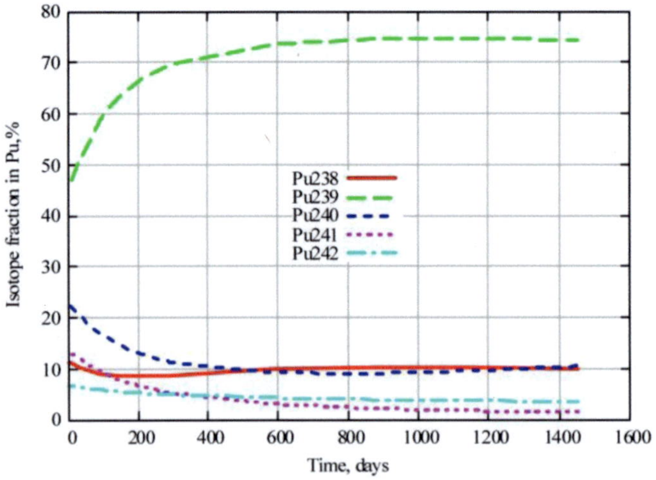

Figure 13.14: Fractions (%) of the different plutonium isotopes as a function of time in the radial blanket fuel elements. The initial plutonium composition with 11.4% Pu-238 remains always above 8% Pu-238 (proliferation-proof) over the full irradiation time period in the FBR [30].

For proliferation-proof plutonium of non-proliferation level Ia or Ib (Section 14) the excess percentage in the initial proliferation-proof plutonium and the fraction of reactor-grade americium of the fresh blanket fuel can be lower.

13.8.2 Plutonium incineration and breeding [17]

The incineration of the denatured plutonium (reduction of its content in the fuel) is shown in Fig. 13.15. In case of operating the FR as burner, the denatured plutonium is incinerated, the reduction being about 33 kg/t of fuel over a burnup of 160 GWd/t. In case of operating the FR as breeder, some additional amount of proliferation-proof plutonium is generated, just about 3 kg/t of fuel over the burnup period of 113 GWd/t.

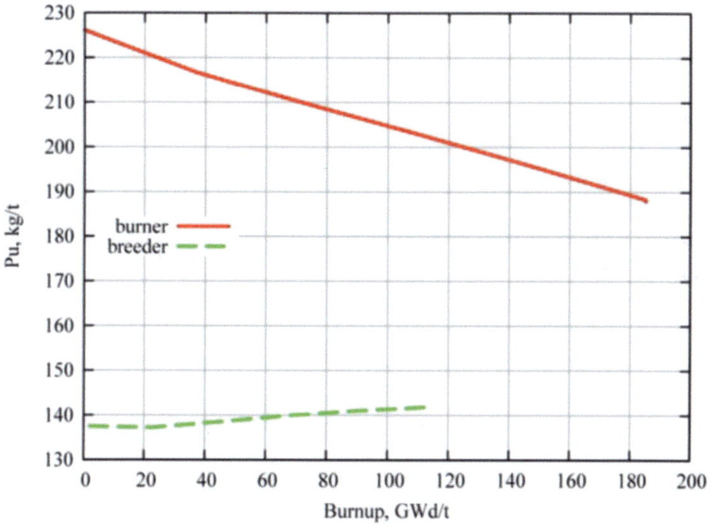

Figure 13.15: Plutonium content per t_{HM} fuel in the FR burner and in the FR breeder (reactor average) during reactor operation [17].

13.8.3 Americium incineration and production of curium

The results for the incineration of americium and the production of curium are shown in Fig. 13.16. and 13.17. At EOL, the fraction of Am-241 in americium is about 70%, this fraction does not vary appreciably after cooling over 2 years.

As indicated earlier, the results of Fig. 13.16 and 13.17 also show the effect of the decay of short-lived actinides, e.g. Pu-241 and Cm-242 for a cooling period of 5 years after unloading of the fuel.

The neptunium content at EOL and after 5 years cooling time is minor. It is only about 0.06 kg/t for the FR burner and for the FR breeder.

Similar to the PWR options, this example of a BN-600-type design loaded with proliferation-proof plutonium and americium fuel is considered conceptually and technically feasible.

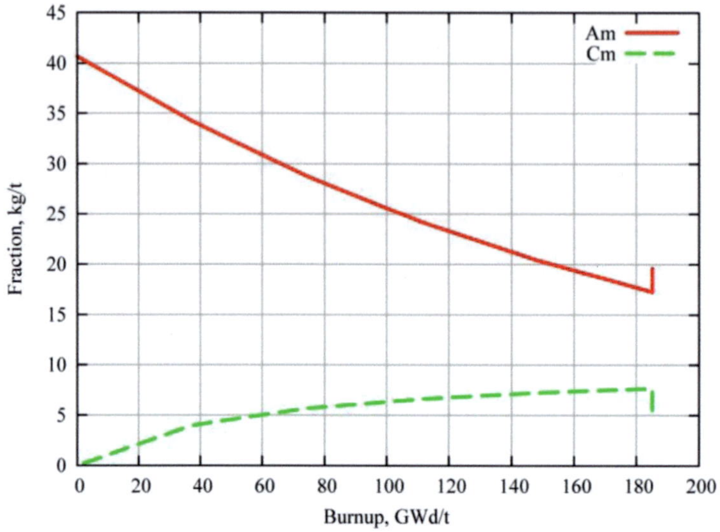

Figure 13.16: Americium and curium content in the burner during reactor operation [17].

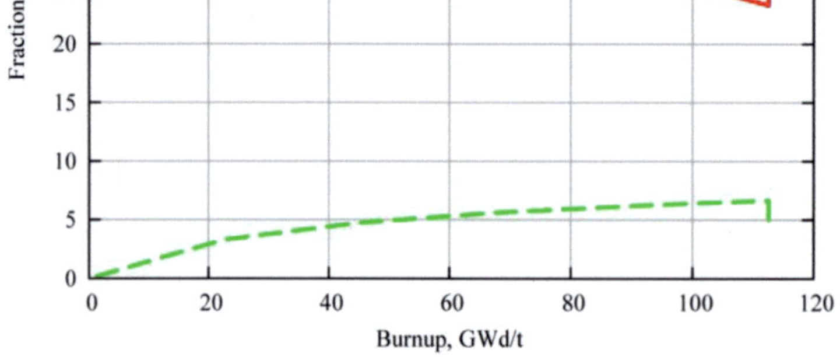

Figure 13.17: Am and Cm content in the breeder during reactor operation [17].

13.9 Conclusion for FRs operating with proliferation-proof plutonium and americium, but avoiding neptunium

The above results show that FRs can be operated with proliferation-proof plutonium and americium. The fresh FR fuel contains no curium. It is assumed that the chemical separation of americium and curium can be developed to technical scale [20,25].

It was shown that FR-Burners can incinerate the proliferation-proof plutonium at a relatively high incineration rate of 33 kg/t of fuel over a burnup of 160 GWd/t. Also breeding is possible with axial and radial blankets. The new plutonium generated by the breeding process becomes also proliferation-proof either already during reactor operation or during the necessary cooling process of the spent blanket fuel elements. The conceivable objection that there exists a short period of time where the blanket plutonium would not be proliferation-proof can be resolved by adding a small amount of proliferation-proof plutonium and a small fraction of reactor-grade americium to the fresh depleted uranium blanket fuel – if considered necessary.

The essential result is that only proliferation-proof plutonium exists in such an FR fuel cycle, i.e. in the reprocessing plant, refabrication plant and in the reactor. **This opens the possibility of future civil proliferation-proof nuclear applications having MOX-PWRs operating in the neptunium-free fuel cycle and FRs operating with proliferation-proof plutonium and proliferation-proof americium creating their own proliferation-proof plutonium and proliferation-proof americium. In such a way depleted uranium can be utilized over thousands of years.**

References Section 13

[1] Loaiza, D.J. et al., Criticality data for spherical U-235, Pu-239 and Np-237 systems reflector-moderated by low capturing-moderator materials, Nucl. Techn., 146, 143-154 (2004).

[2] Loaiza, D.J. et al., Results and analysis of the spherical Np-237 critical experiment surrounded by highly enriched uranium hemispherical shells, Nucl. Sci. Eng., 152, 65-75 (2006).

[3] Albright, D. et al., Troubles Tomorrow? Separated Neptunium-237 and Americium, Chap. V in "The Callenges of Fissile Material Control", Institute for Science and International Security (1999).

[4] Sanchez, R., et al., Criticality of a Np-237 sphere, Nucl. Sci. Eng. 158, 1-14 (2008).

[5] Holden, N.E., et al., Spontaneous fission half-lives for ground-state nuclides, Pure Appl. Chem., 72, 8, 1525-1562 (2000).

[6] Ottmar, H. et al., Demonstration of measurement technologies for neptunium and americium verification in reprocessing, IAEA-SM-367/14/07/P, International Atomic Energy Agency (2001).

[7] Morgenstern, A. et al., Analysis of Np-237 in spent fuel solutions, Radiochimica Acta, 90, 389-393 (2002).

[8] Fukuda, K. et al., Prospects of inventories of uranium, plutonium and minor actinides and mass balance. 2nd consultancy Meeting on Protected Plutonium Production (PPP)-Project, IAEA, Vienna (2006).

[9] Broeders, C.H.M. and Kessler, G., Fuel cycle options for the production and utilization of denatured plutonium, Nucl. Sci. Eng., 156, 1-23 (2007).

[10] Kessler, G., Requirements for nuclear energy in the 21st Century. Progress in Nuclear Energy, 40, No. 3-4, 309-325 (2002).

[11] Wigeland, R.A. et al., Waste management aspects of various fuel cycle options, Technical Meeting on Fissile Material Management Strategies for Sustainable Nuclear Energy, Sept. 12-15, 2005, Vienna, Austria (2005).

[12] Galperin, A. et al., A pressurized water reactor plutonium incinerator based on thorium fuel and seed-blanket assembly geometry, Nucl. Techn., 132, 214-226 (2000).

[13] Sagara, H., Plutonium denaturing by PPP Technology, Second consultancy meeting on Protected Plutonium Production (PPP)-Project, IAEA, Vienna (2006).

[14] Walker, C.T. et al., Transmutation of neptunium and americium in a fast neutron flux: EPMA results and KORIGEN predictions for the superfact fuels, J. of Nucl. Mat., 218, 129-138 (1995).

[15] Sagara, H., et al., Numerical analysis of irradiated Am samples in Experimental Fast Reactor JOYO, Materials Science and Engineering, 9 (2010).

[16] Kessler, G., Proliferation resistance of americium originating from spent irradiated reactor fuel of pressurized water reactors, fast reactors and accelerator driven systems with different fuel cycle options, Nucl. Sci. Eng., 159, 56-82 (2008).

[17] Rineiski, A., Kessler G., Proliferation-resistant fuel options for thermal and fast reactors avoiding neptunium production. Nucl. Eng. Design, 240, 500-510 (2010).

[18] IAEA, Multilateral approaches to the nuclear fuel cycle: Expert group report submitted to the Director General of the IAEA, INFCIRC/640 (2005).

[19] Broeders, C.H.M., Investigations related to the build-up of transurania in pressurized water reactors, FZKA 5784, Karlsruhe, Germany (1996).

[20] CLEFS, Nr. 33, Le Futur de Retraitement, L'Energie Atomique, Paris, France (1996).

[21] Tommasi, J., et al., Long-lived waste transmutation in reactors. Nucl. Techn. 111, 133-148 (1995).

[22] Kloosterman, J.L., et al., Plutonium recycling in pressurized water reactors: influence of the moderator-to-fuel ratio. Nucl. Techn. 130, 227-241 (2000).

[23] Barbrault, P., A plutonium-fueled high-moderated pressurized water reactor for the next century, Nucl. Sci. Eng., 122, 240-246 (1996).

[24] Kim, Y.J., et al., BN-600 full MOX core benchmark analysis. In: Proc. PHYSOR, Chicago, Illinois. American Nuclear Society, Lagrange Park, IL (2004).

[25] Warin, D., Minor actinide partitioning, 1st ACSEPT Workshop, Lisbon, March 31, 2010.

[26] Hill, R., Fuel cycle subcommittee, Overview and Status, Fusion-Fission Hybrid Workshop, Gaithersburg, Md, September 30, 2009, see also: http://web.mit.edu/fusion-fission/WorkshopTalks/Hill.ppt

[27] Glaser, A. et al., Weapon-grade plutonium production potential in the Indian fast breeder reactor, Science and Global Security, 15, p. 85 (2007).

[28] Pandikumar, G. et al., Multi recycling of fuel in prototype fast breeder reactor, PRAMANA-Journal of Physics, Indian Academy of Sciences, 73, 5, p. 819 (2009).

[29] National Academy of Sciences, Management and deposition of excess weapons plutonium, related options, National Academic Press, Washington, D.C. (1985).

[30] A. Rineiski, Personal communication (2010).

14 Future civil proliferation-proof fuel cycles

14.1 Introduction

This Section describes a potential long term strategy for producing and utilizing denatured, proliferation-proof reactor-grade plutonium during a transition period in existing PWRs. This proliferation-proof reactor-grade plutonium can then be incinerated by recycling it once or twice in PWRs and subsequently loading it into FR cores for further multirecycling.

It is proposed that future civil proliferation-proof plutonium-uranium or plutonium-uranium-thorium fuel cycles **shall only utilize proliferation-proof fissile materials**, i.e.

- proliferation-proof reactor-grade plutonium (Sections 9, 10 and 12)
- proliferation-proof reactor-grade americium (Section 11)
- neither produce nor utilize neptunium (only negligible amounts shall be present). The reason is: neptunium cannot be denatured, but could be used for nuclear explosive devices (Section 13).

The nuclear reactors envisaged for this proposal of a long term non-proliferation strategy shall be PWRs and FRs (BWRs can certainly also be used, but all detailed results applied here were obtained in Sections 9 through 14 for PWRs only). FRs will probably be introduced on a large scale at a later time (Section 2.7, Fig. 2.4). This is compatible with this proposal. PWRs are well suited for the conversion of currently existing plutonium into proliferation-safe plutonium during an initial transition phase.

CANDU reactors are not considered here (low burnup fuel see Section 9 with Table 9.3 and Fig. 9.47). Similarly, high temperature gas cooled reactors are not considered for this proposal, since their coated particle fuel is up to now not well suited for reprocessing and recycling.

14.2 Plutonium incineration by a multi-recycling strategy

Because of their high contribution to the long term radiotoxicity of spent fuel in the high level waste repository and for non-proliferation reasons (long term accumulation in the repository and problems of human intrusion) the reactor-grade plutonium and the minor actinides, neptunium and americium shall be destroyed with first priority and as soon as possible in existing PWRs and in FRs later. It was shown (Section 7) that multi-recycling of reactor-grade plutonium and minor actinides (neptunium and americium) is feasible with currently available technology.

Recycling of reactor-grade plutonium in PWRs leads to a higher production of minor actinides. Therefore, recycling of reactor-grade plutonium in PWRs will probably be done only once or twice [1,2,3,4]. Subsequent multi-recycling of reactor-grade plutonium in FR burners (CAPRA type) [3,5,7] or in IFRs [6] leads to more efficient destruction of reactor-grade plutonium, neptunium and americium. Only the inevitable losses during reprocessing and refabrication go into the high level waste (HLW) together with the fission products. The HLW is finally disposed to a deep geological repository.

Cm-243 and Cm-244 are decaying with a half-life of about 24 y and 18 y respectively. Cm-242 decays quickly practically already completely during the usual fuel cooling period. Curium should be chemically separated and stored for further decay of Cm-243 into Pu-239 and Cm-244 into Pu-240, respectively [3,7]. The remaining tiny parts of the Curium isotope (Cm-245), half-life 8500 y would have to be disposed to a deep geological repository or be recycled together with plutonium.

(Multi-recycling of curium together with neptunium and americium appears to be only possible in an FR, e.g. in combination with pyroprocessing. Multi-recycling of curium in thermal spectrum reactors, e.g. PWRs can eventually lead to Cf-252 production, thus requiring longer intermediate storage and making refabrication of actinide doped fuel extremely difficult [7]).

14.3 Needed capacity of reprocessing and Pu/U refabrication plants

The incineration of reactor-grade plutonium and minor actinides by multi-recycling needs sufficient reprocessing and refabrication plant capacity and suitable reactors (MOX-PWRs and CAPRA type FRs or IFRs) to be built up. The optimum strategy is to convert the present reactor-grade plutonium in PWR cores to denatured, proliferation-proof reactor-grade plutonium first (Sections 12 and 13) and then incinerate this proliferation-proof plutonium and the minor actinides mainly in FR burners. This would be similar to the present strategy followed by France [3] and proposed by the USA [6]. Incineration of reactor-grade plutonium as well as neptunium and americium and separation of those fission products (cesium an strontium) causing the highest heat loads would alleviate considerably the needed volume capacities for future deep geological repositories [8].

The build up of the needed reprocessing and refabrication capacity would certainly take time. If the 2300 t of plutonium in spent fuel elements in the world in 2010 [9] after reprocessing would be used for PWR MOX fuel elements with, e.g. for fuel type D of Section 12, this would be sufficient for about 38500 t of MOX fuel elements with 6.5% plutonium correponding to the first core loading of about 300 PWRs of 1.3 GW(e) [10].

14.4 Fuel cycle plant capacity in the world in 2010

Out of the 439 nuclear reactors operating in the world in 2008, there were 265 PWRs and 54 BWRs. Additional PWRs and BWRs were under construction in 2010 (Section 2.1).

The civil reprocessing plant capacity for UO_2 fuel in the world was about 4200 (t_{HM}/y) in 2008, mainly located in NWSs. It consisted of the large scale reprocessing plant at LaHague (France) with 1700 (t_{HM}/y), Sellafield in the UK with 1200 (t_{HM}/y), Mayak in Russia with 500 (t_{HM}/y) and of the smaller reprocessing facilities in India and China (Table 7.2 in Section 7). Japan with its reprocessing plant in Rokkasho-mura with 800 (t_{HM}/y) is up to now the only NNWS operating a large scale civil reprocessing plant. Its safeguards concept was developed in close cooperation with IAEA already during the design and construction phase (Section 8.4 through 8.6). The accompanying critique and discussion was assessed in Sections 8.7 and 8.8.

The USA has no civil reprocessing capacity available anymore. It decided in 1982 (Nuclear Waste Policy Act) to refrain from chemical reprocessing of spent nuclear fuel and from plutonium recycling. As a consequence, only the direct spent fuel disposal concept was pursued. However, this direct high-level waste disposal concept is being reconsidered after the US-DOE withdraw the license application for the Yucca Mountain high level waste repository in 2010 [11]. This was a consequence of the temperature design limits required by USEPA for the vicinity of the waste packages within the deep geological repository (Section 7).

For economical reasons civil spent fuel reprocessing plants have a plant capacity of about 1200 (t_{HM}/y) (Sellafield) or 1700 (t_{HM}/y) (LaHague) in Europe. Such a reprocessing plant can serve about 50 or 70 GW(e) of PWRs (on the basis of 24 t_{HM} unloaded spent fuel per GW(e)·y). Such a plant capacity exceeds the own needs for reprocessing of spent fuel of countries like France and the UK. Therefore, these reprocessing plants also offer reprocessing of spent nuclear fuel to other countries, e.g. European countries or Japan.

The IAEA proposed (INFCIRC/640) the multilateral approach for the nuclear fuel cycle in 2005 [12]. This would open the possibility that several countries in certain regions of the world build common large scale fuel cycle facilities together in multi-partner and multi-ownership under IAEA control. Also the US government announced the Global Nuclear Energy Partnership (GNEP) in 2006 [13] and Russia proposed to create a Global Nuclear Power Infrastructure in 2007 (Section 1) [14].

The capacity of MOX fuel fabrication plants in the world was about 500 (t_{HM}/y) in 2008 (Table 7.3, Section 7.6). These MOX fuel refabrication plants were mainly located in NWSs, e.g. MELOX (195 t_{HM}/y) at Marcoule (France), Sellafield (120 t_{HM}/y) in the UK, Zheleznogorsk 60 (t_{HM}/y) in Russia. Japan with its MOX refabrication plant at Rokkasho-mura (130 t_{HM}/y) is the only NNWS currently operating such a MOX refabrication plant. The accompanying critique and discussion was assessed in Sections 8.7 and 8.8.

The above mentioned reprocessing capacity of 4200 (t_{HM}/y) in the world could reprocess the spent fuel of about 175 GW(e) of PWRs (based on 24 t_{HM} of spent nuclear fuel being unloaded per GW(e)·y). The corresponding MOX refabrication capacity needed would be about 500 t_{HM}/y based on 0.6% or 0.8% plutonium in the UOX spent fuel and 5 to 7% plutonium enrichment of the MOX fuel to be loaded in LWRs.

Also, the capacity of uranium enrichment plants is almost entirely located in NWSs, e.g. in USA (14.3 million kg SWU/y), in Russia (20 million kg SWU/y), in France (10.8 million kg SWU/y) and in China (0.2 million kg SWU/y) (Section 3). Japan as a NNWS has an enrichment capacity of 0.3 million kg SWU/y.

Urenco, a multinational enrichment company (UK, Netherlands and Germany) can be considered already as a multilateral enrichment company (MLEC) along the proposal of IAEA for Multilateral (multinational) enrichment companies. Also, the new LASER enrichment facility owned by Hitachi (Japan) and General Electric (USA) or the Russian offer for a Global Nuclear Power Infrastructure [14] belong to this category of multilateral or multinational fuel cycle companies.

Multilateral uranium enrichment centers (MLECs) and multilateral reprocessing centers (MLRCs) in collocation with MOX fuel refabrication plants – so called MLRCs – as well as waste treatment plants must be large fuel cycle facilities for economical reasons. Therefore, they are well suited to be multilateral, multinational fuel cycle centers as proposed by IAEA (INFCIRC/640) [12].

14.5 Transition phase for the production of proliferation-proof plutonium

Already Massey et al. [16] had pointed out in 1982 that the existing reactor-grade plutonium from LWRs could be converted into proliferation-proof reactor-grade plutonium within several decades. Indeed the results of Section 12 show that proliferation-proof plutonium can be obtained using fresh fuel of types A, B, D or E after one full burnup period of 60 GWd/t_{HM} or 5 to 6 years PWR operation. The large scale conversion of reactor-grade plutonium into proliferation-proof plutonium, therefore, becomes only a question of reprocessing plant and MOX fuel refabrication plant capacities. These capacities determine how fast the present reactor-grade plutonium in spent fuel elements can be converted into proliferation-proof reactor-grade plutonium. **This conversion of present plutonium into proliferation-proof plutonium shall only be done in NWS as neptunium cannot be avoided during this transition phase.**

Fig. 14.1 shows a scheme of the transition phase for PWRs. The UOX-PWRs of NWs receive their LEU fuel from multilateral enrichment centers (MLECs) in NWs. After one burnup period of 60 GWd/t_{HM} their spent nuclear fuel – after intermediate storage – is transported to a multilateral reprocessing center and MOX-Pu/U refabrication center (MLRC) in NWS.

Reprocessing centers are located presently in NWSs (except the Rokkasho-mura reprocessing plant in Japan (see Sections 8.7 and 8.8)). It is proposed that these reprocessing centers in NWSs apply coprocessing of plutonium and neptunium. This can be done by slight modification of the PUREX reprocessing flow sheet (PUREX/COEX) and is already considered for the LaHague reprocessing plant in France [17]. The refabricated MOX fuel shall contain plutonium, neptunium and reenriched uranium (similarly to fuel types A, B, D in Section 12 but without curium).

This MOX Pu/U fuel will be converted in the NWSs into proliferation-proof reactor-grade plutonium after a burnup period of 60 GWd/t_{HM} in MOX PWRs. A simpler and faster way to produce directly proliferation-safe plutonium in the NWSs is the use of re-enriched reprocessed uranium (RRU) as fresh PWR fuel. This leads directly to proliferation-proof reactor-grade plutonium after a burnup period of 60 GWd/t_{HM} [10]. RRU-fuel was already used in PWRs in France [18].

The amount of civil reprocessed uranium in the world was about 120,000 t_{HM} in 2008 [9]. A factor of two more of such uranium was still stored in spent nuclear fuel and would become available after further reprocessing.

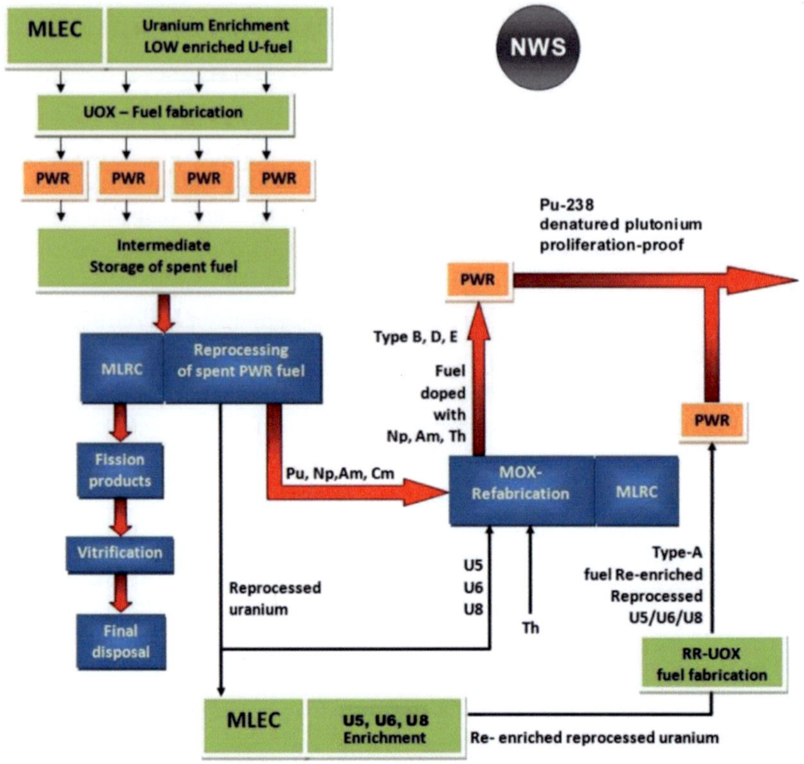

Figure 14.1: Transition phase for the production of proliferation-proof plutonium in NWSs (Reprocessing and Refabrication will in reality be collocated but shown separately here).

The re-enrichment of reprocessed uranium is already performed in centrifuge enrichment plants in Russia and Europe. The presently operating large scale reprocessing plants in the NWSs are able (with slight modifications) to reprocess the fuel type A, B, D for generating proliferation-safe reactor-grade plutonium. Their capacities would be sufficient to serve the initial cores about 170 GW(e) MOX PWRs (Section 14.4) which is close to one half of the PWRs which were operating in the world in 2008.

At a later stage and along with the present development efforts for the separation of americium and curium, for the transmutation and incineration of americium and storage of curium for decay, the present large scale reprocessing plants could be supplemented by additional separation stages. These additional separation stages would apply e.g. the SANEX-GANEX or EXAm processes in France to separate americium and curium [17]. Americium could then be used later in the FR fuel cycle to keep the Pu-238 percentage in the plutonium isotopic composition at the required level for proliferation-proof plutonium (Section 13.8).

14.6 Different levels for non-proliferation criteria of reactor-grade plutonium

In Section 12 it was shown that several options, e.g. fuel type A, B, D or E can be used to produce plutonium with Pu-238 isotopic contents up to about 6-12% in one burnup period of 60 GWd/t_{HM} in a PWR core [10]. Similar results had been obtained by Campbell et al. [19] already in 1978.

In Section 9 and 10 it was demonstrated by nuclear explosion yield calculations and by thermal analyses, that HNEDs become technically unfeasible for reactor-grade plutonium with Pu-238 isotopic contents above certain thermal limits. These thermal limits are listed in Table 14.1 together with the corresponding alpha-particle heat power of HNEDs and the Pu-238 contents of the reactor-grade plutonium of PWR spent fuel. Different possibilities of cooling the HNEDs were also investigated in Section 10. They are also listed as category I and category II in Table 14.1.

As already detailed in Section 10, very high technology HNEDs in terms of the size of the HNEDs, applied chemical explosives, art of implosion technology etc. could only be mastered by advanced NWSs. They do possess already NEDs on the basis of weapon-grade plutonium and highly enriched uranium. According to Garwin [20], deVolpi [21] and Grizzle [22] NWSs have never applied reactor-grade plutonium for plutonium nuclear weapons (Section 10). Therefore, very high technology cases are not included in the present proposal for levels of non-proliferation (Table 14.1).

Table 14.1: Proposed levels for non-proliferation of reactor-grade plutonium for different limits for alpha-particle heat power and corresponding Pu-isotopic composition.

Thermal analysis	Technology assumption	Alpha-particle heat power (kW) of HNED	Pu-238 isotopic content (%)	Proposed safe level for non-proliferation
Outside cooling of HNED by radiation and natural convection by air	low medium	0.12 0.24	1.8 3.6	Ia Ib
Cooling of HNED by liquid helium	low medium	0.24 – 0.46 0.39 – 0.46	3.5 – 7.0 6.0 – 7.0	II
Cooling by internal rods of high conductivity	low	0.4	6.2	II

14.6.1 Scientific proposal for level I criterion for non-proliferation

The level I criterion of non-proliferation shall be split into level Ia (low-technology HNEDs) and level Ib (medium technology HNEDs) for PWR spent fuel in Table 14.1. The level Ia non-proliferation criterion (Table 14.1) shall be valid for reactor-grade plutonium with a Pu-238 content above 1.8% which corresponds to an alpha-particle heat power of the low technology HNED above 0.12 kW. Such reactor-grade plutonium of more than 1.8% Pu-238 is proliferation-proof for all HNEDs of low technology cooled at the outside by radiation and natural convection by air.

Similarly, the safety level Ib criterion for non-proliferation shall be valid for reactor-grade plutonium from PWR spent fuel with a Pu-238 content above 3.6 % which corresponds to an alpha-particle heat power of the medium-technology HNED with an alpha-particle heat above 0.24 kW. Such reactor-grade plutonium with more than 3.6 % Pu-238 is proliferation-proof for all HNEDs, of medium-technology cooled at the outside by radiation and natural convection by air.

The buildup of Am-241 as a function of time shall be neglected in the discussed proposal for the level Ia and Ib non-proliferation criteria in order to remain conservative.

14.6.2 Scientific proposal for level II criterion for non-proliferation

The level II criterion of non-proliferation (Table 14.1) shall cover cooling of the low-and medium-technology HNEDs by liquid helium to -270 °C as well as the possibility of cooling of low technology HNEDs by internal rods of high conductivity.

From Table 14.1 it can be understood that a level II criterion of non-proliferation of 0.46 kW alpha-particle heat power of the HNED (thermal limit) covers cooling of the low- and medium-technology HNEDs down to -270 °C as well as cooling by rods of high conductivity for low-technology HNEDs. This thermal limit corresponds to 7% Pu-238 in the reactor plutonium (Table 9.6b). Such reactor-grade plutonium with more than 7% Pu-238 is considered proliferation-proof.

The buildup of Am-241 as a function of time is neglected in this proposal for the level II non-proliferation criterion in order to remain conservative.

(This safety level II would even cover the very high-technology case with the limit of 0.37 kW for cooling the outside of HNEDs by radiation and natural convection by air (Section 10)).

Proliferation-proof reactor-grade plutonium satisfying level II with about 7% Pu-238 would require slight adaptions of the chemical reprocessing methods and MOX refabrication methods. Present chemical reprocessing plants allow Pu-238 contents of the reactor grade plutonium up to about 6% [17]. For the MOX fuel fabrication also more advanced methods like vibro-compaction or sol-gel processes could be applied (Section 7).

14.6.3 Alpha-particle decay of Pu-238 in proliferation-proof reactor-grade plutonium

Pu-238 decays with a half-life of 87.7 years. This must be accounted for in the management as well as safeguards survey and control of IAEA of a future civil fuel cycle with proliferation-proof reactor-grade plutonium. The fuel cycle turn-around time of the PWR MOX recycle case is 10 years. For future FR recycle times only two years are projected. For these time periods the decay of Pu-238 will be rather small, but should be accounted for. However, proliferation-proof plutonium should not be stored over many decades because the Pu-238 could decay to such percentage that the reactor-grade plutonium would no longer be proliferation-proof any more. Therefore, the proliferation-proof reactor-grade plutonium must be utilized without long delays in the fuel cycle. This must be controlled and verified by IAEA inspectors (Section 8).

14.7 Can proliferation-proof plutonium be converted to weapon-grade plutonium

In this subsection the question is discussed whether such proliferation-proof, reactor-grade plutonium, as , e.g., given in Table 13.3 with 7.7% Pu-238, 44% Pu-239, 31% Pu-240, 10.3% Pu-241, and 7% Pu-242, could be modified by enrichment techniques, to weapon-grade plutonium. This is discussed below for the possibilities of applying either centrifuge or LASER enrichment technology. Present gas ultra-centrifuge designs would have to be changed (section 3.8), since present gas ultra-centrifuges enrich the lower weight U-235 in a U-235/U-238 mixture. This would correspond to a Pu-238 enrichment instead of a desired depletion of Pu-238 in the Pu-238, Pu-239, Pu-240, Pu- 241, Pu-242 mixture.

14.7.1 Centrifuge enrichment technology

For the utilization of centrifuge enrichment technology, the reactor-grade plutonium must be converted into plutonium hexafluoride PuF_6, the only gaseous component of plutonium with a sublimation temperature of 51 °C [26].

14.7.2 Decomposition of PuF_6 by alpha-particle radiation during re-enrichement

Gaseous PuF_6 is formed by reaction between plutonium tetrafluoride PuF_4 with fluorine gas at elevated temperature (750 °C). Plutonium hexafluoride is a very powerful fluorinating agent. Its rate of thermal dissociation is fairly small, but PuF_6 decomposes as a consequence of the high specific alpha radiation activity of its different isotopes. The decomposition product is solid plutonium tetrafluoride PuF_4.

$$PuF_6 \rightarrow PuF_4 + F_2$$

which plates out on the surrounding surfaces. According to Weinstock et al. [26] an average energy of 31 eV is necessary for the decomposition of one PuF_6 molecule. When the PuF_6 is kept as a solid material at temperatures lower than 51 °C a destruction rate of the PuF_6 of 1.5% per day was measured. On the other side, if the PuF_6 is stored as a gas in small cylinders the destruction rate was only 0.1% [26]. In this case, dependent on the geometry of the container and the pressure of the gas a large fraction of the alpha particles from the decay of plutonium isotopes is absorbed by the walls of the container. This difference between the destruction losses of the solid plutonium hexafluoride and the gaseous plutonium hexafluoride by a factor of 15 will be applied in the considerations below.

Somewhat smaller destruction rates as in the PuF_6 gas were reported for uranium hexafluoride [27]. UF_6 has a decomposition energy of about 100 eV per 0.9 molecule. Kryuchkov et al. [27] proposed to use this effect for non-proliferation purposes by admixing, e.g. 1% of U-232 hexafluoride to the U-235/U-238 uranium hexafluoride mixture. U-232 is one of the strongest alpha-particle emitters with a half-life of about 69 years and an average energy of the alpha particles of 5.3 MeV. Kryuchkov et al. [27] showed that the admixture of 1% U-232 would destroy about 48% of all UF_6 molecules within a time period of 1.2 months.

If these calculational procedures of Kryuchkov et al. [27] are applied to different proliferation-proof reactor-grade plutonium isotopic mixtures of plutonium-hexafluoride the following data must be considered. Table 14.2 shows the half-lives, the decay constants, the alpha-decay activities and the energies of the alpha-particles emitted from the different plutonium isotopes. Pu-241 decays via beta-decay into Am-241 with a half-life of 14.4 years. Am-241 itself is an alpha-emitter. However, Am-241 is not considered for the further considerations.

From the activities of the different alpha-particle emitting plutonium isotopes the number of emitted alpha-particles can be determined e.g. for a time period of one year. Each alpha-particle with an energy, e.g. for Pu-239 with 5.5 MeV, can destroy 1.77×10^5 molecules of PuF_6, as the average energy to destroy one PuF_6 molecule into PuF_4 is 31 eV [26]. The measured data of [26] for the case of PuF_6 gas in a container resulted in a loss of PuF_6 molecules by destruction which was by a factor of 15 smaller than in solid PuF_6 material. Applying this factor 15 allows calculating the decomposition losses over a time period of one month for the three different plutonium isotopic compositions of Table 14.3. The isotopic compositions correspond to the levels Ia, Ib and II for non-proliferation which were defined in the previous Section 14.6.

Table 14.2: Data for alpha decay of plutonium isotopes.

Pu-isotope	Pu-238	Pu-239	Pu-240	Pu-241	Pu-242
Half-life for α-decay (y)	87.7	$2.41x10^4$	6563	β-decay	$3.75x10^5$
Decay constant λ $[s^{-1}]$	$2.5x10^{-10}$	$0.91x10^{-10}$	$3.3x10^{-12}$	---*	$0.58x10^{-13}$
Activity of α-decay $(s{\cdot}g)^{-1}$	$6.3x10^{11}$	$0.23x10^{10}$	$0.88x10^{10}$	---*	$0.15x10^9$
Energy of alpha-particles (MeV)	5.5	5.2	5.2	---*	4.9

*Pu-241 decays to Am-241 with a half-life of 14.4 years.

Table 14.3: Different proliferation-proof reactor-grade plutonium compositions.

Pu-isotope	Pu-238	Pu-239	Pu-240	Pu-241	Pu-242
Type Ia composition [%]	1.8	55.2	23.8	12.8	5.4
Type Ib composition [%]	3.6	52.0	23.1	14.1	7.2
Type II composition [%]	7.0	39.6	25.8	18.0	9.6

A plutonium hexafluoride mixture of non-proliferation level Ia (Table 14.3) would lose within one month about 18% of its PuF_6 molecules by alpha-particle decomposition. A plutonium hexafluoride mixture of non-proliferation level Ib (Table 14.3) would lose within one month about 32% of its PuF_6 molecules by alpha-particle induced decomposition. A plutonium hexafluoride mixture of non-proliferation level II (Table 14.3) would lose within one month about 58% of its PuF_6 molecules by alpha-particle induced decomposition.

The results for Pu-238 (highest enrichment factor of all plutonium isotopes and the above high decomposition rates for PuF_6) show that all enrichment methods (gas-ultra-centrifuges, molecular LASER-enrichement) using plutoniumhexafluoride, PuF_6, lead to technically almost insurmountable problems.

14.7.3 Atomic vapor Laser isotope enrichment

For the LASER isotope separation based on plutonium vapor (AVLIS process) insufficient scientific information is available in the open literature to answer this question directly. But it is known that the different isotopes can be selectively excited by LASER beams and be separated. However, if the argumentation of [28] is applied to proliferation-proof reactor- grade plutonium containing Am-241 from the decay of Pu-241, it can be concluded that LASER enrichment with the AVLIS process would become very difficult, if technically possible at all. All earlier efforts of enriching uranium by atomic vapor LASER technology were not pursued to technical scale up to now.

14.8 Future civil Pu/U fuel with proliferation-proof, reactor-grade plutonium

At the end of the transition phase (Fig. 14.1) proliferation-safe, reactor-grade plutonium would be available for MOX Pu/U fuel refabrication. This proliferation-proof reactor-grade plutonium cannot be misused any more for nuclear explosive devices. It can be incinerated in PWRs or FRs.

This must occur under the prerequisites of the reactor-grade plutonium remaining proliferation-proof over the burnup phase of about 60 GWd/t.

Neptunium-237 is a proliferation problem as it can be misused for building nuclear explosive devices (Section 13). Therefore, it must be made sure that only extremely small amounts of neptunium are produced. (A very small amount will be produced by alpha-decay of Pu-241 and neutron capture processes in U-235 and U-236 of the 0.2% U-235 of depleted uranium (Section 13)).

14.8.1 Incineration of proliferation-proof, reactor-grade plutonium in PWRs

In Section 13 it was shown that proliferation-proof reactor-grade plutonium with 7.7% Pu-238 (being slightly above the level II criterion for non-proliferation) can be incinerated by MOX-PWRs (Fig. 14.2). Only tiny amounts of neptuniun would be generated. The fresh MOX fuel would contain proliferation-proof, reactor-grade plutonium and either 0% or 0.5% proliferation-proof americium. In addition the MOX fuel must contain 54.67%

thorium and either 1.69% or 2.17% U-233 to assure acceptable safety-related reactivity coefficients. This design option would allow the incineration of about 40 kg/t_{HM} of proliferation-proof plutonium over a burnup phase of 60 GWd/t_{HM}. The Pu-238 content would even slightly increase during burnup, whereas the Pu-239 content would decrease from 44% to 20% during this burnup phase.

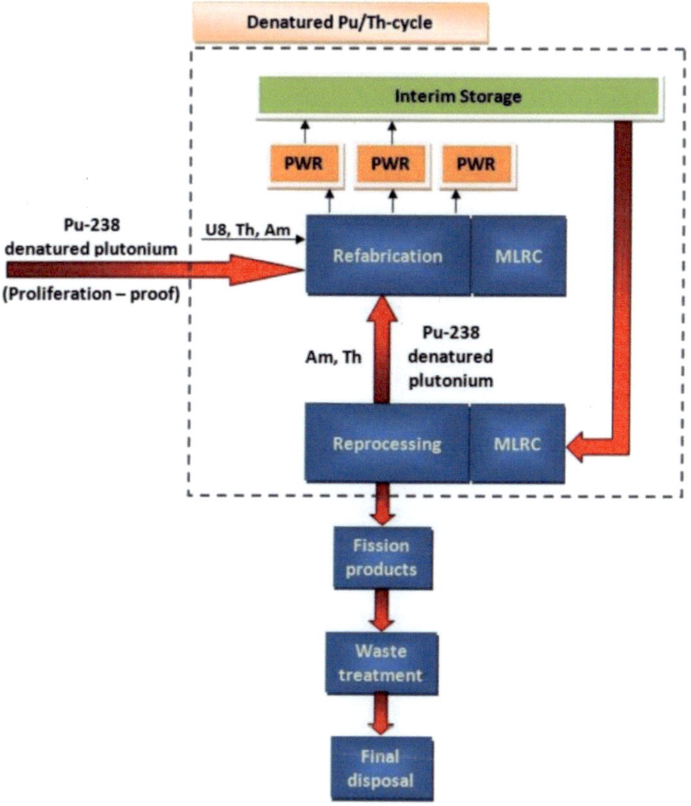

Figure 14.2: Incineration of proliferation-proof plutonium in PWRs in a future civil international proliferation-proof fuel cycle. (Reprocessing and Refabrication step will in reality be collocated in one MLRC but shown separately).

Nevertheless another recycling of this proliferation-safe plutonium would still be possible. For this case the proliferation-proof reactor-grade plutonium after a first recycle containing only about 20% Pu-239 would have to be mixed with the reactor-grade plutonium coming from the transition phase which contains 44% Pu-239. This would enable recycling over a number of decades. This method is similar to the SGR Pu-recycling, described in Section 7.

The PWR core design would have to be modified slightly and a mixed thorium/plutonium/uranium reprocessing scheme would have to be deployed.

For proliferation-proof, reactor-grade MOX fuel corresponding to level Ia or Ib for non-proliferation with only more than 1.8% or 3.6% Pu-238 isotopic content the above described constraints regarding safety related reactivity coefficients would be easier to cope with.

14.8.2 Incineration of proliferation-proof reactor-grade plutonium in FRs

The proliferation-proof plutonium could also be loaded into FR cores, either directly after a transition phase (Section 14.5) or after a recycling phase of a number of decades in PWRs (Section 14.8.1). This is shown in Fig. 14.3.

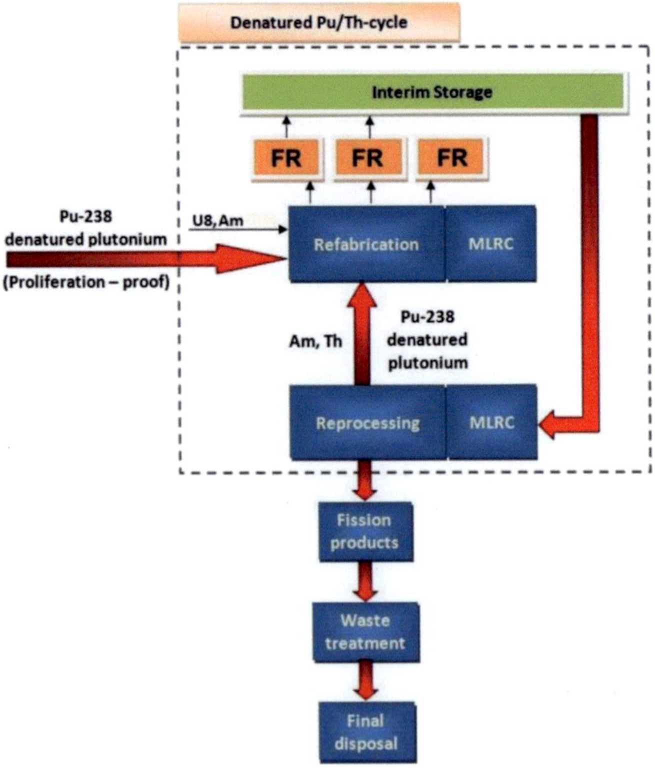

Figure 14.3: Incineration of proliferation-proof plutonium in FRs. Future civil international proliferation-proof fuel cycle.

In this case about 4% proliferation-proof americium must be admixed to the proliferation-proof plutonium in the fresh fuel, in order to keep the Pu-238 sustainable (Section 13.8) over the whole burnup cycle of about 130-150 GWd/t. Only depleted uranium (e.g. containing about 99.8% of U-238 and about 0.2% of U-235) may be used to keep the neptunium production minimal (buildup of neptunium by neutron capture in U-235 and subsequently in U-236).

Also the blanket fuel of FRs would have to consist of depleted uranium with about 4% proliferation-proof americium and about 1% proliferation-proof reactor-grade plutonium if the level II non-proliferation criterion should be fulfilled. The 1% proliferation-proof plutonium should have a Pu-238 content which is about to 4% higher than the proposed level for non-proliferation, e.g. it should be (7+4)% Pu-238 for non-proliferation level II in order to assure proliferation-proof plutonium from the very beginning of irradiation in the blanket fuel (see Fig. 13.15 in Section 13.8.1.2). The Pu-239 from neutron capture in U-238 would be born in the proliferation-proof plutonium admixed to the fresh fuel. In addition Pu-238 would be generated from the decay of Cm-242 (half-life 180 days).

A sodium-cooled BN-600-type fast reactor core was analyzed in Section 13. The results show that for the two considered options the system can operate either as a burner of proliferation-proof plutonium or as a breeder. In case of an FR burner without blankets a conversion ratio of CR = 0.78 and an incineration rate for proliferation-resistant plutonium of 33 kg/t during the burnup phase of 150 GWd/t_{HM} were obtained. If the FR would be operated as a breeder with blankets a breeding ratio of BR = 1.07 would be possible for the BN-600-type FR (Section 13).

However, FR burners would have to operate first for many decades in order to incinerate all plutonium (2500 t in 2010) which was generated by UOX fuelled thermal spectrum reactors and by MOX-PWRs operating during the transition phase. Only when natural uranium will become scarce, FR breeding will have to take over. However, then a breeding ratio of about BR = 1.07 will be sufficient.

FR reactors operating in the thorium/plutonium/uranium fuel cycle with proliferation-proof plutonium would be feasible as well as using fuel with U-233 from option E (Section 12).

The important result is that after a transition phase, in which proliferation-proof plutonium is produced by PWRs, a civil proliferation-proof fuel cycle becomes possible.

The transition phase could be started already with present fuel cycle centers of the NWSs, e.g. LaHague in France, Sellafield in the UK, or Mayak in Russia. These fuel cycle centers could later be complemented by more international MLRCs following the proposal of [23] and IAEA [12].

14.8.2.1 Availability of americium for sustainability of proliferation-proof plutonium in FR burners or-breeders

As shown in Section 14.8.1 and Section 14.8.2 above, americium would be required to keep the proliferation-proof plutonium sustainable in both future FR-burners and FR-breeders. The question then comes up whether there will be sufficient americium available to operate a proliferation-proof civil nuclear fuel cycle.

Fig. 1.3 in Section 1 shows the projections of the IAEA for the generation of americium by nuclear reactors in the world up to 2030. About 450 tons of americium would be available worldwide until 2030.

A fast reactor, e.g. BN-800 with 800 MW(e) power has a mass of 16 tons of PuO_2/UO_2 and 25 tons of UO_2 in the blankets. Assuming 4% of americium in the core and blanket fuel – as assumed above and in Section 13.8 – about 2 to 2.5 tons of americium would be needed for loading such a future FR (somewhat less for FR burners without blankets somewhat more for FR breeders with blankets). On the basis of 2.5 tons of americium per GWe and an availability of 450 tons americium by 2030 this would be sufficient to start about 180 FRs of 1 GW(e) each. The americium will be incinerated (20 kg/ton, see Section 13.8.3) and provide the Pu-238 to keep plutonium proliferation-proof.

A large scale introduction of FR-burners will probably not start before 2040-2050 (Section 2). Even then LWRs and FRs must continue to operate together in symbiosis and MOX-PWRs produce more americium than UOX-PWR (Section 7).

14.8.3 Future international proliferation-proof nuclear fuel cycles

The proliferation-proof plutonium and proliferation-proof americium together with thorium and depleted uranium can then be utilized in a later proliferation-proof international civil nuclear fuel cycle which is open for all countries (Fig. 14.4). Advanced aqueous or pyrochemical reprocessing for plutonium/thorium/uranium fuel and related fuel refabrication technology applying remote handling may become necessary. IAEA safeguards would still be required.

An increase of the present reprocessing and MOX refabrication plant capacity by a factor of three would be sufficient for 525 GWe PWRs which is about a factor of two more than the 265 GWe PWRs operating in 2008. This additional needed reprocessing and MOX refabrication plant capacity should be located where the PWRs are operated (Fig. 14.4).

The NWSs should open their MLRCs for multi-partnership with other countries according to the initiatives of the USA and Russia [13,14].

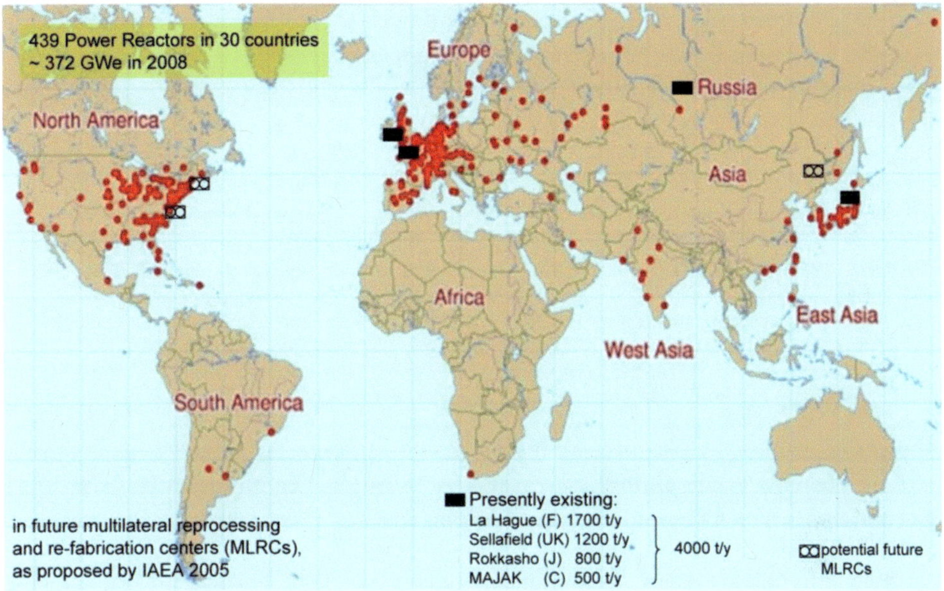

Figure 14.4: Future multilateral reprocessing and refabrication centers (MLRCs) in the world (adapted from IAEA).

14.9 Effect on Safeguards and Non-proliferation Issues in Future Civil Uses of Nuclear Power of the Proposed Concept of Upper Limits for Non-proliferation

Sections 8.3 and 8.4 above had shown the main reason for the existing mistrust of any kind of chemical reprocessing and plutonium recycling in LWRs and FRs to be the ultra-conservative requirements of INFCE/153 [15]. The requirements in INFCE/153 and internal U.S. safeguards conditions and classifications, respectively, are practically identical.

The rule in INFCE/153 [15] that all reactor-grade plutonium, other than plutonium with a Pu-238 isotopic content of ≥80%, must be treated like weapon-grade plutonium cannot be made consistent with the results of the thermal analysis discussed in Section 10, where it was shown that critical spheres without reflectors and without chemical implosion lenses ($k_{eff} = 1$) of reactor-grade plutonium metal are completely molten already above a Pu-238 isotopic content of about 23%. If these spheres of reactor-grade plutonium metal are surrounded by a reflector and tamper of natural uranium metal of 5 cm thickness

(Fig. 10.2, chapter 10, p. 252), as required for a nuclear explosive device, and with the necessary explosive lenses for implosion, this thermal limit drops as follows for reactor-grade plutonium of spent PWR fuel:

1.8% Pu-238 (low-technology HNEDs with 0.12 kW) 3.6% Pu-238 (medium-technology HNEDs with 0.24 kW)	} cooling by radiation and natural convection by air
7% Pu-238 (low-and medium-technology HNEDs with 0.46 kW)	cooling by liquid helium to –270 °C
6.2% Pu-238 (low-technology HNEDs with 0.4 kW)	cooling by internal rods of high thermal conductivity

Above these limits the chemical explosives of the explosive lenses either melt or self-explode. **Reactor-grade plutonium whose Pu-238 isotopic composition is above these thermal limits may be considered proliferation-proof.**

As such proliferation-proof reactor-grade plutonium cannot be misused for building nuclear explosive devices , **it must not be equated** with weapon-grade plutonium as required in INFCE/153 [15].

Reprocessing plants and Pu/U refabrication plants processing such proliferation-proof reactor-grade plutonium no longer fall under the risk described in sections 8.3 and 8.4 which is caused by the measurement error of 1% of the safeguards instrumentation in large reprocessing plants. Hence, there is **no reason either to fear abrupt or protracted diversion**.

Opting out of the NPT and subsequently making use of the proliferation-proof reactor-grade plutonium to build nuclear weapons **would make no sense**. The proliferation-proof reactor-grade plutonium then in possession of the government could not be used to make nuclear weapons.

A specific attempt to produce weapon-grade plutonium by shutting the nuclear reactor down and unloading parts of the fuel elements or the total core at an early point in time (after some weeks) and reprocessing the low-irradiated LEU UOX fuel elements or special U-238 fuel elements must be detected by IAEA safeguards, e.g., anti-neutrino detectors combined with electronic data transmission to the IAEA headquarters (section 8.3). If, however, the reactor is fueled with MOX fuel from proliferation-proof reactor-grade plutonium, this too cannot be used any more for nuclear weapons at any point in time.

If an NNWS does not accede to the NPT and builds and runs on its own all nuclear plants required for making weapon-grade plutonium, it can be prevented from doing so only by diplomatic measures or other deterrents.

14.9.1 Not Included In This Proposal

It should be emphasized again at this point that the following reactor lines cannot be included in the above proposal of an international proliferation-proof civil fuel cycle:

- Gas cooled graphite reactors, e.g. of MAGNOX type and
- CANDU reactors, which attain a maximum fuel burnup of only 7 GWd/t. This is also valid for the more recent versions (CANDU-ATR), with fuel burnup up to 20 GWd/t_{HM}.
- Research reactors, which attain a maximum fuel burnup of only 7 GWd/t.
- Breeding blanket elements with depleted uranium or natural uranium fuel of today's prototype fast breeder reactors.
- The latter are required, in the future to contain about 4% of americium for the production of Pu-238 and about 1% proliferation-proof plutonium with an initial Pu-238 composition of about 11% (non-proliferation level II), in the fresh blanket elements. In this way, the plutonium of the blanket elements will always remain proliferation-proof (section 13).

14.10 Reprocessing and Re-enrichment by Gas Ultra-centrifuges

Whereas chapters 9-14 discussed the proliferation problems of reactor plutonium, reactor americium and reactor neptunium in connection with chemical reprocessing, the following chapters 15-18 will deal with the re-enrichment of U-233/U-235 fuel by gas ulta-centrifuges.

References Section 14

[1] Kessler, G. et al., Moderne Strategien zur Beseitigung von Plutonium, Atomwirtschaft, 46, 132 (2001).

[2] Kessler, G., Requirements for nuclear energy in the 21st century, Progress Nuclear Energy, 40, 3-4, 309 (2002).

[3] Bouchard, J., The enclosed fuel cycle and nonproliferation issues, Proc. Int. Conf. Nuclear Systems for the Future Generation and Global Sustainability (GLOBAL 2005), Tsukuba, Japan, October 9-13, 2005, Atomic Energy Society of Japan (2005).

[4] Aniel-Buchheit, S. et al., Plutonium recycling in a full-MOX 900-MW(electric) PWR: physical analysis of accident behaviors, Nucl. Technol., 128, 245-256 (1999).

[5] Tommasi, J. et al., Long-lived waste transmutation in reactors, Nuclear Technology, 111, 133 (1995).

[6] Wade, D.C, Hill, R.N., The design rationale for the IFR, Progress in Nuclear Energy, 31, 1-2, 13 (1997).

[7] Salvatores, M., Nuclear fuel cycle strategies including partitioning and transmutation, Nucl. Eng. and Design, 235 805-816 (2005).

[8] Wigeland, R.A. et al., Waste management aspects of various fuel cycle options, presented at Technical Mtg. on Fissile Material Management Strategies for Sustainable Nuclear Energy, Vienna, Austria, September 12-15, 2005.

[9] Fukuda, K., IAEA Scenario of MA transmutation in LWR, presented at 1-st International Science and Techology (Forum) on Protected Plutonium Utilization for Peace and Sustainable Prosperity, Tokyo, Japan, March 1-3, 2004.

[10] Kessler, G., Broeders, C.H.M., Fuel cycle options for the production and utilization of denatured plutonium, Nucl. Sci. Eng., 156, 1-23 (2007).

[11] US-DOE withdraws repository license application, Nuclear News, 53, 4, p. 63 (2010).

[12] IAEA, Multilateral approaches to the nuclear fuel cycle: Expert group report submitted to the director general of the IAEA, INFCIRC/640 (2005).

[13] Blake, E.M., GNEP rollout means big jump for fuel cycle, Nucl. News, 49, 3, 64 (2006); see also: "Advanced Fuel Cycle Initiative", available on the Internet at www.ne.anl.gov/research.

[14] Russian initiative for creation of global nuclear power infrastructure. INFCIRC/708 (2007).

[15] The Structure and Content of Agreements between the Agency and States Required in Connection with the Treaty on the Non-Proliferation of Nuclear Weapons. Vienna: International Atomic Energy Agency, INFCIRC/153 Corrected (1972).

[16] Massey, J.V., Schneider A., The role of plutonium-238 in nuclear fuel cycles, Nucl. Technol., 56, 55 (1982).

[17] Warin, D., Minor actinide partitioning, 1st ACSEPT Workshop Lisbon, March 31, 2010.

[18] Vouinou, G. et al., A neutronic analysis of TRU recycling in PWRs loaded with MOX-UE fuel (MOX with U-235 enriched U support), INEL-09, 16091 (2009).

[19] Campbell, D.O., Gift, E.M., Proliferation resistant nuclear fuel cycles, ORNL/TM-6392, Oak Ridge National Laboratory (1978).

[20] Garwin, R.L., http://www.fas.org/rl/90-96.htm.

[21] C.D. Heising-Goodman, An Evaluation of the Plutonium Denaturing Concept in an Effective Safeguards Method, Nucl. Technol., 50, 242 (1980).

[22] Grizzle, S., DOE FACTS, Additional information concerning underground nuclear weapon test of reactor-grade plutonium. http://www.ccnr.org/plute_bomb.html

[23] Häfele, W., Energy in a finite world, A global systems analysis, Harpers & Row, Ltd. UK (1981).

[24] Hoppe, H.O., Urenco Personal Communication (Oct. 2005).

[25] Management and Disposition of Excess Weapons Plutonium: reactor-related options, National Academy of Sciences, National Academics Press, Washington, D.C. (1995).

[26] Weinstock, B. et al., The properties of plutonium hexafluoride, Proc. of the Int. Conf. on the Peaceful Uses of Atomic Energy, Vol. 7, p. 377, United Nations, New York (1956).

[27] Kryuchkov, E.F. et al., Evaluation of self-protection of 20% uranium denatured with ^{232}U against unauthorized reenrichment, Nucl. Sci. and Eng., 162, p. 208 (2009).

[28] Solarz, R.W., A physics overview of AVLIS, UCID-20343, Lawrence Livermore National Laboratory (1985).

15 Uranium fuel with <20% U-235 enrichment as a proliferation problem

15.1 Introduction

Uranium dioxide fuels for modern LWRs have a U-235 enrichment of 4-5% for high fuel burnup of about 50-60 GWd/t. Research reactors which were originally equipped with high enriched uranium (HEU) shall now operate with <20% U-235 enriched fuel. This was recommended during the International Nuclear Fuel Cycle Evaluation, (INFCE) program [1] and is in accordance with IAEA safeguards requirements (section 8.1.1). Uranium fuel with <20% U-235 fuel, also called low enriched fuel (LEU), is not directly usable to build nuclear explosive devices [2]. This is explained by Fig. 15.1 which displays the critical mass of a U-235/U-238 metal uranium sphere with a 4 cm thick beryllium reflector [3]. It can be seen that a metal uranium sphere with U-235 enrichment of <20% U-235 reflected by 4 cm beryllium has a critical mass of more than about 450 kg. This would lead to **a nuclear explosive device (NED)** with all necessary technical components around the fissile areas which is considered too heavy and bulky and therefore unfeasible [2,3,16]. For the uranium isotope U-233 the equivalent enrichment limit would be <12% U-233 (see chapter 17).

The critical mass of a 93% U-235 enriched metallic sphere reflected by 5 cm beryllium is about 22 kg [2,3]. For thicker beryllium reflectors this critical mass is even smaller [2].

The INFCE recommendation [1] in 1980 of isotopic dilution of high enriched uranium (HEU) by natural, depleted or low enriched uranium (LEU) and the proposal of a dividing line at 20% U-235 enriched uranium in weapons-usable (≥20% U-235 enriched uranium) and non-weapons-usable uranium (<20% U-235 enriched uranium) were implemented by IAEA in (INFCIRC/225/Revision 5) [16,17]. In these recommendations IAEA distinguishes - among others – three different enrichment levels. In addition, it defines categories and quantities of material for physical protection measures (see also Tables 16.6 and 17.6). These are discussed in section 16.9.

Table 15.1: Recommendations for U-235 enrichment levels [1,16,17].

IAEA U-235 enrichment levels [17]	Definitions by Forsberg et al. [16]
uranium unirradiated ≥20% U-235 enriched	Weapons-usable (HEU)
uranium unirradiated >10% U-235 <20% enriched	Non-weapons-usable (LEU), can be re-enriched to >20% U-235
uranium unirradiated enriched above natural, but <10% U-235	Non-weapons-usable (LEU)

The dividing line of 20% U-235 enrichment between weapons-usable and non-weapons-usable uranium is based on calculations and technical arguments [2,3,16]. It has been codified, e.g. in US-Department of Energy orders and US-Nuclear Regulatory Commission NRC) regulations [16,18].

15.2 Uranium fuel enrichment for research reactors

Most research reactors were originally operated with 93% U-235 uranium fuel, delivered by the USA or the Soviet Union. During the INFCE program [1] it was recommended to convert this fuel to <20% U-235 enrichment. The Reduced Enrichment for Research and Test Reactors (RERTR) Program of Argonne National Laboratories, USA [5] and the Global Threat Reduction Initiative (GRTS) [20] aim specifically at this conversion of HEU fuel to LEU fuel of research and test reactors. Also Russia initiated a similar program [6] which was followed by the US-Russian RERTR Program [20]. Recommendations by IAEA on reprocessing of HEU fuel from research and test reactors are given in [19].

By the year 2003 there still remained about 20 t of HEU (although partly irradiated) in research reactors of 43 countries [4]. Table 15.2 shows characteristic data of different research reactors. There are fears that this 93% U-235 enriched (HEU) fuel could be stolen or diverted and then be misused to build nuclear explosive devices [2,4]. In addition there are also concerns that <20% U-235 enriched uranium could be re-enriched to 80-93% U-235 uranium using gas ultra-centrifuges (chapter 3) [2,7]. The reasons are the relatively high enrichment factors of gas ultra-centrifuges and the fact that the separative work to enrich 20% U-235 enriched uranium to an enrichment level of 80-93% U-235 is by an order of magnitude lower than in case of enrichment from natural uranium.

Table 15.2: U-235 fissile inventories of research and test reactors [4].

Size	Small reactor	Medium size	Large size
Power level	0.1 – 1 MW	1 - 10 MW	10 - 250 MW
Fissile inventory	5 kg U-235	10 kg U-235	10 – 40 kg U-235
Burnup MWd/t	5,000	20,000 – 50,000	20,000 – 50,000

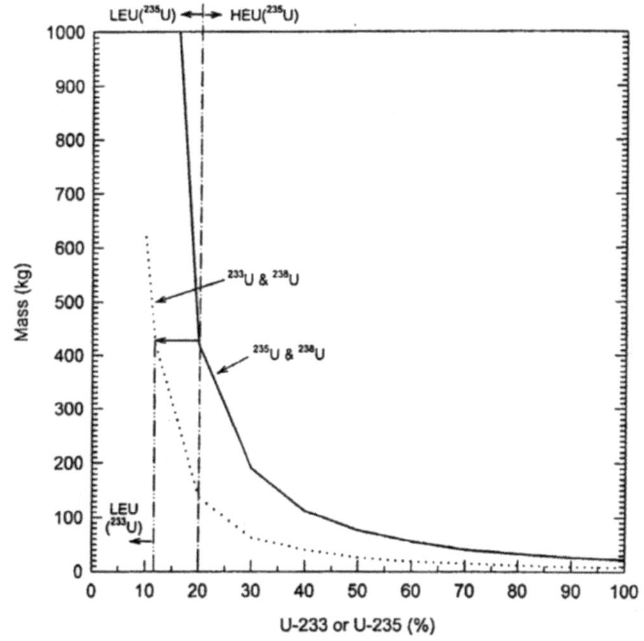

Figure 15.1: Reflected critical masses as a function of percentage of U-233 or U-235 in isotopic mixture with U-238 [3]. (Reflector 4 cm beryllium).

15.3 <20% U-235 Enriched Uranium Fuel for Thorium Fueled Reactor

In addition to <20% U-235 enriched research and test reactor fuel there will in the future also be <20% U-235 enriched uranium fuel needed for reactors operating in the uranium/thorium fuel cycle (chapter 17). For this <20% U-235 enriched uranium fuel the same proliferation concerns can be raised as for the <20% U-235 enriched uranium fuel of research and test reactors. However, it might be possible to re-enrich this <20% U-235 uranium fuel – which probably will exist in large amounts in the civil fuel cycle – by gas ultra-centrifuges in unauthorized enrichment plants not supervised by the IAEA.

15.4 Separative work Requirement for Enriched Uranium

Chapter 3 explained the different characteristics of the enrichment process and the different enrichment methods. In this section we first need equation 3.9 (chapter 3) for the amount of natural uranium needed to obtain 1 kg of enriched uranium product, enriched to a certain percentage of U-235. This Eq. 3.9 is repeated here as Eq. 15.1.

The mass input to the enrichment process is the uranium feed F of natural uranium (atom fraction $x_F = 0.72\%$) or e.g. <20% enriched uranium in the form of uranium hexafluoride. The mass output in the enriched uranium is the product P with atom fraction x_P and the waste stream or tails W with atom fraction x_W of, e.g. 0.2% or 5% in later examples.

The ratio of the feed mass and the product mass is

$$\frac{F}{P} = \frac{x_P - x_W}{x_F - x_W} \tag{15.1}$$

x_F atom fraction of feed material

x_P atom fraction of final enriched product

x_W atom fraction of tails or depleted uranium

The separative work is a measure of the amount of work necessary in the enrichment plant to produce a certain amount of enriched uranium. It has the dimension of mass and is indicated in kg SWU or ton SWU (separation work unit).

The separative work S in SWU/kg U can be expressed in terms of the different mass streams for feed, product and waste [10]. Applying the relations between these mass streams and their atom fractions, the relation for the separative work unit per product mass is given by Eq. 3.11 in chapter 3. It is repeated here as Eq. 15.2

$$\frac{S}{P} = (2x_P - 1)\ln\frac{x_P}{1-x_P} + \frac{x_P - x_F}{x_F - x_W}(2x_W - 1)\ln\frac{x_W}{1-x_W} - \frac{x_P - x_W}{x_F - x_W}(2x_F - 1)\ln\frac{x_F}{1-x_F} \tag{15.2}$$

This relation is evaluated for $P = 1$ kg of product mass in Table 15.3 (extension of Table 3.5) for different enrichment levels x_p [10,11] up to 93% U-235 enrichment. In the left column the different enrichment levels for enriched uranium fuel are shown.

The second column indicates how many kg of natural uranium feed as UF_6 are needed to produce 1 kg of enriched uranium (UF_6). The third column gives the separative work S in kg SWU needed to enrich the natural uranium as feed mass to a certain enrichment.

example: for $P = 1$ kg of 4% U-235 enriched uranium, a feed of $F = 7.43$ kg natural uranium UF_6 and 6.54 kg SWU must be provided.

for 1 kg of 20% U-235 enriched uranium a feed of 38.74 kg of natural uranium as UF_6 and 45.37 kg SWU are needed starting from natural uranium.

for 1 kg of 93% U-235 enriched uranium a feed of 181.6 kg natural uranium as UF_6 and 233.79 kg SWU are needed starting from natural uranium.

The tails assay in all cases of Table 15.3 is 0.2% U-235 as in Table 3.5 (chapter 3).

Table 15.3: Characteristic data: kg of natural uranium feed material as UF_6 required for 1 kg of enriched uranium product and corresponding kg SWU needed.

Enrichment, wt% ^{235}U	kg of Natural U Feed Material as UF_6 to Enrichment Plant	kg Separative Work[a]
	Per kg of Enriched Uranium Product	
Nat. 0.72	1.000	0.000
4.0	7.43	6.54
5.0	9.39	8.85
10.0	19.18	20.86
20.0	38.74	45.37
80.0	156.17	199.09
93.0	181.60	233.79

[a] Tails assay at 0.2 wt% U-235

Table 15.4a: Characteristic data: kg of 15% U-235 enriched uranium feed material as UF_6 required for 1kg of high enriched uranium (HEU) product (80% or 93%) and corresponding kg SWU needed.

Enrichment, wt% ^{235}U	kg of 15% U-235 Uranium Feed Material as UF_6 to Enrichment Plant	kg Separative Work[a]
	Per kg of Enriched Uranium Product	
15	1.000	0.000
80	5.49	21.46
93	6.38	27.22

[a] Tails assay at 0.2 wt. % U-235

However, if the enrichment process is started from 15 or 19.95% U-235 enriched uranium for enrichment up to 93% U-235 there is only 27.2 kg separative work (starting from 15% enriched uranium, Table 15.4a) or 21.1 kg separative work needed (starting from 19.95% enriched uranium, Table 15.4b). This is about one order of magnitude less compared to Table 15.3 (start from natural uranium). The quantities in kg of uranium feed as UF_6 needed are even by a factor of 28 less (starting from 15% enriched uranium) and 38 less (starting from 19.95% enriched uranium) [8].

Table 15.4b: Characteristic data: kg of 19.95% U-235 enriched uranium feed material as UF_6 required for 1 kg of high enriched uranium (HEU) product 80% or 93% and corresponding kg SWU needed.

Enrichment, wt% ^{235}U	kg of 19.95% U-235 Uranium Feed Material as UF_6 to Enrichment Plant	kg Separative Work[a]
	Per kg of Enriched Uranium Product	
19.95	1.000	0.000
80.0	4.10	16.23
93.0	4.76	21.14

[a] Tails assay at 0.2 wt% U-235

If the tails assay or the number of stages in the stripping section are changed from 0.2% tails assay to 1% or 5% tails assay (Tables 15.5a and 15.5b) then the separative work could even decrease to 10.24 kg SWU (in case of 5% tails assay, Table 15.5b). This would be another factor of 2 less in SWU requirement. However the quantities for feed material would increase slightly.

Table 15.5a: Characteristic data: kg of 19.95% U-235 enriched uranium feed material as UF_6 required for 1 kg of high enriched uranium (HEU) 80% or 93% product and corresponding kg SWU needed (tails assay 1% U-235).

Enrichment, wt% ^{235}U	kg of 19.95% U-235 Uranium Feed Material as UF_6 to Enrichment Plant	kg Separative Work[a]
	Per kg of Enriched Uranium Product	
19.95	1.000	0.000
80.0	4.23	11.59
93.0	4.93	15.52

[a] Tails assay at 1 wt% U-235

Table 15.5b: Characteristic data: kg of 19.95% U-235 enriched uranium feed material as UF_6 required for 1kg of high enriched uranium (HEU) product and corresponding kg SWU needed (tails assay 5% U-235).

Enrichment, wt% ^{235}U	kg of 19.95% U-235 Uranium Feed Material as UF_6 to Enrichment Plant	kg Separative Work[a]
	Per kg of Enriched Uranium Product	
19.95	1.000	0.000
80.0	5.09	8.65
93.0	5.97	10.24

[a] Tails assay at 5 wt% U-235

15.5 Number of cascade stages needed to obtain 93% U-235 (HEU)

In chapter 3 it was shown that among present commercially installed enrichment technologies gas ultra-centrifuges (Fig. 15.2) have the highest enrichment factors per stage of a cascade. We apply Eqs. 3.6 through 3.8 of chapter 3 and list them here as Eqs.15.3 through 15.5.

The total number n of stages for an ideal cascade is given by

$$n = 2\,\frac{\ln \dfrac{x_P\,(1-x_W)}{(1-x_P)\cdot x_W}}{\ln \alpha} - 1 \tag{15.3}$$

with

x_P atom fraction of final product

x_W atom fraction of tails depleted uranium

α stage separation factor in a cascade

The number of stages in the stripping section is

$$n_W = \frac{\ln \dfrac{x_F\,(1-x_W)}{(1-x_F)\cdot x_W}}{\ln \beta} - 1 \tag{15.4}$$

x_F atom fraction of feed

β head separation factor in a cascade

For the ideal cascade the head separation factor is:

$$\beta = \sqrt{\alpha}$$

The number of stages in the enriching section

$$n - n_W = \frac{\ln \frac{x_P(1-x_F)}{(1-x_P)\cdot x_F}}{\ln \beta} \tag{15.5}$$

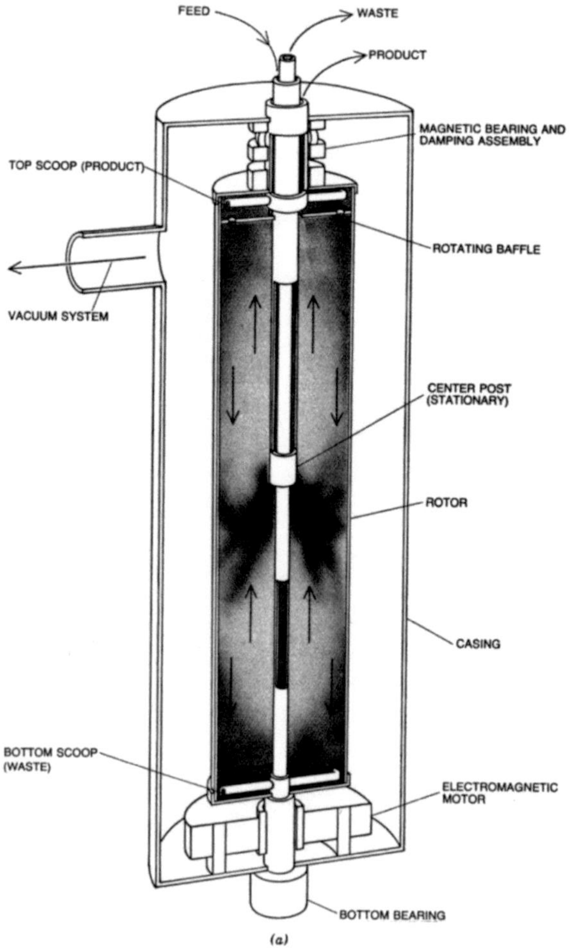

Figure 15.2: Gas Ultra-Centrifuge design [15].

Applying Eqs. 15.3 through 15.5 leads to Table 15.6 for stage enrichment factors of gas ultra-centrifuges of $\alpha = 1.1$ and $\alpha = 1.2$. The lower enrichment factor is often discussed as a first generation commercial gas ultra-centrifuge characteristic which would be procured by illicit trade and be used by countries or by subnational groups (terrorists) for unauthorized re-enrichment [7,8]. Although different cascade arrangements are possible, it is rather simple to describe the number of stages for an **ideal** cascade (section 3.4, Fig. 15.3) and for a **minimum number of stages** cascade. The ideal cascade has twice as much stages than the minimum number of stages cascade minus 1.

$$N_{ideal} = 2\ N_{min} - 1 \qquad N_{min} = 1/2\ (N_{ideal} + 1)$$

The ideal cascade is the most economic cascade with a maximum inventory. There is no mixing of unequal concentrations [11].

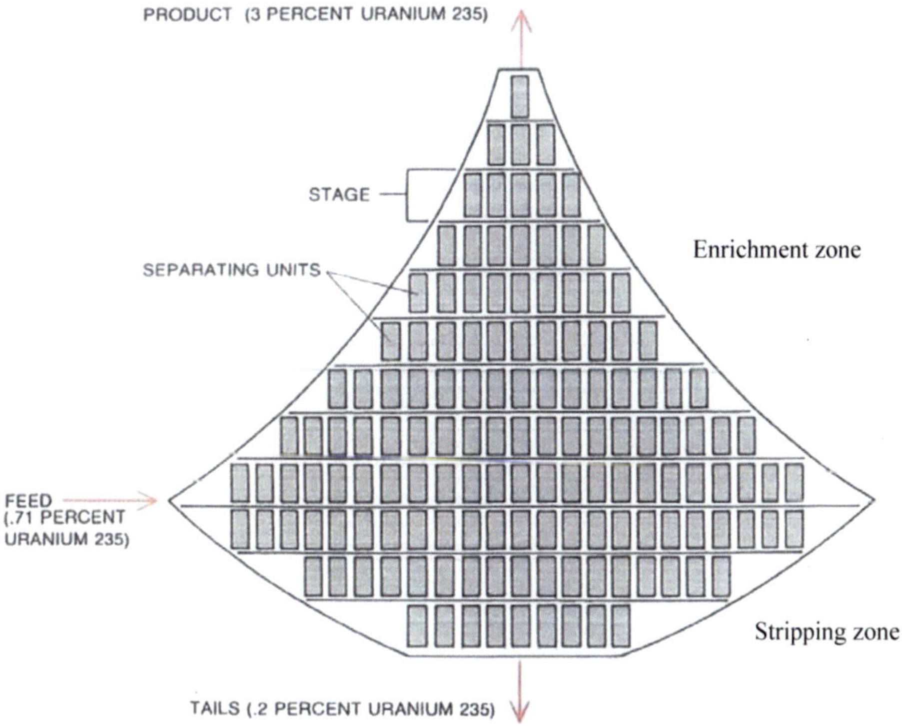

Figure 15.3: Ideal cascade; is made up of stages, each one consisting of a number of gas ultra-centrifuges operating in parallel to provide the required flow of material [15].

Stage enrichment factor in the cascade : $\alpha = 1.1$:

Table 15.6a: Total number of stages as well as number of stages in the enrichment and in the stripping section of an ideal cascade (in brackets) and of a minimum number of stages cascade (without brackets) for an enrichment factor $\alpha = 1.1$, for product enrichment $x_p = 0.93$, for feed enrichments 0.15 and 0.1995 as well as different tails assays of $x_w = 0.003$, $x_w = 0.05$ and $x_w = 0.1$.

Number of stages	Gas centrifuge $\alpha = 1.1$, $x_F = 0.15$				Gas centrifuge $\alpha = 1.1$, $x_F = 0.1995$		
	Tails Assay	x_w= 0.003	x_w= 0.05	x_w= 0.1	x_w= 0.003	x_w= 0.05	x_w= 0.1
Enrichment section		(90) 45	(90) 45	(90) 45	(83) 42	(83) 42	(83) 42
Stripping section		(85) 43	(25) 13	(9) 5	(91) 46	(32) 16	(16) 8
Total number of stages		(175) 88	(115) 58	(99) 50	(174) 88	(115) 59	(99) 50

Table 15.6a and 15.6b show – as an example – the total number of stages required as well as the number of stages needed in the enrichment and in the stripping zone of an ideal cascade (in brackets) and minimum numbers of stages cascade (without brackets). Unauthorized enrichment of 15% or 19.95% enriched uranium ($x_F = 0.15$ and $x_F = 0.1995$) up to 93% U-235 uranium ($x_P = 0,93$) and different tails assays $x_w = 0.3\%$, 5% and 10% results in (99) to (175) and 50 to 88 stages for a stage enrichment factor of $a = 1.1$ (Table 15.6a) as well as (52) or 26 to (91) or 46 stages for a stage enrichment factor $\alpha = 1.2$, respectively (Table 15.6b).

Stage enrichment factor of heads to tails: $\alpha = 1.2$:

Table: 15.6b: Total number of stages as well as number of stages in the enrichment and in the stripping section of an ideal cascade (in brackets) or of a minimum number of stages cascade (without brackets) for stage enrichment factor $\alpha = 1.2$, for product enrichment x_p =0.93, for feed enrichments 0.15 and 0.1995 as well as different tails assays of $x_w = 0.003$, $x_w = 0.05$ and $x_w = 0.1$.

Number of stages	Gas centrifuge $\alpha = 1.2$, $x_F = 0.15$				Gas centrifuge $\alpha = 1.2$, $x_F = 0.1995$		
	Tails Assay	x_w= 0.003	x_w= 0.05	x_w= 0.1	x_w= 0.003	x_w= 0.05	x_w= 0.1
Enrichment section		(47) 24	(47) 24	(47) 24	(44) 22	(44) 22	(44) 22
Stripping section		(44) 22	(13) 7	(5) 3	(47) 24	(16) 8	(8) 4
Total number of stages		(91) 46	(60) 30	(52) 27	(91) 46	(61) 31	(52) 26

Tables 15.6a and 15.6b show the following trends:

- the higher the enrichment factor α, the lower is the required number of stages in the enrichment and in the stripping section
- an enrichment starting from 15% enriched feed material increases the number of stages in the enrichment sections barely, if compared to 19.95% enriched feed material
- the number of stages in the stripping zone is drastically decreased if the tails assay is increased from $x_w = 0.003$ to 0.1.

In addition to the number of stages needed, also the number of machines (gas ultra-centrifuges) working in parallel is interesting to know, if a certain quantity of HEU shall be enriched within a certain time period. For a first generation of commercial gas ultra-centrifuges a number of 250 to 350 machines (gas ultra-centrifuges) was estimated by Kryuchkov et al. [8].This allows to produce one significant quantity (22 kg of 90% U-235 enriched uranium) (x_w = 0.2-0.5%) within a time period of about one month.

Glaser [7] analyzes different break-out scenarios (abrogation from IAEA safeguards by a state) starting from a feedstock of 16.31% U-235 enriched uranium. The scenario is based on information made public by officials from Pakistan and Iran [7]. The analysis bases on a cascade block of 164 centrifuges with 15 cascade stages (Fig. 15.4). The feed stream is introduced in stage number 5. There are 10 enriching stations and 4 stripping stages. The product stream feeds in the next stage and the tails stream feeds back into the previous stage.

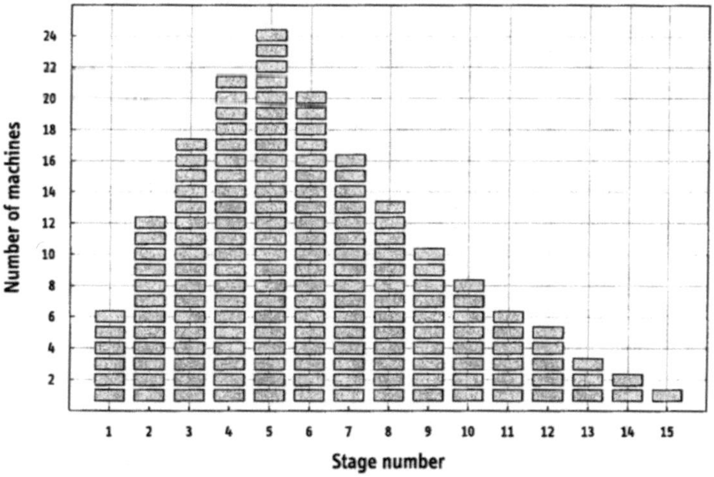

Figure 15.4: Possible arrangement of 164 machines in a 15-stage cascade [7].

This 15 stage cascade block has less cascade stages than given in Tables 15.6a and 15.6b. The gas ultra-centrifuges are so called P1-type centrifuges (early Dutch design of the initial phase of the URENCO project) with 2.4 SWU/yr per gas ultra-centrifuge.

The P1-044 centrifuge design has a specific productivity of 16.9 mg/s of UF_6 for feed rate and 1.73 mg/s UF_6 for product rate and an enrichment factor of a=1.1215. With these design characteristics the 164 machines, 15 stages cascade block starting from x_F = 16.309% feed enrichment achieves x_P = 91.089% U-235 maximum product enrichment with a tails assay of x_W = 7.528%. The main characteristics are given in Table 15.7.

Table 15.7: Characteristics data of the 164 machines, 15 stages cascade block of Fig. 15.4 [7].

x_F = 16.309%	x_P = 91.089%
x_W = 7.568%	α = 1.1215
Separative Work Units of 164 machines: 164 x 2.4 = 393.6 SWU/yr	
Centrifuge rotor diameter: 10 cm	Centrifuge height: 180 cm
Peripheral velocity: 320 m/s	Material: aluminum

The separative work to enrich 1 kg of HEU (x_P = 91.089% U-235 uranium) from x_F = 16.309% uranium enrichment is about 10.6 kg SWU. This leads to the production of 37 kg HEU (91.089% U-235 uranium) or about 1.7 significant quantities of about 22 kg per year. This is by a factor of about 7 less than the value mentioned before as the one proposed by Kryuchkov et al. [8]. The large tails assay of about 7.568% leads to a certain inefficiency and a higher requirement of feed UF_6 with 16.309% U-235 enriched uranium if compared with a so-called re-configurated cascade arrangement [7].

According to Glaser [7] the above 164 machines cascade arrangement becomes very attractive if pre-enriched uranium of 3.5% U-235 uranium is used as feedstock. In this case three parallel blocks of the 164 machines cascade block arrangements would be used to enrich the 3.5% U-235 uranium to 16.309% U-235 uranium. Subsequent enrichment of the 16.309% U-235 uranium in one 164 machines cascade block would then lead to 91.089% U-235 uranium as described above. In total 656 machines would, therefore, be needed.

The production of 37 kg of about 91% U-235 enriched HEU by the above 15 stages 164 machines cascade block can of course be improved by gas ultra-centrifuges of more modern design with a higher performance of e.g. 5 SWU/yr.

15.5.1 Summary of Results

Considering the above given characteristic data it must be concluded:

Uranium with enrichments of 15 or <20% U-235 represents a considerable proliferation risk as it can be used for unauthorized re-enrichment by gas ultra-centrifuges up to weapons uranium (93% U-235).

The technical solution to this proliferation problem will be the **admixing of the uranium isotope U-232** to the <20% U-235 enriched uranium [8,12]. This uranium isotope U-232 should be admixed in small amounts, e.g. 0.1% or somewhat less, to the e.g. 15 or <20% U-235 enriched uranium (denaturing or doping) in order to make the use of such 93% U-235 enriched uranium for weapons application technically **unfeasible**. In the following chapter 16 the nuclear characteristics of U-232 will be summarized. The effects of an admixture of small amounts of U-232 will also be analyzed.

References Section 15

[1] INFCE (International Nuclear Fuel Cycle Evaluation), Report of INFCE Working Group 8: Advanced Fuel Cycle and Reactor Concepts, published by the International Atomic Energy Agency, Vienna, 1980.

[2] Glaser A., On the Proliferation Potential of Uranium Fuel for Research Reactors at Various Enrichment Levels, Science and Global Security, 14:1-24, Copyright © Taylor & Francis Group, LLC, 2006.

[3] Kang J. et al., U-232 and the Proliferation-Resistance of U-233 in Spent Fuel, Science and Global Security, 9:1-32, Copyright © Taylor & Francis Group, LLC, 2001.

[4] Bunn G. et al., Research Reactor Vulnerability to Sabotage by Terrorists, Science and Global Security, 11:85-107, Copyright © Taylor & Francis Group, LLC, 2003.

[5] RERTR: Reduced Enrichment for Research and Test Reactors, Nuclear Engineering Division at Argonne National Laboratory, USA. http://www.rertr.anl.gov/

[6] Bukharin O. et al., US-Russian Reduced Enrichment for Research and Test Reactors (RERTR) Cooperation RANSAC, 2002

[7] Glaser A., Characteristics of the Gas Centrifuge for Uranium Enrichment and their Relevance for Nuclear Weapon Proliferation (corrected), Science and Global Security, 16:1-25, Copyright ©Taylor & Francis Group, LLC, 2008.

[8] Kryuchkov E. F. et al., Evaluation of Self-Protection of 20% Uranium Denatured with ^{232}U against Unauthorized Re-enrichment, Nuclear Science and Engineering: 162, 208-213, 2009.

[9] Kutchesfahani S., Case Study: The Khan Network, Doyle J. E., Nuclear Safeguards, Security and Nonproliferation, Butterworth - Heinemann, an imprint of Elsevier, Amsterdam, 2008.

[10] Rahn F. I. et al., A Guide to Nuclear Power Technology, John Wiley & Sons, New York, 1984

[11] Benedikt M. et al., Nuclear Chemical Engineering, New York, Mc Graw-Hill, 1981.

[12] Moir R., U232 Nonproliferation Features, LLNL-TR-438648, June 2010.

[13] Rosenstock W. et al., Thorium for Nuclear Energy - a Proliferation Risk, DPG-AGA Physics and Disarmament, Dresden. Germany, 2013.

[14] Ganesan S. et al., New Investigations of the Criticality Property of Pure ^{232}U. Annals of Nuclear Energy 29, 1085—1104, 2002.

[15] Olander D.R., The Gas Centrifuge, Scientific American, Vol. 239, Number 2 pp. 27-33, Aug. 1978.

[16] Forsberg C. W. et al., Definition of Weapons-Usable Uranium-233. ORNL/TM-13517, Oak Ridge National Laboratory, USA, March 1998.

[17] Nuclear Security Recommendations on Physical Protection of Nuclear Material and Nuclear Facilities (INFCIRC/225/Revision 5), IAEA, Vienna, 2011.

[18] Alvarez R., Managing the Uranium-233 Stockpile of the United States, Science and Global Security, 21:53-69, Copyright © Taylor & Francis Group, LLC, 2013.

[19] Tozser S. M. et al., Available Reprocessing and Recycling Services for Research Reactor Spent Nuclear Fuel, Atomwirtschaft Vol. 60, Issue 5, page 303-313, 2015.

[20] Past and Current Civilian HEU Reduction Efforts, NTI, 2014.

[21] http://www.nti.org/analysis/articles/past-and-current-civilian-heu-reduction-efforts/

16 Admixture of U-232 to <20% U-235 enriched Uranium

16.1 Nuclear characteristic data of U-232 and other uranium isotopes

In Table 16.1 the main nuclear characteristic data of U-232 are compared to those of other important uranium isotopes: U-234, U-235 and U-238. (The nuclear characteristic data of U-233 will be discussed separately in chapter 17).

Table 16.1: Basic nuclear properties of uranium isotopes in uranium fuel [5].

	^{232}U	^{234}U	^{235}U	^{238}U
Half-life (y)	68.9	$2.45x10^5$	$7.04x10^8$	$4.47x10^9$
Specific yield of α-particles [1/(g·s)]	$8x10^{11}$	$2.3x10^8$	$7.9x10^4$	$1.2x10^4$
Mean energy of α.particles (MeV)	5.3	4.76	4.4	4.19
Specific yield of spontaneous fission neutrons [1/(g·s)]	1.3	$5.02x10^{-3}$	$2.99x10^{-4}$	$1.36x10^{-2}$
Fission cross section (E_n=0.0253 eV) [barn]	77.15	0.465	583.2	$1.2x10^{-5}$

U-232 decays with a half-life of 69 years to Th-228 which decays in four subsequent α-decays to Pb-212 (Fig. 16.1 and Table 16.2). This Pb-212 isotope decays emitting β-particles to Bi-212 and then to Po-212. Also Bi-212 decays emitting α-particles to Tl-208 which again decays emitting β^--particles to stable Pb-208. Tl-208 emits high energy γ-radiation of 2.6 MeV. Finally, Po-212 decays by emitting α-particles to stable Pb-208.

- U-232 is a very strong α-particle emitter with $8x10^{11}$ α-particles per g·s [5]. This causes an α-heat production of 6809 W/kg after chemical separation. This very strong α-heat production is even increasing to 5,322 W/kg after about 9 years by the α-decay of Th-228 and the other daughter products (Fig. 16.13)

- The α-particles emitted by U-232 are able to dissociate the UF_6 molecules and can make it practically impossible to enrich uranium to weapons-grade level [5].

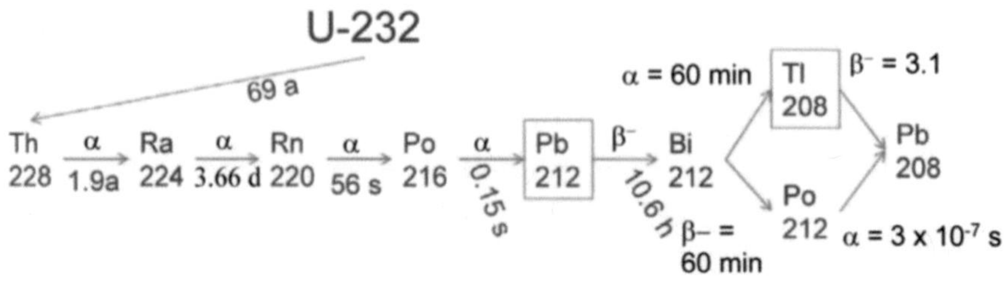

Figure 16.1: Decay chain of U-232 [22].

Table 16.2: Decay products, emitters of α-particles [5].

	Decay Products of ^{232}U					
	^{228}Th	^{224}Ra	^{220}Rn	^{216}Po	^{212}Bi	^{212}Po
Half-life	1.91 y	3.66 d	55.6 s	0.145 s	1.06 h	$3x10^{-7}s$
Energy of α-particles (relative intensity) (MeV)	5.42 (71.7%) 5.34 (27.6%)	5.69 (94.9%) 5.45 (5.1%)	6.29 (100%)	6.78 (100%)	6.09 (9.7%) 6.05 (25.2%)	8.78 (100%)

- Compared to other uranium isotopes U-232 is a very strong spontaneous fission neutron source of 1.3 n/(g·s). However, compared to e.g. Pu-240 and Pu-242 it is a much weaker spontaneous fission neutron source.
- U-232 with its very strong α-particle source is able to initiate (α,n)-reactions with light elements existing as impurities such as boron, carbon etc. in uranium metal. This can magnify the spontaneous fission neutron source considerably (section 16.4).
- The decay products Tl-208 and Bi-212 emit a high-energy γ-radiation (2.6 MeV and 0.7 MeV, respectively) which can give a certain protection of uranium denatured by sufficient percentage of U-232 against the misuse for weapons purposes.

U-232 has a relatively high fission cross section and a low critical mass. Different cross section sets lead to a relatively wide range of critical mass of 3.7 to 13.6 kg (there are no critical experiments available yet) [17].

16.2 Production routes of U-232

U-232 originates from nuclear reactions with Th-230 and Th-232 in reactor cores operating in the Th/U fuel cycle.

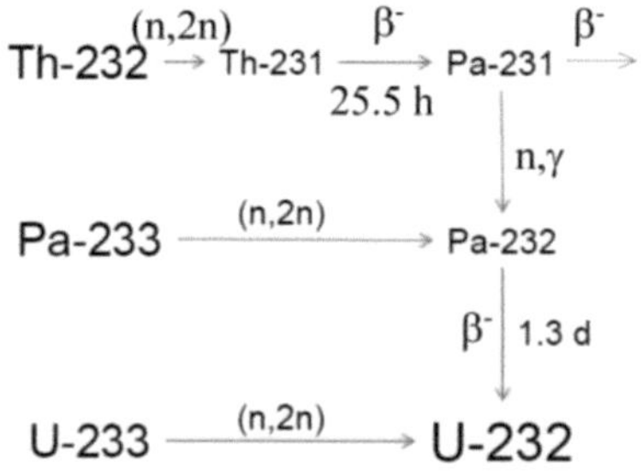

Figure 16.2: Production routes of U-232 via (n,γ) and (n,2n)-reactions [22] (adapted).

The intense γ-radiation of U-232 daughter products e.g., Tl-208 (2.6 MeV) represent an obstacle to essential parts of the Th/U fuel cycle. U-233 fuel will always be contaminated by U-232 and is, therefore, difficult to handle. According to the above nuclear reactions, U-232 originates after (n,2n)-reactions with U-233 and Pa-233. The fraction of U-232 increases with burnup of the fuel in the reactor core and can attain up to several 1000 ppm for high burnup Th/U fuel. The lowest U-232 impurity levels are attained in CANDU-type fuel elements at low burnup [1]

U-232 can be produced as a relatively pure material for radioisotope radiation sources if the following production scheme (Fig. 16.3) is followed [2,3]:

An amount of 10 kg thorium (consisting of about 1 kg Th-230 and 9 kg of Th-232) is put into an irradiation capsule which is loaded into a somewhat under-moderated reactor core with a partly epithermal neutron spectrum and a neutron flux of 10^{14}n/cm^{2}s. Th-230 is a daughter product of U-238 and can be separated from uranium refinery wastes. It is also found in several percent concentrations in thorium ores.

After an irradiation period of 400 d the irradiation capsule is unloaded and cooled for 270 d. This cooling period is essential. During the irradiation period Th-231 and Th-233 are formed by neutron capture in Th-230 and Th-232.

Th-231 and Th-233 have different half-lives for β–decay into Pa-231 (half-life: 32,800 y) and Pa-233 (half-life: 27d). After 10 half-lives for the decay of Pa-233

$$10x27 = 270 \text{ d}$$

this isotope has essentially decayed into U-233, whereas Pa-231 with its long half-life remains.

Chemical separation results in thorium (Th-230 and Th-232), fission products, Pa-231 and the uranium isotopes (U-232, U-233, U-234, U-235), (see Fig. 16.3).

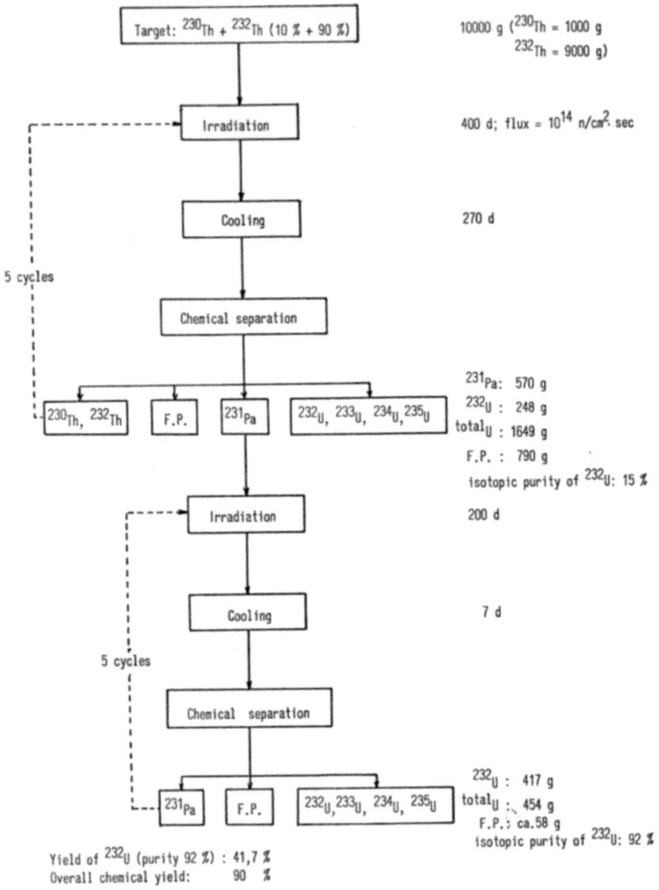

Figure 16.3: Production scheme for U-232 after irradiation of Th- 230 and Th- 232 [2,3].

The thorium (Th-230 and Th-232) and Pa-231 are then recycled back in different capsules into the reactor core. The thorium is irradiated 5 times each over 400 d, whereas the Pa-231 is irradiated 5 times each over 200 d. Neutron capture in Pa-231 leads to Pa-232 which decays to U-232.The cooling time after irradiation of the Pa-231 is only 7d. Chemical separation after irradiation and cooling leads to remaining Pa-231, fission products and the uranium isotopes U-232, U-233, U-234 and U-235.

The second five irradiation cycles of Pa-231 finally lead to 417 g of U-232 in within 454 g uranium isotopes (U-232, U-233, U-234, U-235) which is an enrichment of 92% U-232. This irradiation and chemical separation scheme was proposed for the production of U-232-beryllium neutron sources. The 15% U-232 enriched in uranium isotopes (248 g in 1,649 g uranium isotopes) resulting after the first five irradiation cycles can also be used for other applications than neutron irradiation sources.

Admixing of this U-232 to 19.95% U-235 enriched uranium and its effects on re-enrichment by gas ultra-centrifuges will be discussed in the following sections.

16.3 Destruction of Uranium Hexafluoride UF_6 by α-Radiation of U-232

Bernhard et al. [4] reported that the gaseous uranium hexafluoride used for uranium enrichment e.g. in gaseous diffusion plants or gas ultra-centrifuge plants will be ionized and destroyed by α-radiation. This α-radiation arises from U-234, U-235 and U-238 during enrichment of natural uranium. When natural uranium is enriched in a diffusion cascade more than 100-fold to 93% enriched weapons type uranium this self-irradiation can lead to about 0.2% of the UF_6 molecules to be ionized and destroyed over the time period of enrichment procedures.

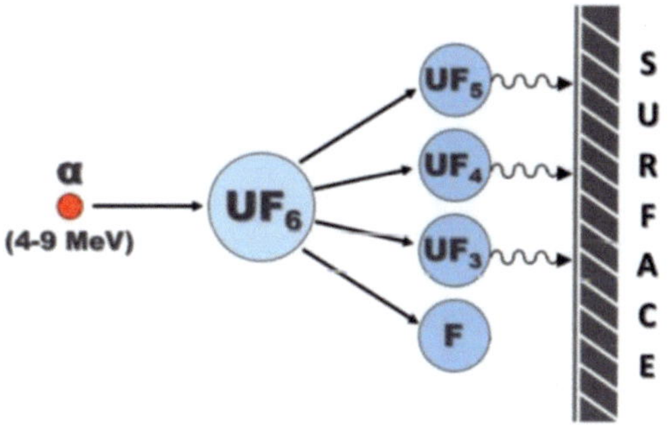

Figure 16.4: Destruction of UF_6 by α-particles.

Kryuchkov et al. [5], therefore, proposed to intensify this destruction process by admixing U-232 with its very high α-particle irradiation to e.g. 19.95% enriched uranium (see also section 14.7.2). UF_6 is the only gaseous composition of chemical uranium fluoride compounds. All other uranium fluoride compounds, originating after destruction of UF_6 e.g., UF_5, UF_4, UF_3, U_2F_9 and U_4F_{12} [4,5,6] are either low-volatile or non-volatile. In addition, free fluorine is released.

Experiments showed that e.g., UF_3 and UF_4 are adsorbed at inner surfaces, whereas UF_5 can also partially recombine with fluorine. However, if the reservoir for UF_6 is small and at low pressure this recombination process can slow down substantially and non-volatile uranium fluorides have enough time to plate out at inner walls.

According to experimental results [4,5,6] about 0.9 molecules of UF_6 are destroyed per each 100 eV of absorbed energy of α-particles. One α-particle of 5.3 MeV from the decay of U-232 (Table 16.1), therefore, destroys 47,700 UF_6 molecules.

Kryuchkov et al. [5] showed in an example calculation -similarly as in section 14.7.2 for reactor plutonium- that a 1% U-232 concentration in 19.95% U-235 enriched uranium would destroy 48% of the UF_6 molecules within 1.2 months during enrichment in a gas ultra-centrifuge cascade arrangement. The assumption of 1% is only an example, as in a number of stages with higher enrichment the U-232 concentration will be higher than 1% during re-enrichment. In addition UF_6 molecules will also be destroyed during transport.

Kryuchkov et al. [5] also showed that an admixture of 0.1% U-232 would increase to 11% (10 enriching stages, stage enriching factor 1.6 for U-232) at the end of the gas ultra-centrifuge cascade, due to the lower atomic weight of U-232 (three atomic mass units) compared to U-235, (see section 16.5).The higher content of U-232 in the higher enrichment stages of the ultra-centrifuge cascade arrangement would intensify the α-particle radiation to such an extent that almost all of the UF_6 molecules might be destroyed.

In addition, the lower uranium fluorides and the free fluorine constitute a light fraction in the gaseous mixture which will disturb the enrichment process and violate the proper order of enrichment [5].

16.4 α-Particle Reactions with Light Elements as Impurities in Uranium Metal

α-particles react with light elements producing neutrons along the following nuclear reactions [26]:

$$^{7}Li + \alpha \rightarrow {}^{10}B + n$$

$$^{9}Be + \alpha \rightarrow {}^{12}C + n$$

$$^{11}B + \alpha \rightarrow {}^{14}N + n$$

$$^{19}F + \alpha \rightarrow {}^{22}Na + n$$

$$^{27}Al + \alpha \rightarrow {}^{30}P + n$$

U-232 is a very strong α-emitter (Table 16.1). It can -if admixed to U-235/U-238- increase considerably the total fission neutron source of the fuel by additional (α,n)-reactions with light elements. Such light elements do exist as impurities in U-235/U-238 metal or dioxide fuel.

Kryuchkov et al. [5], therefore proposed to admix 0.1% U-232 to the 19.95% U-235 enriched uranium in order to make it **proliferation-proof** against unauthorized re-enrichment by gas ultra-centrifuges. They calculated the (α,n)-nuclear reaction rates by using the nuclear cross section sets JENDL [7] and EXFOR [8] and applying the ′′the Stopping and Range of Ions in Matter,′′ code SRIM [9]. This is done for the enrichment available after re-enriching the 19.95% U-235 uranium doped with 0.1% U-232 up to weapons type level. This number of neutrons from (α,n)-reactions with light elements is finally added to the number of neutrons from spontaneous fission neutron emitters.

Kryuchkov et al. [5] used the information on the content of light element impurities from publications of the Novosibirsk Chemical Concentration Plant and the Ulba Metallurgical Plant in Kazakhstan [11,12]. As an example the data on light element impurities in uranium metal of the Ulba plant in Kazakhstan are given in Table 16.3. Other light element impurity data by Yemel'yanov et al. [15] and URENCO [16] are similar. The data of URENCO are given in Table 16.4. They are categorized differently and are valid for UF_6 which makes a direct comparison with Table 16.3 difficult. In any case the data on impurities could also be adapted as needed during the metallurgical process by adding light elements.

Table 16.3: Light elements as impurities found in uranium metal of the Ulba plant at Kazakhstan [11,12].

Light element	Impurity in ppm in uranium metal
Lithium	1
Beryllium	1
Boron	0.2
Carbon	550
Sodium	30
Magnesium	10
Aluminum	50
Silicon	100
Phosphorous	100
Potassium	50
Calcium	100
Manganese	10
Iron	200

Table 16.4: Light element impurities found in UF_6 in the German URENCO plant at Gronau, Germany [16].

Light Element	Impurity in ppm in UF_6
Boron	≤0.4
Carbon	≤1200
Lithium, Beryllium	
Magnesium, Calcium	≤100
Phosphorous	≤20
Silicon	≤50
Bromine	≤5

16.5 Intensified Enrichment of U-232 during Enrichment of Multicomponent Uranium Isotope Mixtures in Gas Ultra-Centrifuges

As U-232 is lighter by three atomic mass units than U-235 it will be higher enriched in a mixture of U-232, U-234, U-235 and U-238 isotopes than e.g., U-235 [13,14]. During unauthorized re-enrichment of 19.95% U-235 enriched uranium with admixture of e.g. 0.1% U-232, therefore, the latter will be enriched by a factor of about 110 to 11% U-232, whereas U-235 is only enriched by a factor of 3.624 to 72.3% U-235. This is achieved in 10 equal gas ultra-centrifuge cascade enriching stages with a stage enrichment factor of 1.6 for U-232 and 1.1374 for U-235 [13,14]. Table 16.5 shows the different enrichment levels prior and after unauthorized re-enrichment to weapons type level enrichment:

Table 16.5: Characteristic data after re-enrichment of 19.95% enriched U-235 uranium with admixture of 0.1% U-232 [5].

Prior to re-enrichment	Content of U-232 in 19.95% U-235 enriched uranium is 0.1%
	Content of U-238 in uranium: 79.95%
After re-enrichment	Content of U-232 in weapons type enriched uranium is 11%
	Content of U-235 in weapons type enriched uranium is 72.3.0%
	Content of U-238 in weapons type enriched uranium is 16.7%

The U-232 replaces partly the U-235. It also has a lower critical mass than U-235. (The isotope U-234 is neglected here). The (α,n)- and spontaneous fission neutrons are determined by Kryuchkov et al. [5] to be 0.89 x 10^6 n/s for a content of 11% U-232 after re-enrichment (initial admixture of 0.1% of U.232).

In the following the admixture of 0.1% U-232 will be discussed in detail. It will be shown that 0.1% admixture of U-232 will satisfy all requirements to render re-enriched 19.95% uranium **proliferation-proof**.

16.6 Hypothetical Nuclear Explosive Devices

16.6.1 Gun-type system Hypothetical Nuclear Explosive Device

Kryuchkov et al. [5] assumed an internal neutron source of S_1= 0.89 x 10^6 n/s for a gun-type system arrangement of a hypothetical nuclear explosive device (HNED). As this number cannot be easily recalculated from all other data given in their paper a second number for the internal neutron source S_2 will be used here for the further analysis. The difference between both numbers S_1 and S_2 can be explained probably by the uncertainties of the critical mass which might have been calculated with different cross section sets and the fact that no critical experiments are available yet for U-232. This was discussed in detail by Ganesan et al. [17] who reported critical masses between 3 to 13 kg for U-232 depending on different cross section sets. The highly enriched weapons type uranium analyzed here contains 11% U-232 (previous section).

The second value of S_2= 0.43 x 10^6 n/s is based on a critical mass of 23.14 kg,(6.15 cm radius and 8.2 cm length) of a cylinder. It was calculated by applying the Monte Carlo code MCNP [24] and the nuclear cross section set ENDF/B VII [23]. In addition, a spontaneous fission yield of 1.3 n/(g·s) (Table 16.1) and an (α,n)-yield from light element impurities of 8.024 n/(g·s) (calculated in [5], see also Fig. 16.5) were applied.

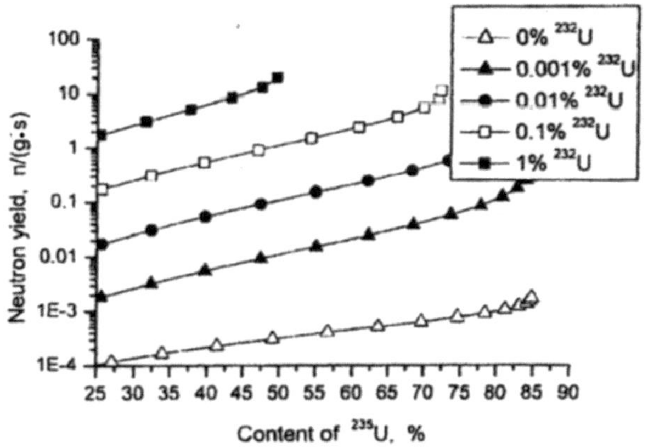

Figure 16.5: Neutron yield from (α,n)-reactions of light elements in 19.95% re-enriched uranium and different U-232 contents (decay chain of U-232 not taken into account [5]).

According to sections 11.5.1 and 11.7.1 the first gun-type nuclear explosion device had about 2 or more critical masses. Therefore, a factor of 2.3 is chosen here.

$S_2 = 2.314x10^4$ x 2.3 x (8.024 + 0.11 x 1.3) = 0.43 x 10^6 n/s

16.6.2 Spherical Implosion Hypothetical Nuclear Explosion Devices

For comparison also an HNED with an initial $k_{eff} = 0.98$ will be analyzed. Its critical mass was determined to be

18.51 kg (near critical radius 6.15 cm for $k_{eff} = 0.98$))

applying the Monte Carlo code MCNP and the nuclear cross section set ENDF/B –VII [23]. This leads to an internal neutron source [24] of

$S_3 = 1.851x10^4$ x (8.024 + 0.11 x 1.3) = 0.15 x 10^6 n/ s

This subcritical neutron source has to be multiplied by $1/(k_{eff}-1)$ which results in 1/(1-0.98) = 50 and finally to

$S_3 = 7.6$ x 10^6 n/s

16.7 Analysis of the Explosion Potential of HNEDs from Unauthorized Enrichment of 19.95% Enriched U-235 Doped with U-232

For the analysis of the explosion potential of HNEDs from unauthorized re-enrichment of 19.95% U-235 enriched uranium doped with U-232 the methods used in sections 9.11 and 11.7 are applied.

16.7.1 Probability of Pre-ignition by Spontaneous Fission Neutrons and (α,n)-neutrons

The theory of pre-ignition of a nuclear explosion by spontaneous fission neutrons was described in chapter 9.11. It is applied here to spontaneous fission neutron and (α,n)-neutrons sources.

It is presented in a simplified way for a ramp increase, where the criticality factor k_{eff} or $\Delta k = k_{eff}-1$ increases from prompt criticality to a maximum $k_{eff,max}$ or to $\Delta k_{max} = k_{eff,max} - 1$ during a short time period to (Fig. 16.6a).

Similarly, the Rossi alpha

$$\alpha(t) = \frac{\Delta k}{\tau} \qquad (16.1)$$

with τ as the neutron lifetime can be presented as a ramp function (Fig. 16.6b).

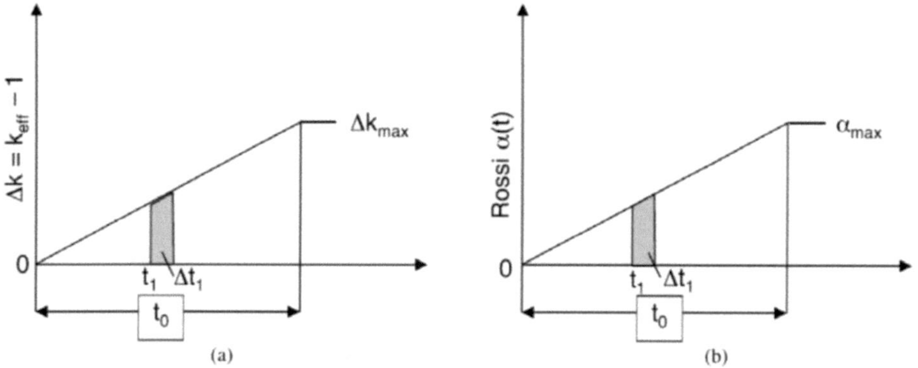

Figure 16.6: Fig. 16.6: (a) $\Delta k(t)$ as a ramp function of time t_1 and (b) Rossi $\alpha(t)$ as a ramp function of time t_1.

The differential probability $p(t_1)$ of a first persistent fission chain being sponsored at time t_1 in the time interval dt_1 is then

$$p(t_1)\cdot dt_1 = \frac{2\Delta k_{max}}{\bar{\nu}\cdot\Gamma_2}\cdot\frac{S}{t_0}\cdot t_1 \times \exp\left[\frac{-\Delta k_{max}}{\bar{\nu}\Gamma_2}\cdot\frac{S}{t_0}\cdot t_1^2\right] dt_1 \quad (s^{-1}) \qquad (16.2)$$

and the cumulative probability $P(t_1)$ that persistent fission chains have occurred until time t_1 (integral of the differential probability between zero and t_1) is given by

$$P(t_1) = 1 - \exp\left[-\frac{\Delta k_{max}}{\bar{\nu}\Gamma_2}\cdot\frac{S}{t_0}\cdot t_1^2\right] \qquad (16.3)$$

where

$\Delta k_{\max}$ = maximum $k_{eff,\max} - 1$ at full compression

$\bar{\nu}$ = average number of neutrons per spontaneous fission

$$\Gamma_2 = \nu \cdot (\nu - 1)/\bar{\nu}^2 \approx 0.8$$

S = spontaneous fission neutrons per second.

For Δk_{max} of ~1, ν ~2.5, and $\Gamma_2 \approx 0.8$ the factor $\Delta k_{max}/(\nu \cdot \Gamma_2)$ becomes ½ and Eqs. 16.2 and 16.3 become Eqs. 16.4 and 16.5 as they were derived in section 9.11.1 of chapter 9.

$$p(t_1) \cdot dt_1 = S \cdot \frac{t_1}{t_0} \cdot \exp\left[-\frac{1}{2} \cdot S \cdot \frac{t_1^2}{t_0}\right] \cdot dt_1 \quad (s^{-1}) \tag{16.4}$$

and

$$P(t_1) = 1 - \exp\left[-\frac{1}{2} \cdot S \cdot \frac{t_1^2}{t_0}\right] . \tag{16.5}$$

The maximum nominal nuclear explosive yield Y_o would theoretically be attained if $P(t_o)$= 0; i.e., no persistent fission chains would have occurred up to t_0 (no pre-ignition up to Δk_{max}). In practical reality, $P(t_0)$ can only be minimized.

The minimum nuclear explosive yield is called minimum fizzle yield $Y_{F,min}$. It is determined by the criterion that pre-ignition is initiated at the earliest possible point in time t_1=0 and that approximately e^{45} fissions will have occurred during the chain reaction (section 9.12 and 9.13). The internal energy accumulated will then be high enough such that the nuclear fissile material will have vaporized and begun to exert a pressure in the mega bar range on its surroundings. This will initiate a rapid expansion. It also marks the maximum excess reactivity introduced for this case.

The minimum ratio of the fizzle yield can then be determined to be

$$x_{F,\min} = (Y/Y_0)_{F,\min} = \left(\frac{90\tau}{t_0}\right)^{3/2} \tag{16.6}$$

by applying Serber's relation (section 9.9).

In a more general way, a relation between the cumulative probability of pre-ignition (pre-detonation) and the yield fractions $x = Y/Y_0$ (with Y as the explosive yield achieved after pre-ignition at some point in time and Y_0 as the maximum nominal yield) was also derived as

$$P(x) = P\left(\frac{Y}{Y_0}\right) = 1 - \exp\left[-\frac{1}{2} S \cdot t_0 \left(\frac{Y}{Y_0}\right)^{2/3} + 45 \cdot S \cdot \tau\right] \tag{16.7}$$

16.7.2 Numerical Examples

For a further evaluation Eqs. (16.4 through 16.7) require representative numerical values for the spontaneous fission and (α,n)-neutron source S, the effective neutron life time τ, and for the time period t_0.

16.7.2.1 Spontaneous Fission Neutron Sources

From section 16.5 the total neutron source (spontaneous fission neutron and (α,n)-neutron source) is known to be:

$S_1 = 0.89 \times 10^6$ n/s (gun-type device [5])

$S_2 = 0.43 \times 10^6$ n/s (gun-type device, section 16.5.3)

$S_3 = 7.6 \times 10^6$ n/s (spherical implosion device, section 16.5.4)

16.7.2.2 Effective Neutron Lifetime

The effective neutron lifetime τ can be assumed to be $\sim 10^{-8}$ s for enriched U-232, U-235 and U-238 (section 9.8.2)

16.7.2.3 Time Period t_0 from Reaching Prompt Criticality to Δkmax

The time period t_0 for the increase from $\Delta k = 0$ up to Δk_{max} depends on the method of compaction of the nuclear fissile material. Two methods are known: the gun-type system and the implosion method.

16.7.2.4 Gun-Type-System

For a gun-type system with highly U-235 enriched uranium an assembly velocity of 300 m/s in the gun barrel can be assumed. This leads to $t_0 = 10^{-3}$ s which is a factor of 100 higher for the time period t_0 than for the implosion system with $t_0 = 10^{-5}$ s.

For the subsequent evaluation, the time period t_0 for the gun-type system is treated in a parametric way with two values $t_0 = 5 \times 10^{-4}$ s and $t_0 = 10^{-3}$ s.

16.7.2.5 Implosion Method

For the implosion method, a time period $t_0 = 10^{-5}$ s is assumed, which would correspond to a particle velocity of 6 km/s during shock compression of a 6-cm-radius solid uranium metal sphere.

16.7.3 Results for the Parametric Evaluation

Figures 16.7 through 16.10 show the evaluation of Eqs. (16.5) and (16.7) for the above defined parameters.

16.7.3.1 Gun-Type System

Figures 16.7 and 16.8 show the cumulative probabilities of pre-ignition for $t_0 = 10^{-3}$ s and $t_0 = 5\times10^{-4}$ s as a function of the ratio t_1/t_0. The time derivative of this curve represents the differential probability of pre-ignition at t_1 within the time interval dt_1. Figure 16.7 shows that pre-ignition would occur early within the compaction phase up to $t_1/t_0 = 0.1$-0.15. For a shorter compaction time $t_0 = 5\times10^{-4}$ s (Fig. 16.8), this time interval for early pre-ignition is $t_1/t_0 = 0.15$-0.2.

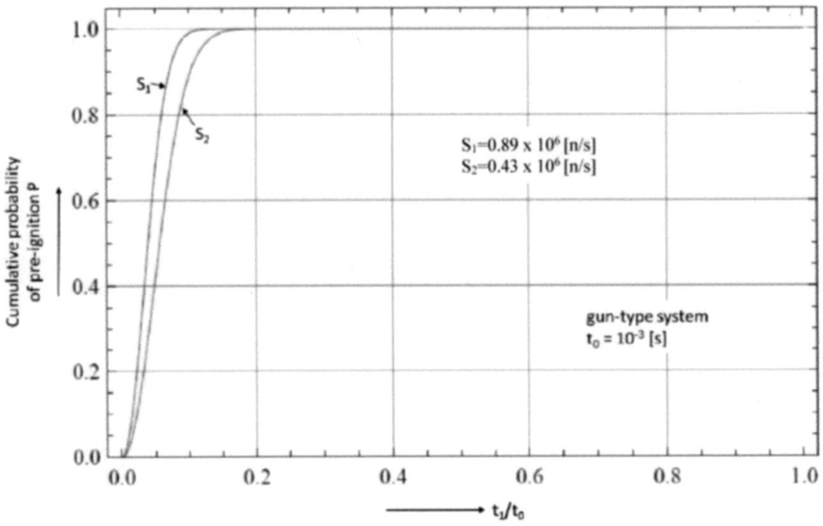

Figure 16.7: Cumulative probability of pre-ignition for a time period of $t_0 = 10^{-3}$ s of compaction by a gun-type system as a function of t_1/t_0. Neutron source $S_1 = 0.89\times10^6$ n/s and $S_2 = 0.43\times10^6$ n/s.

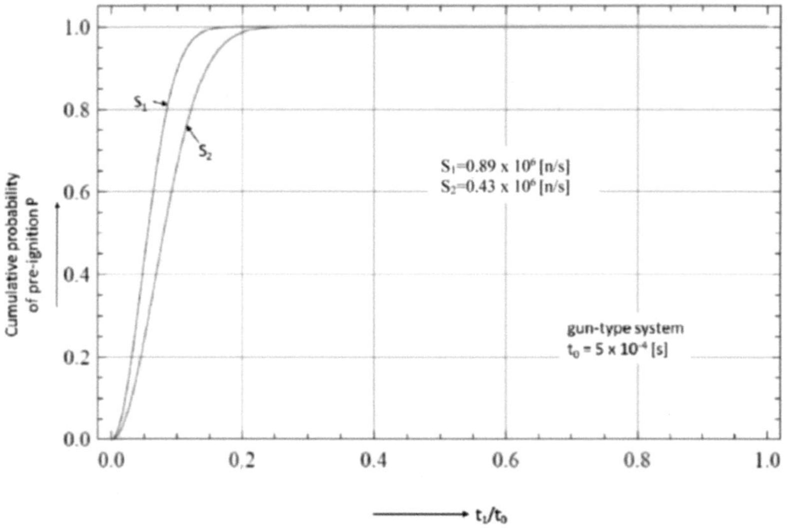

Figure 16.8: Cumulative probability of pre-ignition for a time period of $t_0 = 5\times10^{-4}$ s of compaction by gun-type system as a function of t_1/t_0. Spontaneous neutron source $S_1 = 0.89\times10^6$ n/s and $S_2 = 0.43\times10^6$ n/s.

Figures 16.9 and 16.10 show the cumulative probability of pre-ignition as a function of the ratio $x = Y/Y_0$ of the attainable nuclear explosive yield Y to the maximum nominal yield Y_0 for the time period $t_0 = 10^{-3}$ s and 5×10^{-4} s (gun-type system). Both curves for S_1 and S_2 coincide into one single rectangular like curve. Each value of the cumulative probability for pre-ignition belongs to a certain value of the ratio x: On the right hand side $(1-P(Y/Y_0))$ shows the cumulative probability to attain Y/Y_0. The smallest ratios of $x=Y/Y_0$ or minimum fizzle yield ratio $x_{F,min}$ will occur at the intersection of the curves with the abscissa or $P(Y/Y_0)=0$ (cumulative probability for pre-ignition=0, see also section 9.13.2). They are

$$x_{F,\min} = (Y/Y_0)_{F,\min}$$
$$= 2.7 \times 10^{-5} \text{ for the time period } t_0 = 10^{-3} \text{ s}$$

and

$$x_{F,\min} = (Y/Y_0)_{F,\min}$$
$$= 7.6 \times 10^{-5} \text{ for the time period } t_0 = 5\times10^{-4} \text{ s}.$$

For Y_0, a value of, e.g., 20 kt TNT can be considered. This leads to:

$Y_{F,min} = 2.7\times10^{-5}$ x $2\times10^4 = 0.54$ t TNT for $t_0 = 10^{-3}$ s

$Y_{F,min} = 7.6\times10^{-5}$ x $2\times10^4 = 1.52$ t TNT for $t_0 = 5$ x 10^{-4} s

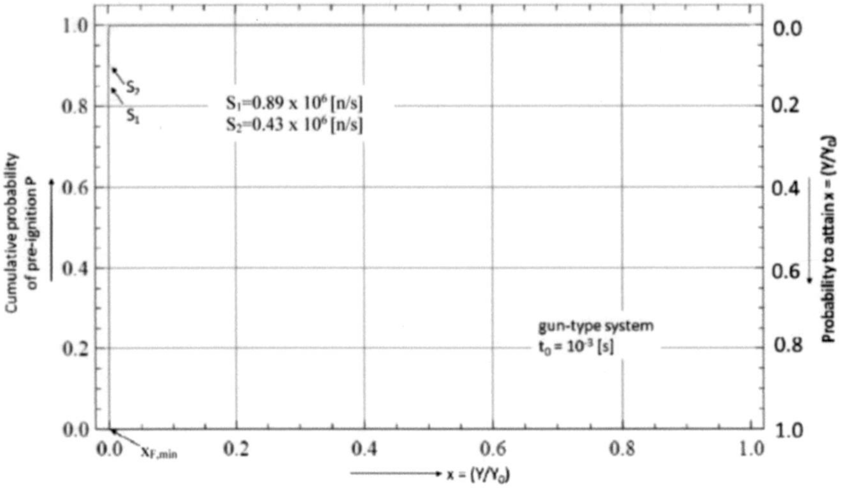

Figure 16.9: Cumulative probability of pre-ignition as a function of the ratio x of the attainable nuclear explosive yield Y to the maximum nominal yield Y_0 for the time period of compaction $t_0 = 10^{-3}$ s. Neutron source $S_1 = 0.89\times10^6$ n/s and $S_2 = 0.43\times10^6$ n/s.

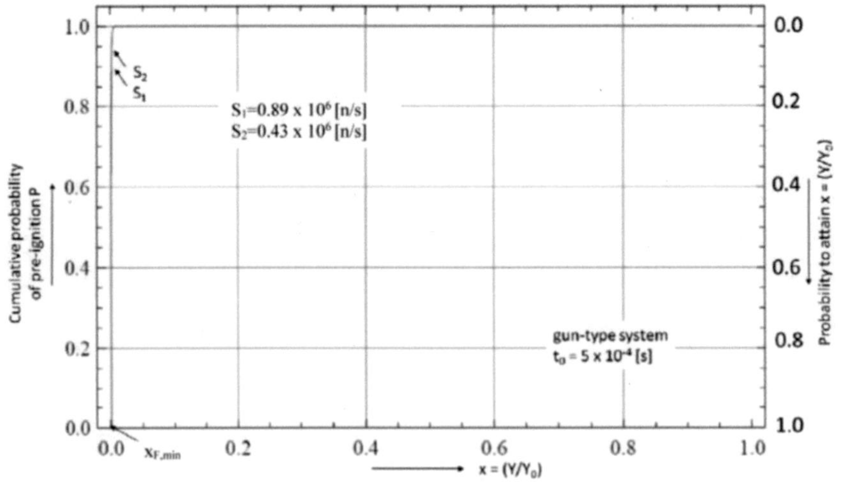

Figure 16.10: Cumulative probability of pre-ignition as a function of the ratio x of the attainable nuclear explosive yield Y to the maximum nominal yield Y_o for the time period of compaction $t_0 = 5\times10^{-4}$ s. Spontaneous fission neutron source $S_1=0.89\times10^6$ n/s and $S_2=0.43\times10^6$ n/s.

It can be concluded that for the gun-type system with 11% U-232, 72.3% U-235 and 16.7% U-238 the attainable nuclear explosive yield Y would be extremely low and of no interest for NEDs (0.54 t TNT for $t_0 = 10^{-3}$ s and 1.5 t TNT for $t_0 = 5 \times 10^{-4}$ s, if $Y_0 = 20$ kt TNT is applied.

16.7.4 Spherical Implosion System

Fig. 16.11 shows the cumulative probabilities of pre-ignition for $t_0 = 10^{-5}$ s as a function of the ratio t_1/t_0. As the geometry of the spherical implosion system is relatively compact and has a $k_{eff} \approx 0.98$, it must be considered as a subcritical system in which the internal neutron source is subject to a source multiplication (section 11.7.8).

$$M = \frac{1}{1 - k_{eff}} \quad \text{for } k_{eff} = 0.98 \rightarrow M = 50 \ .$$

The neutron sources 0.15×10^6 n/s (section 16.6.2) must be multiplied by $M = 50$, which results in value of 7.6×10^6 n/s. The evaluation of Eqs. 16.5 and 16.7 leads then to Figs. 16.11 and 16.12.

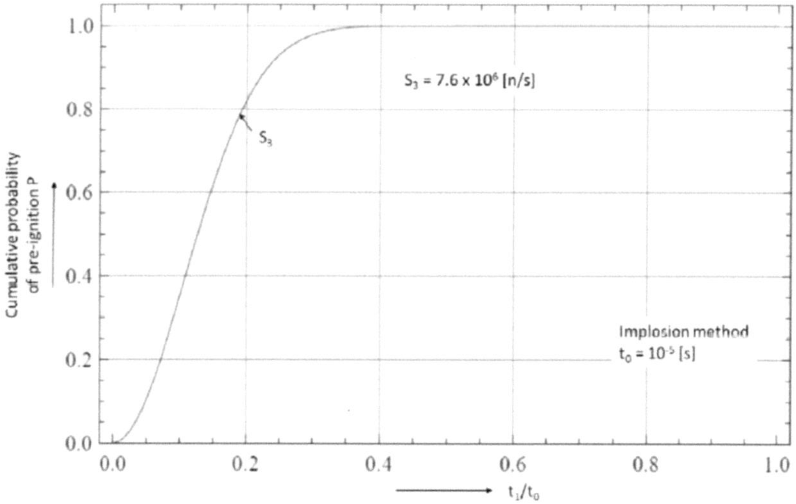

Figure 16.11: Cumulative probability of pre-ignition for a time period 10^{-5}s and spontaneous fission neutron source $S_3 = 7.6 \times 10^6$ n/s.

Fig. 16.11 shows the cumulative probabilities of pre-ignition for the time period $t_0 = 10^{-5}$ s and the neutron source $S_3 = 7.6 \times 10^6$ n/s. For a time period $t_0 = 10^{-5}$ s, Fig. 16.11 shows that pre-ignition already occurs between $t_1/t_0 = 0$ and $t_1/t_0 = 0.3$ for a spontaneous fission neutron source, $S_3 = 7.6 \times 10^6$ n/s. Fig. 16.12 shows the cumulative probability $(1-P(Y/Y_0)$ to attain $x = Y/Y_0$ for $t_0 = 10^{-5}$ s and $S_3 = 7.6 \times 10^6$ n/s. The ratio $x_{F,min}$, which depends on t_0 (see Eq. 16.6) is higher for the spherical implosion system than for the gun system.

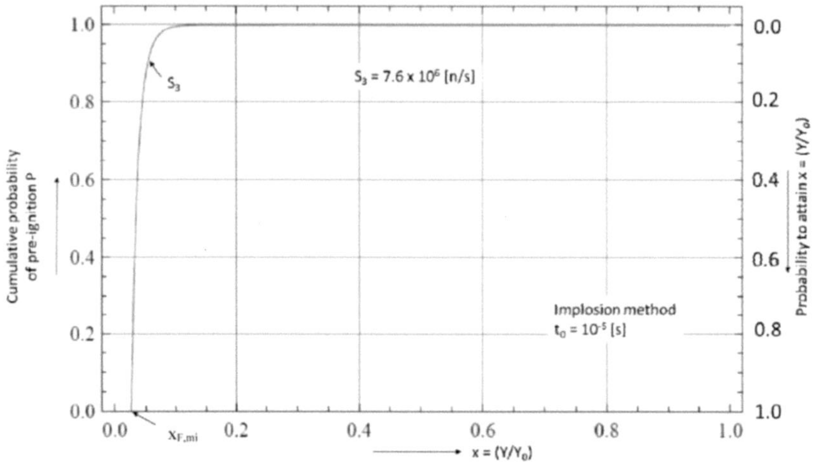

Figure 16.12: Cumulative probability of pre-ignition as a function of the ratio x of the attainable explosive yield Y and the maximum nominal yield Y_0 for the time period $t_0 = 10^{-5}$ s (implosion method) and spontaneous fission and (α,n)-neutron source of $S_3 = 7.6 \times 10^6$ n/s.

For $t_0 = 10^{-5}$ s, the ratio $x_{F,\min}$ of the minimum fizzle yield (Eq. 16.6) is

$$x_{F,\min} = \left(\frac{Y}{Y_0}\right)_{F,\min} = 0.027 \quad \text{for } t_0 = 10^{-5}\ \text{s}\ .$$

It leads to the conclusion that for a compaction time $t_1 = 10^{-5}$ s, a minimum nuclear explosive yield of 0.54 kt TNT could be attained (relative to a maximum nominal yield of $Y_0 = 20$ kt TNT). For lower probabilities (right ordinate scale) even higher explosive yields could be attained.

16.8 Thermal Analysis of a high-technology HNED based on 11% U-232, 72.3% U-235 and 16.7% U-238 Uranium

The highly enriched uranium with 11% U-232, 72.3% U-235, and 16.7% U-238 resulting as the end product of the gas ultra-centrifuge cascade (initial feed enrichment 0.1% U-232 and 19.95% U-235 enriched uranium) would contain

18.51 x 0.11 = 2.036 kg U-232

16.8.1 Heat rate of U-232 as a function of time after reprocessing

The number of atoms per kg mass as a function of time after reprocessing [20] follows the relations:

$$N_{U\text{-}232}(t) = N_{U\text{-}232,0} \cdot e^{-t/\tau_{U\text{-}232}} \tag{16.8}$$

$N_{U\text{-}232,0}$: number of U-232 atoms per kg mass just after reprocessing

$$\frac{dN(t)_{Th\text{-}228}}{dt} = \frac{N(t)_{U\text{-}232}}{\tau_{U\text{-}232}} - \frac{N(t)_{Th\text{-}228}}{\tau_{Th\text{-}228}} \tag{16.9}$$

Initial conditions:

$t = 0 \; N(0)_{Th\text{-}228} = 0$ (16.9a)

$N(0)_{U\text{-}232} = N_{U\text{-}232,0}$

half-life of Th-228 = 1.9 years

The decay chain of U-232 (half-life =70 years) altogether follows 6 α-decays and 2 β-decays to stable Pb-208. The energy release of the decaying U-232 to Th-228 is 5.4 MeV. The half-life of Th-228 is 1.9 years followed by decays with much shorter half-life (Fig. 16.1) and a total energy release of 40.8 MeV including a penetrating γ-radiation of 2.6 MeV and some γ-radiation with much lower energy [20].

The heat rate of U-232 follows the relation:

$$heat\ rate = \frac{N_{U-232,0}}{\tau_{U-232}} E_{U-232} \cdot e^{-t/\tau_{U-232}} + \frac{N_{TH-228}}{\tau_{Th-228}} (E_{tot} - E_{U-232}) \tag{16.10}$$

With $E_{U\text{-}232} = 5.4$ MeV $\quad E_{tot} - 40.8$ MeV

The heat rate in W/kg as shown in Fig. 16.13 is the obtained by applying the conversion factor of $1.6021 \cdot 10^{13}$ J/MeV.

The rate of γ-radiation is:

$$rate\ of\ \gamma - radiation = \frac{N_{Th-228}}{\tau_{Th-228}} \cdot E_{\gamma} \qquad (16.11)$$

With $E\gamma = 2.6$ MeV

In the first week after reprocessing the content of Th-228 is close to zero. Ra-224 with its half-life of 3.7 days and its following decay products with even a shorter half-life are dominating. Therefore, the γ-radiation is still fairly low [20]. The rate of γ-radiation starts at zero (just after reprocessing) and the heat rate is finite as shown by Fig. 16.13.

16.8.2 Heat rate and temperature profile in a high-technology HNED

U-232 produces 0.679 W/kg just after reprocessing (section 16.1 and Fig. 16.13). This results in:

2.036 kg x 0.680 kW/kg = 1.383 kW

α-heat rate or power, after reprocessing for a high-technology HNED. This heat rate-power will increase due to the α-particle heat rate and absorbed radiation of the decay products of U-232 (Fig. 16.1) after one year to about: 2.1 kW/kg U-232 and after 9 years to: 5.322 kW/kg U-232 resulting in a total heat power for the high-technology HNED

after one year: 4.275 kW and after 9 years: 10.835 kW

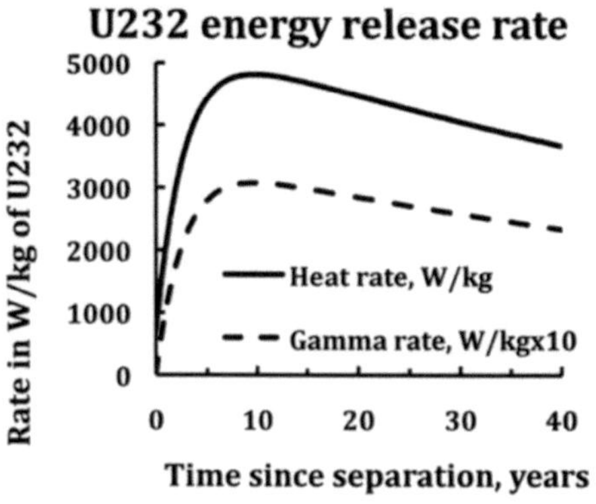

Figure 16.13: Heat rate and γ-radiation rate of U-232 as a function of time after reprocessing according to Moir [20].

This heat power of 1.383 kW will lead to very high temperatures within the HNED. Therefore, in the following only a high-technology HNED with an outer radius of 21 cm (section 10.1, Figs. 10.1 and 10.2) and a α-heat power of 1,383 W will be analyzed. Low-technology and medium-technology HNEDs would lead to even higher temperatures because of their thicker high-explosive lenses (section 10.1). Fig. 16.14 shows the geometry of the high-technology HNED to be analyzed.

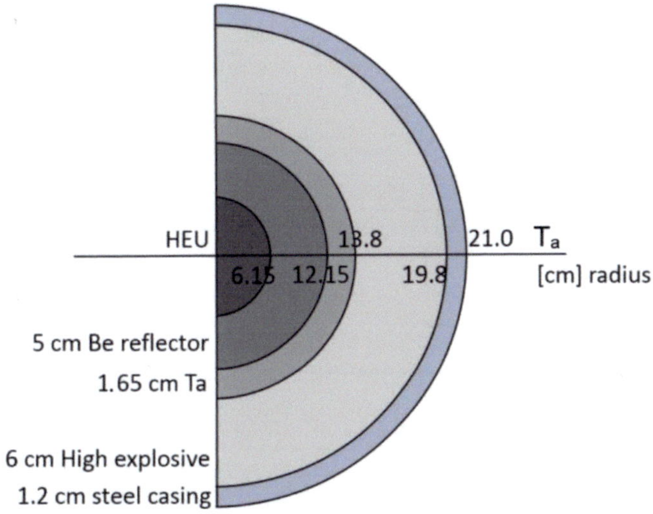

Figure 16.14: Geometric dimensions of the high-technology HNED.

PBX 9502 and TATB will be assumed as high explosives in the HNED. A layer of tantalum of 1.65 cm thickness shall smoothen so called Rayleigh-Taylor instabilities of the spherical shock waves (section 9.5). The characteristic materials data of these high explosives are given in Table 10.2 of section 10.3.3. The thermal conductivities for the different materials of the HNED are given in Table 10.7 of section 10.9.

The outer steel casing temperature T_a of the high-technology HNED with an outer radius of 21 cm is obtained by interpolation in the last column of Table 10.5 of section 10.6:

$$T_a = 197\ °C$$

for a α-heat power of 1,383 W.

Applying Eqs. 10.2 through 10.4 of section 10.7 together with the thermal conductivities for the different materials used in the HNED results in the radial temperature profile shown by Fig. 16.15

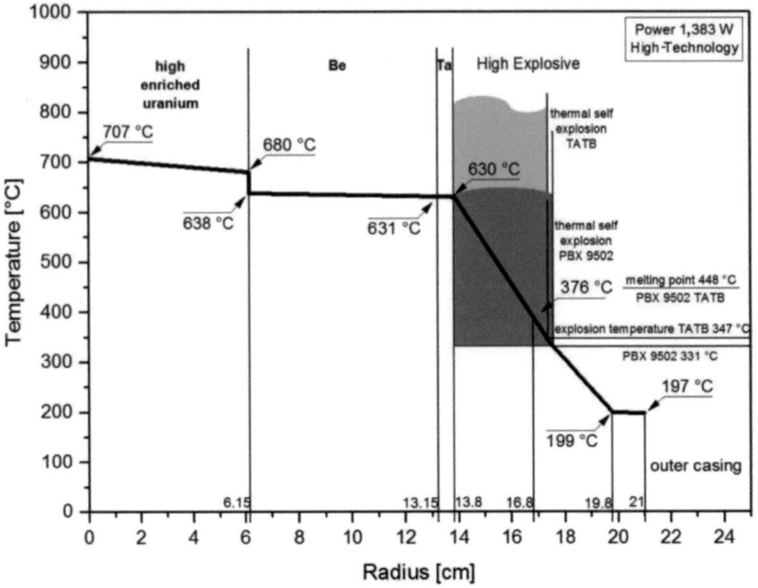

Figure 16.15: Radial temperature profile in the high-technology HNED for an α-heat power of 1,383 W.

It shows that the limiting temperatures (melting points of 448 °C for PBX 9502 and TATB as well as the temperatures for self-explosion 331 °C for PBX 9502 and 347 °C for TATB) are exceeded within the inner one half of the explosive lenses. The high-technology HNED would not work and is **technically unfeasible**. As for the medium- and low-technology HNEDs (chapter 10) the high explosives have lower thermal heat conductivities, lower limiting temperatures and thicker explosive lenses. They would **not work either at all**. The above **conclusions**, therefore, are **very conservative**, as the α-heat power increases drastically as a function of time (Fig. 16.13)

16.9 Radiation Exposure during Manufacturing due to U-232 Content in HNEDs

The γ-radiation exposure in 0.5 m distance of a sphere of uranium one year after chemical separation containing different amounts of U-232 was reported by Kang et al. [1] and Forsberg et al. [18], see Table 16.6. It lists the γ-radiation exposure due to a sphere of U-233 with different contents of U-232. It is assumed here that the γ-radiation exposure of a 5 kg U-233 sphere (4 cm radius) and an 18.5 kg U-235/U-238 sphere (radius 6.15 cm) with the same contents of U-232 is about equal in 0.5 m distance. In 1 m distance this γ-radiation is by a factor of 4 less.

Table 16.6: Radiation exposure due to a 5 kg U-233 sphere or a 18.5 kg U-235/U-238 sphere with different U-232 content in ppm or % in a distance of 0.5 m (adapted from Kang et al. [1]).

Content of U-232	Radiation Exposure
100 ppm (0.01%)	1.27 rem/h (0.0127 Sv/h)
1000 ppm (0.1%)	12.7 rem/h (0.127 Sv/h)
10,000 ppm (1%)	127 rem/h (1.27 Sv/h)
24,000 ppm (2.4%)	305 rem/h (3.05 Sv/h)

An unshielded re-enriched uranium sphere of 18.5 kg with an isotopic composition of

11% U-232, 72.3% U-235, 16.7% U-238

leads to: 11 x 127 = 1,397 rem/h or 13.97 Gray/h at 0.5 m distance

or 329 rem/h or 3.29 Gray/h at 1 m distance

(γ-radiation factor = 1)

An appreciation of this high level of γ-radiation can be obtained from a discussion of:

- The Nuclear Security Recommendations on Physical Protection of Nuclear Material and Nuclear Facilities (INFCIRC/225/Revision 5) [19].
- The radiobiological effects to human beings (acute radiation syndrome).

The IAEA, in its Recommendations of INFCIRC/225/Revision 5 [19] (Table 16.7 footnote b) sets a dividing line of 1 Gray/h or 100 rem/h radiation exposure at a distance of 1 m (unshielded) from the fissile material. This dividing line distinguishes between **unirradiated** and **irradiated** fissile material. The IAEA recognizes thereby that the diversion by national or subnational groups of **unirradiated** weapons-usable fissile material with less or equal 100 rem/h at 1 m distance (unshielded) is easier than for **irradiated** weapons-usable material which has a self-protection of >100 rem/h (unshielded).

Table 16.7 shows 3 safeguards categories for the **unirradiated** materials:

- Plutonium
- Uranium enriched to 20% U-235 or more
- Uranium enriched to 10% U-235, but less than 20% U-235
- Uranium enriched above natural, but less than 10% U-235
- U-233

Table 16.7: 16.7 Safeguards characterization of nuclear materials [19].

Material	Form	Category I	Category II	Category III[c]
1 Plutonium[a]	Unirradiated[b]	2 kg or more	Less than 2 kg but more than 500 g	500 g or less but more than 15 g
2 Uranium-235(^{235}U)	Unirradiated[b]			
	- Uranium enriched to 20% 235U or more	- 5 kg or more	- Less than 5 kg but more than 1 kg	- 1 kg or less but more than 15 g
	- Uranium enriched to 10% ^{235}U but less than 20% ^{235}U		- 10 kg or more	- Less than 10 kg but more than 1 kg
	- Uranium enriched above natural, but less than 10% ^{235}U			- 10 kg or more
3 Uranium-233(^{233}U)	Unirradiated[b]	2 kg or more	Less than 2 kg but more than 500 g	500 g or less but more than 15 g
4 Irradiated fuel (The categorization of irradiated fuel in the tables is based on international *transport* considerations. The State may assign a different category for domestic use, storage and *transport* taking all relevant factors into account)			Depleted or natural uranium, thorium or low enriched fuel (less than 10% fissile content)[d,e]	

[a] All plutonium except that with isotopic concentration exceeding 80% in Pu-238.

[b] Material not irradiated in a reactor or material irradiated in a reactor but with a radiation level equal to or less than 1 Gy/h. (100 rad/h) at 1 m unshielded.

[c] Quantities not falling in Category III and natural uranium, depleted uranium and thorium should be protected at least in accordance with prudent management practice.

[d] Although this level or protection is recommended, it would be open to States, upon evaluation of the specific circumstances, to assign a different category of physical protection.

[e] Other fuel which by virtue of its original fissile material content is classified as Category I or II before irradiation may be reduced one category level while the radiation level from the fuel exceeds 1 Gy/h(100 rad/h) at 1 m unshielded.

The three categories define different safeguards-relevant quantities for the unirradiated fissile materials, e.g. 2 kg or more for plutonium as well as U-233 and 5 kg or more for uranium enriched to 20% or more in category I. In category II these safeguards-relevant quantities are less, e.g. <2 kg, but more than 500 g for plutonium and U-233 and <5 kg but >1 kg for uranium enriched to 20% and more etc. For category III these safeguards quantities are again smaller.

Footnote e of Table 16.7 allows material classified in category I or II to be reduced by one category if the radiation level of this material exceeds a radiation exposure of 1 Gray/h or 100 rem/h at 1 m distance (unshielded).

The IAEA requirements for physical protection against unauthorized removal of fissile material in use or storage is most stringent for material of category I and somewhat lower for material of category II. As an example:

Material classified in category II should be used and stored in a protected area which should be located in a limited access area equipped with physical barriers, intrusion detection etc.

Material classified in category I should have in addition an inner area which provides an additional layer for detection and control of access. Vehicle barriers should be installed etc.

In conclusion: footnote e of Table 16.7 is applicable to the above 18.5 kg sphere with a radiation exposure of 329 rem/h in a distance of 1 m (unshielded). This means: fissile material belonging to safeguards category I (the above 18.5 kg sphere with >100 rem/h radiation in a distance of 1 m (unshielded)) can be shifted to safeguards category II.

Manufacturing of an HNED with such high γ-radiation would require remote handling techniques in hot cells.

If subnational groups or terrorists would do handling, transport and manufacturing of the HNED without the proper shielding and remote handling techniques the following radiobiological effect become valid:

- Gastro-intestinal disorders are produced at >2 Gray (nausea, vomiting, hyposthenia).
- At more than 6-8 Gray, there are practically no chances of survival.

It has been proposed, e.g. by [25] to solve the γ-radiation problem caused by some isotopes (mainly Bi-212 and Tl-208) of the decay chain of the contaminant U-232 by chemical separation of these isotopes. This is technically feasible. However, the U-232 itself cannot be separated chemically from the uranium isotopes and the γ-radiation of its daughter products (mostly Bi-212 and Tl-208) rises again to 100 rem/h or 1 Sv/h, e.g. for a U-232 concentration of 3.15% in 1 m distance of a 5 kg U-233 sphere within 1 year (Fig. 16.16). This was shown by Kang et al. [1], Forsberg et al. [18] and Moir [20]. (It is assumed here that the above 18.5 kg U-232/U-235/U-238 sphere and the 5 kg

U-232/U-233 sphere with the same U-232 concentration lead to about equal radiation exposure). In addition, the saturation value of the γ-radiation of a 5 kg unshielded sphere of U-233 with 3.15% U-232 would be about 300 rem/h or 3 Sv/h at 1 m distance after about 9 years since chemical separation (Fig. 16.16). At γ-radiation exposure of more than 6-8 Sv there is practically no chance of survival.

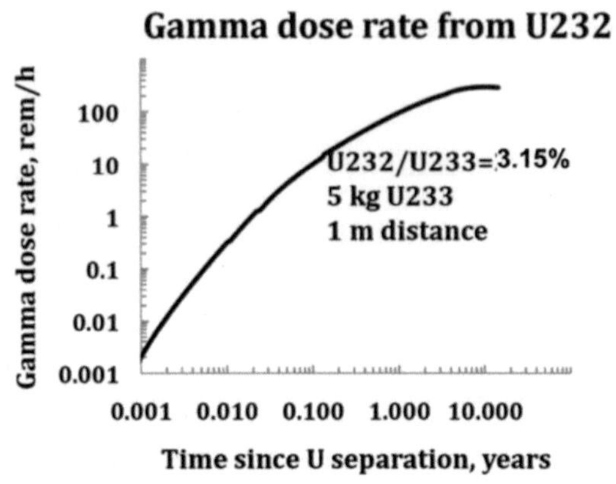

Figure 16.16: γ- dose rate from U-232 (adapted from [20]).

16.10 Radiation Damage to High Explosives

The high explosives used in the explosive lenses of HNEDs can withstand only a γ-radiation exposure of 8.77 x 10^5 Gray for medium-technology explosives e.g. HMX, RDX and 8.77 x 10^6 Gray for high-technology explosives, e.g. TATB and DATB [21].

At a distance of 13.8 cm (inner radius of the high explosive lenses, Fig. 16.14) this γ-radiation increases to

$$13.97 \times 0.5^2 / 0.138^2 = 183 \text{ Gray/h.}$$

With this γ-radiation of 183 Gray/h the limit of 8.77 x 10^5 Gray is reached within 4,792 hours or about 200 days.

Again it should be mentioned that this γ-radiation increases considerably during the first 9 years since chemical separation (Fig. 16.16) due to the decay products of U-232. At this high γ-radiation exposure limit of 8.77 x 10^5 Gray several radiation damage effects, e.g. gas evolution, weight loss or crumbling etc. of the high explosives occur which make the proper functioning of the explosive lenses **unfeasible**. Also the electronic components are very sensitive to such high γ-radiation exposure [20,21].

16.11 Summarizing Remarks

The addition or doping of 0.1% of U-232, to the 19.95% enriched U-235/U-238 uranium would make the re-enrichment and manufacturing or functioning of HNEDs **unfeasible** on several independent ways:

- The high α-particle radiation of the 11% U-232 would destroy most of the UF_6 gas during short time within the cascade of gas ultra-centrifuges (Kryuchkov et al. [5]).
- The spontaneous fission neutrons of U-232 and the neutrons produced from (α,n)-reactions of the α-particle radiation of U-232 with light element impurities lead to a high neutron background source.
- a gun-type system HNED would have an extremely low explosion yield of 0.54 t to 1.5 t of TNT which would be of **no interest.**
- for an implosion-type HNED the minimum fizzle yield would be 0.54 kt TNT (relative to a nominal maximum yield of 20 kt TNT). However, the thermal analysis of such an implosion-type HNED -even for high-technology designs- results in such high temperatures within the high explosive lenses that any type of **HNED becomes unfeasible.**
- the high γ-radiation produced by the 11% U-232 content after re-enrichment would destroy the function of the high explosive lenses within about 200 days for medium-technology explosives. High-technology explosives are not available for NNWSs or subnational groups.

If one compares the above results which all would make the functioning of HNEDs with 11% U-232 high enriched uranium **unfeasible**, one finds that the heat production of U-232 apart from the destruction of UF_6 has the strongest influence.

Now, the addition of 0.1% U-232 to 19.95% U-235 enriched uranium had been chosen somewhat arbitrarily by Kryuchkov et al. [5]. One can suppose that even a lower percentage of U-232 could also be sufficient to produce such a high heat generation within the HNED that still the temperature limits of the high explosive lenses would be exceeded. This is the more valid if one takes the longer time periods for transport and manufacturing into consideration.

In summary it can be concluded that the present **RERTR** program (section 15.1) for the <20% U-235 enriched fuel of present research and test reactors **should be complemented by adding about 0.1% or 1000ppm U-232 (doping)**. In this case unauthorized re-enrichment of the <20% U-235 research reactor fuel would make the functioning of **HNEDs technically unfeasible or proliferation-proof.** It is obvious that adding (doping with) U-232 will complicate the manufacturing process for the research reactor fuel. But this is the price it takes to make research and test reactor fuel **proliferation-proof**.

References Section 16

[1] Kang J. et al., U-232 and the Proliferation-Resistance of U-233 in Spent Fuel, Science and Global Security, 9:1-32, Copyright © Taylor & Francis Group, LLC, 2001.

[2] Kim J. I., The Production of U-232 in Large Quantities, Proceedings of the 2nd International Symposium on Power from Radioisotopes, p. 121-138, Madrid, 28th May-1st June 1972.

[3] Bechtel R. D., Uranium-232 Beryllide Neutron Source, In Partial Fulfillment of the Requirements for the Degree Master of Science in Nuclear and Radiological Engineering, Georgia Institute of Technology, May 2007.

[4] Bernhardt H. A. et al., Radiation Effects of Alpha Particles on Uranium Hexafluoride, Paper No. 552, Proceedings 2nd Int. Conference. On the Peaceful Uses of Atomic Energy, Geneva, 1958.

[5] Kryuchkov E. F. et al., Evaluation of Self-Protection of 20% Uranium Denatured with 232-U against Unauthorized Re-enrichment, Nuclear Science and Engineering: 162, 208-213, 2009.

[6] Katz J. J. et al., The Chemistry of Uranium, Mc Graw- Hill Company, New York, 1951.

[7] Kikoin I. K. et al., Experimental Reactor with Gaseous Fissionable Substance (UF_6), Paper 2502, Proceedings 2nd Int. Conf. On the Peaceful Uses of Atomic Energy, Geneva, 1958.

[8] Nuclear Reaction Data Center (JCPRG), Web Site at: http://www.jcprg.org.

[9] Experimental Nuclear Reaction (EXFOR) Data Base, Version of Dec. 19, 2008; available on the INTERNET at: http://www-nds.iaea.org/exfor/exfor.htm.

[10] SRIM-The Stopping and Range of Ions in Matter: available on the Particle Interactions with Matter Web Site at: http://www.srim.org.

[11] Russia's Nuclear Fuel Cycle (updated June 2016) Web Site at: http://www.world-nuclear.org/information-library/country-profiles/countries-o-s/russia-nuclear-fuel-cycle.aspx

[12] ''Uranium Products'' available on the Kazatomprom Web Site at: http://ulba.kz.//en/production12.htm.

[13] Donedu F. et al., Optimization Studies for Gas Centrifuges, Separation Science and Technology, 35:8, 1207-1221, Febr. 2007.

[14] Wood H. G., Effects of Separation Processes on Minor Uranium Isotopes Enrichment Cascades, Science and Global Security, 16:26-36, Copyright © Taylor & Francis Group, LLC, 2008.

[15] Yemel'yanov V. S. et al., The Metallurgy of Nuclear Fuel, Properties and Principles of the Technology of Uranium, Thorium and Plutonium, Pergamon Press, Oxford, 1969.

[16] Lammert R., Personal Communication, Urenco, Germany, May 2009.

[17] Ganesan S. et al., New Investigations of the Criticality Property of Pure U-233, Annals of Nuclear Energy 29, 1085-1104, 2002.

[18] Forsberg C. W. et al., Uses for Uranium-233: What Should Be Kept For Future Needs?, ORNL-6952, Oak Ridge National Laboratory, USA, September 1999.

[19] Nuclear Security Recommendations on Physical Protection of Nuclear Material and Nuclear Facilities (INFCIRC/225/Revision 5), IAEA, Vienna, 2011.

[20] Moir R., U232 Nonproliferation Features, LLNL-TR-438648, June 2010.

[21] Avrami L. et al., Radiation-induced changes in explosive material, AD-775371, Dec 1973, distributed by NTIS National Technical Information Service US Department of Commerce, 582 Fort Royal Road, Springfield, Va. 22131.

[22] Rosenstock W. et al., Thorium for Nuclear Energy - a Proliferation Risk, DPG-AGA Physics and Disarmament, Dresden, Germany, 2013

[23] MacFarlane R. E. et al., Method of Processing ENDF/B-VII with NJOY. Science Direct, Nuclear Data Sheets 111 (2010) 2739-2890. www.elsevier.com/locate/nds.

[24] Briesmeister J. F., editor, MCNP-A General Monte-Carlo N-Particle Transport Code, Version 4C, Technical Report, LA-13709-M, Los Alamos National Laboratory, USA, 2000.

[25] Technical white paper (pdf) March 2014, V1.0.1 Transatomic Power. http://www.transatomicpower.com/wp-content/uploads/2015/04/transatomic-white-paper.pdf

[26] Soppera, N. et al., JANIS Book of alpha-induced cross sections, comparison of evaluated and experimental data from JENDL/AN-2005, TENDL-2011 and EXFOR, OECD NEA DATA Bank, 2012.

[27] Kessler G., Sustainable and Safe Nuclear Fission Energy, Technology and Safety of Fast and Thermal Reactors, Springer Verlag, Berlin, Heidelberg, 2012.

17 The U-233/Thorium Fuel Cycle and Nuclear Characteristics of U-233

17.1 Introduction

Unlike U-235 -which is a naturally occurring fissile isotope- U-233 is an artificial isotope which is produced by neutron reactions with Th-232 (conversion or breeding process, section 4.8).

Nuclear reactor deployment, therefore, had to be started in a first step with natural or enriched uranium leading to the production of plutonium. In a second step enriched uranium or plutonium mixed with thorium as fertile material as well as thorium in the cores of thermal neutron reactors or in the blanket of FBRs then lead to U-233 (conversion or breeding process) as an artificial fissile isotope (Fig. 17.1) [1,2].

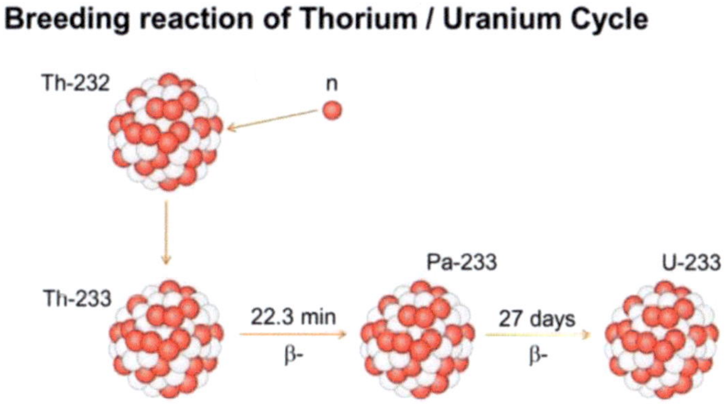

Figure 17.1: Conversion or breeding process of U-233 [1].

U-233 diluted in U-238 (denaturing) and mixed with thorium can be used in all reactor types (thermal and fast spectrum reactors) to utilize also thorium as an abundant nuclear energy source [4].

17.2 U-233/Thorium Fuel in Thermal Spectrum Reactors and Fast Breeders

U-233/thorium fuel can be loaded as mixed oxide fuel UO_2/ThO_2 in the cores of PWRs, BWRs, CANDUs or Fast Breeders. However, for the production of U-233 the thorium must be mixed, in a first step, with U-235 enriched uranium or plutonium, as long as not sufficient quantities of U-233 are available. This allows breeding of U-233 by neutron capture in Th-232. After a certain burnup time in the core the spent fuel must be unloaded and after a certain cooling time be disassembled for chemical reprocessing in order to obtain the U-233. U-233/thorium can then be fabricated as mixed oxide fuel and be inserted as fuel subassemblies into the cores of PWRs, BWRs, CANDUs, Fast Breeders or as molten salt fuel in MSRs.

17.3 Worldwide available Thorium Reserves

Unlike natural uranium, thorium contains only fertile material, no fissile isotope. Therefore, thorium can be compared with U-238. Thorium reserves are considered to be more abundant in the world than uranium reserves. Thorium can be recovered as a by-product from minerals mined for the extraction of titanium, tin and zirconium. Monazite in placer (black sand) is the main thorium-bearing mineral ($ThSiO_4$). Most of the reasonably assured reserves are located in India, China, Brazil and the USA, with large estimated additional reserves in Canada, Egypt, Australia and the USA. Monazite sands in India, Brazil, Australia and Egypt contain 4.6-7% thorium. In the United States and Canada, thorium reserves are also found in vein deposits. Table 17.1 indicates the major thorium deposits and resources in the world. Worldwide thorium resources are estimated to total about 6.2 million tons of thorium including undiscovered resources. Opportunities exist to expand known reserves and discover new areas. However, due to a limited market for thorium there has been little exploration to discover new reserves [3].

Table 17.1: Major thorium deposit types and resources [3].

Deposit type	Resources (1000 t Th)
Carbonatite	1,800
Placer (glacial, alluvial)	2,200
Vein-type	1,500
Alkaline rocks	0,600
Other	0,100
Total	6,200

In addition to monazite sand, thorium also occurs in uranium ores exploited for uranium production. The thorium recoverable from such uranium mines in Canada has reached a production level of several thousand tons per annum. Further increases should be attainable from uranium deposits specially mined for their thorium value.

17.4 Concentration, purification and conversion of thorium

Thorium ores usually contain up to a few percent of thorium and must be separated from by-products to reduce the weight for subsequent shipment. This concentration stage is predominately performed by leaching processes similarly as for uranium (section 2.4). However, physical concentration methods are also applied including crushing and sizing, gravity, magnetic, electrostatic and flotation types of separation. Roasting may be employed to improve the solubility of thorium [4].

17.5 Uranium-233/Thorium Fuel Fabrication

$^{233}UO_2/ThO_2$ pellet type fuel for PWRs, LWRs or CANDUs or Fast Breeders must be fabricated by remote fabrication lines in shielded concrete cells. The fabrication process follows similar steps of powder mixing, pressing, sintering and grinding of the pellets, as described for the PuO_2/UO_2 mixed oxide fuel fabrication in section 7.6. Remote fabrication in hot cells is required because of the U-232 contamination of U-233 and the Th-228 contamination of thorium. U-232 is formed according to Fig. 16.1 (section 16.1). Th-228 is a decay product (α-decay) of Th-232. Some decay products of Th-228 are emitters of very high-energy γ-radiation (Tl-208 with 2.6 MeV and Bi-212 with 0.7 MeV). Recycled uranium in the U-233/Th fuel cycle will contain a few 100 ppm of U-232. Its high-energy γ-radiation together with neutrons produced from (α,n) reactions (α-decay of uranium and thorium isotopes) with light elements as impurities, such as oxygen or aluminum, fluorine and carbon, require shielding during re-fabrication. Most of the experience in U-233 UO_2/ThO_2 pellet fuel fabrication has been gained in the US LWBR project. Fabrication processes for gas cooled reactors are not discussed here, since this kind of fuel can hardly be reprocessed [4].

17.6 The THOREX REPROCESSING PROCESS

The U/Th fuel is dissolved in very highly concentrated 13 M nitric acid, 0.05 M hydrofluoric acid and 0.1 M aluminum nitrate held at boiling temperature. The residual solids are removed from the solution by centrifuging. The solution with $Th(NO_3)_4$ and $UO_2(NO_3)_2$

then enters the first extraction column (Fig. 17.2) and is moved in a counter current flow against TBP dissolved in a hydrocarbon solvent. TBP selectively dissolves thorium nitrate and uranyl nitrate while moving upward in the column. The fission products, protactinium and aluminum nitrate leave the column at the bottom together with the scrub solution, which is added at the top of the column. Careful adjustment of these chemical processes is necessary to separate fission products, especially Zr-95, from thorium. In the tetravalent state, thorium is chemically very similar to zirconium. Pa-233 with a half-life of 27.0 d is the precursor of U-233. The fertile fuel can be cooled until Pa-233 has decayed into U-233. In the second column, $Th(NO_3)_4$ is recovered from the TBP by being moved in a counter current flow against diluted nitric acid. $Th(NO_3)_4$ and nitric acid leave the second column at the bottom. The organic solution together with $UO_2(NO_3)_2$ flows into the third column, where uranium is re-extracted. Uranium is then purified further in additional solvent extraction separation steps. Small traces of plutonium and neptunium may be separated by additional extraction chromatography. In case more plutonium is built up e.g., in 19.95% enriched U-235/U-238/Th fuel, the separation from plutonium and neptunium by extraction chromatography is not sufficient. In these cases the plutonium must be co-extracted with uranium and thorium. This may also be achieved by solvent extraction in contact with TBP. However, this PUREX/THOREX process is more complicated than the THOREX process [4].

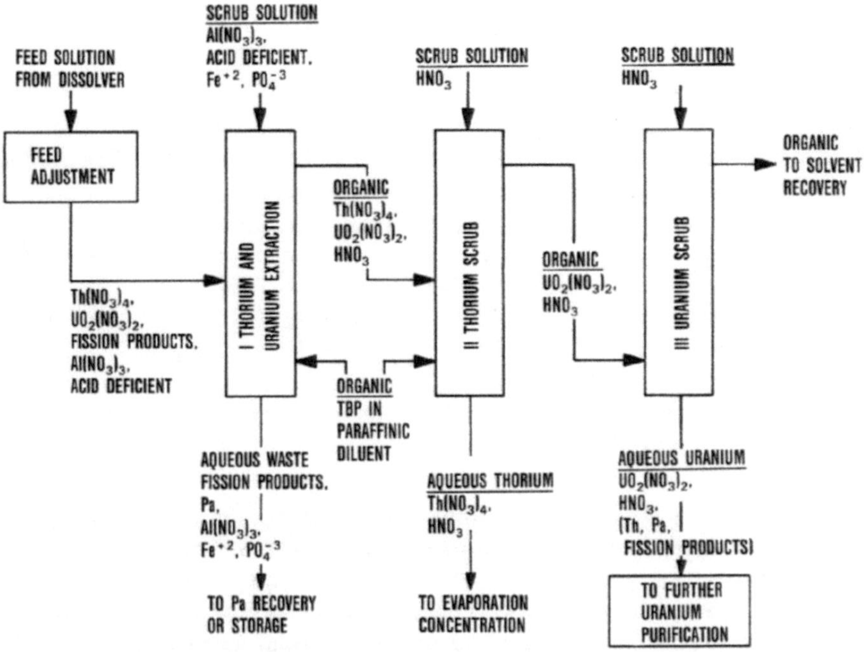

Figure 17.2: The THOREX reprocessing flowsheet [4].

17.7 PWRs and HWRs operating in the uranium/thorium recycle mode

The nuclear characteristics of the U-233/Th-232 fuel cycle show that conversion ratios near 1 (near breeding) could be attained in reactor cores with a well thermalized neutron spectrum. PWRs operating in the U/Pu fuel cycle have only conversion ratios of 0.6-0.72. Therefore, research and demonstration projects were started to demonstrate the potential viability of neutron breeders with a thermalized neutron spectrum. These were the Elk River reactor in Minnesota, the Molten Salt reactor at Oak Ridge National Laboratory and the Light Water Breeder reactor at Shippingport, Pennsylvania, USA. By 1977, however, pursuit of the thorium fuel cycle was abandoned in the USA [2,4,9,10,11].

The first commercial nuclear plants to utilize thorium were:

- the Indian Point Unit 1, a PWR that began operation in 1962. However the costs of recovering U-233 have been so extremely high, that the utility switched to uranium fuel.
- the Peach Bottom Unit, a prototype 40 MW(e) high temperature reactor using thorium fuel. It operated from 1967 to 1974.
- the Fort St. Vrain reactor plant a 330 MW(e) high temperature reactor. It operated from 1979 to 1983. [4,9,11]

At present, essentially, only a small prototype test reactor is going to be started in India. India also plans to construct and operate a 300 MW(e) advanced heavy water reactor (AHWR) using U-233/Th and Pu/Th mixed oxide fuel [9]. Thorium/uranium fuel cycle facilities (reprocessing and re-fabrication plants) are not yet available on a technical, commercial scale. In the following, therefore, only reactor design proposals can be discussed

Table 17.2 shows the fuel cycle design data of PWRs operating in the U-235/U-238/ Th-232 recycle mode or in the U-233/U-238/Th-232 recycle mode. In addition fuel cycle design data for HWRs (CANDU) operating in the U-233/Th-232 recycle mode are presented. The PWR reactor operates with 19.95% U-235/U-238/Th-232 fuel or with 11.95% U-233/U-238/Th-232 fuel. Although the fissile U-233 produced plus the unused U-235 are recycled into the reactor core after chemical separation, an additional 455 kg of U-235 must be fed annually as 19.95% enriched uranium fuel (load factor 0.85) [4].

Table 17.2 also contains data for LWRs operated in a U-233/Th-232 fuel cycle. The initial fuel has a U-233 enrichment of 11.95% U-233 in depleted uranium. This U-233 recycle LWR has an initial core inventory of 1,904 kg U-233 and, in addition to the 580 kg of fissionable uranium recycled, requires an annual 358 kg of U-233 which is obtained by chemical reprocessing of the spent fuel from other nuclear reactors run on thorium as fertile fuel (load factor 0.85) [4]. The HWR recycle reactor with 11.95% enriched U-233 uranium fuel has a core inventory of 1,859 kg U-233. It needs no natural uranium. It requires an annual makeup of 129 kg of U-233 (load factor 0.85). However, its initial inventory of 1,859 kg of U-233 and the annual makeup must come from other nuclear reactors which, in turn, use the fissile U-235. Due to the nuclear characteristics of U-235/U-233 and Th-232 in a thermalized neutron spectrum an enrichment of either slightly below 20% U-235 or slightly below 12% U-233 (see section 15.1 and 17.11) is needed for all reactor designs with thorium of Table 17.2. The requirements of 19.95% U-235 enriched uranium for proliferation and safeguards reasons was explained in section 15.1. The safequards requirement of 11.95% U-233 will be explained in section 17.11.

Table 17.2: Fuel cycle design data for PWRs and HWRs operating with U-235/U-238/Th-232 or U-233/Th-232 fuel (30 year operating time, load factor 0.85) [4].

Reactor type		PWR- recycle	PWR-recycle	HWR-recycle
Total thermal power	MW(th)	3,765	3,800	4,029
Net Electric Power	MW(e)	1,229	1,270	1,260
Fuel cycle		U-235/Th	U-233/Th	U-233/Th
Equilibrium, recycle enrichment	%	19.95%(U-235)	11.95%(U-233)	11.95%(U-233)
Fraction of core replaced per year		0.33	0.33	0.5
Average burn-up	MWd(th)/kg	33.5	33.4	14
Fissile requirements	kg/GW(e)	2,544	1,904	1,859
Annual recycled Eq.	kg/GW(e)	677	580	1,039
Annual makeup Eq.	kg/GW(e)	455	358	129
Natural uranium Requirements: - Initial core - 30 years cumulative	t/GW(e)	493 2,685	not adequate	not adequate
Thorium requirements: - Initial core - 30 years cumulative	t/GW(e)	63 827	57.9 655	105.4 1,030
Fissile material in spent fuel: Enrichment	%	9.3	7.9	11.2
Annual discharge - U_{fiss} Eq. - Pu_{fiss}	kg/GW(e)	530 68	478 65	848 22.8

17.8 The Thorium Seed – Blanket PWR Core Design Concept

Radkowsky et al. [5,6,7] proposed the thorium based seed-blanket assembly core design concept, shown by Fig. 17.3

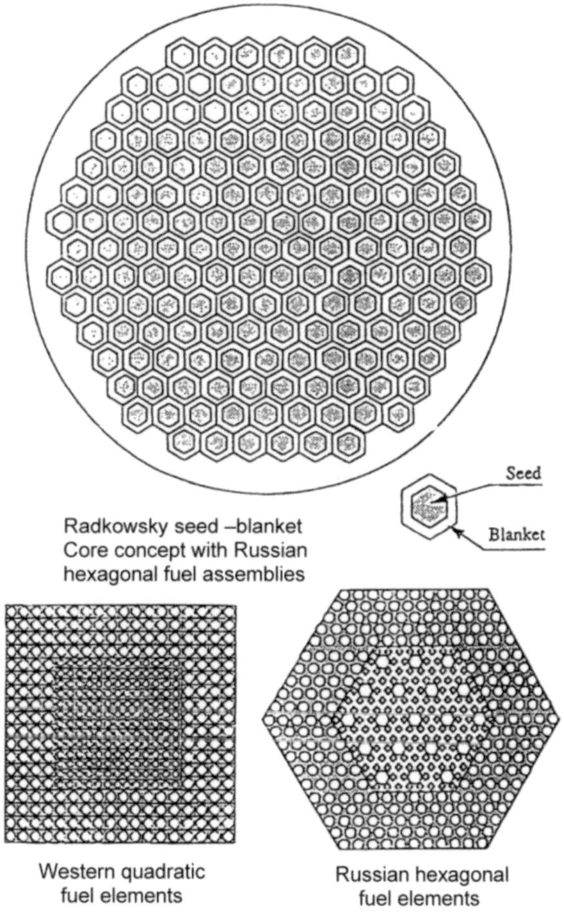

Figure 17.3: This design concept can operate with 19,95% U-235 enriched uranium[1] [5,6] or it allows the incineration of plutonium (weapons plutonium [6,7]) in the seed part of the seed-blanket assembly and it can achieve conversion of U-233 via neutron capture in thorium-232 in its blanket part.

[1] The U-235 enrichment of the U-235 enriched uranium is 20% in [5,6,7]. It is slightly modified to 19.95% here to be in accordance with section 15.1

17.8.1 The U-235/thorium Burner PWR [5,6]

The seed-blanket assemblies can directly replace the quadratic fuel assemblies of present Western PWRs or the hexagonal fuel assemblies of present Russian PWRs without any further design modifications. The separation in a seed and blanket region allows enough flexibility for designing the seed region to be an efficient source of well thermalized neutrons to the subcritical thorium/uranium blanket region. Within the blanket region the thorium is spiked with a small amount (about 10%) of 19.95% U-235 enriched uranium. The U-233 originating from the conversion process with Th-232 is born into U-235/ U-238 and becomes denatured. The seed region is designed with such lattice parameters that the neutron spectrum becomes over-moderated ($V_m/V_f = 3.3$, V_m being the moderator volume and V_f is the fuel volume), whereas the lattice parameters of the blanket region ($V_m/V_f = 1.8$) lead to an under-moderated neutron spectrum. The high fuel enrichment in the fuel rods (19.95% U-235 uranium or weapons plutonium) in the seed region results in a relatively high power density in the fuel. Therefore, a U/Zr alloy or U/Pu/Zr alloy metallic fuel is chosen. This leads to fuel temperatures below 500 °C.

Table 17.3: Characteristic core design data of a seed-blanket core design operating with 19.95% U-235 enriched uranium (U-Burner) or burning weapons-Pu (Weapons-Pu-Burner) [5,6,7].

Reactor type	PWR seed-blanket core design [3,4,5]	
	19.95% U-235 U-Burner	Weapons-Pu-Burner
Total Power MW(th)	3,400	3,400
Net Electric Power MW(el)	1,100	1,100
Fuel	19.95% -235U/Th	Weapons-Pu/Th/U
Fraction of seeds Replaced per year	0.33	0.33
Blanket regions replaced after (years)	10	6
Fissile fuel reload per year (kg)	729-935 U-235	1,131 W-Pu
Annual Fissile Fuel Dis-charge (kg)	128 (U-235)	267-468 (Pu)
U-233 discharge (kg)	630 (after 10 years)	725 (after 6 years)

The blanket region contains 90% thorium and 10% of 19.95% enriched uranium in the U-235-burner version. In this case the discharged uranium of the blanket region contains the uranium isotopes U-232, U-233, U-234, U-235, U-236 and U-238. The fissile isotope content of U-233 an U-235 remains well below 17% which satisfies the safeguards requirements, Eq. 17.1 of section 17.11

The plutonium originating from neutron capture in U-238 etc. can be considered proliferation-proof under certain conditions. In case of the 19.95% U-235 U-Burner it has the following isotopic composition [6] for:

seed: 0.059 Pu-238, 0.466 Pu-239, 0.236 Pu-240, 0.151 Pu 241, 0.088 Pu-242

blanket: 0.114 Pu-238, 0.386 Pu-239, 0.159 Pu-240, 0.144 Pu-241, 0.197 Pu-242

The Pu-238 content of 0.059 in the seed region is slightly lower than the number of 0.070 which is determined in section 14.9 as proliferation-proof for low- and medium-technology. This can be changed easily by admixing slight amounts of reprocessed uranium to the fresh seed fuel (keeping the 19.95% U-235 fixed). Neutron capture in the higher contents of U-236 and Np-237 leads then to higher contents of Pu-238 in the unloaded plutonium.

The unloaded plutonium in the case of the Weapons-Pu-Burner [7] is not proliferation-resistant. This is not necessary as such Weapons-Pu-Burners would only be allowed to operate in NWSs.

The U-232 content in the unloaded U-233 is about 593 ppm or roughly 600 ppm (Table IV of [7])

17.9 Nuclear Characteristics of U-233

U-233 is build up after neutron capture in Th-232 followed by two subsequent beta-decays of Th-233 and Pa-233 (Fig. 17.1). In a thermal neutron spectrum U-233 is the best fissile material because it has the best neutron yield.

Fig. 17.4 shows the neutron yield η, i.e. the total number of fission neutrons generated per neutron absorbed for U-233 and other fissile isotopes U-235, Pu-239 and Pu-241.

Some nuclear characteristics of U-233 are compared in Table 17.4. The decay chain of U-233 is shown in Fig. 17.5 [11].

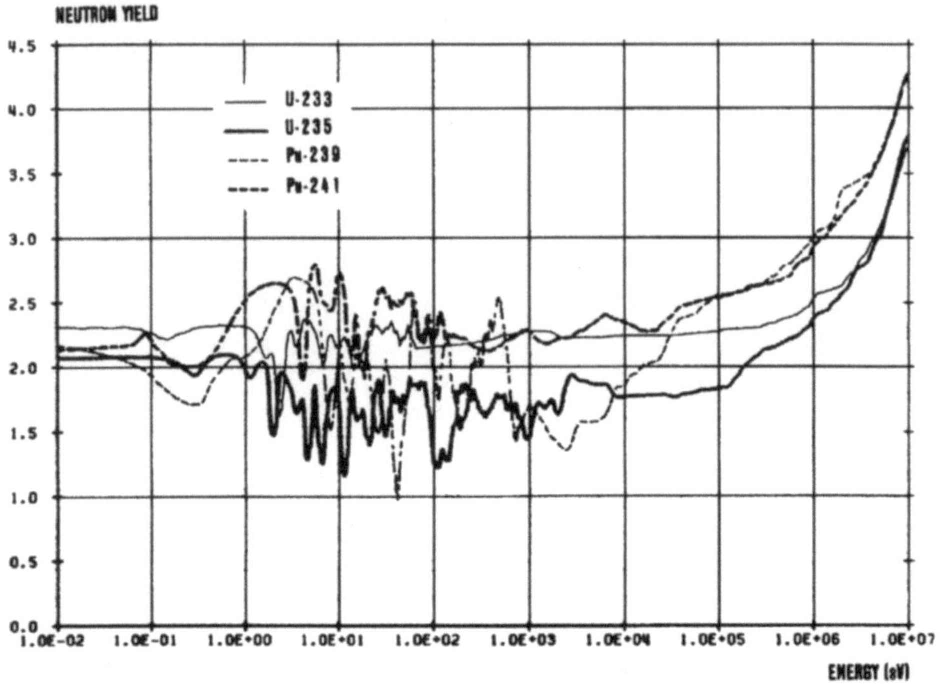

Figure 17.4: Neutron yield as a function of neutron energy for various fissile fuel isotopes [4].

Table 17.4: Comparison of some nuclear characteristics of U-233 and U-235 [8].

Nuclear Properties	U-233	U-235
Half-life (years)	1.6×10^5	7×10^8
Critical mass (kg) Reflector	8.4 (93.8% U-233) 3.7 cm Be	21 (93.5% U-235) 5.1 cm Be
Neutrons released per fission (1 MeV) (0.025 eV)	 2.5 2.28	 2.3 2.07
Spontaneous fission rate per g and s	 0.5	 0.6
Decay heat (W/kg)	0.3	0.0001
α-particle Activity (Ci/g)	 9.6×10^{-3}	 2.2×10^{-6}

U-233 metal is difficult to handle for manufacturing because of its own radiation and of the associated α-radiation level caused by decay products of U-232.

The specific α-particle activity of U-233 is 9.6 x 10^{-3} Ci/g plus rapid build-up of α-particle activity of short lived α-decay products. It is about three orders of magnitude higher than that of weapons type uranium. This necessitates glove box handling techniques similar to those for weapons plutonium metal [11].

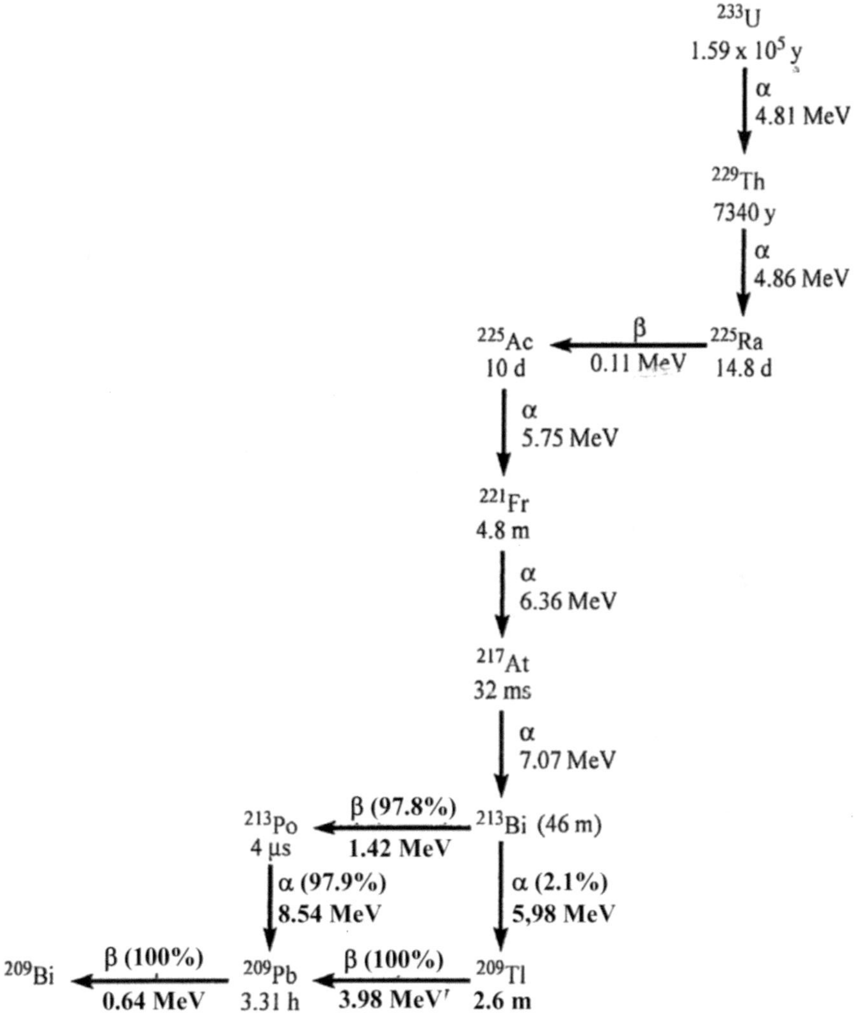

Figure 17.5: Decay chain of U-233 [15, slightly adapted].

17.10 γ-radiation of U-233 contaminated by U-232

U-233 contains always trace amounts of the isotope U-232. The content of U-232 depends on the production techniques (spectrum of the reactor core, irradiation time) [8,9,11]. Relatively pure U-233 contains about 5 to 10 ppm U-232 [10] and requires already special handling techniques according to international radiation protection standards (Table17.5) [8,11]. The radiation exposure coming from U-232 is predominantly caused by Tl-208 with its 2.6 MeV γ-radiation.

Table 17.5: Unshielded working hours required to accumulate a 5 rem or 0.05 Sv dose (5 kg metal U-233 sphere at 0.5 m distance one year after chemical separation) (Kang et al. [8] adapted).

U-232 content	Dose rate (Sv/hr)	Hours
Weapons-grade Plutonium	0.000013	3,800
U-233 (1 ppm U-232)	0.00013	380
U-233 (5 ppm U-232)	0.00064	76
U-233 (100 ppm U-232)	0.0127	3.8
U-233 (1% U-232)	1.27	0.04
U-233 (3.15% U-232)	4	0.0121

Table 17.5 shows the radiation exposure at 0.5 m distance from a 5 kg metal sphere of U-233 with different contents of U-232. In 1 m distance the dose rate would be a factor of 4 lower. In addition, it indicates the possible number of working hours allowed before the occupationally limited radiation exposure limit of 5 rem for workers is accumulated [8].

The chemical separation of the daughter products of U-232 (mainly Tl-208) with the aim of decreasing the γ-radiation was discussed already in section 16.9. After chemical separation the α-radiation would again rise within one year to considerably high values. For weapons use the U-233, therefore, must be produced as relatively pure U-233.

The spontaneous fission neutron rate of pure U-233 is small and even somewhat less than that of U-235 (Tables 8.15 and 17.4). Relatively pure U-233, therefore, can be used to build a gun-type or implosion-type nuclear weapon. Nearly pure U-233 has been used for nuclear weapons tests by several nations (USA, USSR, India). Ranges for U-232 contents for both gun-type and implosion-type nuclear weapons are given in [2].

In the following section only commercial type reactors, e.g. LWRs and LMFBRs operating in the U/Th fuel cycle will be considered.

17.11 Dilution of U-233 by U-238

INFCE working group 8 [12] proposed the dilution or denaturing of U-233 by U-238 (natural or depleted uranium) as a solution of the proliferation problem. It also proposed the limit of 12% U-233 (diluted by U-238) as the dividing line between weapons-usable and non-weapons-usable U-233. In addition, it defined Eq. 17.1

$$\frac{\text{weight of U-233} + 0.6 \text{ x weight of U-235}}{\text{weight of total uranium}} < 12\% \qquad (17.1)$$

Eq. 17.1 leads for: Weight U-235 = 0 to <12% for U-233

The physics and technical basis for Eq. 17.1 was not given by INFCE working group 8 [12], but provided by Forsberg et al. [11,15] and later also by Kang et al. [8] and Glaser [17]. Forsberg et al. [11] also proposed a safeguards categorization for U-233 (Table 17.6), similar to that for U-235 (see section 15.1) defined by INFCIRC/225/Revision 5 [13]. In this proposal the limits of ≥12% as well as the ranges of 6% to 12% and 0.66% to 6% are defined together with material quantities (see also section 16.9). In the open literature it is commonly accepted that 12% U-233 diluted by U-238 is equivalent to 20% U-235 diluted by U-238 [5,11,16].

Table 17.6: Definition of enrichment categories for U-233 by Forsberg et al. [11].

Definition by Forsberg et al. [11]	
uranium-233 unirradiated ≥12% U-233 enriched	Weapons-usable (HEU)
uranium-233 unirradiated ≥6% U-233 <12% enriched	Non-weapons-usable (LEU, Can be re-enriched to >12% U-233
uranium-233 unirradiated >0,66% but <6% U-233	Non weapons-usable (LEU)

IAEA and US regulatory organizations have not endorsed the 12% U-233 dividing line yet. Instead, IAEA, e.g. in INFCIRC/225/Revision 5 [13] in 2011 defined un-irradiated U-233 of ≥2 kg as safeguards category I and 0.5 to 2 kg as safeguards category II. US regulatory organizations [10] apply the same safeguards standards for separated U-233 as INFCIRC/225/Revision 5 [13]. A more detailed safeguards definition by IAEA, as proposed by Forsberg et al. [11], will only become necessary when reactors operating in the U/Th fuel cycle will be deployed on a large scale in the future.

In this chapter 17 and the following chapter 18 it is assumed that the definitions proposed by INFCE working group 8 [12] and Forsberg et al. [11,15] for U-233 will become valid as safeguards standard in the future. They appear scientifically justified and are applied already in the open literature [5,6,7,16,18].

U-233 diluted by U-238 with an enrichment of <12% U-233 will be considered Low Enriched Uranium U-233 (LEU) equivalent to <20% U-235. U-233 enriched between 6% and <12% is not weapons-usable, but could be converted to weapons-usable U-233/U-238 by re-enrichment in a relatively small enrichment plant [18]. It is assumed that the α-radiation of the contaminant U-232 would not prevent terrorists from handling separated U-233. For such persons it might not matter whether they receive high radiation exposures. Radiation exposure even at high levels would not result in immediate death [4,11].

In this connection it is interesting to analyze how much the contaminant U-232 can protect the U-233 - enriched between 6% and <12% - against re-enrichment in gas ultra-centrifuge cascades and **whether it can be demonstrated** - as in chapters 15 and 16 for <20% enriched U-235 doped with U-232 - that such re-enriched U-233 with a certain concentration of U-232 is **proliferation-proof.**

References Section 17

[1] Rosenstock W. et al., Thorium for Nuclear Energy - A Proliferation Risk, DPG - AGA Physics and Disarmament, Dresden, Germany, 2013.

[2] Uranium-233 -wikipedia, the free encyclopedia http://en.wikipedia.org/wiki/Uranium-233

[3] Barthel F. H. et al., THORIUM Occurrences, Geological Deposits and Resources, IAEA International Symposium on Thorium Raw Material for Nuclear Energy (URAM 2014), 23-27 June 2014, Vienna, Austria. http://www-pub.iaea.org/iaeameetings/cn216pn/Thursday/Session10/164-Barthel.pdf

[4] Kessler G., Sustainable and Safe Nuclear Energy, Technology and Safety of Fast and Thermal Reactors, Springer, Heidelberg, 2012.

[5] Radkowsky A. et al., The Nonproliferative Light Water Thorium Reactor: A New Approach To Light Water Reactor Core Technology, Nuclear Technology, Vol. 124, 215-222, Dec. 1998.

[6] Galperin A. et al., A Pressurized Water Reactor Plutonium Incinerator based on Thorium Fuel and Seed-Blanket Assembly Geometry, Nuclear Technology, Vol. 132, 214-226, Nov. 2000.

[7] Galperin A. et al., A Thorium-Based Seed-Blanket Assembly Concept to Enhance PWR Proliferation Resistance, Proceedings of the International Conference on Future Nuclear Systems-Global'99, August 29-September 3, Jackson Hole, Wyoming, 1999.

[8] Kang J. et al., U-232 and the Proliferation-Resistance of U-233 in Spent Fuel, Science and Global Security, 9:1-32, Copyright © Taylor & Francis Group, LLC, 2001.

[9] IAEA, Thorium Fuel Cycle-Potential Benefits and Challenges, IAEA-TECDOC-1450, Vienna, Austria, May 2005.

[10] Alvarez R., Managing the Uranium-233 Stockpile of the United States, Science and Global Security, 21:53-69, Copyright © Taylor & Francis Group, LLC, 2013.

[11] Forsberg C. W. et al., Definition of Weapons-usable Uranium-233, ORNL/TM-13517, Oak Ridge National Laboratory, USA, March 1998.

[12] Report of INFCE Working Group 8, Advanced Fuel Cycle and Reactor Concepts, IAEA, Vienna, Austria (1980).

[13] Nuclear Security Recommendations on Physical Protection of Nuclear Material and Nuclear Facilities (INFCIRC/225/Revision 5), IAEA, Vienna, 2011.

[14] U.S.NRC, Protecting people and the environment, Subpart C--Occupational Dose Limits. www.nrc.gov/

[15] US Nuclear Regulatory Commission, NRC Dose Limits (http://www.nrc.gov/NRC/EDUCATE/REACTOR/09-DOSESTANDARD/index).

[16] Forsberg C. W. et al., Uses for Uranium-233: What Should Be Kept For Future Needs?, ORNL-6952, Oak Ridge National Laboratory, USA, September 1999.

[17] Shwageraus E. et al., Use of Thorium for Transmutation of Plutonium and Minor Actinides in PWRs, Nuclear Technology, 147, 53, 2004.

[18] Glaser A., On the proliferation potential of uranium fuel for research reactors at various enrichment levels, Science and Global Security, 14:1-34, Copyright © Taylor & Francis Group, LLC, 2006.

[19] Glaser A., Characteristics of the gas centrifuge for uranium enrichment and their relevance to nuclear weapons proliferation (corrected), Science and Global Security, 16:1-25, Copyright © Taylor & Francis Group, LLC, 2008.

18 Proliferation-Resistance of U-233 Contaminated by U-232

18.1 U-232 Produced in Reactor Cores as Contaminant of U-233 in the U/Th Fuel Cycle

The production of U-233 in the U/Th fuel cycle in the cores of different reactor types (PWRs, LMFBRs or ADSs) leads inevitably to the production of U-232 as contaminant in a few hundred ppm range. This production depends on the neutron spectrum in the reactor core and on the burnup of the U/Th fuel [1,2,3]. Table 18.1 shows some results for the production of U-232 in different fuel compositions and different reactor types. A wide range of U-232 production is associated with the production of U-233.

Table 18.1: Different U-232 contamination in different reactor types at different fuel burnups [2,3,7,10].

Reactor Type	Fuel Composition	Burn-up MWd/t	U-232 ppm
Seed-Blanket	Weapons-Pu (seed) and U/Th (blanket)	160,000 63,000	 597
PWR	19.95% U-235/Th	50,000	1,436
PWR	19.95% U-235/Th	100,000	2,052
LMFBR	11.95% U-233/Th	100,000	1,727
ADS	11.95% U-233/Th	40,000	3,000
CANDU	See Fig. 18.1		

The results of Table 18.1 can be explained by Fig. 18.1 which is based on analyses by Kang et al. [1]. Fig. 18.1 shows the ratio of (U-232+Pa-232)/(U-233+Pa-233) as a function of (U-233+Pa-233)/Th-232) for CANDUs, LMFBRs, PWRs and different burn-up of the fuel according to Kang et al. [1].

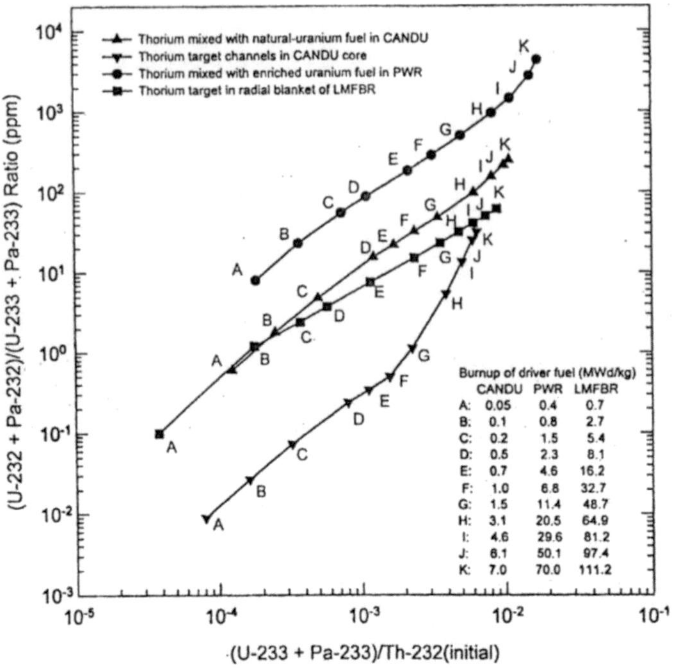

Figure 18.1: (U-232+Pa-232)/(U-233+Pa-233) ratio as a function of (U-233+Pa-233)/Th-232 ratio in different reactor types (CANDU, PWR and LMBFR) [1].

In case of CANDU reactors the (U-233+Pa-233) is produced either in cores with fuel containing thorium and natural uranium or in channels located in the reactor core. These channels are loaded with 19.5% U-235 enriched uranium and thorium. The highest U-232 concentrations result from irradiation in a homogeneous PWR core, e.g. above a burn-up of about 20,000 MWd/t the U-232 concentration is >1,000 ppm. The lowest U-232 concentration can be achieved in a CANDU type reactor with thorium and natural uranium fuel as well as very low fuel burn-up or with target channels of CANDU reactors as described above or in radial blanket subassemblies of LFMBRs. **Such loadings of CANDU reactor cores or radial blankets of thorium fueled LMFBRs should not be allowed by IAEA.**

An exception is the Radkowsky seed and blanket PWR core design of section 17.8 with about 600 ppm U-232 concentration. This is probably due to the special seed- blanket design [5,7].

U-232 contamination is often mentioned to represent a high degree of proliferation-resistance. It will be shown in the following sections that similar results as in sections 15 and 16 can only be achieved above a certain concentration of U-232. This is due to the fact that the UF_6 molecules of U-233 are only one atomic mass unit (amu) heavier than U-232. In the following section only PWR thorium fuel mixed with enriched uranium will be discussed. CANDU reactors and LMFBR radial blanket fuel as described in Fig. 18.1 will not be considered (see also section 18.9.6).

18.2 Re-enrichment of Denatured U-233

Re-enrichment of denatured U-233 in relatively small gas ultra-centrifuge cascades can lead to weapons-usable U-233. This will be analyzed in the following sections for three representative examples:

- a homogeneous PWR core design with fresh fuel mixed of 76% thorium and 24% of 19.95% enriched U-235/U-238 uranium (isotopic composition: 0.199% U-234, 19.95% U-235, 0.092% U-236, 79.759% U-238), (example A) [4].
- the Radkowsky seed-blanket core design concept with fresh mixed fuel of 49% weapons type plutonium and 51% natural uranium in the seed region as well as 91% thorium and 9% natural uranium in the blanket region (example B) [5,7].
- a homogeneous PWR core with the composition of the fresh fuel as fuel type example A, but with a U-232 concentration according to Kang et al. [1], see Fig. 18.1, (example C)

Fuel type example A

After a burn-up of the fuel of 60,200 MWd/t in the PWR core of example A the spent fuel will have the following isotopic composition of the uranium [4], shown by Table 18.2. It is assumed that the isotopic composition of the uranium hardly changes between the times of discharge of the spent fuel and the chemical reprocessing procedure. The data for U-232 were not given in [4] but are taken over from Table 18.1 for 50,000 MWd/t burn-up of the fuel in a PWR and extrapolated to 60,200 MWd/t burn-up. The values given by Kang et al. [1] are higher (see fuel type example C)

Table 18.2: Isotopic composition of uranium for fuel type example A (homogeneous PWR core) after a burn-up of 60,200 MWd/t [4].

Isotopic	composition	
U-232	0.172%	
U-233	4.830%	
U-234	0.740%	
U-235	6.470%	Fuel type example A
U-236	2.940%	
U-238	84.848%	

Fuel type example B

Fuel composition example B is derived from Table IV of [7] and corresponds to a burn-up of the blanket region of 63,000 MWd/t after 6 years of reactor operation. The uranium fuel composition is shown by Table 18.3.

Table 18.3: Isotopic composition of uranium for fuel type example B (seed and blanket core design [7]) after a burn-up of 63,000 MWd/t in the blanket region.

Isotopic	composition	
U-232	0.06%	
U-233	9.33%	
U-234	1.92%	
U-235	0.58%	Fuel type example B
U-236	0.18%	
U-238	87.93%	

Fuel type example C

Fuel composition example C will have the same composition as fuel type example A, but will have a U-232 contamination of 3,500 ppm taken from Kang et al. [1] and Fig. 18.1 For fuel type example B the U-233 concentration is almost twice compared to fuel type example A or C, but the U-232 concentration is only one third or one sixth of the U-232 concentration of fuel type example A or C. This difference of the two examples is probably due to the seed-blanket core design concept of Radkowsky [7].

Table 18.4: Isotopic composition of uranium for fuel type example C (homogeneous PWR core after a burnup of 60,200 MWd/t; U-232 content taken from Fig. 18.1 [1]).

Isotopic	composition	
U-232	0.35%	
U-233	4.83%	
U-234	0.74%	Fuel type example C
U-235	6.47%	
U-236	2.94%	
U-238	84.67%	

In the following section it will be analyzed how these fuel compositions can be re-enriched to high enriched, weapons type U-233/U-235 by a cascade of gas ultra-centrifuges. It is again assumed that terrorists or subnational groups would not care for the standard occupational radiation exposure limits for workers of 50 mSv per year [8,10], but would be willing to take much higher total radiation exposures in the range of several Sievert during the re-enrichment process and the manufacturing of a nuclear explosive device.

As an example, gastro-intestinal disorders will be produced at >2 Gray or Sievert (γ-radiation) with nausea or vomiting. At more than 6-8 Gray or Sievert (γ-radiation) there is practically no chance of survival [9].

18.3 Enrichment Factors of Gas Ultra-Centrifuges for Multicomponent Isotopic UF_6-Mixtures

The different molecular weights of the UF_6 molecules of the isotopes U-232, U-233, U-234, U-235, U-236 and U-238 result in different overall stage enrichment factors for these isotopes of the multicomponent isotopic mixture [11,12,13]. For gas ultra-centrifuges the different stage enrichment factors were calculated by Wood [11]. They are given in Table 18.5, related to a stage enrichment factor of 1.6 for U-235. The lighter isotopes U-232, U-233 and U-234 have higher stage enrichment factors than U-235.

Table 18.5: Overall stage enrichment factors for different uranium isotopes in a gas ultra-centrifuge. The different factors are derived from Eq. 9 and Eq. 4 of Wood [11]. For U-233 an average value of the two procedures proposed by Wood [11] is assumed.

Isotope	stage enrichment factor
U-232	2.2
U-233	2.1
U-234	1.8
U-235	1.6
U-236	1.4

These stage enrichment factors of Wood [11] are now related (keeping the same ratios) to different lower stage enrichment factors for U-235, e.g. 1.16 for fuel type examples A and C, 1.09 for fuel type example B. These lower stage enrichment factors shall represent gas ultra-centrifuge designs of the first commercial generation [13,18].

With 6 stages in the enriching cascade section, 3 stages in the stripping section and one stage for the feed stream of a gas ultra-centrifuge cascade we arrive at 10 stages in total (see section 15.5). With such a 6 stage enriching section in the cascade the following isotopic compositions after re-enrichment are achieved:

Fuel type example A

6 enriching stages, stage enrichment factor for U-235 = 1.16

Table 18.6: Isotopic composition of fuel type example A after re-enrichment in a 10 stages gas ultra-centrifuge cascade with 6 enriching stages.

Isotopic	composition	
U-232	2.83%	
U-233	60.15%	
U-234	3.65%	Fuel type example A
U-235	15.76%	after re-enrichment
U-236	3.21%	
U-238	14.40%	

Fuel type example B

6 enriching stages stage enrichment factor for U-235 = 1.09

Table 18.7: Isotopic composition of fuel type example B after re-enrichment in a 10 stages gas ultra-centrifuge cascade with 6 enriching stages.

Isotopic	composition	
U-232	0.68%	
U-233	80.00%	
U-234	6.45%	Fuel type example B
U-235	0.96%	after re-enrichment
U-236	0.13%	
U-238	11.78%	

Comparing the isotopic compositions of fuel type example A and B one finds a difference of about a factor 4 in the concentrations of U-232.

Fuel type example C

Table 18.8: Isotopic composition of fuel type example C after re-enrichment in a 10 stages gas ultra-centrifuge cascade with 6 enriching stages.

Isotopic	composition	
U-232	5.76%	
U-233	60.15%	
U-234	3.65%	Fuel type example C
U-235	15.76%	after re-enrichment
U-236	3.21%	
U-238	11.47%	

In the next section the critical mass and critical dimensions of HNEDs for the three fuel types examples A, B and C are determined.

18.4 Critical Masses and Dimensions for Fuel Type Examples A, B, and C

Applying the Monte Carlo code MCNP and the cross section set ENDF/B VII.1 (see chapter 16) one finds the following critical dimensions for the three compositions of fuel types examples A, B and C. It is assumed that the critical sphere will be surrounded by a 5 cm thick reflector of beryllium.

Fuel type example A and C: critical radius = 5.2 cm

For k_{eff} = 0.98 and a uranium density of 18.9 g/cm^3

Critical mass =11.13 kg

Fuel type example B: critical radius = 4.9 cm

For k_{eff} = 0.98 and a uranium density of 18.9 g/cm^3

Critical mass = 9.31 kg

In the following sections we will analyze:

- the destruction of uranium hexafluoride by the α-radiation of U-232, U-233 and U-234
- the production of neutrons by (α,n)-reactions with light element impurities in uranium metal
- the nuclear explosive yield of a hypothetical nuclear explosive device (HNED)
- the temperatures in an HNED as a consequence of the α-heat production of U-232 in an HNED
- the radiation exposure during manufacturing of an HNED caused by U-232 daughter products
- the radiation damage caused in the high explosives.

This will be done similar to the analysis performed in chapter 16 for 19.95% U-235 enriched uranium of research and test reactors, doped with U-232.

18.5 Destruction of Uranium Hexafluoride UF_6 by α-Radiation of U-232

Following section 16.2 of chapter 16 the UF_6 molecules in gas ultra-centrifuge cascades are destroyed by α-radiation of the uranium isotopes. The strongest α-radiation emitters of the uranium isotope mixtures A, B and C are U-232 with an activity of 8×10^{11} (1/g·s) and U-233 with an activity of 3.35×10^{8} (1/g·s) as well as U-234 with an activity of 2.3×10^{8} (1/g·s) (see Table16.1). All other uranium isotopes can be neglected. In addition, the activities of the isotopes U-233 and U-234 contribute only about 1% or 6% to the total α-activity of the isotope mixtures examples A, B, and C. The α-energies are 5.3 MeV for U-232, 4.8 MeV for U-233 and 4.76 MeV for U-234.

Kryuchkov et al. [13] showed that for a concentration of 1% U-232 about 48% of the UF_6 molecules in the cascade of gas ultra-centrifuges would be destroyed within 1.2 months. These results of Kryuchkov et al. [13] can be directly used for the calculation of the amount of destruction of UF_6 molecules by the α-radiation of the uranium isotopes U-232, U-233 and U-234 of this analysis. The destruction of UF_6 molecules is directly proportional to the number of α-particles arising from the decay of the α-emitters U-232, U-233, and U-234. The α-activities in 1/(g·s) are given in Table 18.9 for the feed stage and stages number 4 and 6 of the enriching section (stage number 6 is the end enriching stage of the 10 stages cascade).

Table 18.9: α-activities (1/(g·s)) due to the isotopes U-232, U-233, and U-234 in different stages (feed stage, fourth enriching stage and end enriching stage).

	Fuel type example		
	A	B	C
Feed stage	$1.39x10^9$	$0.52x10^9$	$2.82x10^9$
Fourth stage	$8.99x10^9$	$2.48x10^9$	$1.81x10^{10}$
End stage	$2.28x10^{10}$	$5.72x10^9$	$4.63x10^{10}$

According to Kryuchkov et al. [13] 1% U-232 or $8x10^9$ (1/(g·s)) lead to 48% destruction of UF_6 in 1.2 months or 36 days [13]

During transport through the gas ultra-centrifuge cascade the UF_6 gas molecules remain equal time periods in each stage. With the above assumptions and the calculations for the α-activities of U-232, U-233, and U-234 in the feed stage and stages number 4 and 6 of the enriching section (Table 18.9) these results are related to the number of α-particles emitted by the 1% U-232 reported by Kryuchkov et al. [13]. This leads to the following results for the fuel types examples A, B, and C.

Table 18.10: Time periods for destruction of 48% UF_6 within the gas ultra-centrifuge cascade for fuel type examples A, B, and C.

	Destruction of 48% UF_6 in days or months Fuel type example		
	A	B	C
Feed stage	6.9 months	18.5 months	3.4 months
Fourth stage	32 days	3.8 months	16 days
End stage	12.6 days	1.7 months	6 days

An appreciation of the time periods of Table 18.10 is possible if they are compared to the results reported by Glaser [18] in section 15.5. With the 164 machines cascade discussed by Glaser a quantity of 37 kg of HEU with 91% U-235 enrichment could be produced within a time period of one year. This would be one significant quantity of about 21 kg within 7 months.

On the other hand Kryuchkov et al. [13] estimate that with a 250 to 350 machines cascade of gas ultra-centrifuges with much higher separation power per machine a significant quantity of 90% U-235 enriched uranium (about 21kg) could be produced within about one month (their data, however, do not allow a detailed recalculation).

Kryuchkov et al. [13] did only consider the α-particles of the α-decay of U-232 to Th-228 with an energy of 5.3 MeV, but not the following decay chain with α-particles. This is justified as these α-particle emitting daughter products of U-232 could be chemically separated prior to the re-enrichement process. In addition, these α-particles will be delayed mostly by the half-life of **1.91 years** of Th-228. This could shorten somewhat the time periods in Table 18.10 (only important for fuel type example B).

Whereas the α-particles will have destroyed a certain part of the UF_6 molecules when being transported through each cascade to the end cascade, this destruction process will also continue during the subsequent time periods for transportation and conversion of the UF_6 to uranium metal. Therefore it can be concluded:

Although there are differences of about a factor of 4 or 8 in the time periods shown in Table 18.10 between fuel type example B and A or C during re-enrichment with such a gas ultra-centrifuge cascade arrangement most of the UF_6 molecules would be destroyed in all cases. This would make the construction of an HNED impossible.

18.6 The Production of Neutrons by (α,n)-Reactions with Light Element Impurities in Uranium Metal

Following section 16.3 and assuming the results of Kryuchkov et al. [13] who applied the SRIM code [15] and the nuclear data of [16,17] as well as the light element impurities data of the Ulba plant at Kazakhstan (Table 16.3), one can start from Fig. 16.5. The lead isotope for re-enrichment is U-233 now instead of U-235. The neutron yield at begin of re-enrichment (ordinate of Fig. 16.5) is proportional to the concentration of U-232 in the uranium. Only the stage enrichment factors are slightly different from those shown in Fig. 16.5.

With these initial data we obtain the following neutron yields (Fig. 18.2) at the begin (feed stage) and at the end of the re-enrichment cascade for the three fuel type examples A, B, and C.

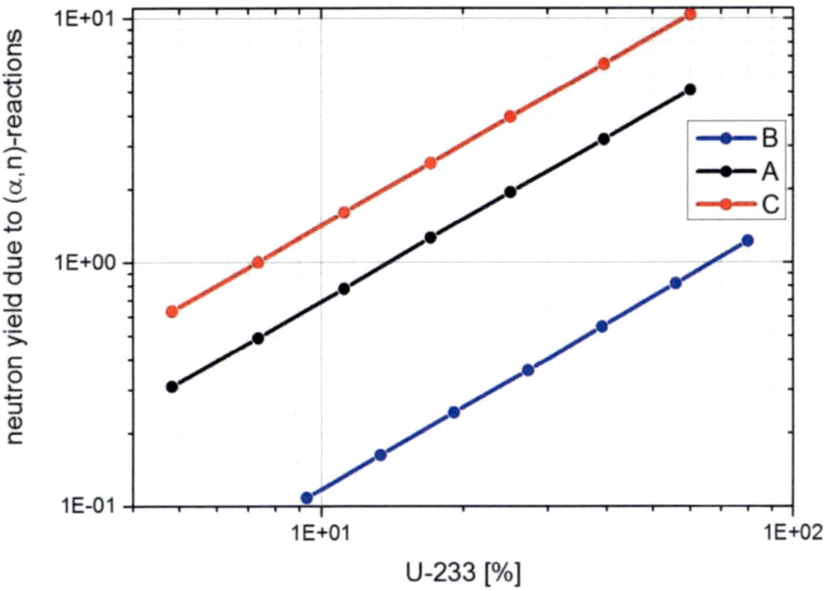

Figure 18.2: Neutron Yield from (α,n)-reactions with light elements as impurities in uranium during enrichment in the 6 enriching stages of the 10 stages gas ultra-centrifuge cascade (for fuel type examples A, B, and C).

The neutron yields in Fig. 18.2 have the following values at begin and end of the re-enrichment

Fuel type example	A	B	C
Begin of re-enrichment:			
neutron yield:	0.31	0.11	0.63
Enrichment of U-233:	4.83%	9.33%	4.83%
End of re-enrichment:			
neutron yield	5.1	1.22	10.4
Enrichment of U-233:	60.15%	80%	60.15%

These data show that the neutron yield caused by (α,n)-reactions with light element impurities increases considerably within the 6 stages of the enriching section.

The neutron yield at the end of re-enrichment will also be the neutron yield within the HNED after manufacturing. It will be the basis for the analysis of the explosion yield of HNEDs in the next section.

18.7 Internal Neutron Sources of Hypothetical Nuclear Explosive Devices with Fuel Type Examples A, B, and C

Both gun-type and implosion-type hypothetical nuclear explosive devices (HNEDs) will be analyzed for the fuel type examples A, B, and C to compare their potential explosive yields. For the internal neutron sources both the neutron yield from (α,n)-reactions with light element impurities and the spontaneous fission yields for the fuel isotopes U-232, U-233, and U- 235 must be considered.

For fuel type example A the following internal neutron source is calculated:

Implosion-type:

$S_{A,impl.}$= 11,130 x (5.1+0.0283 x 1.3+0.6015 x 0.5+0.1576 x 0.6)=61,572 (n/s)

For an initial k_{eff}=0.98 and an implosion-type HNED this value has to be multiplied

by 1/(1-0.98)=50 (see section 8.9.10) which leads to

$S_{A,impl.}=3.08x10^6$ (n/s)

For a **gun-type** HNED with a diameter of 8.25 cm and a length of 11 cm as critical dimensions the following value for the internal neutron source is determined. The first gun-type nuclear explosion device had about 2.3 times critical masses (section 11.5).

$S_{A,gun}$=11,130 x 2.3x (5.1+0.0283 x 1.3+0.6015 x 0.5+0.1675 x 0.6)

$S_{A,gun}=1.42x10^5$ (n/s)

Similarly the neutron source values for gun-type and implosion-type devices for example B and C were calculated.

Table 18.11: Internal neutron sources for gun-type and implosion-type HNEDs with fuel type examples A, B, and C.

	Internal neutron sources S_A, S_B, and S_C (n/s)		
Fuel types example	A	B	C
Gun-type	$1.42x10^5$	$3.49x10^4$	$2.78x10^5$
Implosion-type	$3.08x10^6$	$0.76x10^6$	$6.05x10^6$

Fuel type example B has the lowest internal neutron source for gun-type and implosion-type HNEDs. It also has the lowest U-232 isotope concentration of only 0.06% at begin of the re-enrichment (section 18.2)

18.8 Analysis of the Explosion Potential of HNEDs with Re-enriched U-233/U-235

For the analysis of the explosion potential of HNEDs of re-enriched U-233/U-235 contaminated by U-232 the methods used in section 9.11 and 9.13 are applied.

18.8.1 Probability of Pre-ignition by Spontaneous Fission Neutrons n and (α,n)- Neutrons in an HNED

The theory of pre-ignition probability is always presented in a simplified way for a ramp increase, where the criticality factor k_{eff} or $\Delta k = k_{eff}- 1$ increases from prompt criticality to a maximum $k_{eff,max}$ or to $\Delta k_{max} = k_{eff,max} -1$ during a short time period of t_0 (Fig. 18.3a).

Similarly, the Rossi alpha

$$\alpha(t) = \frac{\Delta k}{\tau} \tag{18.1}$$

with τ as the neutron lifetime can be presented as a ramp function (Fig. 18.3b).

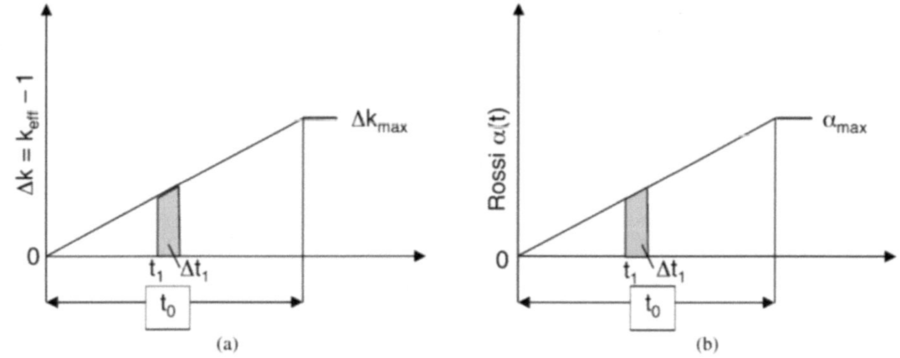

Figure 18.3: *Δk(t)* as a ramp function of time t_1 and (b) Rossi α(t) as a ramp function of time t_1.

The differential probability of a first persistent fission chain being sponsored at time t_1 in the time interval dt_1 is then

$$p(t_1)\cdot dt_1 = \frac{2\Delta k_{max}}{\bar{\nu}\cdot\Gamma_2}\cdot\frac{S}{t_0}\cdot t_1 \times \exp\left[\frac{-\Delta k_{max}}{\bar{\nu}\Gamma_2}\cdot\frac{S}{t_0}\cdot t_1^2\right] dt_1 \qquad (18.2)$$

and the cumulative probability that persistent fission chains have occurred until time t_1 (integral of the differential probability between zero and t_1) is given by

$$P(t_1) = 1 - \exp\left[-\frac{\Delta k_{max}}{\bar{\nu}\Gamma_2}\cdot\frac{S}{t_0}\cdot t_1^2\right] \qquad (18.3)$$

where

Δk_{max} = maximum $k_{eff,max}$ -1 at full compression
$\bar{\nu}$ = average number of neutrons per spontaneous fission

$\Gamma_2 = \nu\cdot(\nu - 1)/\bar{\nu}^2 \approx 0.8$

S = spontaneous fission source and (α,n)-neutrons per second.

For Δk_{max} of ~1, ν of ~2.5, and $\Gamma_2 \approx 0.8$ the factor $\Delta k_{max}/(\nu\cdot\Gamma_2)$ becomes ½ and Eqs. 18.2 and 18.3 become

$$p(t_1)\cdot dt_1 = S\cdot\frac{t_1}{t_0}\cdot\exp\left[-\frac{1}{2}\cdot S\cdot\frac{t_1^2}{t_0}\right]\cdot dt_1 \qquad (18.4)$$

and

$$P(t_1) = 1 - \exp\left[-\frac{1}{2}\cdot S\cdot\frac{t_1^2}{t_0}\right] \qquad (18.5)$$

The maximum nominal nuclear explosive yield Y_o would theoretically be attained if $P(t_0) = 0$; i.e., no persistent fission chains would have occurred up to t_0 (no pre-ignition up to Δk_{max}). In practical reality, $P(t_0)$ can only be minimized.

The minimum nuclear explosive yield is called minimum fizzle yield $Y_{F,min}$. It is determined by the criterion that pre-ignition is initiated at the earliest possible point in time t_1=0 and that approximately e^{45} fissions will have occurred during the chain reaction. The internal energy accumulated will then be high enough such that the nuclear fissile material will have vaporized and begun to exert a pressure in the megabar range on its surroundings. This will initiate a rapid expansion. It also marks the maximum excess reactivity introduced for this case (see section 9.12).

The minimum ratio of the fizzle yield can then be determined to be

$$x_{F,\min} = (Y/Y_0)_{F,\min} = \left(\frac{90\tau}{t_0}\right)^{3/2} \quad (18.6)$$

by applying Serber's relation (see section 9.9)

In a more general way, a relation between the cumulative probability of pre-ignition (pre-detonation) and the yield fractions $x = Y/Y_0$ (with Y as the explosive yield achieved after pre-ignition at some point in time and Y_0 as the maximum nominal yield) can also be derived (see section 9.1.3.1) as:

$$P(x) = P\left(\frac{Y}{Y_0}\right) = 1 - \exp\left[-\frac{1}{2} S \cdot t_0 \left(\frac{Y}{Y_0}\right)^{2/3} + 45 \cdot S \cdot \tau\right] \quad (18.7)$$

18.8.2 Numerical Examples

A further evaluation of Eqs. 18.4 through 18.7 requires representative numerical values for the spontaneous fission and (α,n)-neutron source S, the effective neutron life time τ, and for the time period t_0.

18.8.2.1 Spontaneous Fission Neutron Sources

The spontaneous fission neutron sources for the fuel types examples:

A, B, and C are listed in Table 18.11 for gun-type and implosion-type HNED

18.8.2.2 Effective Neutron Lifetime

The effective neutron lifetime τ can be assumed to be about ~10^{-8} s for high enriched U-232/U-233/U-235 and U-238 metal in an HNED

18.8.2.3 Time Period t_0 from Reaching Prompt Criticality to Δk_{max}

The time period t_0 for the increase from $\Delta k = 0$ up to Δk_{max} depends on the method of compaction of the nuclear fissile material. Two methods are known: the gun-type-system and the implosion method (see section 11.5).

18.8.2.4 Gun-Type-System

For a gun-type system with high enriched U-233/U-235 uranium an assembling velocity of 300 (m/s) in the gun barrel can be assumed. This leads to $t_0 = 10^{-3}$s which is by a factor of about 100 higher for the time period t_0 than for the implosion method with $t_0 = 10^{-5}$ s.

For the subsequent evaluation, the time period t_0 for the gun system is treated in a parametric way with two values $t_0 = 5\times10^{-4}$ s and $t_0 = 10^{-3}$ s.

18.8.2.5 Implosion Method

For the implosion method, two time periods $t_0 = 2x10^{-5}$ s and $t_0 = 10^{-5}$ s are assumed. The latter would correspond to a particle velocity of 5 km/s during shock compression of a 5-cm-radius metallic uranium metal sphere.

18.8.3 Results for the Parametric Evaluation

Figures 18.4 through 18.7 show the evaluation of Eqs. 18.5 for the above defined parameters.

18.8.3.1 Gun-Type System

Figures 18.4 and 18.5 show the cumulative probabilities of pre-ignition for $t_0 = 10^{-3}$ s and $t_0 = 5\times10^{-4}$ s as a function of the ratio t_1/t_0. The time derivative of this curve represents the differential probability of pre-ignition at t_1 within the time interval dt_1. Figure 18.4 shows that pre-ignition would occur early within the compaction phase up to $t_1/t_0 = 0.2\text{-}0.3$ for S_C and S_A and 0.6 for S_B. For a shorter compaction time $t_0 = 5\times10^{-4}$ s (Fig. 16.5), this time interval for early pre-ignition is $t_1/t_0 = 0.3\text{-}0.4$ for S_C and S_A and 0.80 for S_B

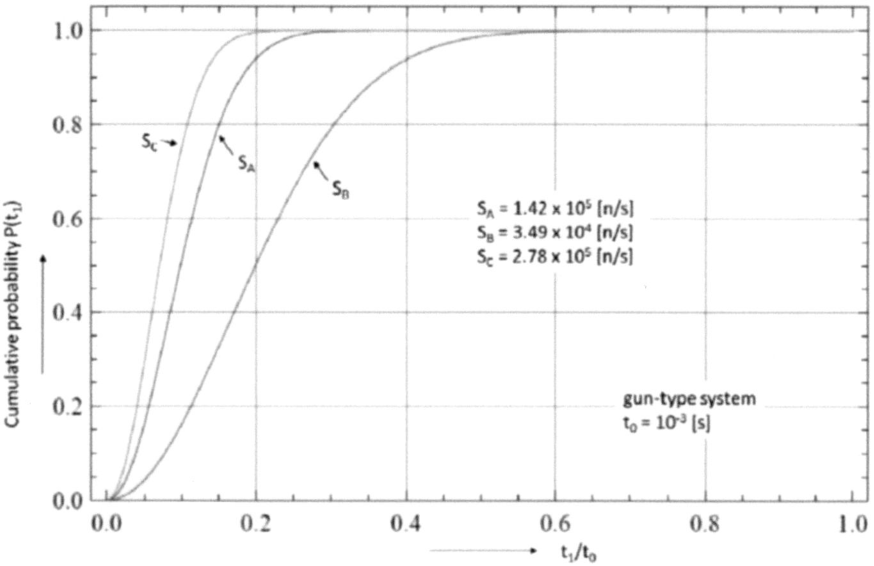

Figure 18.4: Cumulative probability of pre-ignition for a time period of t_0=10^{-3} s of compaction by gun-type system as a function of t_1/t_0. Spontaneous and (α,n)-neutron sources S_A, S_B, and S_C.

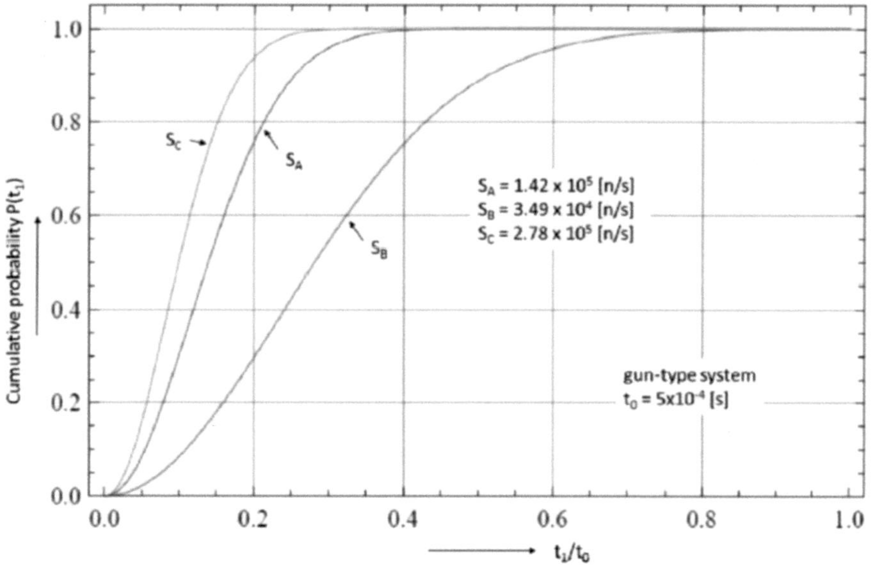

Figure 18.5: Cumulative probability of pre-ignition for a time period of $t_0 = 5\times10^{-4}$ s of compaction by gun-type system as a function of t_1/t_0. Spontaneous and (α,n)-neutron source, S_A, S_B, and S_C.

Figures 18.6 and 18.7 show the cumulative probability of pre-ignition as a function of the ratio x of the attainable nuclear explosive yield Y to the maximum nominal yield Y_0 for the time periods $t_0 = 10^{-3}$ s and 5×10^{-4} s (gun-type system). The curves for S_A, S_B, and S_C coincide into one single curve. Each value of the cumulative probability for pre-ignition belongs to a certain value of the ratio x: On the right hand *(1-P(Y/Y₀))* shows the cumulative probability of no pre-ignition or the probability to attain $x=Y/Y_0$. The smallest ratios of $x=Y/Y_0$ or minimum fizzle yield ratio $x_{F,min}$ would occur at the intersection of the curves with the abscissa or the cumulative probability for pre-ignition=0. They are (Eq. 18.5)

$$x_{F,\,min} = (Y/Y_0)_{F,\,min}$$
$$= 2.7 \times 10^{-5} \text{ for the time period } t_0 = 10^{-3} \text{ s}$$

and

$$x_{F,\,min} = (Y/Y_0)_{F,\,min}$$
$$= 7.6 \times 10^{-5} \text{ for the time period } t_0 = 5\times10^{-4} \text{ s}.$$

For Y_0, a value of, e.g., 20 kt TNT can be considered (see section 9.1.3).

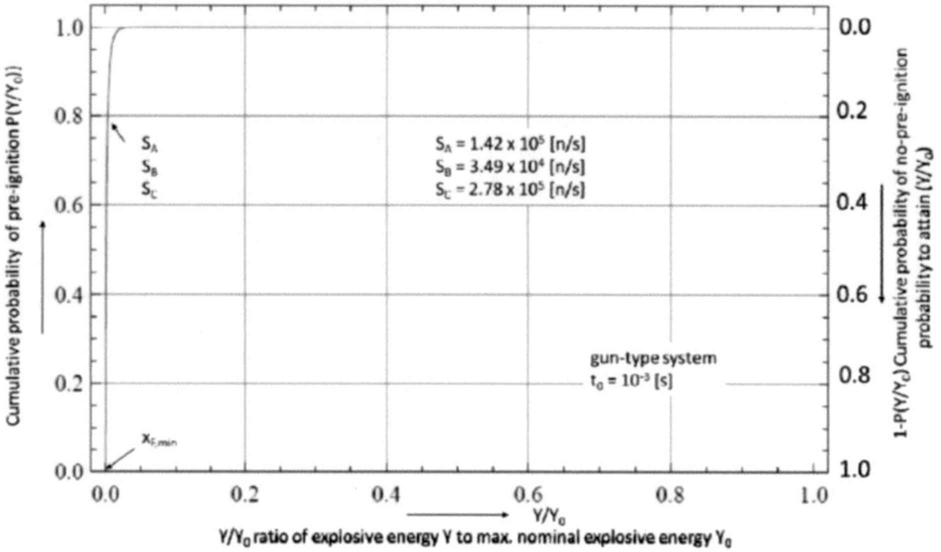

Figure 18.6: Cumulative probability of pre-ignition as a function of the ratio x of the attainable nuclear explosive yield Y to the maximum nominal yield Y_o for the time period of compaction $t_0 = 10^{-3}$ s. Spontaneous fission and (α,n)-neutron source S_A, S_B, and S_C.

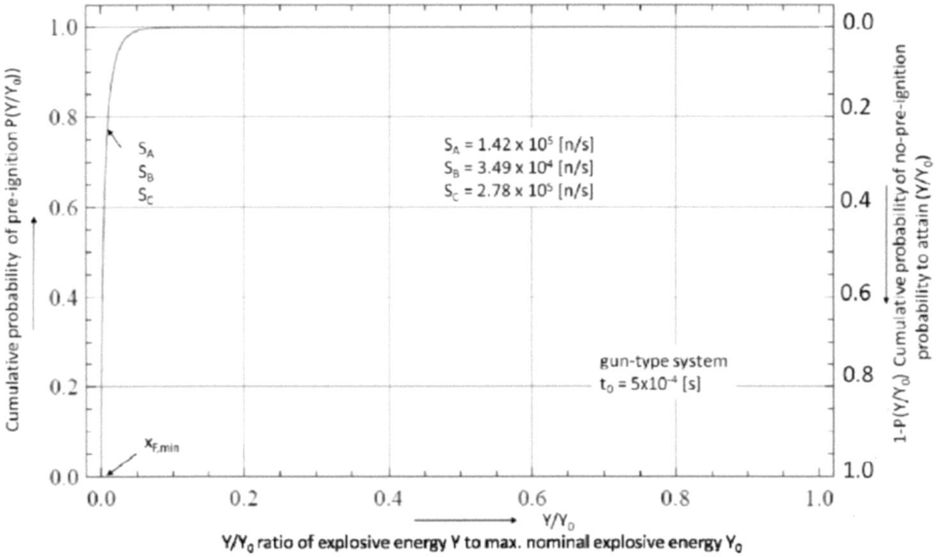

Figure 18.7: Cumulative probability of pre-ignition as a function of the ratio x of the attainable nuclear explosive yield Y to the maximum nominal yield Yo for the time period of compaction $t_0 = 5{\times}10^{-4}$ s. Spontaneous fission and (α,n)-neutron sources S_A, S_B, and S_C.

It can be concluded that for the gun-type system with fuel type example A, B, and C that the attainable nuclear explosive yield Y would be extremely low and of no interest for NEDs (0.54 t TNT for $t_0 = 10^{-3}$ s and 1.5 t TNT for $t_0 = 5 \times 10^{-4}$ s, if $Y_0 = 20$ kt TNT is applied).

18.8.4 Spherical Implosion System

Fig. 18.8 and 18.9 show the cumulative probabilities of pre-ignition for $t_0 = 2 \text{ x } 10^{-5}$ s and $t_0 = 10^{-5}$ s as a function of the ratio t_1/t_0.

As the geometry of the spherical implosion system is relatively compact and has a $k_{eff} \approx$ 0.98, it must be considered as a subcritical system in which the internal neutron source is subject to a source multiplication. This is accounted for in Table 18.11 for S_A, S_B, and S_C.

Due to the faster compaction velocities of the implosion method the pre-ignition starts later, compared to the gun-type system velocities This is shown in Fig. 18.8 for t_0= 2x10^{-5} s and in Fig. 18.9 for t_0=10^{-5} s and three spontaneous fission and (α,n)-neutron sources S_A, S_B, and S_C

The ratio $x_{F,min}$, which depends on t_0 (see Eq. 18.5) is higher for the spherical implosion system than for the gun system (Figs. 18.10 and 18.11). .

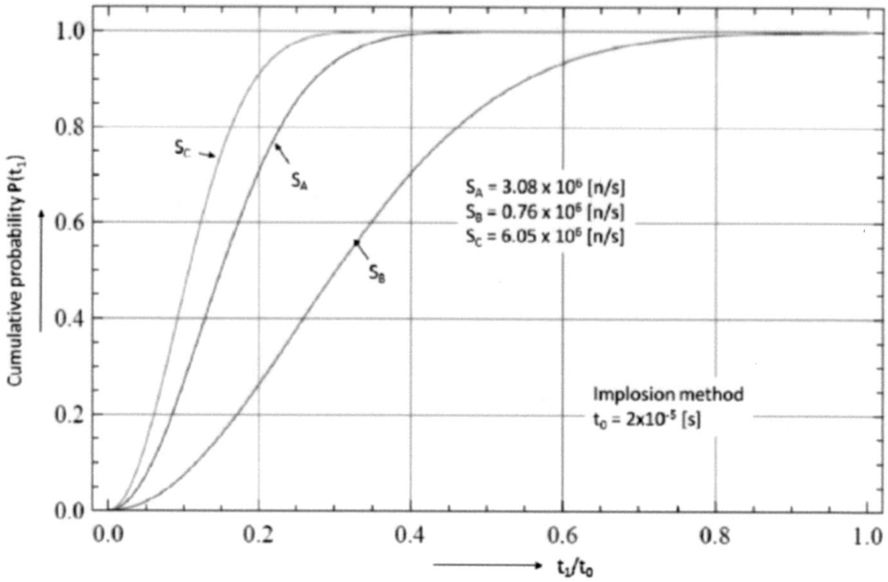

Figure 18.8: Cumulative probability of pre-ignition for a time period t_0= 2x10^{-5}s and internal spontaneous and (α,n)-neutron sources S_A, S_B, and S_C as a function of t_1/t_0.

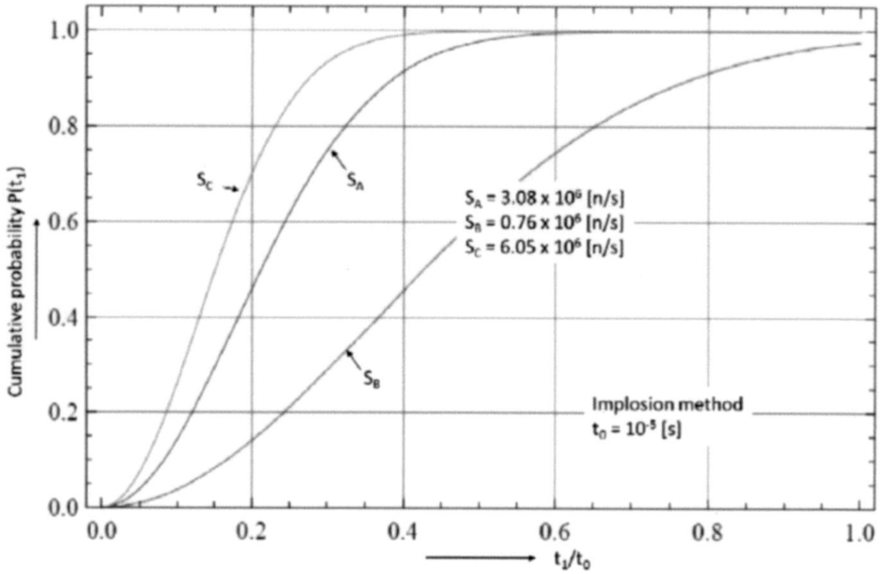

Figure 18.9: Cumulative probability of pre-ignition for a time period t_0=10^{-5}s and internal spontaneous and (α,n)-neutron source S_A, S_B, and S_C as a function of t_1/t_0.

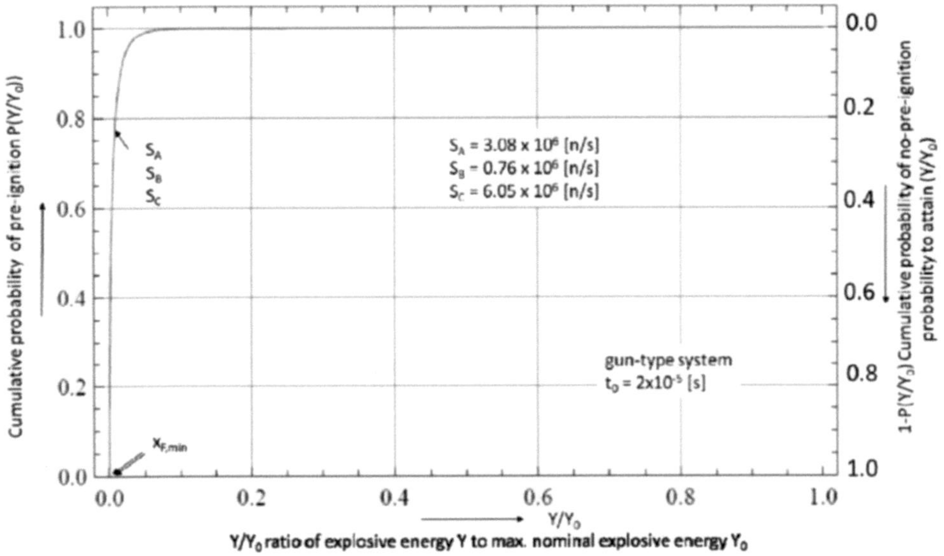

Figure 18.10: Cumulative probability of pre-ignition as a function of the ratio x of the attainable explosive yield Y and the maximum nominal yield Y_0 for the time period t_0=2x10^{-5}s (implosion method) as well as spontaneous fission and (α,n)-neutron sources of S_A, S_B, and S_C.

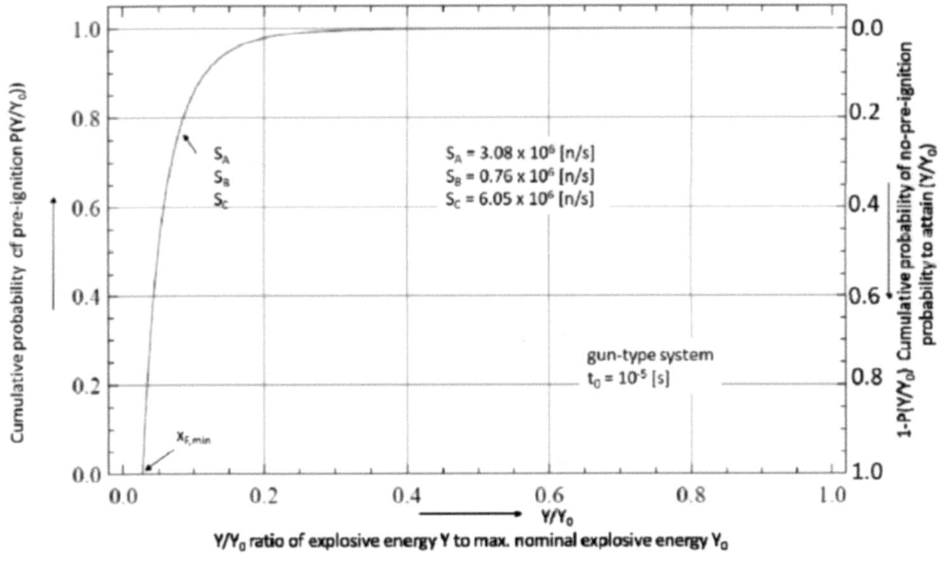

Figure 18.11: Cumulative probability of pre-ignition as a function of the ratio x of the attainable explosive yield Y and the maximum nominal yield Y_0 for the time period t_0=10^{-5}s (implosion method) as well as spontaneous fission and (α,n)-neutron sources of S_A, S_B, and S_C.

The curves for S_A, S_B, and S_C coincide in one curve in Figs. 18.10 and 18.11. For $t_0 = 10^{-5}$ s, the ratio $x_{F,min}$ of the minimum fizzle yield (Eq. 16.5) is highest.

$$x_{F,\min} = \left(\frac{Y}{Y_0}\right)_{F,\min} = 0.027 \quad \text{for } t_0 = 10^{-5}\text{ s} .$$

This is the same number as derived for reactor plutonium in section 9.12 and leads to the same conclusion that for a compaction time $t_0 = 10^{-5}$ s, a minimum nuclear explosive yield of 0.54 kt TNT could be attained (relative to a maximum nominal yield of Y_0 = 20 kt TNT). For lower probabilities to attain x these values can even be higher than $x_{F,min}$.

18.9 Internal α-Heat Production of HNEDs Based on Fuel Type Examples A, B, and C

The fuel type examples A, B, and C resulting as the end product of re-enrichment would contain the following masses of U-232 in their critical mass:

Example A	11,130 x 0.0283 = 314.9 g U-232
Example B	9,310 x 0.0068 = 63.3 g U-232
Example C	11,130 x 0.0576 = 641.1 g U-232

U-232 produces 680 W/kg just after reprocessing due to the α-decay of U-232. After one year this α-heat production has increased to about 2,100 W/kg due to the α-particle decay chain (Fig. 16.6) and increases even further as a function of time. It is assumed here that conversion to UF_6, re-enrichment and manufacturing would last only one year. This is extremely conservative.

The α-heat power of an HNED with uranium of the different fuel types are:

Example A	0.3419 x 2,100 = 661 W
Example B	0.0633 x 2,100 = 133 W
Example C	0.6411 x 2,100 = 1,346 W

In the following sections the temperature profiles in such HNEDs of implosion-type will be discussed.

18.9.1 Geometric Dimensions for HNEDs with Fuel of Type A, B, and C

The critical radius for k_{eff}=0.98 of fuel type examples A and C with 5,2 cm and fuel type example B with 4.9 cm differ only slightly from the examples for HNEDs for reactor-grade plutonium with 5.8 cm critical radius analyzed already in chapter 10. Therefore, the same geometric dimensions for the outer casing and the high explosive lenses can be chosen for the thermal analysis. Only the thickness of the Al-layer for the damping of the Rayleigh-Taylor instabilities of the shock waves are slightly increased.

The outer casing radius is:

- for high-technology HNEDs: 21 cm
- for medium-technology HNEDs: 42 cm
- for low-technology HNEDs: 65 cm

The thickness of the outer casing is 1.2 cm for all cases.

The thickness of the explosive lenses is:

- for high-technology HNEDs: 6 cm
- for medium-technology HNEDs: 25 cm
- for low-technology HNEDs: 43 cm

The characteristic materials data for the high explosives of the three different technologies are taken from Tables 10.1, 10.2 and 10.3. The thermal conductivity data of the different materials are taken from Tables 10.1, 10.2, 10.3 and 10.7.

The results of chapter 10 and Kessler [19] show that fuel type examples A and C with 661 W and 1,346 W internal α-heat power lead for low- and medium-technology HNEDs to too high temperatures in the high explosive lenses (Figs. 10.16 and 10.17). Such low- and medium-technology HNEDs, therefore become **technically unfeasible**. Only high-technology HNEDs are of interest for the thermal analysis. Fuel type example A is with an internal α-heat power of 661W slightly above the limit of 640 W for high-technology HNEDs cooled at the outside by air and thermal radiation (Fig. 10.16). Fuel type example C is with 1,346 W even above the limit of 950 W for high-technology HNEDs cooled at the outside by liquid helium (Fig. 10.17).

On the other hand fuel type example B is with 133 W internal α-heat production technically unfeasible for low-technology HNEDs cooled at the outside by air and thermal radiation, but feasible for medium-technology HNEDs cooled at the outside by air and thermal radiation (Fig. 10.16).

18.9.2 Thermal Analysis for HNEDs with Fuel Type Examples A and C

As explained above only high-technology HNEDS have to be considered for fuel type example A with 661 W and fuel type C with 1,346 W. The geometric dimensions for HNEDs fuel type A and C are shown by Fig. 18.12.

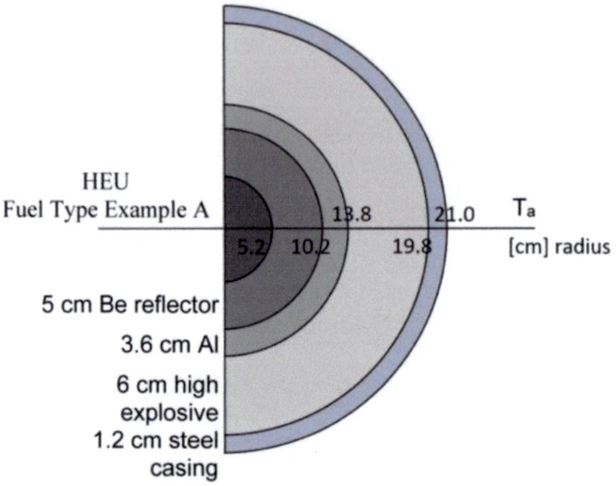

Figure 18.12: Geometric dimensions for a high-technology HNED with fuel type examples A and C.

The outside temperature of HNED fuel type example A with 661 W and cooled at the outside by air and thermal radiation can be determined from Table 10.5, by interpolation.This leads to an outside temperature of T_a = 134 °C

Fig. 18.13 shows the radial temperature profile within HNED fuel type example A . The thermal conductivities of TATB and PBX 9502 are almost equal. Therefore, only one curve is shown. The radial temperature increases from 134 °C at the outside to 341 °C at the inner border of the chemical explosive lenses. This is just below the limiting temperature for self-explosion of TATB (347°C), but above the self-explosion temperature of PBX 9502 (331°C). The radial temperature increases then further to 388°C in the center of the U-232/U-233/U-235/U-238 part of the HNED. No melting point of the various materials is attained. The high-technology HNED fuel type example A would just be **technically unfeasible**, because the limiting temperature for start of self-explosion of PBX 9502 would be exceeded.

As a high-technology HNED with fuel type C has an internal α-heat production of 1,346 W it would be **technically unfeasible**, even if it would be cooled at the outside by liquid helium. The 1,346 W is above the limit of 950 W given in Fig. 10.17.

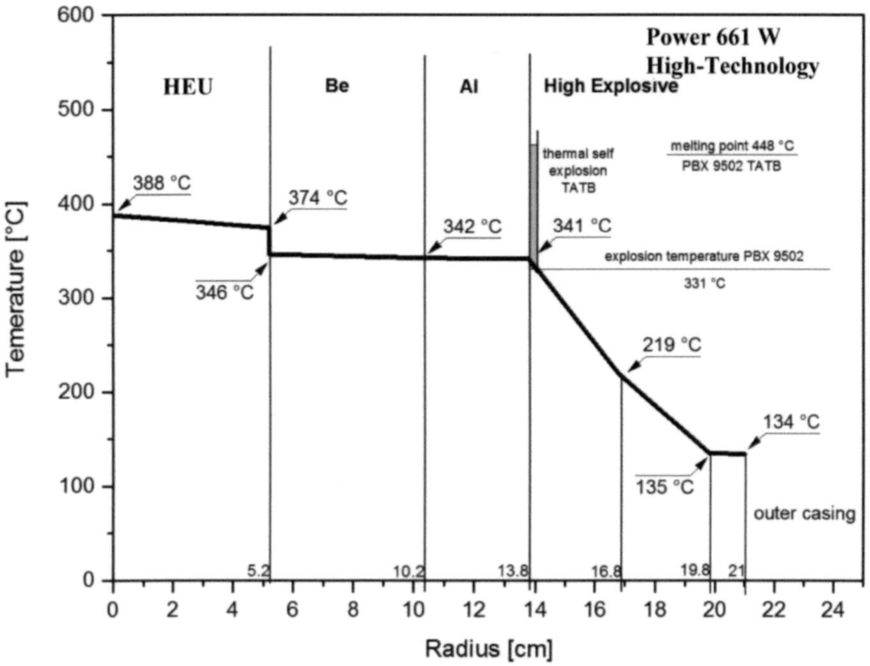

Figure 18.13: Radial temperature profile in a high-technology HNED with an internal α-heat power of 661 W (fuel type example A).

18.9.3 Thermal Analysis of an HNED with Fuel Type Example B

As explained above an HNED with fuel type example B having an internal α-heat power of 133 W is not technically feasible as a low-technology design (see Fig. 10.16). However it is technically feasible as medium and high-technology design. The radial temperature profile of a medium-technology design will, therefore, be analyzed below. The geometric dimensions of a medium-technology design with fuel type example B are shown by Fig. 18.14

The outside temperature of an HNED with fuel type example B and an internal α-heat power of 133 W cooled on the outside by air and thermal radiation can be determined from Table 10.5 by interpolation.

Outside temperature $T_a = 35$ °C

The chemical explosive lenses contain – as in chapter 10 – the different chemical explosives: HMX and DATB. The thermal conductivities and limiting temperatures of these explosives can be found in Table 10.2.

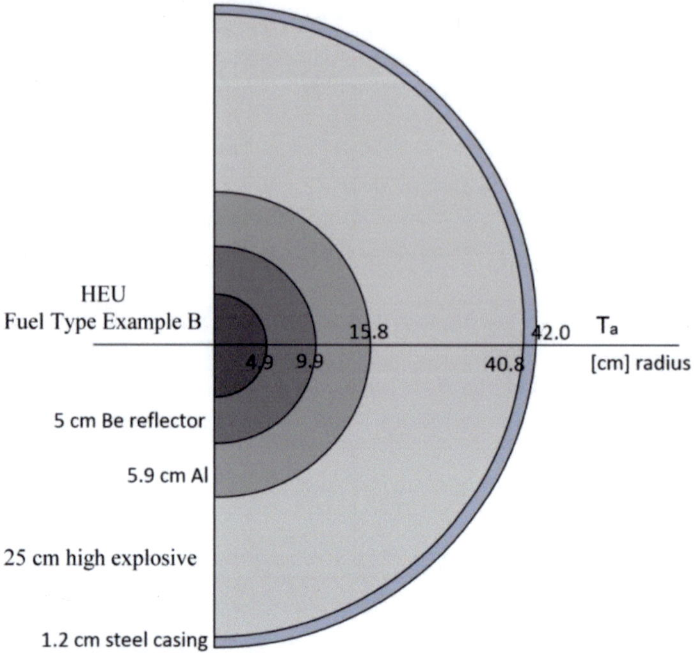

Figure 18.14: Geometric dimensions of a medium-technology HNED with fuel type example B.

Fig. 18.15 shows the radial temperature profile within such a medium-technology HNED. The radial temperatures rise up from 35°C outside temperature to 141°C for the explosive HMX and to 193°C for DATB at the inner surface of the explosive lenses. No limiting temperatures (melting point for HMX is 256-286°C and for DATB it is 286 °C; limiting temperatures for onset of pyrolysis and onset of self-explosion are above 259°C, see Table 10.2) are exceeded. The radial temperatures increase further within the Al-layer, the beryllium reflector and in the U-232/ U-233/U-335/U-238 sphere up to 203°C, respect-tively, but no melting point is exceeded.

The medium-technology and high-technology HNED for fuel type example B are **technically feasible**. For cooling at the outside of the HNED by liquid helium, even a low-technology HNED would be technically feasible (see Fig. 10.17).

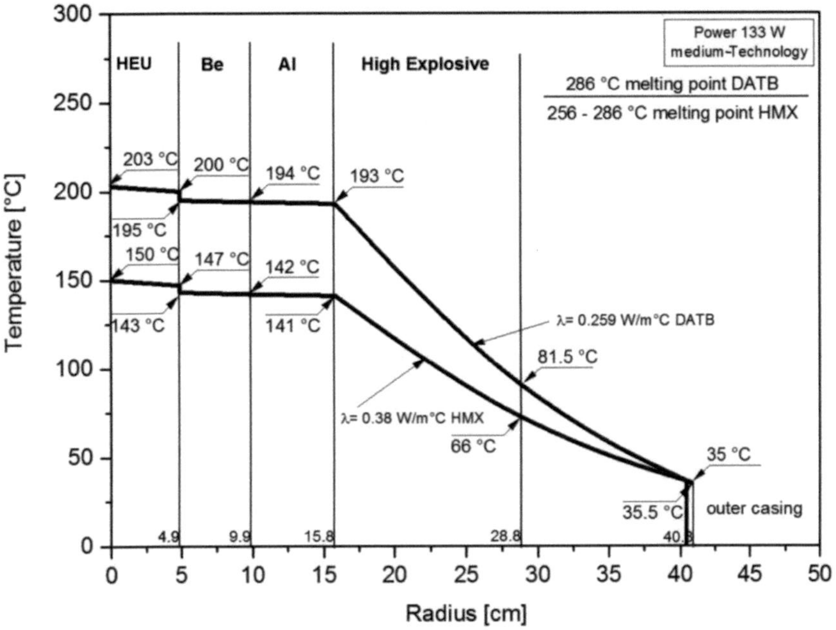

Figure 18.15: Radial temperature in a medium-technology HNED and fuel type example B with an α-heat power of 133 W and DATB and HMX as high explosives.

18.9.4 Radiation Exposure due to U-232 Content in Fuel Types Examples A, B, and C

The γ-radiation exposure in 0.5 m distance of a U-232/U-233/U-235/U-238 metal sphere one year after reprocessing was already discussed in section 16.9 applying data from Kang et al. [1] and Forsberg et al. [10]. For metal spheres with the U-232 contents of the fuel types A, B, and C after re-enrichment the following radiation exposures can be determined at 0.5 m distance.

Table 18.12: γ-radiation exposure from U-232/U-233/U-235/U-238 metal spheres of fuel type examples A, B, and C in 0.5 m distance one year after reprocessing.

U-232 content	Dose rate (Sv/h)	Working hours
U-233/U-235 (1% U-232)	1.27	0.04
U-233/U-235 (3.15% U-232)	4.00	0.0127
Fuel type example A (2.83% U-232)	3.59	0.14
Fuel type example B (0.68% U-232)	0.86	0.059
Fuel type example C (5.76% U-232)	7.31	0.0069

Manufacturing of such HNEDs with very high γ-radiation would require remote handling techniques in hot cells. The recommendations of IAEA [21] on physical protection for >1 Sv/h radiation dose at 1 m distance (see section 16.9) or 4 Sv/h at 0.5 m distance is almost attained by fuel type example A and exceeded by almost a factor of 2 in case of fuel type example C. For fuel type example B, however, the γ-radiation exposure would be lower by more than a factor 4 if compared to the IAEA recommendations on physical protection [21].

According to Moir [22] the γ-radiation increases further by a factor of 2.9 during the next 10-20 years. As already explained in section 16.9 the daughter products of U-232 can be chemically separated, but the γ-radiation of the daughter products Tl-208 would rise again to considerable values within a year.

18.9.5 Radiation Damage to High Explosives

The high explosives used in the explosive lenses of HNEDs can withstand only a certain γ-radiation exposure. For the medium-technology high explosive, e.g. HMX this exposure limit is 8.77×10^5 Gray. For high-technology explosives, e.g. TATB this exposure limit is 8.77×10^6 Gray [22,23].

At this high γ-radiation exposure limit several radiation damage effects, e.g. gas evolution, weight loss or crumbling of the high explosives occur [22,23]. Assuming a radiation weighting factor of 1 for γ-radiation the above dose rates (Table 18.12) of the different fuel types examples A, B, and C at 0.5 m distance correspond to:

Fuel type example	γ-radiation exposure (Gray/hr)
A (2.83% U-232)	3.59
B (0.68% U-232)	0.86
C (5.76% U-232)	7.31

At the inner surface of the high explosive lenses the γ-radiation dose will be higher than at 0.5 m distance. For fuel type examples A and C the γ-radiation dose at the inner surface of the high explosive lenses at 13.8 cm will be:

$$3.59 \times 0.5^2/0.138^2 = 47.1 \text{ Gray/h}$$

For fuel type B the γ-radiation dose at the inner surface of the high explosive lenses at 15.8 cm will be:

$$0.86 \times 0.5^2/0.158^2 = 8.6 \text{ Gray/h}$$

The radiation exposure limits will be reached after the following time periods for the medium-technology high explosive HMX and for the high-technology high explosive TATB (Table 18.13)

Table 18.13: Time periods for reaching the radiation exposure limits of high explosives in HNEDs with different fuel types examples A, B, and C.

Fuel type example	8.77×10^5 (HMX)	8.77×10^6 (TATB)
A (2.83% U-232)	26 months	260 months
B (0.68% U-232)	122 months	1220 months
C (5.76% U-232)	13 months	130 months

If it is assumed that high-technology high explosives are not available to terrorist and subnational groups, it can be concluded that fuel type example C would produce sufficient γ-radiation to make an HNED with fuel type example C **technically unfeasible**. One year or 13 months can certainly be assumed for production of the metal sphere and assembling the HNED. After 13 months the high explosive would not work any more properly. For fuel type example B this is not the case at all. For fuel type example A the HNEDs would have to be assembled and used within 26 months.

18.9.6 Summarizing Remarks

The initial 19.95% U-235 enrichment in fresh U/Th fuel needed for conversion of Th-232 to U-233 in e.g. PWRs has to be treated like the 19.95% U-235 uranium fuel of research and test reactors. It should be doped with U-232 as suggested in chapter 16 to become **proliferation-proof**. If already recycled, the U-233 would already contain U-232 contamination from previous irradiations in a PWR core doping with **U-232** would then probably not be necessary.

Another possibility would be to use as reactor fuel **proliferation-proof reactor plutonium** with more than 7% Pu-238 (see chapter 14) instead of 19.95% U-235 enriched uranium mixed with thorium. This is possible as was shown by Neuhaus et al. [24] when investigating the possibility of Pu-incineration in the Pu/Th fuel cycle. The isotope U-232 is only one atomic mass unit (amu) lighter than U-233. Therefore, the enrichment of U-232 in a U-232/U-233 mixture by a gas ultra-centrifuge arrangement is not as efficient as it was shown for U-232/U-235 fuel of research and test reactors in chapters 15 and 16.

Radial blanket elements in LMFBRs containing only thorium should not be allowed by IAEA (see Fig. 18.1 by Kang et al. [1] as they will produce U-233 with only a few 10 ppm U-232).

The most efficient influence characteristic of the contaminant U-232 is found to be the destruction of the UF_6 molecules by (α,n)-reactions. In this case even the relatively low 600 ppm contamination with U-232 in fresh fuel of fuel type example B can be considered to be sufficient to make the fresh U-232/U-233/U-235/U-238 fuel **proliferation-proof** (Table 18.14).

Table 18.14: Is the concentration of the contaminant U-232 sufficiently high to make fuel type example A, B or C **proliferation-proof**?

Fuel type example	A	B	C
U-232 ppm	1,700	600	3,500
48% destruction by α-particles in	yes 13 days	yes 1.7 months	yes 6 days
Gun-type HNED	(negligible nuclear explosion yield)		
Implosion-type HNED	Explosion yield (0.45 kt) and more	Explosion yield (0.45 kt) and more	Explosion yield (0.45 kt) and more
Thermal analysis	yes	no (low- technology)	yes
α-radiation	almost	no	yes
Destruction of chemical Explosives within	yes (2 years)	no (122 months)	yes (1 year)

Because of the high internal neutron sources and of the pre-ignition problem in a gun-type HNED with fuel type examples A, B, and C could produce only negligible (0.54 to 1.5 t of TNT) nuclear explosive yields. However, implosion-type HNEDs would have a minimum explosion yield of 0.45 kt TNT. Higher explosion yields would be possible at lower probabilities. The thermal analysis, shows that medium-technology HNEDs could be built with fuel type example B. On the other hand even high-technology HNEDs would be **technically unfeasible** for fuel type example A if the HNED would be cooled at the outside by air and thermal radiation. Fuel type example C would be **proliferation-proof** for high-technology HNEDs and all types of outside cooling.

The γ-radiation in 1 m distance would not be sufficient for fuel type example B to fulfill all IAEA recommendations of 1 Sv/h for physical protection [21]. For fuel type example A the γ-radiation of 0.86 Sv/h in 1 m distance would be close to fulfilling the IAEA recommendations for physical protection [21]. Fuel type example C could well satisfy the IAEA recommendations for physical protection. The γ-radiation exposure in the high explosive lenses necessary to destroy their chemical and physical properties would be high enough after 13 months for fuel type example C (**proliferation-proof**) and after 26 months for fuel type example A (**close to be proliferation-proof**). Fuel type example B with 106 months would be technically feasible. For a gas ultra-centrifuge cascade as described in section 15.5 with an overall stage enrichment factor of 1.1215 and 164 machines with 2.4 SWU/year the U-232 contamination of 600 ppm (fuel type example B) would be sufficient to destroy almost all UF_6 gas during re-enrichment (**proliferation-proof**). From this chapter and Table 18.14 it can be concluded that a U-232 contamination of about 1,700 ppm and above would satisfy even more criteria of Table 18.14 for **being proliferation-proof**. According to Kang et al. [1] such values of >1,700 ppm are reached **in PWRs** with a fuel burnup above about 30,000 MWd/t.

References Section 18

[1] Kang J. et al., U-232 and the Proliferation-Resistance of U-233 in Spent Fuel, Science and Global Security, 9:1-32, Copyright © Taylor & Francis Group, LLC, 2001.

[2] IAEA, Thorium Fuel Cycle-Potential Benefits and Challenges, IAEA-TECDOC-1450, Vienna, Austria, May 2005.

[3] Forsberg C. W. et al., Uses for Uranium-233: What Should Be Kept For Future Needs?, ORNL-6952, Oak Ridge National Laboratory, USA, September 1999.

[4] Ade B. et al., Safety and Regulatory Issues of the Thorium Fuel Cycle, NUREG/ CR-7176, ORNL/TM-2013/543, February 2014.

[5] Radkowsky A. et al., The Nonproliferative Light Water Thorium Reactor: A New Approach To Light Water Reactor Core Technology, Nuclear Technology, Vol. 124, 215-222, Dec. 1998.

[6] Galperin A. et al., A Pressurized Water Reactor Plutonium Incinerator Based on Thorium Fuel and Seed-Blanket Assembly Geometry, Nuclear Technology, Vol. 132, 214-226, Nov. 2000.

[7] Galperin A. et al., A Thorium-Based Seed-Blanket Fuel Assembly Concept to Enhance PWR Proliferation Resistance, Proceedings of the International Conference on Future Nuclear Systems-Global'99, August 29-September 3, Jackson Hole, Wyoming, 1999.

[8] International Commission on Radiation Protection (ICRP)), Annals of the ICRP, Publication 103- The 2007 Recommendations of the International Commission on Radiological Protection, Elsevier, Amsterdam, 2007.

[9] Kessler G., Sustainable and Safe Nuclear Energy, Technology and Safety of Fast and Thermal Nuclear Reactors, Springer, Heidelberg, 2012.

[10] Forsberg C. W. et al., Definition of Weapons-Usable Uranium-233, ORNL/TM-13517, Oak Ridge National Laboratory, USA, March 1998.

[11] Wood H. G., Effects of Separation Processes on Minor Uranium Isotopes in Enrichment Cascades, Science and Global Security, 16:26-36, Copyright © Taylor & Francis Group, LLC, 2008.

[12] Von Halle E., Multicomponent Isotope Separation in Matched Abundance Ratio Cascades Composed of Stages with Large Separation Factors, K/ITP-131; Workshop on Separation Phenomena in Liquids and Gases, Technische Hochschule Darmstadt, Germany, July 20-23, 1987.

[13] Kryuchkov E. F. et al., Evaluation of Self-Protection of 20% Uranium Denatured with ^{232}U Against Unauthorized Reenrichment. Nuclear Science and Engineering 162, 208-213, 2009.

[14] Glaser A., On the proliferation potential of uranium fuel for research reactors at various enrichment levels, Science and Global Security, 14:1-34, Copyright © Taylor & Francis Group, LLC, 2006.

[15] SRIM - The Stopping and Range of Ions in Matter: available on the Particle Interactions with Matter Website at: http://www.srim.org.

[16] Nuclear Reaction Data Center (JCPRG) Website at: http://www.jcprg.org.

[17] Experimental Nuclear Reaction (EXFOR) Data Base, Version of Dec.19 2008; available on the INTERNET: http://www-nds.iaea.org/exfor/exfor.htm.

[18] Glaser A., Characteristics of the Gas Centrifuge for Uranium Enrichment and Their Relevance for Nuclear Weapons Proliferation (corrected), Science and Global Security, 16:1-25, Copyright © Taylor & Francis Group, LLC, 2008.

[19] Kessler G,. Plutonium Denaturing by Pu-238, Nuclear Science and Engineering 155, 53-73, 2007.

[20] IAEA, Management of Reprocessed Uranium, Current Status and Future Prospects, IAEA-TECDOC-1529, Vienna, Austria, February 2007.

[21] Nuclear Security Recommendations on Physical Protection of Nuclear Material and Nuclear Facilities (INFCIRC/225/Revision 5), IAEA, Vienna, 2011.

[22] Moir R., U232 Nonproliferation Features, LLNL-TR-438648, June 2010.

[23] Avrami L. et al., Radiation-Induced Changes In Explosive Materials, AD-775371, Picatinny Arsenal, Dover, New Jersey, Dec. 1973.

[24] Neuhaus I. et al., Comparison of Uranium- and Thorium-based Plutonium-Recycling in a Pressurized Water Reactor, Juel-3690, Forschungszentrum Juelich, Germany, 1999.

List of Abbreviations

ADS	Accelerator Driven Systems
AGR	Advanced gas cooled, graphite moderated reactor
ANL	Argonne National Laboratory (USA)
ANS	American Nuclear Society
APS	American Physical Society
AREVA	French company (reactor manufacturer)
AUC	ammonium uranyl carbonate
AUPuC	ammonium uranyl plutonyl carbonate
AVLIS	Atomic vapour isotope separation
AVR	Arbeitsgemeinschaft Versuchsreaktor, experimental HTR plant (Federal Republic of Germany)
BN 350	prototype fast breeder reactor (USSR)
BN 600	Advanced prototype fast breeder reactor (USSR)
BOR 60	Soviet experimental fast reactors
BR	breeding ratio
BR-1, BR-2, BR-5	Soviet experimental fast reactors
BWR	Boiling water reactor, cooled and moderated by light water
CANDU	Canadian deuterium uranium reactor
CANDU-PHWR	Canadian deuterium uranium pressurized heavy water reactor
CAPRA	Consommation Améliorée du Plutonium dans les Réacteur Avancés
CASTOR	German fuel transport cask
CEMO	Continous enrichment monitoring
CHEMO	Cascade header enrichment monitor
CR	Conversion ratio
CRDM	Control rod drive mechanism
C/S	Containment/Surveillance
DA	Destructive analysis

DFR	Dounreay fast reactor, experimental fast test reactor (UK)
DIV	Design information verification
DOE	Department of Energy
EBR-I, EBR-II	experimental breeder reactors (USA)
ECCS	Emergency core cooling system
EFWS	Emergency feed water system
ENDF	Evaluated nuclear data file
EOSS	Electronic optical sealing system
EPR	European Pressurized Water Reactor
EPRI	Electric Power Research Institute (USA)
EPSS	Emergency power supply system
EURATOM	European Organisation
EURODIF	derived from European diffusion, international company for uranium isotope separation plants
FE	Fuel element
FR	Fast reactors
FBR	Fast breeder reactor
FOC	Fiber optic cable
GGH	Generalized geometry hold up
GGR	Gas cooled, graphite moderated reactor
GNEP	Global Nuclear Energy Partnership
GTRI	Global Threat Reduction Initiative
HEU	Highly enriched uranium
HELIKON	Vortex tube enrichment process developed in South Africa
HLW	High level waste
HLWC	High level waste concentrate
HEPA	High efficiency particulate air filter
HF	Hydrogen fluoride
HLNC	High level coincidence counter
HM	Heavy metal

HNED	Hypothetical Nuclear Explosive Devise
HP	High pressure
HRG	High resolution gamma spectrometer
HTGR	High temperature gas cooled, graphite moderated reactor with prismatic fuel elements
HTR	High temperature gas cooled, graphite moderated reactor with spherical fuel elements (pebble bed reactor)
HTR-Th	HTR with thorium fuel
HWR	Heavy water reactor
IAEA	International Atomic energy Agency
ICRP	International Commission on Radiological Protection
IRWST	In-containment refueling water storage tank
IHX	Intermediate heat exchanger
IFR	Integral Fast Reactor
INFCE	International Nuclear Fuel Cycle Evaluation
INFCIRC	Reports or provisions of IAEA
ISIS	Institute for Science and International Security
IUREP	International Uranium Resource Evaluation Project
JAEA	Japanese Atomic energy Agency
JOYO	Japanese experimental fast breeder reactor
JSFR	Japanese sodium cooled fast reactor
KFA	Kernforschungs-Anlage Jülich, Nuclear Research Center, Jülich (Federal Republic of Germany)
KfK	Kernforschungszentrum Karlsruhe, Nuclear Research Center, Karlsruhe (Federal Republic of Germany)
KMP	Key measurement point
LBE	Liquid lead bismuth eutectic alloy
LEU	Low enriched uranium
LFUA	Limited frequency unanounced access
LGR	Light water cooled graphite moderated reactor
LIDAR	Light detection and ranging

LLW	Low level waste
LMFBR	Liquid metal cooled fast breeder reactor
LOCA	Loss-of-coolant accident
LOF	Loss-of-flow accident
LWBR	Light water breeder reactor
LWR	Light water cooled and light water moderated reactor
LWGR	Light water cooled graphite moderated reactor
LWR-Pu	LWR with plutonium fuel
MAGNOX	derived from magnesium alloy; gas cooled graphite moderated reactor using a Mg alloy as the cladding material and uranium dioxide as fuel
MBA	Material balance area
MDC	Moderator density coefficient
MELOX	French MOX fabrication plant
MEU	Medium enriched uranium
MLIS	Molecular laser isotope separation
MLW	Medium level waste
MTC	Moderator temperature coefficient
MLEC	Multilateral Enrichment Center
MLRC	Multilateral Reprocessing Center
MF	Moderator fuel ratio
MONJU	fast breeder prototype reactor (Japan)
MOX	PuO_2/UO_2 mixed oxide fuel
MSBR	Molten salt breeder reactor
MUF	Material unaccounted for
NAPA	US Nuclear Nonproliferation Act
NAS	National Academy of Sciences
NEA	Nuclear energy Agency of OECD
NED	Nuclear explosive device
NDA	Non destructive analysis

NWS	Nuclear Weapon State
NNWS	Non-Nuclear Weapon State
NNPA	Nuclear nonproliferation act
NPT	Non-Proliferation Treaty
NRTA	Near real time accountancy
NRC	see USNRC
OECD	Organization for Economic Cooperation and Development OT Once-through fuel cycle
PFR	Prototype Fast Reactor (UK)
PHENIX	fast breeder prototype reactor (France)
LP	Low pressure
PCRV	Prestressed concrete reactor vessel
PIV	Physical inventory verification
PWR	Pressurized ligh water cooled and light water moderated reactor
PUREX	Plutonium and uranium recovery by extraction
RAR	Reasonably assured resources
rem	roentgen equivalend man
RPV	Reactor pressure vessel
RRU	Reenriched reprocessed uranium
RETRR	Reduced Enrichement for Reasearch and Test Reactors
SAGSI	Standing Advisory Group on Safeguards Implementation
SCRAM	Fast Reactor shut down
SEFOR	Southwest experimental fast oxide reactor (USA)
SFR	Sodium cooled fast reactor
SGHWR	Steam generating heavy water reactor
SGR	self-generated recycling
SILEX	Separation of isotopes in Laser exitation enrichment process
SNR 300	schneller natriumgekühlter Reaktor, protoype fast breeder reactor (Federal Republic of Germany/Belgium/Netherlands)
SS	stainless steel

SSAC	State System of Accountancy and Control
SQ	Significant quantity
SUPERPHENIX	fast breeder reactor power plant following after PHENIX in the French breeder reactor program
SWU	Separative work unit
TMT	Trinitrotolol High explosive
TBP	tri-n-butyl phosphate
TDLS	Tunable diode laser spectroscopy
THOREX	Thorium oxide recovery by extraction
THTR	Thorium fueled high temperature gas cooled reactor
TLD	Thermoluminicent dosimeter
UKAEA	United Kingdom Atomic energy Authority
U/Pu	Uanium/plutonium
UMS	Unattended monitoring system
UOX	Uranium dioxide fuel
URENCO	Uranium Enrichment Company Ltd. (UK)
USDOE	United States Department of Energy
USEPA	United States Environmental Protection Agency
USNRC	United States Nuclear Regulatory Commission
VACOSS	Variable coding sealing system
VHTGR	Very high temperature gas cooled reactor
VVER	Russian pressurized water reactor

Subject Index